SRA
Connecting Math Concepts

Level B Presentation Book 2

COMPREHENSIVE EDITION

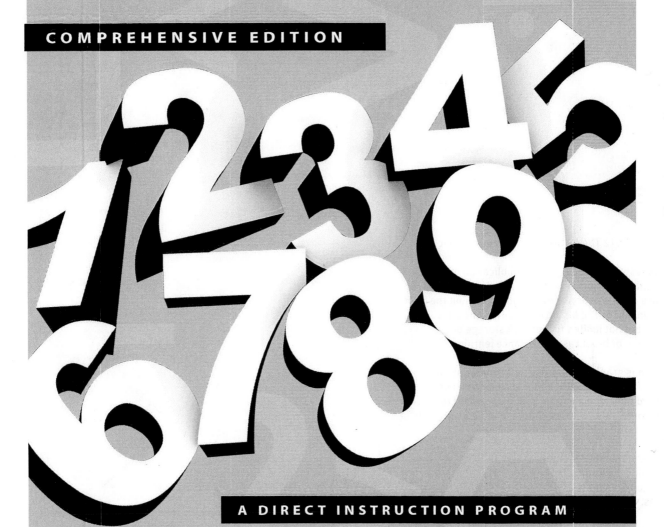

A DIRECT INSTRUCTION PROGRAM

 **Education**

Bothell, WA • Chicago, IL • Columbus, OH • New York, NY

MHEonline.com

 **Education**

Copyright © 2012 The McGraw-Hill Companies, Inc.

Send all inquiries to:
McGraw-Hill Education
4400 Easton Commons
Columbus, OH 43219

ISBN: 978-0-02-103589-2
MHID: 0-02-103589-X

Printed in the United States of America.

6 7 8 9 10 MER 16 15 14

Table of Contents

Lessons 41-45 Planning Page

	Lesson 41	**Lesson 42**	**Lesson 43**	**Lesson 44**	**Lesson 45**
Student Learning Objectives	**Exercises** 1. Find missing numbers in number families 2. Count forward or backward from a given number 3. **Identify and count pennies** 4. Say addition and subtraction facts with a small number of 2 5. **Identify and count dimes and pennies** 6. Say and write addition and subtraction facts with a small number of 1 7. Say and write place-value equations for 3-digit numbers 8. Solve 2-digit/3-digit addition and subtraction problems 9. Add 10 to 2-digit numbers 10. Identify place value and write numbers 11. Solve problems with three addends 12. Complete work independently	**Exercises** 1. Say two addition facts and two subtraction facts for number families 2. Count forward or backward from a given number 3. Add 10 to 2-digit numbers 4. Find missing numbers in number families 5. Identify and count dimes and pennies 6. Say and write addition and subtraction facts with a small number of 1 7. Say and write place-value equations for 3-digit numbers 8. Solve 2-digit/3-digit addition and subtraction problems 9. **Align symbols for two-column problems** 10. Solve problems with three addends 11. Complete work independently	**Exercises** 1. Say addition and subtraction facts with a small number of 2 2. Identify and count dimes and pennies 3. **Count forward or backward from a given number and plus 10** 4. Solve problems with three addends 5. Say and write place-value equations for 3-digit numbers 6. Find and write missing numbers in number families 7. Align symbols for two-column problems 8. Say and write addition and subtraction facts with a small number of 2 9. Write and solve word problems 10. Complete work independently	**Exercises** 1. **Say the value of dimes and pennies** 2. **Say and write addition and subtraction facts with a small number of 1 or 2** 3. Count forward or backward from a given number 4. **Count the value of dimes and pennies** 5. Find missing numbers in number families 6. Solve problems with three addends 7. **Write the symbols for word problems in columns and solve** 8. **Find and write the big number and write a subtraction fact for number families with a small number of 2** 9. Say and write place-value equations for 3-digit numbers 10. Say and write addition and subtraction facts with a small number of 2 11. Complete work independently	**Exercises** 1. Say the value of dimes and pennies 2. **Say addition facts with a small number of 1 or 2** 3. Count forward or backward from a given number 4. **Count the value of dimes and pennies in mixed groups** 5. Find missing numbers in number families 6. Solve problems with three addends 7. Write the symbols for word problems in columns and solve 8. Find and write missing numbers in number families 9. Say and write place-value equations for 2- or 3-digit numbers 10. **Say and write subtraction facts with a small number of 1 or 2** 11. Complete work independently

Common Core State Standards for Mathematics

	Lesson 41	Lesson 42	Lesson 43	Lesson 44	Lesson 45
1.OA 1			✔		
1.OA 3	✔	✔	✔	✔	
1.OA 5	✔	✔			
1.OA 6	✔	✔	✔	✔	✔
1.OA 8	✔	✔	✔	✔	✔
1.NBT 1	✔	✔	✔	✔	✔
1.NBT 2	✔	✔	✔	✔	✔
1.NBT 4	✔	✔	✔	✔	✔
1.NBT 5	✔	✔	✔	✔	✔

Teacher Materials	**Presentation Book 2,** Board Displays CD or chalkboard
Student Materials	Workbook 1, Pencil
Additional Practice	• Student Practice Software: Block 2: Activity 1 (1.NBT.1), Activity 2 (1.NBT.1 and 1.NBT.2), Activity 3 (1.NBT.1), Activity 4 (1.NBT.4 and 1.NBT.5), Activity 5 (1.NBT.2 and 1.NBT.4), Activity 6 (1.NBT.2) • Math Fact Worksheets 9–10 (After Lesson 41), 11 (After Lesson 43), 12–16 (After Lesson 44)
Mastery Test	

Lesson

EXERCISE 1: NUMBER FAMILIES
PLUS/MINUS DISCRIMINATION

a. (Display:) [41:1A]

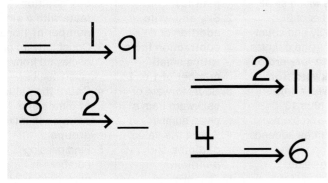

All of these families have a missing number. Some of these families have a small number of 1. Some have a small number of 2. You're going to say the problem for the missing number. Remember, if a small number is missing, you minus.

- (Point to ⟶9.) What's missing, a small number or the big number? (Signal.) *A small number.*
- Say the problem for the numbers that are shown. (Signal.) *9 minus 1.*
    (To correct:)
    - (Point to ⟶9.) Is a small number missing? (Signal.) *Yes.*
    - So do you minus? (Signal.) *Yes.*
    - Start with the big number and say the minus problem for the numbers that are shown. (Signal.) *9 minus 1.*
- What's the missing small number? (Signal.) *8.*
- Say the fact. (Signal.) *9 – 1 = 8.*
b. (Point to 8 ⟶2_.) What's missing, a small number or the big number? (Signal.) *The big number.*
- Say the problem for the numbers that are shown. (Signal.) *8 plus 2.*
- What's the missing big number? (Signal.) *10.*
- Say the fact. (Signal.) *8 + 2 = 10.*
c. (Point to ⟶2 11.) What's missing, a small number or the big number? (Signal.) *A small number.*
- Say the problem for the numbers that are shown. (Signal.) *11 minus 2.*
- What's the missing small number? (Signal.) *9.*
- Say the fact. (Signal.) *11 – 2 = 9.*

d. (Point to 4 ⟶6.) What's missing, a small number or the big number? (Signal.) *A small number.*
- Say the problem for the numbers that are shown. (Signal.) *6 minus 4.*
- What's the missing small number? (Signal.) *2.*
- Say the fact. (Signal.) *6 – 4 = 2.*
(Repeat steps a through d that were not firm.)

EXERCISE 2: MIXED COUNTING

a. Your turn to count by hundreds to 1000. Get ready. (Tap.) *100, 200, 300, 400, 500, 600, 700, 800, 900, 1000.*
b. Count backward from 350 to 340. Get 350 going. *Three hundred fiftyyy.* Count backward. (Tap.) *349, 348, 347, 346, 345, 344, 343, 342, 341, 340.*
c. Start with 100 and count by tens to 200. Get 100 going. *One huuundred.* Count. (Tap.) *110, 120, 130, 140, 150, 160, 170, 180, 190, 200.*
- Start with 100 and count backward by tens. Get 100 going. *One huuundred.* Count backward. (Tap.) *90, 80, 70, 60, 50, 40, 30, 20, 10.*
d. Start with 120 and count backward to 110. Get 120 going. *One hundred twentyyy.* Count backward. (Tap.) *119, 118, 117, 116, 115, 114, 113, 112, 111, 110.*
e. Start with 37 and plus tens to 107. Get 37 going. *Thirty sevennn.* Plus tens. (Tap.) *47, 57, 67, 77, 87, 97, 107.*
f. You have 307. What number? (Signal.) *307.*
- When you plus tens, what's the next number? (Signal.) *317.*
- So what's 307 plus 10? (Signal.) *317.*
- What's 307 plus 100? (Signal.) *407.*
- Listen: 301 plus 1. What's 301 plus 1? (Signal.) *302.*

EXERCISE 3: COINS
PENNIES

a. (Display:) [41:3A]

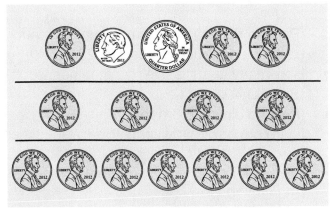

- (Point to penny in 1st row.) This coin is a penny. What kind of coin is this? (Touch.) *(A) penny.*
b. (Point to 1st row.) Some of these coins are pennies.
- (Point to penny.) Is this a penny? (Touch.) *Yes.*
- (Point to dime.) Is this a penny? (Touch.) *No.*
- (Repeat for:) Quarter, *No;* Penny, *Yes;* Penny, *Yes.* (Repeat until firm.)
c. Listen: A penny is worth 1 cent. How many cents is a penny worth? (Signal.) *1.*
- So when you count pennies, you count by ones. What do you count by for pennies? (Signal.) *Ones.*
- How many cents is each penny worth? (Signal.) *1.*
d. (Point to penny in 2nd row.) What kind of coins are in this row? (Touch.) *Pennies.*
- When you count for pennies, do you count by ones or by tens? (Signal.) *Ones.*
  I'll touch each coin in this row. You'll count to figure out how many cents the row is worth.
- Count by one for each penny I touch. Get ready. (Touch.) *1, 2, 3, 4.*
- How many cents are in this row? (Signal.) *4.*
e. (Point to penny in 3rd row.) What kind of coins are in this row? (Touch.) *Pennies.*
- What do you count by for pennies? (Signal.) *Ones.*
  I'll touch each coin in this row. You'll count to figure out how many cents the row is worth.
- Count by one for each penny I touch. Get ready. (Touch.) *1, 2, 3, 4, 5, 6, 7.*
- How many cents are in this row? (Signal.) *7.*

EXERCISE 4: NUMBER FAMILIES
SMALL NUMBER OF 2

a. (Display:) [41:4A]

2 + 8	10 − 8
6 − 4	4 + 2
6 + 2	8 − 2

All of these problems are from families with a small number of 2. Some of these problems minus. So a small number is missing.
- (Point to **2 + 8**.) Read the problem. (Touch.) *2 plus 8.*
- Does that problem minus? (Signal.) *No.*
- So is a small number missing? (Signal.) *No.* Right, a small number is not missing. So the big number is missing.
- Is the answer a small number or the big number? (Signal.) *The big number.*
b. (Point to **6 − 4**.) Read the problem. (Touch.) *6 minus 4.*
- Does that problem minus? (Signal.) *Yes.*
- So is a small number missing? (Signal.) *Yes.* That's the answer.
- Is the answer a small number or the big number? (Signal.) *A small number.*
c. (Repeat the following tasks for the remaining problems:)

(Point to __.) Read the problem.	Does the problem minus?	So is the answer a small number?	Is the answer a small number or the big number?
6 + 2	No	No	The big number.
10 − 8	Yes	Yes	A small number.
4 + 2	No	No	The big number.
8 − 2	Yes	Yes	A small number.

d. This time, you'll tell me the missing numbers.
- (Point to **2 + 8**.) Read the problem. (Touch.) *2 plus 8.*
- Is the answer a small number or the big number? (Signal.) *The big number.*
- What's 2 plus 8? (Signal.) *10.*
e. (Point to **6 − 4**.) Read the problem. (Touch.) *6 minus 4.*
- Is the answer a small number or the big number? (Signal.) *A small number.*
- What's 6 minus 4? (Signal.) *2.*

f. (Repeat the following tasks for the remaining problems:)

(Point to __.) Read the problem.	Is the answer a small number or the big number?	What's __?	
6 + 2	The big number.	6 + 2	8
10 – 8	A small number.	10 – 8	2
4 + 2	The big number.	4 + 2	6
8 – 2	A small number.	8 – 2	6

(Repeat problems that were not firm.)

EXERCISE 5: COINS
DIMES AND PENNIES

a. (Display:) [41:5A]

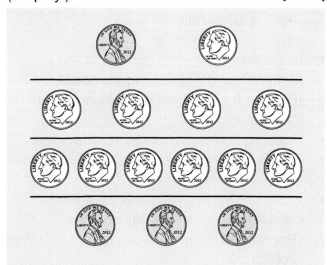

These are coins.
- (Point to penny in 1st row.) What's this coin? (Touch.) *(A) penny.*
- How many cents is a penny worth? (Signal.) *1.*
- So what do you count by for each penny? (Signal.) *One.*

b. (Point to dime.) This coin is a dime. What coin? (Touch.) *(A) dime.*
- Listen: Each dime is worth 10 cents. How many cents is each dime worth? (Signal.) *10.*
- So when you count for dimes, do you count by ones or by tens? (Touch.) *Tens.*
(Repeat step b until firm.)

c. (Point to dimes in 2nd row.) What are these coins? (Touch.) *Dimes.*
You're going to figure out how many cents this row of dimes is worth.
- How many cents is each dime worth? (Signal.) *10.*
- What do you count by for each dime? (Signal.) *Ten.*

d. My turn to count the cents for this row. (Touch each dime.) 10, 20, 30, 40.
- How many cents is this row of dimes worth? (Signal.) *40.*
Yes, 40.

e. (Point to dimes.) Your turn: Count by ten for each dime. (Touch dimes.) *10, 20, 30, 40.*
- How many cents is this row of dimes worth? (Signal.) *40.*

f. (Point to dimes in 3rd row.) You're going to figure out how many cents this row of dimes is worth.
- How many cents is each dime worth? (Signal.) *10.*
- So what do you count by for each dime? (Signal.) *Ten.*
- Count by ten for each dime. Get ready. (Touch.) *10, 20, 30, 40, 50, 60.*
- How many cents is this row of dimes worth? (Signal.) *60.*
Remember, you count by ten for each dime.

g. (Point to pennies in 4th row.) What kind of coins are in this row? (Touch.) *Pennies.*
- How many cents is each penny worth? (Touch.) *1.*
- What do you count by for each penny? (Signal.) *One.*
- Count by one for each penny. Get ready. (Touch each penny.) *1, 2, 3.*
- How many cents is this row of pennies worth? (Signal.) *3.*
Remember: You count by one for each penny.

EXERCISE 6: FACTS
ADDITION/SUBTRACTION (SMALL NUMBER OF 1)

a. (Distribute unopened workbooks to students.)
- Open your workbook to Lesson 41 and find part 1.
(Observe children and give feedback.)
(Teacher reference:)

a. 10 – 1   c. 5 – 4   e. 8 – 1
b. 1 + 7   d. 9 + 1   f. 7 – 6

All of these problems are from number families with a small number of 1.

b. Touch and read problem A. (Signal.) *10 minus 1.*
- What's the answer? (Signal.) *9.*
- Touch and read problem B. (Signal.) *1 plus 7.*
- What's the answer? (Signal.) *8.*

c. (Repeat the following tasks for problems C through F:)
- Touch and read problem __.
- What's the answer?
d. Read each problem to yourself and write the answer.
(Observe children and give feedback.)
e. Check your work.
- Problem A: 10 minus 1. What's the answer? (Signal.) *9.*
- Problem B: 1 plus 7. What's the answer? (Signal.) *8.*
- Problem C: 5 minus 4. What's the answer? (Signal.) *1.*
- Problem D: 9 plus 1. What's the answer? (Signal.) *10.*
- Problem E: 8 minus 1. What's the answer? (Signal.) *7.*
- Problem F: 7 minus 6. What's the answer? (Signal.) *1.*

EXERCISE 7: PLACE-VALUE ADDITION
3-DIGIT NUMBERS

`REMEDY`

a. Find part 2 on worksheet 41. ✔
(Teacher reference:)

`R` `Part N`

a. ___ + ___ + ___ = 813
b. ___ + ___ + ___ = 500
c. ___ + ___ + ___ = 406
d. ___ + ___ + ___ = 170

- Touch the number shown in problem A. ✔
- What number does the problem end up with? (Signal.) *813.*
- Listen: 13. Say the place-value addition for 13. Get ready. (Signal.) *10 plus 3.*
- Now say the place-value addition for 813. Get ready. (Signal.) *800 + 10 + 3 = 813.*
- Complete equation A.
(Observe children and give feedback.)
b. For the rest of the place-value equations, you'll say zero. Touch the number shown in problem B. ✔
- What number does the problem end up with? (Signal.) *500.*
Two of the numbers in the place-value equation are zero.
- Say the place-value addition for 500. (Signal.) *500 + 0 + 0 = 500.*
- Complete equation B.
(Observe children and give feedback.)

c. Touch the number shown in problem C. ✔
- What number does the problem end up with? (Signal.) *406.*
- Say the place-value addition for 406. (Signal.) *400 + 0 + 6 = 406.*
- Complete equation C.
(Observe children and give feedback.)
d. Touch the number shown in problem D. ✔
- What number does the problem end up with? (Signal.) *170.*
- Say the place-value addition for 170. (Signal.) *100 + 70 + 0 = 170.*
- Complete equation D.
(Observe children and give feedback.)
(Answer key:)

a. $\underline{800} + \underline{10} + \underline{3} = 813$
b. $\underline{500} + \underline{0} + \underline{0} = 500$
c. $\underline{400} + \underline{0} + \underline{6} = 406$
d. $\underline{100} + \underline{70} + \underline{0} = 170$

EXERCISE 8: COLUMN PROBLEMS

a. Find part 3 on worksheet 41. ✔
(Teacher reference:)

$$
\begin{array}{r} a.\quad 1\,0\,6 \\ -\quad\ \ 5 \\ \hline \end{array}
$$

$$
\begin{array}{r} b.\quad 6\,0 \\ +\ 1\,5 \\ \hline \end{array}
$$

- Touch and read problem A. Get ready. (Signal.) *106 minus 5.*
- Touch and read the problem for the ones. Get ready. (Signal.) *6 minus 5.*
- Look at the tens column. What will you write in the answer for the tens? (Signal.) *Zero.*
- Look at the hundreds column. What will you write in the answer for the hundreds? (Signal.) *1.*
b. Touch and read problem B. Get ready. (Signal.) *60 plus 15.*
- Touch and read the problem for the ones. Get ready. (Signal.) *Zero plus 5.*
- Touch and read the problem for the tens. Get ready. (Signal.) *6 plus 1.*

c. Go back to problem A and write the correct digits for the answer in the right places. Then work problem B. Put your pencil down when you've finished part 3.
(Observe children and give feedback.)
- Check your work. You'll touch and read each problem and the answer.
- Problem A. (Signal.) *106 – 5 = 101.*
- Problem B. (Signal.) *60 + 15 = 75.*

EXERCISE 9: PLUS 10

a. Find part 4 on worksheet 41. ✔
(Teacher reference:)

a. 38 + 10
b. 26 + 10
c. 15 + 10
d. 13 + 10

You're going to say answers to problems that plus 10. Some of the problems start with a teen number. Remember, a teen number plus 10 equals a 20s number.
You'll read each problem, tell me the answer, and then you'll write the answers.
- Touch and read problem A. Get ready. (Signal.) *38 plus 10.*
- What's the answer to 38 plus 10? (Signal.) *48.*
b. Touch and read problem B. Get ready. (Signal.) *26 plus 10.*
- What's the answer to 26 plus 10? (Signal.) *36.*
c. Touch and read problem C. Get ready. (Signal.) *15 plus 10.*
- What's the answer to 15 plus 10? (Signal.) *25.*
d. Touch and read problem D. Get ready. (Signal.) *13 plus 10.*
- What's the answer to 13 plus 10? (Signal.) *23.*
e. Complete the equations in part 4. Put your pencil down when you're finished.
(Observe children and give feedback.)
(Answer key:)

a. 38 + 10 = 48
b. 26 + 10 = 36
c. 15 + 10 = 25
d. 13 + 10 = 23

EXERCISE 10: NUMBERS IN COLUMNS
HUNDREDS, TENS, ONES

REMEDY

a. Find part 5 on worksheet 41. ✔
(Teacher reference:)

R **Part B**

You're going to write numbers in a column.
- Look at the spaces where you'll write the numbers for part 5. ✔
These columns don't have letters for hundreds, tens, and ones.
- Touch the space for the hundreds digit in row A. ✔
- Touch the space for the tens digit in row A. ✔
- Touch the space for the ones digit in row A. ✔
(Repeat until firm.)
b. Listen: You're going to write 3 in row A. What number? (Signal.) *3.*
- Does 3 start with hundreds, tens, or ones? (Signal.) *Ones.*
- Write 3 in row A.
(Observe children and give feedback.)
c. You're going to write 165 in row B. What number? (Signal.) *165.*
- Does 165 start with hundreds, tens, or ones? (Signal.) *Hundreds.*
- Write 165 in row B.
(Observe children and give feedback.)
d. You're going to write 12 in row C. What number? (Signal.) *12.*
- Does 12 start with hundreds, tens, or ones? (Signal.) *Tens.*
- Write 12 in row C.
(Observe children and give feedback.)
e. You're going to write 470 in row D. What number? (Signal.) *470.*
- Does 470 start with hundreds, tens, or ones? (Signal.) *Hundreds.*
- Write 470 in row D.
(Observe children and give feedback.)
f. You're going to write 39 in row E. What number? (Signal.) *39.*
- Does 39 start with hundreds, tens, or ones? (Signal.) *Tens.*
- Write 39 in row E.
(Observe children and give feedback.)

g. Now you'll touch and read the numbers you just wrote.
- Touch and read the number in row A. Get ready. (Signal.) *3.*
- Row B. Get ready. (Signal.) *165.*
- Row C. Get ready. (Signal.) *12.*
- Row D. Get ready. (Signal.) *470.*
- Row E. Get ready. (Signal.) *39.*
(Repeat step g until firm.)
(Answer key:)

a.			3
b.	1	6	5
c.		1	2
d.	4	7	0
e.		3	9

EXERCISE 11: ADDITION
3 ADDENDS

a. Find part 6 on worksheet 41. ✔
(Teacher reference:)

a. $10 + 3 + 1$ b. $2 + 6 + 1$

- Touch and read problem A. (Signal.) *10 plus 3 plus 1.*
- Say the problem for the underlined part. (Signal.) *10 plus 3.*
- What's the answer? (Signal.) *13.*
- Say the next problem. (Signal.) *13 plus 1.* (Repeat until firm.)
- What's the answer to 13 plus 1? (Signal.) *14.*
- Complete equation A. ✔
- Everybody, touch and read the whole equation. Get ready. (Signal.) *10 + 3 + 1 = 14.*

b. Touch and read problem B. (Signal.) *2 plus 6 plus 1.*
- Say the problem for the underlined part. (Signal.) *2 plus 6.*
- What's the answer? (Signal.) *8.*
- Say the next problem. (Signal.) *8 plus 1.* (Repeat until firm.)
- What's the answer to 8 plus 1? (Signal.) *9.*
- Complete equation B. ✔
- Everybody, touch and read the whole equation. Get ready. (Signal.) *2 + 6 + 1 = 9.*

EXERCISE 12: INDEPENDENT WORK

a. Turn to the other side of worksheet 41 and find part 7. ✔
(Teacher reference:)

Part 7
a. $\underline{6 \quad 2} \to 8$ b. $\underline{3 \quad 2} \to 5$

Part 8
a. $7 + 1$ c. $1 + 1$ e. $8 + 1$
$7 + 2$ $1 + 2$ $8 + 2$
$97 + 2$ $51 + 2$ $88 + 2$

b. $5 + 1$ d. $3 + 1$ f. $4 + 1$
$5 + 2$ $3 + 2$ $4 + 2$
$15 + 2$ $33 + 2$ $64 + 2$

Part 9
a. $400 + 50 + 0$ d. $200 + 0 + 9$
b. $100 + 10 + 3$ e. $700 + 10 + 6$
c. $\quad 70 + 2$ f. $600 + 0 + 0$

You'll write two plus facts and two minus facts below each number family.
You'll complete the equations in parts 8 and 9.

b. Your turn: Complete worksheet 41.
(Observe children and mark incorrect responses on children's worksheets as you give feedback.)

Lesson 42

EXERCISE 1: NUMBER FAMILIES
COMMUTATIVE AND INVERSE RELATIONSHIPS

a. (Display:)   [42:1A]

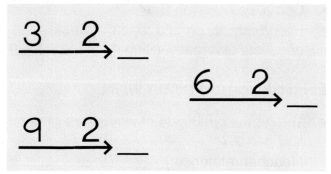

All of these families have a small number of 2. You'll tell me the big number for each family but I won't write it. You have to remember it.

- (Point to $\overset{3\quad 2}{\longrightarrow}$_.) What are the small numbers for this family? (Touch.) *3 and 2.*
- What's the big number? (Signal.) *5.*

b. (Point to $\overset{9\quad 2}{\longrightarrow}$_.) What are the small numbers for this family? (Touch.) *9 and 2.*
- What's the big number? (Signal.) *11.*

c. (Point to $\overset{6\quad 2}{\longrightarrow}$_.) What are the small numbers for this family? (Touch.) *6 and 2.*
- What's the big number? (Signal.) *8.*
(Repeat families that were not firm.)

d. Now you'll say both plus facts and both minus facts for each family.
- (Point to $\overset{3\quad 2}{\longrightarrow}$_.) Say the fact that starts with the first small number. Get ready. (Touch.) *3 + 2 = 5.*
- Say the other plus fact. (Signal.) *2 + 3 = 5.*
- Say the fact that goes back along the arrow. (Signal.) *5 − 2 = 3.*
- Say the other minus fact. (Signal.) *5 − 3 = 2.*

e. (Point to $\overset{9\quad 2}{\longrightarrow}$_.) Say the fact that starts with the first small number. Get ready. (Touch.) *9 + 2 = 11.*
- Say the other plus fact. (Signal.) *2 + 9 = 11.*
- Say the fact that goes back along the arrow. (Signal.) *11 − 2 = 9.*
- Say the other minus fact. (Signal.) *11 − 9 = 2.*

f. (Point to $\overset{6\quad 2}{\longrightarrow}$_.) Say the fact that starts with the first small number. Get ready. (Touch.) *6 + 2 = 8.*
- Say the other plus fact. (Signal.) *2 + 6 = 8.*
- Say the fact that goes back along the arrow. (Signal.) *8 − 2 = 6.*
- Say the other minus fact. (Signal.) *8 − 6 = 2.*
(Repeat steps d through f that were not firm.)

EXERCISE 2: MIXED COUNTING

a. Your turn to count by hundreds to 1000. Get ready. (Tap.) *100, 200, 300, 400, 500, 600, 700, 800, 900, 1000.*

b. Count backward from 250 to 240. Get 250 going. *Two hundred fiftyyy.* Count backward. (Tap.) *249, 248, 247, 246, 245, 244, 243, 242, 241, 240.*
- Start with 100 and count backward by tens. Get 100 going. *One huuundred.* Count backward. (Tap.) *90, 80, 70, 60, 50, 40, 30, 20, 10.*

c. Start with 100 and count by tens to 200. Get 100 going. *One huuundred.* Count. (Tap.) *110, 120, 130, 140, 150, 160, 170, 180, 190, 200.*

d. Start with 120 and count backward to 110. Get 120 going. *One hundred twentyyy.* Count backward. (Tap.) *119, 118, 117, 116, 115, 114, 113, 112, 111, 110.*

e. Start with 32 and plus tens to 102. Get 32 going. *Thirty-twooo.* Plus tens. (Tap.) *42, 52, 62, 72, 82, 92, 102.*

f. You have 501. What number? (Signal.) *501.*
- When you plus ones, what's the next number? (Signal.) *502.*
- You have 501. When you plus tens, what's the next number? (Signal.) *511.*
- So what's 501 plus 10? (Signal.) *511.*
- You have 501. What's 501 plus 100. (Signal.) *601.*

EXERCISE 3: PLUS 10

a. (Display:) [42:3A]

53 + 10	41 + 10
12 + 10	15 + 10
29 + 10	11 + 10

These problems plus 10. Some of these problems start with a teen number. Remember, a teen number plus 10 equals a 20s number. You'll read each problem. Then you'll say each fact.

- (Point to **53 + 10.**) Read this problem. (Touch.) *53 plus 10.*
- What's 53 plus 10? (Signal.) *63.*
- Say the fact for 53 plus 10. (Signal.) *53 + 10 = 63.*

b. (Point to **12 + 10.**) Read this problem. (Touch.) *12 plus 10.*
- What's 12 plus 10? (Signal.) *22.*
- Say the fact for 12 plus 10. (Signal.) *12 + 10 = 22.*

c. (Point to **29 + 10.**) Read this problem. (Touch.) *29 plus 10.*
- What's 29 plus 10? (Signal.) *39.*
- Say the fact for 29 plus 10. (Signal.) *29 + 10 = 39.*

d. (Point to **41 + 10.**) Read this problem. (Touch.) *41 plus 10.*
- What's 41 plus 10? (Signal.) *51.*
- Say the fact for 41 plus 10. (Signal.) *41 + 10 = 51.*

e. (Point to **15 + 10.**) Read this problem. (Touch.) *15 plus 10.*
- What's 15 plus 10? (Signal.) *25.*
- Say the fact for 15 plus 10. (Signal.) *15 + 10 = 25.*

f. (Point to **11 + 10.**) Read this problem. (Touch.) *11 plus 10.*
- What's 11 plus 10? (Signal.) *21.*
- Say the fact for 11 plus 10. (Signal.) *11 + 10 = 21.*
(Repeat problems that were not firm.)

━━━━━━━━ **INDIVIDUAL TURNS** ━━━━━━━━

(Call on individual students to perform one of the following tasks.)

- (Point to **15 + 10.**) Say the fact. (Call on a student.) *15 + 10 = 25.*
- (Point to **41 + 10.**) Say the fact. (Call on a student.) *41 + 10 = 51.*

EXERCISE 4: NUMBER FAMILIES
PLUS/MINUS DISCRIMINATION

REMEDY

a. (Display:) [42:4A]

All of these families have a missing number. Some of these families have a small number of 1. Some have a small number of 2. You're going to say the problem for the missing number.
Remember, if a small number is missing, you minus.

- (Point to $\xrightarrow{9 \quad 2}$_.) What's missing, a small number or the big number? (Signal.) *The big number.*
- Say the problem for the numbers that are shown. (Signal.) *9 plus 2.*
- What's the missing big number? (Signal.) *11.*
- Say the fact. (Signal.) *9 + 2 = 11.*

b. (Point to $\xrightarrow{4}$5.) What's missing, a small number or the big number? (Signal.) *A small number.*
- Say the problem for the numbers that are shown. (Signal.) *5 minus 4.*
- What's the missing small number? (Signal.) *1.*
- Say the fact. (Signal.) *5 − 4 = 1.*

c. (Point to $\xrightarrow{1}$9.) What's missing, a small number or the big number? (Signal.) *A small number.*
- Say the problem for the numbers that are shown. (Signal.) *9 minus 1.*
- What's the missing small number? (Signal.) *8.*
- Say the fact. (Signal.) *9 − 1 = 8.*

d. (Point to =—²→10.) What's missing, a small number or the big number? (Signal.) *A small number.*
- Say the problem for the numbers that are shown. (Signal.) *10 minus 2.*
- What's the missing small number? (Signal.) *8.*
- Say the fact. (Signal.) *10 – 2 = 8.*

e. (Point to ⁶—→8.) What's missing, a small number or the big number? (Signal.) *A small number.*
- Say the problem for the numbers that are shown. (Signal.) *8 minus 6.*
- What's the missing small number? (Signal.) *2.*
- Say the fact. (Signal.) *8 – 6 = 2.*

f. (Point to ⁷—²→__.) What's missing, a small number or the big number? (Signal.) *The big number.*
- Say the problem for the numbers that are shown. (Signal.) *7 plus 2.*
- What's the missing big number? (Signal.) *9.*
- Say the fact. (Signal.) *7 + 2 = 9.*

g. (Point to ³—→4.) What's missing, a small number or the big number? (Signal.) *A small number.*
- Say the problem for the numbers that are shown. (Signal.) *4 minus 3.*
- What's the missing small number? (Signal.) *1.*
- Say the fact. (Signal.) *4 – 3 = 1.*

h. (Point to ⁵—¹→__.) What's missing, a small number or the big number? (Signal.) *The big number.*
- Say the problem for the numbers that are shown. (Signal.) *5 plus 1.*
- What's the missing big number? (Signal.) *6.*
- Say the fact. (Signal.) *5 + 1 = 6.*
  (Repeat steps a through h that were not firm.)

a. (Display:) [42:5A]

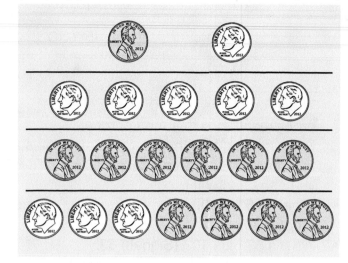

- (Point to penny in 1st row.) What kind of coin is this? (Touch.) *(A) penny.*
- How many cents is a penny worth? (Touch.) *1.*
- So you count by one for each penny. What do you count by for each penny? (Signal.) *One.*

b. (Point to dime.) What kind of coin is this? (Touch.) *(A) dime.*
- How many cents is a dime worth? (Touch.) *10.*
- So what do you count by for each dime? (Signal.) *Ten.*
  (Repeat until firm.)

c. (Point to 2nd row.) What are these coins? (Touch.) *Dimes.*
- How many cents is 1 dime worth? (Signal.) *10.*
- What do you count by for each dime? (Signal.) *Ten.*
- (Point to 1st dime.) Count by ten for each dime. Get ready. (Touch.) *10, 20, 30, 40, 50.*
- How many cents is this row of dimes worth? (Signal.) *50.*

d. (Point to 3rd row.) What are these coins? (Touch.) *Pennies.*
- How many cents is 1 penny worth? (Signal.) *1.*
- What do you count by for each penny? (Signal.) *One.*
- (Point to 1st penny.) Count by one for each penny. Get ready. (Touch.) *1, 2, 3, 4, 5, 6.*
- How many cents is this row of pennies worth? (Signal.) *6.*

e. (Point to 4th row.) This row has dimes and pennies.
- What do you count by for each dime? (Signal.) *Ten.*
- What do you count by for each penny? (Signal.) *One.*
- Count by tens for the dimes. Get ready. (Touch.) *10, 20, 30.* (Keep finger on last dime.) (Repeat until firm.)

f. (Keep touching last dime.) You have 30 cents.
- Get 30 going. *Thirtyyyy.* Count by one for each penny. (Touch pennies.) *31, 32, 33, 34.* (Repeat until firm.)
- How many cents is this row worth? (Touch.) *34.*

g. Let's do it again, the fast way.
- (Point to 1st dime.) Count for the dimes. (Touch dimes.) *10, 20, 30.* Count for the pennies. (Touch pennies.) *31, 32, 33, 34.*
- How many cents is this row of coins worth? (Touch.) *34.*
(Repeat step g until firm.)

EXERCISE 6: FACTS
ADDITION/SUBTRACTION (SMALL NUMBER OF 1)

a. (Distribute unopened workbooks to students.)
- Open your workbook to Lesson 42 and find part 1.
(Observe children and give feedback.)
(Teacher reference:)

a. $11 - 1$	d. $5 - 4$	g. $11 - 10$
b. $10 - 9$	e. $10 + 1$	h. $4 - 1$
c. $1 + 3$	f. $10 - 1$	

All of these problems are from number families with a small number of 1.
- Touch and read problem A. (Signal.) *11 minus 1.*
- What's the answer? (Signal.) *10.*

b. Touch and read problem B. (Signal.) *10 minus 9.*
- What's the answer? (Signal.) *1.*

c. (Repeat the following tasks for problems C through H:)
- Touch and read problem __.
- What's the answer?

d. Read each problem to yourself and complete the equation.
(Observe children and give feedback.)

e. Check your work.
- Problem A: 11 minus 1. What's the answer? (Signal.) *10.*
- Problem B: 10 minus 9. What's the answer? (Signal.) *1.*
- (Repeat for:) C, 1 + 3, *4*; D, 5 − 4, *1*; E, 10 + 1, *11*; F, 10 − 1, *9*; G, 11 − 10, *1*; H, 4 − 1, *3*.

EXERCISE 7: PLACE-VALUE ADDITION
3-DIGIT NUMBERS

a. Find part 2 on worksheet 42. ✔
(Teacher reference:)

a. ___ + ___ + ___ = 825
b. 700 + 0 + 3 = ___
c. ___ + ___ + ___ = 350
d. ___ + ___ + ___ = 405

You'll complete the place-value equations in part 2.
For one of these problems you'll write the number you end up with. For others, you'll show the numbers you plus.
- Touch problem A. ✔
- What number does the problem end up with? (Signal.) *825.*
- Complete the place-value equation. (Observe children and give feedback.)
- Read the place-value equation for A. (Signal.) *800 + 20 + 5 = 825.*

b. Touch and read problem B. (Signal.) *700 plus zero plus 3 equals.*
- Write the number you end up with. ✔
- Say the place-value addition equation for B. (Signal.) *700 + 0 + 3 = 703.*

c. Complete the rest of the place-value equations in part 2.
(Observe children and give feedback.)

d. Check your work.
- Read the place-value equation for C. Get ready. (Signal.) *300 + 50 + 0 = 350.*
- Read the place-value equation for D. Get ready. (Signal.) *400 + 0 + 5 = 405.*

EXERCISE 8: COLUMN PROBLEMS

a. Find part 3 on worksheet 42. ✔
    (Teacher reference:)

a. $\begin{array}{r} 39 \\ +720 \\ \hline \end{array}$

b. $\begin{array}{r} 894 \\ -84 \\ \hline \end{array}$

c. $\begin{array}{r} 402 \\ +7 \\ \hline \end{array}$

d. $\begin{array}{r} 581 \\ +208 \\ \hline \end{array}$

• Touch and read problem A. Get ready. (Signal.) *39 plus 720.*
• Touch and read the problem for the ones. (Signal.) *9 plus zero.*
• Touch and read the problem for the tens. (Signal.) *3 plus 2.*
• Look at the hundreds column. What will you write in the answer for the hundreds? (Signal.) *7.*

b. Touch and read problem B. Get ready. (Signal.) *894 minus 84.*
• Touch and read the problem for the ones. Get ready. (Signal.) *4 minus 4.*
• Touch and read the problem for the tens. Get ready. (Signal.) *9 minus 8.*
• Look at the hundreds column. What will you write in the answer for the hundreds? (Signal.) *8.*

c. Touch and read problem C. Get ready. (Signal.) *402 plus 7.*
• Touch and read the problem for the ones. Get ready. (Signal.) *2 plus 7.*
• Look at the tens column. What will you write in the answer for the tens? (Signal.) *Zero.*
• Look at the hundreds column. What will you write in the answer for the hundreds? (Signal.) *4.*

d. Touch and read problem D. Get ready. (Signal.) *581 plus 208.*
• Touch and read the problem for the ones. Get ready. (Signal.) *1 plus 8.*
• Touch and read the problem for the tens. Get ready. (Signal.) *8 plus zero.*
• Touch and read the problem for the hundreds. Get ready. (Signal.) *5 plus 2.*

e. Go back to problem A and write the correct digits for the answer in the right places. Put your pencil down when you're finished. (Observe children and give feedback.)
• Touch and read the problem and the answer for A. Get ready. (Signal.) *39 + 720 = 759.*
• Work the rest of the problems in part 3. Put your pencil down when you've written the correct answers. (Observe children and give feedback.)

f. Check your work.
• Touch and read the equation for B. Get ready. (Signal.) *894 – 84 = 810.*
• Touch and read the equation for C. Get ready. (Signal.) *402 + 7 = 409.*
• Touch and read the equation for D. Get ready. (Signal.) *581 + 208 = 789.*

EXERCISE 9: WRITING COLUMN PROBLEMS   REMEDY

a. (Display:) W   [42:9A]

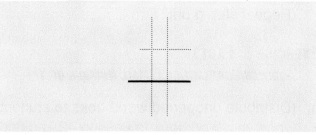

I'm going to write a column problem. Listen to the problem. 15 plus 30.
Listen again: 15 plus 30. Say that problem. (Signal.) *15 plus 30.*
(Repeat until firm.)

b. So I write 15 in the first row. (Touch tens column in top row.) I write the plus sign here. (Touch far left of equals bar.) I write 30 here in this row too, but in the right columns. (Touch tens column above equals bar.)
• (Point to first row.) What do I write here? (Touch.) *15.*
• (Point to where sign goes.) What do I write here? (Touch.) *Plus.*
• (Point where 30 goes.) What do I write here? (Touch.) *30.*
(Repeat step b until firm.)

c. This time I'll write the symbols.
- (Point to first row.) What do I write in this row? (Touch.) *15.*
- Does 15 start with hundreds, tens, or ones? (Signal.) *Tens.*
  (Add to show:) [42:9B]

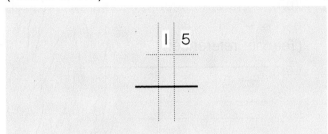

- (Point to where sign goes.) What do I write here? (Signal.) *Plus.*
  (Add to show:) [42:9C]

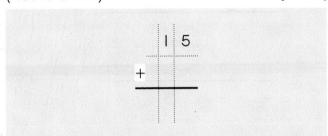

- What number do I write next? (Signal.) *30.*
- Does 30 start in the hundreds, tens, or ones? (Signal.) *Tens.*
  (Add to show:) [42:9D]

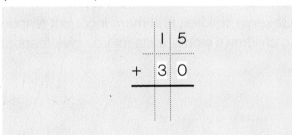

All of the columns are lined up.
d. (Point to **15.**) I'll touch the symbols and you'll read the problem. Get ready. (Touch.) *15 plus 30.*
e. Find part 4 on worksheet 42. ✔
  (Teacher reference:)   R Part G

- Touch the problem 17 plus 62. ✔
  You're going to write 17 plus 62 in the column for A. The equals bar is already shown.
- Touch where you'll write 17. ✔
- Touch where you'll write plus. ✔
- Touch where you'll write 62. ✔
  (Repeat until firm.)

f. Do 17 and 62 start with the hundreds, tens, or ones? (Signal.) *Tens.*
- Write 17.
  (Observe children and give feedback.)
- Write the plus sign.
  (Observe children and give feedback.)
- Write 62.
  (Observe children and give feedback.)
g. Touch the problem you'll write for B.
- Touch and read the problem for B. Get ready. (Signal.) *36 minus 15.*
  Yes, 36 minus 15. The equals bar is already shown.
- Touch where you'll write 36. ✔
- Touch where you'll write minus. ✔
- Touch where you'll write 15. ✔
  (Repeat until firm.)
h. Do 36 and 15 start with the hundreds, tens, or ones? (Signal.) *Tens.*
- Write 36.
  (Observe children and give feedback.)
- Write the minus sign.
  (Observe children and give feedback.)
- Write 15.
  (Observe children and give feedback.)
  (Teacher reference:)

You'll work these problems as part of your independent work.

EXERCISE 10: ADDITION
3 ADDENDS   REMEDY

a. (Display:) W   [42:10A]

$$1 + 5 + 2$$

- (Point to **1.**) Read the whole problem. (Touch.) *1 plus 5 plus 2.*
- Say the problem for the underlined part. (Signal.) *1 plus 5.*
- What's the answer? (Signal.) *6.*
- Say the next problem. (Signal.) *6 plus 2.*
- What's the answer? (Signal.) *8.*
  (Repeat until firm.)
  (Add to show:)   [42:10B]

$$1 + 5 + 2 = 8$$

- Read the equation. (Touch.) *1 + 5 + 2 = 8.*

b. Find part 5 on worksheet 42. ✔
(Teacher reference:)

a. 3 + 2 + 1
b. 7 + 1 + 2

These problems are like the one you just worked.

- Touch and read problem A. (Signal.) *3 plus 2 plus 1.*
- Say the problem for the underlined part. (Signal.) *3 plus 2.*
- What's the answer? (Signal.) *5.*
- Say the next problem. (Signal.) *5 plus 1.* (Repeat until firm.)
- What's the answer? (Signal.) *6.*
- Complete equation A. ✔
- Everybody, touch and read equation A. Get ready. (Signal.) *3 + 2 + 1 = 6.*

c. Touch and read problem B. (Signal.) *7 plus 1 plus 2.*
- Say the problem for the underlined part. (Signal.) *7 plus 1.*
- What's the answer? (Signal.) *8.*
- Say the next problem. (Signal.) *8 plus 2.* (Repeat until firm.)
- What's the answer? (Signal.) *10.*
- Complete equation B. ✔
- Everybody, touch and read equation B. Get ready. (Signal.) *7 + 1 + 2 = 10.*

EXERCISE 11: INDEPENDENT WORK

a. Now we'll go over your independent work.
- Touch part 4 again. ✔
You'll work the column problems you wrote in part 4.
b. Turn to the other side of worksheet 42 and find part 6. ✔
(Teacher reference:)

Part 6   a. 6 2 →8   b. 9 2 →11

Part 7
a. 13 + 10   d. 51 + 10   g. 28 + 10
b. 47 + 10   e. 14 + 10   h. 75 + 10
c. 62 + 10   f. 86 + 10   i. 11 + 10

Part 8
a. 100 + 10 + 1   e. 800 + 20 + 0
b.   30 + 0   f. 800 + 0 + 2
c. 400 + 0 + 8   g. 800 + 10 + 2
d.   10 + 6   h. 700 + 0 + 0

Part 9
a. ___ + ___ = 39   c. ___ + ___ = 15
b. ___ + ___ = 13   d. ___ + ___ = 40

You'll write four facts below each family — two plus facts and two minus facts.
Then you'll complete the equations for parts 7, 8, and 9.

c. Your turn: Complete worksheet 42.
(Observe children and mark incorrect responses on children's worksheets as you give feedback.)

Lesson 43

EXERCISE 1: FACTS
ADDITION/SUBTRACTION (SMALL NUMBER OF 2) [REMEDY]

a. (Display:) [43:1A]

2 + 7	9 − 7
5 − 2	4 − 2
2 + 6	11 − 9
2 + 9	3 + 2

All of these problems are from families with a small number of 2. Some of these problems minus and some of them plus. You'll tell me the missing number. Remember, if a problem minuses, the answer is a small number.

- (Point to **2 + 7.**) Read the problem. (Touch.) *2 plus 7.*
- Does the problem minus? (Signal.) *No.*
- So is the answer a small number? (Signal.) *No.*
- What's 2 plus 7? (Signal.) *9.*

b. (Point to **5 − 2.**) Read the problem. (Touch.) *5 minus 2.*
- Does this problem minus? (Signal.) *Yes.*
- So is the answer a small number? (Signal.) *Yes.*
- What's 5 minus 2? (Signal.) *3.*

c. (Repeat the following tasks for the remaining problems:)

(Point to __.) Read the problem.	Does the problem minus?	So is the answer a small number?	What's __?	
2 + 6	No	No	2 + 6	8
2 + 9	No	No	2 + 9	11
9 − 7	Yes	Yes	9 − 7	2
4 − 2	Yes	Yes	4 − 2	2
11 − 9	Yes	Yes	11 − 9	2
3 + 2	No	No	3 + 2	5

EXERCISE 2: COINS
DIMES AND PENNIES

a. (Display:) [43:2A]

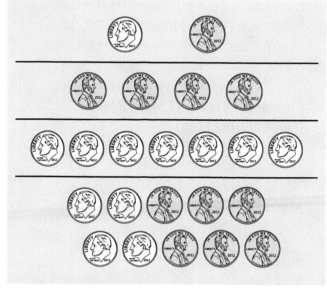

- (Point to dime in 1st row.) What coin is this? (Touch.) *(A) dime.*
- How much is each dime worth? (Signal.) *10 cents.*
- What do you count by for each dime? (Signal.) *Ten.*

b. (Point to penny.) What coin is this? (Touch.) *(A) penny.*
- How much is each penny worth? (Signal.) *1 cent.*
- What do you count by for each penny? (Signal.) *One.*

c. (Point to 2nd row.) What are these coins? (Touch.) *Pennies.*
- What do you count by for each penny? (Signal.) *One.*
- (Point to 1st penny.) Count by one for each penny. Get ready. (Touch.) *1, 2, 3, 4.*
- How many cents is this row of pennies worth? (Signal.) *4.*

d. (Point to 3rd row.) What are these coins? (Touch.) *Dimes.*
- What do you count by for each dime? (Signal.) *Ten.*
- (Point to 1st dime.) Count by ten for each dime. Get ready. (Touch.) *10, 20, 30, 40, 50, 60, 70.*
- How many cents is this row of dimes worth? (Signal.) *70.*

e. (Point to dimes in 4th row.) What are these coins? (Touch.) *Dimes.*
• What do you count by for each dime? (Signal.) *Ten.*
• (Point to pennies.) What are these coins? (Touch.) *Pennies.*
• What do you count by for each penny? (Signal.) *One.*
• (Point to 1st dime.) Count by ten for each dime. Get ready. (Touch.) *10, 20, 30, 40.*
f. (Keep touching the last dime.) You have 40 cents.
• Get 40 going. *Fortyyy.* Count for each penny. (Touch pennies.) *41, 42, 43, 44, 45, 46.* (Repeat until firm.)
• How many cents is this row worth? (Touch.) *46.*

EXERCISE 3: MIXED COUNTING
PLUS 10

a. My turn to start with 94 and plus tens. Ninety-fouuur. 104, 114, 124.
• Your turn: Start with 94 and plus tens to 124. Get 94 going. *Ninety-fouuur.* Plus tens. (Tap.) *104, 114, 124.*
b. Start with 91 and plus tens to 121. Get 91 going. *Ninety-wuuun.* Plus tens. (Tap.) *101, 111, 121.*
c. Start with 99 and plus tens to 129. Get 99 going. *Ninety-niiine.* Plus tens. (Tap.) *109, 119, 129.*
d. Count by hundreds to 1000. Get ready. (Tap.) *100, 200, 300, 400, 500, 600, 700, 800, 900, 1000.*
e. Start with 100 and count by tens to 200. Get 100 going. *One huuundred.* Count. (Tap.) *110, 120, 130, 140, 150, 160, 170, 180, 190, 200.*
f. Start with 100 and count backward by ten. Get 100 going. *One huuundred.* Count backward. (Tap.) *90, 80, 70, 60, 50, 40, 30, 20, 10.*
g. Now you're going to answer problems that plus 10.
• Listen: 16 plus 10. What's 16 plus 10? (Signal.) *26.*
• Say the fact. (Signal.) *16 + 10 = 26.*
• New problem: 11 plus 10. What's 11 plus 10? (Signal.) *21.*
• Say the fact. (Signal.) *11 + 10 = 21.*
• New problem: 25 plus 10. What's 25 plus 10? (Signal.) *35.*
• Say the fact. (Signal.) *25 + 10 = 35.*
• New problem: 23 plus 10. What's 23 plus 10? (Signal.) *33.*
• Say the fact. (Signal.) *23 + 10 = 33.*

EXERCISE 4: ADDITION
3 ADDENDS   REMEDY

a. (Display:) W   [43:4A]

$$4 + 1 + 2$$

This is like the problems you've worked before, but there is not an underlined part.
• (Point to **4 + 1.**) Everybody, read the problem. Get ready. (Touch.) *4 plus 1 plus 2.*
• Say the first problem you'll work. Get ready. (Signal.) *4 plus 1.*
• What's the answer? (Signal.) *5.*
• Say the next problem. (Signal.) *5 plus 2.* (Repeat until firm.)
• What's the answer? (Signal.) *7.* (Add to show:)   [43:4B]

$$4 + 1 + 2 = 7$$

• Read the equation. (Touch.) *4 + 1 + 2 = 7.*
b. (Distribute unopened workbooks to students.)
• Open your workbook to Lesson 43 and find part 1. (Observe children and give feedback.) (Teacher reference:)   R Part R

a. $1 + 9 + 5$ b. $1 + 5 + 1$

These problems are like the ones you just worked.
• Touch and read problem A. Get ready. (Signal.) *1 plus 9 plus 5.*
• Say the first problem you'll work. Get ready. (Signal.) *1 plus 9.*
• What's the answer? (Signal.) *10.*
• Say the next problem. Get ready. (Signal.) *10 plus 5.*
• What's the answer? (Signal.) *15.*
• Complete the equation. (Observe children and give feedback.)
• Touch and read the equation. (Signal.) *1 + 9 + 5 = 15.*

c. Touch and read problem B. Get ready.
(Signal.) *1 plus 5 plus 1.*

- Say the first problem you'll work. Get ready.
(Signal.) *1 plus 5.*
- What's the answer? (Signal.) *6.*
- Say the next problem. Get ready. (Signal.)
6 plus 1.
- What's the answer? (Signal.) *7.*
- Complete the equation.
(Observe children and give feedback.)
- Touch and read the equation. (Signal.)
1 + 5 + 1 = 7.

EXERCISE 5: PLACE-VALUE ADDITION
3-DIGIT NUMBERS

a. Find part 2 on worksheet 43. ✔
(Teacher reference:)

a. $\underline{} + \underline{} + \underline{} = 741$
b. $500 + 60 + 0 = \underline{}$
c. $300 + 0 + 9 = \underline{}$
d. $\underline{} + \underline{} + \underline{} = 217$

You're going to complete the place-value addition for the problems in part 2.
Be careful because there are zeros in some of the equations. You know how to work problem A.

- Touch problem B. ✔
- Read problem B and say the answer. (Signal.)
500 + 60 + 0 = 560.
- Touch problem C. ✔
- Read problem C and say the answer. (Signal.)
300 + 0 + 9 = 309.
- Complete the place-value addition for each problem.
(Observe children and give feedback.)
(Answer key:)

a. $\underline{700} + \underline{40} + \underline{1} = 741$
b. $500 + 60 + 0 = \underline{560}$
c. $300 + 0 + 9 = \underline{309}$
d. $\underline{200} + \underline{10} + \underline{7} = 217$

b. Check your work. You'll touch and read the place-value equation for each problem.

- Problem A. Get ready. (Signal.)
700 + 40 + 1 = 741.
- Problem B. Get ready. (Signal.)
500 + 60 + 0 = 560.
- Problem C. Get ready. (Signal.)
300 + 0 + 9 = 309.
- Problem D. Get ready. (Signal.)
200 + 10 + 7 = 217.

EXERCISE 6: NUMBER FAMILIES
PLUS/MINUS DISCRIMINATION

a. Find part 3 on worksheet 43. ✔
(Teacher reference:)

a. $\underline{} \xrightarrow{2} 5$ d. $6 \xrightarrow{} 7$ g. $\underline{} \xrightarrow{2} 11$
b. $4 \xrightarrow{} 6$ e. $5 \xrightarrow{1} \underline{}$ h. $\underline{} \xrightarrow{2} 9$
c. $8 \xrightarrow{2} \underline{}$ f. $\underline{} \xrightarrow{1} 4$

The families in this part have a small number of 1 or a small number of 2. A number is missing in each of the families. You'll tell me the missing number for each family. Then you'll complete the families.

- Touch family A. ✔
- Is a small number or the big number missing?
(Signal.) *A small number.*
- Say the problem to figure out the missing number. (Signal.) *5 minus 2.*
- So what's the missing number? (Signal.) *3.*

b. Touch family B. ✔

- Is a small number or the big number missing?
(Signal.) *A small number.*
- Say the problem. (Signal.) *6 minus 4.*
- So what's the missing number? (Signal.) *2.*

c. (Repeat the following tasks for problems C through H:)

Touch family __.	Is a small number or the big number missing?	Say the problem.	So what's the missing number?
C	The big number.	8 + 2	10
D	A small number.	7 – 6	1
E	The big number.	5 + 1	6
F	A small number.	4 – 1	3
G	A small number.	11 – 2	9
H	A small number.	9 – 2	7

(Repeat families that were not firm.)

d. Write the missing number for each family. Put your pencil down when you're finished.
(Observe children and give feedback.)

e. Check your work. You'll touch and read the number you wrote for each family.

- Family A. (Signal.) *3.*
- (Repeat for:) B, *2;* C, *10;* D, *1;* E, *6;* F, *3;* G, *9;* H, *7.*

EXERCISE 7: WRITING COLUMN PROBLEMS REMEDY

a. Find part 4 on worksheet 43. ✔
   (Teacher reference:) R Part H

ₐ. 50 + 318   ᵦ. 479 − 61

• Touch the problem you'll write for column A. ✔
• Read it. Get ready. (Signal.) *50 plus 318.*
  You're going to write 50 plus 318 in the column for A. The equals bar is already shown.
b. Touch where you'll write 50. ✔
• Touch where you'll write plus. ✔
• Touch where you'll write 318. ✔
  (Repeat step b until firm.)
c. Does 50 start with hundreds, tens, or ones?
  (Signal.) *Tens.*
• Write 50.
  (Observe children and give feedback.)
• Write the plus sign.
  (Observe children and give feedback.)
• Does 318 start with hundreds, tens, or ones?
  (Signal.) *Hundreds.*
• Write 318.
  (Observe children and give feedback.)
  (Teacher reference:)

ₐ. 50 + 318
```
    50
+ 3 1 8
```

d. Touch and read the problem you'll write for column B. Get ready. (Signal.) *479 minus 61.*
  You're going to write 479 minus 61 in the column for B.
e. Touch where you'll write 479. ✔
• Touch where you'll write minus. ✔
• Touch where you'll write 61. ✔
  (Repeat step e until firm.)
f. Does 479 start with hundreds, tens, or ones?
  (Signal.) *Hundreds.*
• Write 479.
  (Observe children and give feedback.)
• Write the minus sign. ✔
• Does 61 start with hundreds, tens, or ones?
  (Signal.) *Tens.*

• Write 61.
  (Observe children and give feedback.)
  (Teacher reference:)

ₐ. 50 + 318   ᵦ. 479 − 61

You'll work these problems as part of your independent work.

EXERCISE 8: FACTS
ADDITION/SUBTRACTION (SMALL NUMBER OF 2) REMEDY

a. Turn to the other side of worksheet 43 and find part 5. ✔
   (Teacher reference:) R Part S

ₐ. 9 − 7   c. 6 − 2   e. 2 + 7
ᵦ. 5 + 2   d. 7 − 5   f. 8 − 2

You're going to read each problem and tell me if the big number or a small number is missing.
• Touch and read problem A. (Signal.) *9 minus 7.*
• Is the answer the big number or a small number? (Signal.) *A small number.*
• What's 9 minus 7? (Signal.) *2.*
b. Touch and read problem B. (Signal.) *5 plus 2.*
• Is the answer the big number or a small number? (Signal.) *The big number.*
• What's 5 plus 2? (Signal.) *7.*
c. (Repeat the following tasks for problems C through F:)

Touch and read problem __.	Is the answer the big number or a small number?	What's __?	
C	*A small number.*	6 − 2	4
D	*A small number.*	7 − 5	2
E	*The big number.*	2 + 7	9
F	*A small number.*	8 − 2	6

(Repeat problems that were not firm.)
d. Write answer to all the problems in part 5.
   (Observe children and give feedback.)
e. Check your work. You'll touch and read each fact.
• Fact A. (Signal.) *9 − 7 = 2.*
• (Repeat for:) B, *5 + 2 = 7;* C, *6 − 2 = 4;*
  D, *7 − 5 = 2;* E, *2 + 7 = 9;* F, *8 − 2 = 6.*

EXERCISE 9: WORD PROBLEMS
WRITING AND SOLVING

a. Find part 6 on worksheet 43. ✔
(Teacher reference:)

a. _____ b. _____ c. _____

I'll tell you word problems. You'll write the symbols for each problem. Then you'll complete the equation.

- Listen to problem A: A truck had 50 boxes on it. Then men put 7 more boxes on the truck. How many boxes ended up on the truck?
- Listen to the first part again: A truck had 50 boxes on it. What symbol will you write for that part? (Signal.) *50.*
- Write the symbol for the first part of problem A. ✔
- Listen to the next part: Then men put 7 more boxes on the truck. What symbols do you write for that part? (Signal.) *Plus 7.*
- Write the symbols and complete the equation. Raise your hand when you know how many boxes ended up on the truck.
(Observe children and give feedback.)
- Everybody, touch and read the equation for problem A. (Signal.) *50 + 7 = 57.*
- How many boxes ended up on the truck? (Signal.) *57.*

b. Listen to problem B: There were 6 pies in a store. A woman bought 5 of those pies. How many pies ended up in the store?

- Listen to the first part again: There were 6 pies in a store. What symbol will you write for that part? (Signal.) *6.*
- Write the symbol for the first part of problem B. ✔
- Listen to the next part: A woman bought 5 of those pies. What symbols do you write for that part? (Signal.) *Minus 5.*
- Write the symbols and complete the equation. Raise your hand when you know how many pies ended up in the store.
(Observe children and give feedback.)
- Everybody, touch and read the equation for problem B. (Signal.) *6 – 5 = 1.*
- How many pies ended up in the store? (Signal.) *1.*

c. Listen to problem C: Linda put 7 books on a shelf. Dan put 2 more books on the shelf. How many books ended up on the shelf?

- Listen to the first part again: Linda put 7 books on a shelf. What symbol will you write for that part? (Signal.) *7.*
- Write the symbol for the first part of problem C. ✔
- Listen to the next part: Dan put 2 more books on the shelf. What symbols do you write for that part? (Signal.) *Plus 2.*
- Write the symbols and complete the equation. Raise your hand when you know how many books ended up on the shelf.
(Observe children and give feedback.)
- Everybody, touch and read the equation for problem C. (Signal.) *7 + 2 = 9.*
- How many books ended up on the shelf? (Signal.) *9.*

EXERCISE 10: INDEPENDENT WORK

a. Find part 7 on worksheet 43. ✔
(Teacher reference:)

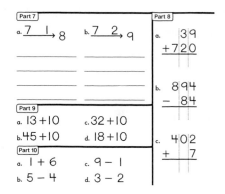

You'll write four facts below each number family.

b. Find part 8. ✔
You'll work the column problems for part 8. Then you'll complete the equations for parts 9 and 10.

c. Turn back to the other side of worksheet 43 and find the column problems you wrote for part 4. ✔
You'll work the problems in part 4.

d. Your turn: Complete worksheet 43.
(Observe children and mark incorrect responses on children's worksheets as you give feedback.)

Lesson

EXERCISE 1: COINS

a. Listen: If you count pennies, you count
   by ones.
- If you count dimes, what do you count by?
  (Signal.) *Tens.*
- If you count pennies, what do you count by?
  (Signal.) *Ones.*
  (Repeat until firm.)

b. You're going to count for dimes. What do you
   count by? (Signal.) *Tens.*
- Listen: You have 30 cents and you count for
  dimes. What's the next number? (Signal.) *40.*
- You have 45 cents and you count for dimes.
  What's the next number? (Signal.) *55.*
- You have 95 cents and you count for dimes.
  What's the next number? (Signal.) *105.*

c. You're going to count for pennies. What do
   you count by? (Signal.) *Ones.*
- Listen: You have 35 cents and you count for
  pennies. What's the next number? (Signal.) *36.*
- You have 50 cents and you count for pennies.
  What's the next number? (Signal.) *51.*

EXERCISE 2: FACTS
ADDITION (SMALL NUMBER OF 1 OR 2)

a. (Display:)   [44:2A]

6 + 1	10 + 1
7 + 2	1 + 8
2 + 7	2 + 5
2 + 10	

Some of these problems are from families with
a small number of 1. Some of these problems
are from families with a small number of 2. All
these problems plus. So all the answers are
big numbers.
- (Point to **6 + 1.**) Read the problem. (Touch.)
  6 plus 1.
- What's the big number? (Signal.) *7.*
- Say the fact for 6 plus 1. (Signal.) *6 + 1 = 7.*

b. (Point to **7 + 2.**) Read the problem. (Touch.)
   7 plus 2.
- What's the big number? (Signal.) *9.*
- Say the fact for 7 plus 2. (Signal.) *7 + 2 = 9.*

c. (Repeat the following tasks for remaining
   problems:)
- (Point to __.) Read the problem.
- What's the big number?
- Say the fact for __.

EXERCISE 3: MIXED COUNTING

a. My turn to start with 92 and plus tens.
   Ninety-twooo, 102, 112, 122.
- Your turn: Start with 92 and plus tens to 122.
  Get 92 going. *Ninety-twooo.* Plus tens. (Tap.)
  102, 112, 122.
- Start with 95 and plus tens to 125. Get 95 going.
  Ninety-fiiive. Plus tens. (Tap.) *105, 115, 125.*
- Start with 91 and plus tens to 121. Get 91 going.
  Ninety-wuuun. Plus tens. (Tap.) *101, 111, 121.*

b. Now you're going to count by other numbers.
- Count by hundreds to 1000. Get 100 going.
  One huuundred. Count. (Tap.) *200, 300, 400,
  500, 600, 700, 800, 900, 1000.*

c. Start with 100 and count by tens to 200. Get
   100 going. *One huuundred.* Count. (Tap.) *110,
   120, 130, 140, 150, 160, 170, 180, 190, 200.*
- Start with 100 and count backward by tens. Get
  100 going. *One huuundred.* Count backward.
  (Tap.) *90, 80, 70, 60, 50, 40, 30, 20, 10.*

d. Listen: 12 plus 10. What's 12 plus 10?
   (Signal.) *22.*
- Listen: 16 plus 10. What's 16 plus 10?
  (Signal.) *26.*
- Listen: 11 plus 10. What's 11 plus 10?
  (Signal.) *21.*
- Listen: 49 plus 10. What's 49 plus 10?
  (Signal.) *59.*

EXERCISE 4: COINS

a. (Display:)   **REMEDY**   [44:4A]

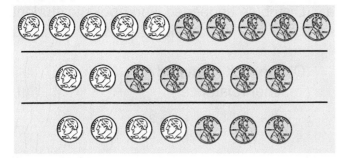

- (Point to dimes in 1st row.) This row has dimes and pennies.
- What do you count by for each dime? (Touch.) *Ten.*
- What do you count by for each penny? (Touch.) *One.*
- Count for the dimes. Get ready. (Touch.) *10, 20, 30, 40, 50.* (Keep touching last dime.)
- How many cents do you have? (Signal.) *50.*
- Get 50 going. *Fiftyyy.* Count for each penny. (Touch pennies.) *51, 52, 53, 54, 55.* (Repeat until firm.)
- How many cents is this row of coins worth? (Signal.) *55.*

b. Let's do it again, the fast way.
- (Point to dimes.) Count for the dimes. (Touch.) *10, 20, 30, 40, 50.* Keep counting for the pennies. (Touch pennies.) *51, 52, 53, 54, 55.*
- How many cents is this row of coins worth? (Signal.) *55.*
(Repeat step b until firm.)

c. (Point to dimes in 2nd row.) You'll count for the coins in this row.
- Count for the dimes. Get ready. (Touch dimes.) *10, 20.* (Keep your finger on the last dime.)
- How many cents do you have? (Signal.) *20.*
- Get 20 going. *Twentyyy.* Count for each penny. (Touch pennies.) *21, 22, 23, 24, 25.* (Repeat until firm.)
- How many cents is this row of coins worth? (Signal.) *25.*

d. Let's do it again, the fast way.
- (Point to 2 dimes.) Count for the dimes. (Touch.) *10, 20.* Keep counting for the pennies. *(Touch pennies.) 21, 22, 23, 24, 25.*
- How many cents is this row of coins worth? (Signal.) *25.*
(Repeat step d until firm.)

e. (Point to dimes in 3rd row.) You'll count for the coins in this row.

- Count for the dimes. Get ready. (Touch dimes.) *10, 20, 30, 40.* (Keep your finger on the last dime.)
- How many cents do you have? (Signal.) *40.*
- Get 40 going. *Fortyyy.* Count for each penny. (Touch pennies.) *41, 42, 43.* (Repeat until firm.)
- How many cents is this row of coins worth? (Signal.) *43.*

f. Let's do it again, the fast way.
- (Point to 4 dimes.) Count for the dimes. (Touch.) *10, 20, 30, 40.* Keep counting for the pennies. (Touch pennies.) *41, 42, 43.*
- How many cents is this row of coins worth? (Signal.) *43.*
(Repeat step f until firm.)

g. I'll start at the other end. Let's see if we end up with 43.
- (Touch pennies.) *1, 2, threee.* (Touch dimes.) *13, 23, 33, 43.*
- Did we end up with 43 cents? (Touch.) *Yes.*

EXERCISE 5: NUMBER FAMILIES
PLUS/MINUS DISCRIMINATION (SMALL NUMBERS 1, 2)

a. (Display:)   [44:5A]

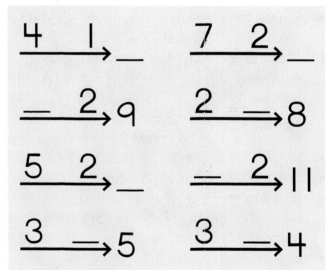

All of these families have a missing number. Some of these families have a small number of 1. Some have a small number of 2. You're going to say the problem for the missing number. Remember, if a small number is missing, you minus.
- (Point to $\xrightarrow{4}$_.) What's missing, a small number or the big number? (Signal.) *The big number.*
- Say the problem for the numbers that are shown. (Signal.) *4 plus 1.*
- What's the missing big number? (Signal.) *5.*
- Say the fact. (Signal.) *4 + 1 = 5.*

b. (Point to =²→9.) What's missing, a small number or the big number? (Signal.) *A small number.*
- Say the problem for the numbers that are shown. (Signal.) *9 minus 2.*
- What's the missing small number? (Signal.) *7.*
- Say the fact. (Signal.) *9 − 2 = 7.*

c. (Point to ⁵—²→_.) What's missing, a small number or the big number? (Signal.) *The big number.*
- Say the problem for the numbers that are shown. (Signal.) *5 plus 2.*
- What's the missing big number? (Signal.) *7.*
- Say the fact. (Signal.) *5 + 2 = 7.*

d. (Point to ³—→5.) What's missing, a small number or the big number? (Signal.) *A small number.*
- Say the problem for the numbers that are shown. (Signal.) *5 minus 3.*
- What's the missing small number? (Signal.) *2.*
- Say the fact. (Signal.) *5 − 3 = 2.*

e. (Point to ⁷—²→_.) What's missing, a small number or the big number? (Signal.) *The big number.*
- Say the problem for the numbers that are shown. (Signal.) *7 plus 2.*
- What's the missing big number? (Signal.) *9.*
- Say the fact. (Signal.) *7 + 2 = 9.*

f. (Point to ²—→8.) What's missing, a small number or the big number? (Signal.) *A small number.*
- Say the problem for the numbers that are shown. (Signal.) *8 minus 2.*
- What's the missing small number? (Signal.) *6.*
- Say the fact. (Signal.) *8 − 6 = 2.*

g. (Point to =²→11.) What's missing, a small number or the big number? (Signal.) *A small number.*
- Say the problem for the numbers that are shown. (Signal.) *11 minus 2.*
- What's the missing small number? (Signal.) *9.*
- Say the fact. (Signal.) *11 − 2 = 9.*

h. (Point to ³—→4.) What's missing, a small number or the big number? (Signal.) *A small number.*
- Say the problem for the numbers that are shown. (Signal.) *4 minus 3.*
- What's the missing small number? (Signal.) *1.*
- Say the fact. (Signal.) *4 − 3 = 1.*
(Repeat steps a through h that were not firm.)

EXERCISE 6: ADDITION
3 ADDENDS

a. (Distribute unopened workbooks to students.)
- Open your workbook to Lesson 44 and find part 1.
(Observe children and give feedback.)
(Teacher reference:)

a. $1 + 10 + 1 =$ c. $1 + 9 + 2 =$
b. $2 + 6 + 1 =$ d. $5 + 2 + 2 =$

- Touch and read problem A. Get ready. (Signal.) *1 plus 10 plus 1 equals.*
- Say the first problem you'll work. Get ready. (Signal.) *1 plus 10.*
- What's the answer? (Signal.) *11.*
- Say the next problem. Get ready. (Signal.) *11 plus 1.*
- What's the answer? (Signal.) *12.*
- Complete the equation. ✔
- Touch and read equation A. Get ready. (Signal.) *1 + 10 + 1 = 12.*

b. Touch and read problem B. Get ready. (Signal.) *2 plus 6 plus 1 equals.*
- Say the first problem you'll work. Get ready. (Signal.) *2 plus 6.*
- What's the answer? (Signal.) *8.*
- Say the next problem. Get ready. (Signal.) *8 plus 1.*
- What's the answer? (Signal.) *9.*

c. Touch and read problem C. Get ready. (Signal.) *1 plus 9 plus 2 equals.*
- Say the first problem you'll work. Get ready. (Signal.) *1 plus 9.*
- What's the answer? (Signal.) *10.*
- Say the next problem. Get ready. (Signal.) *10 plus 2.*
- What's the answer? (Signal.) *12.*

d. Touch and read problem D. Get ready. (Signal.) *5 plus 2 plus 2 equals.*
- Say the first problem you'll work. Get ready. (Signal.) *5 plus 2.*
- What's the answer? (Signal.) *7.*
- Say the next problem. Get ready. (Signal.) *7 plus 2.*
- What's the answer? (Signal.) *9.*
(Repeat steps b through d until firm.)

e. Work problems B, C, and D. Put your pencil down when you've completed the equations for part 1.
(Observe children and give feedback.)

f. Check your work.
• Touch and read equation B. Get ready. (Signal.) *2 + 6 + 1 = 9.*
• Touch and read equation C. Get ready. (Signal.) *1 + 9 + 2 = 12.*
• Touch and read equation D. Get ready. (Signal.) *5 + 2 + 2 = 9.*

EXERCISE 7: WORD PROBLEMS
WRITING IN COLUMNS
 REMEDY

a. Find part 2 on worksheet 44. ✔ (Teacher reference:)

R Part C

You're going to write the symbols for word problems in columns and work them. The equals bars are already shown.
• Touch where you'll write the symbols for problem A. ✔
Listen to problem A: A man started with 62 nails. The man found 14 more nails. How many nails did the man end up with?
• Listen again: A man started with 62 nails. How many nails did the man start out with? (Signal.) *62.*
• So what number do you start with? (Signal.) *62.*
• Does 62 start with hundreds, tens, or ones? (Signal.) *Tens.*
• Write 62.
(Observe children and give feedback.)
b. Listen to the next part: The man found 14 more nails. How many nails did the man find? (Signal.) *14.*
• Tell me the sign and the number you write for: The man found 14 more nails. Get ready. (Signal.) *Plus 14.*
• Write the plus sign. ✔
• Does 14 start with hundreds, tens, or ones? (Signal.) *Tens.*
• Write 14.
(Observe children and give feedback.)

c. Work problem A. Put your pencil down when you know how many nails the man ended up with.
(Observe children and give feedback.)
(Teacher reference:)

• Read the problem and the answer you wrote for A. Get ready. (Signal.) *62 + 14 = 76.*
• How many nails did the man end up with? (Signal.) *76.*
d. Touch where you'll write the symbols for problem B. ✔
Listen to problem B: 45 people were on a bus. Then 21 of those people got off the bus. How many people ended up on the bus?
• Listen again: 45 people were on a bus.
• What number do you write first? (Signal.) *45.*
• Does 45 start with hundreds, tens, or ones? (Signal.) *Tens.*
• Write 45.
(Observe children and give feedback.)
e. Listen to the next part: Then 21 of those people got off the bus.
• Tell me the sign and the number you write for: Then 21 of those people got off the bus. Get ready. (Signal.) *Minus 21.*
• Write the minus sign. ✔
• Does 21 start with hundreds, tens, or ones? (Signal.) *Tens.*
• Write 21.
(Observe children and give feedback.)
f. Work problem B. Put your pencil down when you know how many people ended up on the bus.
(Answer key:)

• Read the problem and the answer you wrote for B. Get ready. (Signal.) *45 – 21 = 24.*
• How many people ended up on the bus? (Signal.) *24.*

EXERCISE 8: NUMBER FAMILIES
SMALL NUMBER OF 2

a. Find part 3 on worksheet 44. ✔
   (Teacher reference:)

   a. 8 2, ___   c. 10 2, ___
   _____   _____

   b. 4 2, ___   d. 7 2, ___
   _____   _____

   All of these families have a small number of 2.
   You'll write the missing number in each family.
   Then you'll write one fact for each family.
   • Start with the first small number and say the
     problem for the missing number in family A.
     (Signal.) *8 plus 2.*
   • What's the answer? (Signal.) *10.*
b. Start with the first small number and say the
   problem for the missing number in family B.
   (Signal.) *4 plus 2.*
   • What's the answer? (Signal.) *6.*
c. (Repeat the following tasks for remaining
   families:)
   • Start with the first small number and say the
     problem for the missing number in family __.
   • What's the answer?
d. Write the big number in each family.
   (Observe children and give feedback.)
e. (Display:)   [44:8A]

   a. 8 ___ 2 →10   c. 10 ___ 2 →12

   b. 4 ___ 2 →6   d. 7 ___ 2 →9

   This shows the big number you should have
   written for each family.
   • Make sure you've written the correct big
     number for each family.
   (Check to make sure students have correct
   big numbers before presenting step f.)
f. Touch family A. ✔
   • Start with 10 and say the fact that goes
     backward along the arrow. (Signal.) *10 – 2 = 8.*
   • Say the fact that starts with 10 but does not
     go straight back along the arrow. (Signal.)
     10 – 8 = 2.
g. Touch family B. ✔
   • Start with 6 and say the fact that goes
     backward along the arrow. (Signal.) *6 – 2 = 4.*

• Say the fact that starts with 6 but does not
  go straight back along the arrow. (Signal.)
  6 – 4 = 2.
• Touch family C. ✔
• Start with 12 and say the fact that goes
  backward along the arrow. (Signal.) *12 – 2 = 10.*
• Say the fact that starts with 12 but does not
  go straight back along the arrow. (Signal.)
  12 – 10 = 2.
• Touch family D. ✔
• Start with 9 and say the fact that goes
  backward along the arrow. (Signal.) *9 – 2 = 7.*
• Say the fact that starts with 9 but does not go
  straight back along the arrow. (Signal.) *9 – 7 = 2.*
h. In the space below each family, you'll write
   the minus fact that does not go straight back
   along the arrow.
• Touch the space below family A. ✔
• Say the fact that does not go straight back
  along the arrow. Get ready. (Signal.) *10 – 8 = 2.*
  That's the fact you'll write for family A.
• Say the fact you'll write for family A again.
  (Signal.) *10 – 8 = 2.*
i. Touch the space below family B. ✔
   You'll write the minus fact that does not go
   straight back along the arrow in that space.
• Say the fact you'll write for family B. Get ready.
  (Signal.) *6 – 4 = 2.*
j. Touch the space below family C. ✔
• Say the fact you'll write for family C. Get ready.
  (Signal.) *12 – 10 = 2.*
k. Touch the space below family D. ✔
• Say the fact you'll write for family D. Get ready.
  (Signal.) *9 – 7 = 2.*
   (Repeat steps i through k that were not firm.)
l. Write the minus fact that does not go straight
   back along the arrow below the families in part
   3. Put your pencil down when you've written
   the fact for each family.
   (Observe children and give feedback.)
   (Answer key:)

   a. 8 2, 10   c. 10 2, 12
      10 – 8 = 2   12 – 10 = 2

   b. 4 2, 6   d. 7 2, 9
      6 – 4 = 2   9 – 7 = 2

m. Check your work. You'll touch and read the
   fact below each family.
   • Family A. Read the fact. (Signal.) *10 – 8 = 2.*
   • (Repeat for remaining problems.)

EXERCISE 9: PLACE-VALUE ADDITION

a. Find part 4 on worksheet 44. ✔
   (Teacher reference:)

   a. $500 + 20 + 9 =$ ____
   b. ____ $+$ ___ $+$ ___ $= 204$
   c. ___ $+$ ___ $= 68$
   d. $700 + 10 + 3 =$ ____
   e. ___ $+$ ___ $= 95$

   You'll work these problems, but be careful.
   Some of them are two-digit numbers.
- Complete each place-value equation. Put your
   pencil down when you've finished.
   (Observe children and give feedback.)
   (Answer key:)

   a. $500 + 20 + 9 = \underline{529}$
   b. $\underline{200} + \underline{0} + \underline{4} = 204$
   c. $\underline{60} + \underline{8} = 68$
   d. $700 + 10 + 3 = \underline{713}$
   e. $\underline{90} + \underline{5} = 95$

b. Check your work. You'll touch and read the
   place-value equation for each problem.
- Problem A. Get ready. (Signal.)
   $500 + 20 + 9 = 529$.
- Problem B. Get ready. (Signal.)
   $200 + 0 + 4 = 204$.
- Problem C. Get ready. (Signal.) $60 + 8 = 68$.
- Problem D. Get ready. (Signal.)
   $700 + 10 + 3 = 713$.
- Problem E. Get ready. (Signal.) $90 + 5 = 95$.

EXERCISE 10: FACTS
ADDITION/SUBTRACTION (SMALL NUMBER OF 2)

a. Find part 5 on worksheet 44. ✔
   (Teacher reference:)

   a. $11 - 9$
   b. $10 - 2$
   c. $4 + 2$
   d. $7 - 2$
   e. $8 - 6$
   f. $2 + 9$

   You're going to tell me if the big number or a
   small number is missing.
- Touch and read problem A. (Signal.) *11 minus 9.*
- Is the answer the big number or a small
   number? (Signal.) *A small number.*
- What's 11 minus 9? (Signal.) *2.*

b. Touch and read problem B. (Signal.) *10 minus 2.*
- Is the answer the big number or a small
   number? (Signal.) *A small number.*
- What's 10 minus 2? (Signal.) *8.*
c. (Repeat the following tasks for remaining
   problems:)
- Touch and read problem ___.
- Is the answer the big number or a small
   number?
- What's ___?
d. Write answers to all the problems in part 5.
   (Observe children and give feedback.)
e. Check your work. Touch and read each fact.
- Fact A. (Signal.) *11 – 9 = 2.*
- Fact B. (Signal.) *10 – 2 = 8.*
- (Repeat for remaining facts.)

EXERCISE 11: INDEPENDENT WORK

a. Turn to the other side of worksheet 44 and find
   part 6. ✔
   (Teacher reference:)

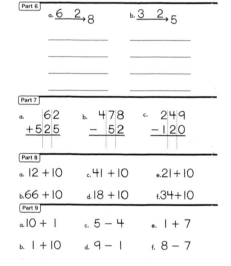

   You'll write four facts below each number
   family.
b. Find part 7. ✔
   You'll work the column problems in part 7.
   Then you'll complete the equations in parts 8
   and 9.
c. Your turn: Complete worksheet 44.
   (Observe children and mark incorrect responses
   on children's worksheets as you give feedback.)

Lesson 45

EXERCISE 1: COINS

a. Listen: If you count pennies, you count by ones.
• If you count dimes, what do you count by? (Signal.) *Tens.*
• If you count pennies, what do you count by? (Signal.) *Ones.*
   (Repeat until firm.)
b. You're going to count for pennies and dimes.
• Listen: You have 35 cents and you count for dimes. What's the next number? (Signal.) *45.*
• Listen: You have 35 cents and you count for pennies. What's the next number? (Signal.) *36.*
• Listen: You have 60 cents and you count for pennies. What's the next number? (Signal.) *61.*
• You have 60 cents and you count for dimes. What's the next number? (Signal.) *70.*

EXERCISE 2: FACTS

ADDITION (SMALL NUMBERS 1 OR 2)

a. (Display:)   [45:2A]

6 + 1	1 + 4
8 + 2	2 + 6
2 + 9	7 + 1

Some of these problems are from families with a small number of 1. Some are from families with a small number of 2.
• All these problems plus. So are all the answers big numbers or small numbers? (Signal.) *Big numbers.*
• (Point to **6 + 1.**) Read the problem. (Touch.) *6 plus 1.*
• What's the big number? (Signal.) *7.*
• Say the fact for 6 plus 1. (Signal.) *6 + 1 = 7.*
b. (Point to **8 + 2.**) Read the problem. (Touch.) *8 plus 2.*
• What's the big number? (Signal.) *10.*
• Say the fact for 8 plus 2. (Signal.) *8 + 2 = 10.*

c. (Repeat the following tasks for remaining problems:)
• (Point to __.) Read the problem.
• What's the big number?
• Say the fact for __.

EXERCISE 3: MIXED COUNTING

a. Your turn: Start with 83 and plus tens to 123. Get 83 going. *Eighty-threee.* Plus tens. (Tap.) *93, 103, 113, 123.*
• Start with 88 and plus tens to 128. Get 88 going. *Eighty-eieieight.* Plus tens. (Tap.) *98, 108, 118, 128.*
• Start with 85 and plus tens to 125. Get 85 going. *Eighty-fiiive.* Plus tens. (Tap.) *95, 105, 115, 125.*
b. Now you're going to count by other numbers.
• Count by hundreds to 1000. Get 100 going. *One huuundred.* Count. (Tap.) *200, 300, 400, 500, 600, 700, 800, 900, 1000.*
c. Start with 100 and count by tens to 200. Get 100 going. *One huuundred.* Count. (Tap.) *110, 120, 130, 140, 150, 160, 170, 180, 190, 200.*
• Start with 100 and count backward by tens. Get 100 going. *One huuundred.* Count backward. (Tap.) *90, 80, 70, 60, 50, 40, 30, 20, 10.*
d. Listen: 11 plus 10. What's 11 plus 10? (Signal.) *21.*
• Listen: 8 plus 10. What's 8 plus 10? (Signal.) *18.*
• Listen: 5 plus 10. What's 5 plus 10? (Signal.) *15.*
• Listen: 26 plus 10. What's 26 plus 10? (Signal.) *36.*

EXERCISE 4: COINS

a. (Display:) [45:4A]

(Point to dimes in 1st row.) This row has dimes and pennies.

- What do you count by for each dime? (Signal.) *Ten.*
- What do you count by for each penny? (Signal.) *One.*
- Count for the dimes. Get ready. (Touch.) *10, 20, 30.* (Keep touching last dime.)
- How many cents do you have? (Signal.) *30.*
- Get 30 going. *Thirtyyy.* Count for each penny. (Touch pennies.) *31, 32, 33.* (Repeat until firm.)
- How many cents is this row of coins worth? (Signal.) *33.*

b. Let's do it again, the fast way.

- (Point to dimes.) Count for the dimes. (Touch.) *10, 20, 30.* Keep counting for the pennies. (Touch pennies.) *31, 32, 33.*
- How many cents is this row of coins worth? (Signal.) *33.* (Repeat step b until firm.)

c. (Point to dimes in 2nd row.) You'll count for the coins in this row.

- Count for the dimes. Get ready. (Touch dimes.) *10, 20, 30, 40, 50, 60.* (Keep touching last dime.)
- How many cents do you have? (Signal.) *60.*
- Get 60 going. *Sixtyyy.* Count for each penny. (Touch pennies.) *61, 62.* (Repeat until firm.)
- How many cents is this row of coins worth? (Signal.) *62.*

d. Let's do it again, the fast way.

- (Point to 6 dimes.) Count for the dimes. (Touch.) *10, 20, 30, 40, 50, 60.* Keep counting for the pennies. (Touch pennies.) *61, 62.*
- How many cents is this row of coins worth? (Signal.) *62.* (Repeat step d until firm.)

EXERCISE 5: NUMBER FAMILIES
PLUS/MINUS DISCRIMINATION (SMALL NUMBER OF 1, 2)

a. (Display:) [45:5A]

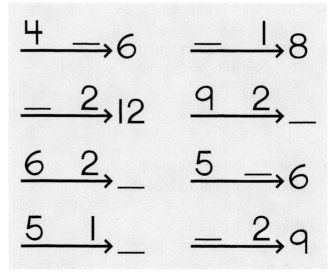

These families have a small number of 2 or a small number of 1. A number is missing in each family. You're going to say the problems for the missing numbers.

- (Point to $\overset{4}{\longrightarrow}6$.) Say the problem for the missing number. Get ready. (Signal.) *6 minus 4.*
- What's the missing number? (Signal.) *2.*
- Say the fact. (Signal.) *6 − 4 = 2.*

b. (Point to $\overset{2}{\longrightarrow}12$.) Say the problem for the missing number. Get ready. (Signal.) *12 minus 2.*

- What's the missing number? (Signal.) *10.*
- Say the fact. (Signal.) *12 − 2 = 10.*

c. (Repeat the following tasks for the remaining problems:)

(Point to __.) Say the problem for the missing number. Get ready.	What's the missing number?	Say the fact.
$\overset{6\quad 2}{\longrightarrow}\_$	6 + 2	8	6 + 2 = 8
$\overset{5\quad 1}{\longrightarrow}\_$	5 + 1	6	5 + 1 = 6
$\overset{1}{\longrightarrow}8$	8 − 1	7	8 − 1 = 7
$\overset{9\quad 2}{\longrightarrow}\_$	9 + 2	11	9 + 2 = 11
$\overset{5}{\longrightarrow}6$	6 − 5	1	6 − 5 = 1
$\overset{2}{\longrightarrow}9$	9 − 2	7	9 − 2 = 7

(Repeat for families that were not firm.)

EXERCISE 6: ADDITION
3 ADDENDS

a. (Distribute unopened workbooks to students.)
*   Open your workbook to Lesson 45 and find part 1.
    (Observe children and give feedback.)
    (Teacher reference:)

    a. 1 + 10 + 2
    b. 2 + 9 + 1
    c. 1 + 8 + 2

*   Touch and read problem A. Get ready.
    (Signal.) *1 plus 10 plus 2.*
*   Say the first problem you'll work. Get ready.
    (Signal.) *1 plus 10.*
*   What's the answer? (Signal.) *11.*
*   Say the next problem. Get ready. (Signal.)
    11 plus 2.
*   What's the answer? (Signal.) *13.*
*   Complete the equation. ✔
*   Touch and read equation A. Get ready.
    (Signal.) *1 + 10 + 2 = 13.*
b. Touch and read problem B. (Signal.) *2 plus 9 plus 1.*
*   Say the first problem you'll work. Get ready.
    (Signal.) *2 plus 9.*
*   What's the answer? (Signal.) *11.*
*   Say the next problem. Get ready. (Signal.)
    11 plus 1.
*   What's the answer? (Signal.) *12.*
c. Touch and read problem C. Get ready.
    (Signal.) *1 plus 8 plus 2.*
*   Say the first problem you'll work. Get ready.
    (Signal.) *1 plus 8.*
*   What's the answer? (Signal.) *9.*
*   Say the next problem. Get ready. (Signal.)
    9 plus 2.
*   What's the answer? (Signal.) *11.*
    (Repeat steps b and c until firm.)
d. Work problems B and C. Put your pencil down when you've completed the equations for part 1.
    (Observe children and give feedback.)
e. Check your work.
*   Touch and read equation B. Get ready.
    (Signal.) *2 + 9 + 1 = 12.*
*   Touch and read equation C. Get ready.
    (Signal.) *1 + 8 + 2 = 11.*

EXERCISE 7: WORD PROBLEMS
WRITING IN COLUMNS

a. Find part 2 on worksheet 45. ✔
    (Teacher reference:)

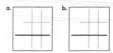

You're going to write the symbols for word problems in columns and work them. The equals bars are already shown.
*   Touch where you'll write the symbols for problem A. ✔
    Listen to problem A: There were 52 birds on a wire. Then 17 more birds landed on the wire. How many birds ended up on the wire?
*   Listen again: There were 52 birds on a wire.
*   What number do you write first? (Signal.) *52.*
*   Does 52 start with hundreds, tens, or ones? (Signal.) *Tens.*
*   Write 52.
    (Observe children and give feedback.)
b. Listen to the next part: Then 17 more birds landed on the wire.
*   Tell me the sign and the number you write for: Then 17 more birds landed on the wire. Get ready. (Signal.) *Plus 17.*
c. Write the plus sign and 17.
    (Observe children and give feedback.)
d. Work problem A. Put your pencil down when you know how many birds ended up on the wire.
    (Observe children and give feedback.)
*   Touch and read the problem and the answer you wrote for A. Get ready. (Signal.) *52 + 17 = 69.*
*   How many birds ended up on the wire? (Signal.) *69.*
e. Touch where you'll write the symbols for problem B. ✔
    Listen to problem B: There were 680 packages on a truck. 170 of those packages were taken off of the truck. How many packages ended up on the truck?
*   Listen again: There were 680 packages on a truck.
*   So what number do you write first? (Signal.) *680.*
*   Does 680 start with hundreds, tens, or ones? (Signal.) *Hundreds.*
*   Write 680.
    (Observe children and give feedback.)

f. Listen to the next part: 170 of those packages were taken off the truck.
• Tell me the sign and the number you write for: 170 of those packages were taken off the truck. Get ready. (Signal.) *Minus 170.*
g. Write the minus sign and 170. (Observe children and give feedback.)
h. Work problem B. Put your pencil down when you know how many packages ended up on the truck. (Observe children and give feedback.)
• Touch and read the problem and the answer you wrote for B. Get ready. (Signal.) *680 – 170 = 510.*
• How many packages ended up on the truck? (Signal.) *510.*
(Answer key:)

```
a.       b.
   5 2      6 8 0
 + 1 7    - 1 7 0
   6 9      5 1 0
```

EXERCISE 8: NUMBER FAMILIES
PLUS/MINUS DISCRIMINATION
(SMALL NUMBER OF 1, 2)

[REMEDY]

a. Find part 3 on worksheet 45. ✔
(Teacher reference:)

[R Part F]

a. 9 → 11
_____

b. __ 1 → 9
_____

c. 9 2 → __
_____

d. __ 2 → 9
_____

These families have a missing number. Some of these families have a small number of 1. Some have a small number of 2. For each family you're going to say the problem for finding the missing number. Then you'll write the fact below the family and complete the family.
b. Family A. The numbers shown are 9 and 11.
• Say the problem for the numbers that are shown. (Signal.) *11 minus 9.*
• What's the missing number? (Signal.) *2.*
• Say the fact. (Signal.) *11 – 9 = 2.*
• Touch the space below family A. ✔
• Write the fact 11 minus 9 equals 2. Then write the missing number in the family. (Observe children and give feedback.)

c. Family B. The numbers shown are 1 and 9.
• Say the problem for the numbers that are shown. (Signal.) *9 minus 1.*
• What's the missing number? (Signal.) *8.*
• Say the fact. (Signal.) *9 – 1 = 8.*
• Write the fact for family B. Then write the missing number in the family. (Observe children and give feedback.)
d. Family C. The numbers shown are 9 and 2.
• Say the problem for the numbers that are shown. (Signal.) *9 plus 2.*
• What's the missing number? (Signal.) *11.*
• Say the fact. (Signal.) *9 + 2 = 11.*
• Write the fact for family C. Then write the missing number in the family. (Observe children and give feedback.)
e. Family D. The numbers shown are 2 and 9.
• Say the problem for the numbers that are shown. (Signal.) *9 minus 2.*
• What's the missing number? (Signal.) *7.*
• Say the fact. (Signal.) *9 – 2 = 7.*
• Write the fact for family D. Then write the missing number in the family. (Observe children and give feedback.)
f. Check your work. You'll read the fact for finding the missing number in each family.
• Family A. Read the fact for the missing number. (Signal.) *11 – 9 = 2.*
• Family B. Read the fact for the missing number. (Signal.) *9 – 1 = 8.*
• Family C. Read the fact for the missing number. (Signal.) *9 + 2 = 11.*
• Family D. Read the fact for the missing number. (Signal.) *9 – 2 = 7.*
You'll write more facts for finding the missing number in families as part of your independent work.

EXERCISE 9: PLACE-VALUE ADDITION

a. Find part 4 on worksheet 45. ✔
(Teacher reference:)

a. ___ + ___ + ___ = 207
b. ___ + ___ + ___ = 200
c.   60 + 1 = ___
d. ___ + ___ + ___ = 416

You'll work these problems, but be careful. One of them is a two-digit number.

- Complete each equation.
  (Observe children and give feedback.)
  (Answer key:)

  a. <u>200</u>+<u>0</u>+<u>7</u>=207

  b. <u>200</u>+<u>0</u>+<u>0</u>=200

  c. 60 + 1 = <u>61</u>

  d. <u>400</u>+<u>10</u>+<u>6</u>=416

b. Check your work.
- Say the place-value addition for problem A.
  (Signal.) *200 + 0 + 7 = 207.*
- Say the place-value addition for problem B.
  (Signal.) *200 + 0 + 0 = 200.*
- Say the place-value addition for problem C.
  (Signal.) *60 + 1 = 61.*
- Say the place-value addition for problem D.
  (Signal.) *400 + 10 + 6 = 416.*

EXERCISE 10: FACTS
SUBTRACTION (SMALL NUMBER OF 1 OR 2)

a. Find part 5 on worksheet 45. ✔
  (Teacher reference:)

  a. 7−2   e. 10−2

  b. 6−1   f. 4−1

  c. 8−6   g. 7−5

  d. 9−8   h. 6−2

  Some of these problems are from families with
  a small number of 1. Some are from families
  with a small number of 2. All these problems
  minus. So are all the answers big numbers or
  small numbers? (Signal.) *Small numbers.*
- Touch and read problem A. (Signal.)
  7 minus 2.
- What numbers are shown? (Signal.) *7 (and) 2.*
- So what's the missing small number?
  (Signal.) *5.*
- Say the fact for 7 minus 2. (Signal.) *7 − 2 = 5.*
b. Touch and read problem B. (Signal.) *6 minus 1.*
- What numbers are shown? (Signal.) *6 (and) 1.*
- So what's the missing small number?
  (Signal.) *5.*
- Say the fact for 6 minus 1. (Signal.) *6 − 1 = 5.*
c. (Repeat the following tasks for problems C
  through H:)
- Touch and read problem __.
- What numbers are shown?
- So what's the missing small number?
- Say the fact for __.
d. Complete the equations for all the problems in
  part 4.
  (Observe children and give feedback.)

(Answer key:)

a. 7−2 =5   e. 10−2 =8

b. 6−1 =5   f. 4−1 =3

c. 8−6 =2   g. 7−5 =2

d. 9−8 =1   h. 6−2 =4

EXERCISE 11: INDEPENDENT WORK

a. Turn to the other side of worksheet 45 and find
  part 6. ✔
  (Teacher reference:)

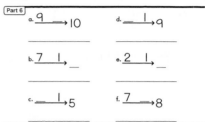

- Touch family A. ✔
- Is a small number missing? (Signal.) *Yes.*
- So do you minus to find the missing number?
  (Signal.) *Yes.*
- Say the problem for finding the missing
  number in family A. (Signal.) *10 minus 9.*
- What's 10 minus 9? (Signal.) *1.*
- That's the missing number. Touch where you'll
  write the missing number. ✔
b. Say the fact for 10 minus 9. (Signal.) *10 − 9 = 1.*
- You'll write 10 minus 9 equals 1 below family
  A. Touch where you'll write the fact. ✔
c. For each family in part 6, you'll write the
  missing number. You'll write the equation for
  finding the missing number below the family.
- You'll start plus facts with the first small
  number. Will you start plus facts with the first
  small number or the other small number?
  (Signal.) *The first small number.*
d. Find part 7. ✔
  You'll work the column problems in part 7.
  Then you'll complete the equations in parts 8
  and 9.
- Your turn: Complete worksheet 45.
  (Observe children and mark incorrect responses
  on children's worksheets as you give feedback.)

Lessons 46–50 Planning Page

	Lesson 46	Lesson 47	Lesson 48	Lesson 49	Lesson 50
Student Learning Objectives	**Exercises** 1. **Say subtraction facts with a small number of 1 or 2** 2. Count forward or backward from a given number 3. Count the value of dimes and pennies in mixed groups 4. Find missing numbers in number families 5. Solve problems with three addends 6. Write the symbols for word problems in columns and solve 7. Find and write missing numbers in number families 8. Say and write place-value equations for 2- or 3-digit numbers 9. Say and write subtraction facts with a small number of 1 or 2 10. Complete work independently	**Exercises** 1. Say addition and subtraction facts with a small number of 1 or 2 2. Count forward or backward from a given number 3. **Say two addition and two subtraction facts for number families with a small number of 10** 4. Count the value of dimes and pennies in mixed groups 5. **Solve addition and subtraction facts** 6. Say and write place-value equations for 3-digit numbers 7. **Find and write the big number for number families with a small number of 10** 8. Write the symbols for word problems in columns and solve 9. Solve problems with three addends 10. Complete work independently	**Exercises** 1. Find the big number and say an addition fact for number families with a small number of 10 2. Count forward or backward from a given number 3. Say addition and subtraction facts with a small number of 1 or 2 4. **Count by fives** 5. Solve problems with three addends 6. **Solve subtraction facts with a small number of 1 or 2** 7. Count the value of dimes and pennies in mixed groups 8. Write the symbols for word problems in columns and solve 9. Find and write the big number for number families with a small number of 10 10. Say and write subtraction facts with a small number of 1 or 2 11. Complete work independently	**Exercises** 1. **Say subtraction facts for number families with a small number of 1 or 2** 2. Count by fives 3. Count forward or backward from a given number 4. Find the big number and say two addition facts and two subtraction facts for number families with a small number of 10 5. **Count a part in a whole and count on for the whole** 6. Say and write addition and subtraction facts with a small number of 1 or 2 7. Count the value of dimes and pennies in mixed groups 8. **Solve problems with three addends in columns** 9. Write the symbols for word problems in columns and solve 10. Complete work independently	**Exercises** 1. Say subtraction facts with a small number of 1 or 2 2. Count forward or backward from a given number 3. Find the big number and say two addition and two subtraction facts for number families with a small number of 10 4. Count by fives 5. **Count by ones and plus and minus 10 and 100** 6. Count a part in a whole and count on for the whole 7. **Find and write missing numbers in number families with a small number of 5** 8. Write the symbols for word problems in columns and solve 9. Solve addition and subtraction facts 10. Solve problems with three addends in columns 11. Complete work independently

Common Core State Standards for Mathematics

1.OA 3	✔	✔		✔	✔
1.OA 5		✔	✔	✔	✔
1.OA 6	✔	✔	✔	✔	✔
1.OA 8	✔	✔	✔	✔	✔
1.NBT 1	✔	✔	✔	✔	✔
1.NBT 2	✔	✔	✔	✔	✔
1.NBT 4	✔	✔	✔	✔	✔
1.NBT 5	✔	✔	✔	✔	✔
Teacher Materials	Presentation Book 2, Board Displays CD or chalkboard				
Student Materials	Workbook 1, Pencil				
Additional Practice	Student Practice Software: Block 2: Activity 1 (1.NBT.1), Activity 2 (1.NBT.1 and 1.NBT.2), Activity 3 (1.NBT.1), Activity 4 (1.NBT.4 and 1.NBT.5), Activity 5 (1.NBT.2 and 1.NBT.4), Activity 6 (1.NBT.2)				
Mastery Test					Student Assessment Book (Present Mastery Test 5 following Lesson 50.)

Lesson 46

EXERCISE 1: NUMBER FAMILIES

MINUS FACTS   REMEDY

a. (Display:)   [46:1A]

11 − 9	7 − 2
9 − 2	10 − 9
5 − 3	5 − 1
8 − 1	6 − 2

Some of these problems are from families with a small number of 1. Some are from families with a small number of 2. All these problems minus. So all the answers are small numbers.

- (Point to **11 − 9**.) Read the problem. (Touch.) *11 minus 9.*
- What numbers are shown? (Signal.) *11 (and) 9.*
- So what's the missing number? (Signal.) *2.*
- Say the fact for 11 minus 9. (Signal.) *11 − 9 = 2.*

b. (Point to **9 − 2**.) Read the problem. (Touch.) *9 minus 2.*
- What numbers are shown? (Signal.) *9 (and) 2.*
- So what's the missing number? (Signal.) *7.*
- Say the fact for 9 minus 2. (Signal.) *9 − 2 = 7.*

c. (Repeat the following tasks for remaining problems:)

(Point to __.) Read the problem.	What numbers are shown?	So what's the missing number?	Say the fact for __.	
5 − 3	5 (and) 3	2	5 − 3	5 − 3 = 2
8 − 1	8 (and) 1	7	8 − 1	8 − 1 = 7
7 − 2	7 (and) 2	5	7 − 2	7 − 2 = 5
10 − 9	10 (and) 9	1	10 − 9	10 − 9 = 1
5 − 1	5 (and) 1	4	5 − 1	5 − 1 = 4
6 − 2	6 (and) 2	4	6 − 2	6 − 2 = 4

(Repeat problems that were not firm.)

EXERCISE 2: MIXED COUNTING

a. Count by hundreds to 1000. (Tap.) *100, 200, 300, 400, 500, 600, 700, 800, 900, 1000.*
- Your turn: Start with 76 and plus tens to 116. Get 76 going. *Seventy-siiix.* Plus tens. (Tap.) *86, 96, 106, 116.*
- Start with 96 and plus tens to 126. Get 96 going. *Ninety-siiix.* Plus tens. (Tap.) *106, 116, 126.*
- Start with 89 and plus tens to 129. Get 89 going. *Eighty-niiine.* Plus tens. (Tap.) *99, 109, 119, 129.*

b. Start with 100 and count by tens to 200. Get 100 going. *One huuundred.* Count. (Tap.) *110, 120, 130, 140, 150, 160, 170, 180, 190, 200.*

c. Start with 100 and count backward by tens. Get 100 going. *One huuundred.* Count backward. (Tap.) *90, 80, 70, 60, 50, 40, 30, 20, 10.*

d. You have 30. When you count by ones, what's the next number? (Signal.) *31.*
- You have 30. When you count by tens, what's the next number? (Signal.) *40.*
- You have 49. When you count by ones, what's the next number? (Signal.) *50.*
- You have 49. When you count by tens, what's the next number? (Signal.) *59.*
- You have 53. When you count by ones, what's the next number? (Signal.) *54.*
- You have 53. When you count by tens, what's the next number? (Signal.) *63.*

EXERCISE 3: COINS

a. (Display:) [46:3A]

- (Point to dimes in 1st row.) This row has dimes and pennies.
- What do you count by for each dime? (Signal.) *Ten.*
- What do you count by for each penny? (Signal.) *One.*
- Count for the dimes. Get ready. (Touch.) *10, 20, 30, 40.* (Keep touching last dime.)
- How many cents do you have? (Signal.) *40.*
- Get 40 going. *Fortyyy.* Count for each penny. (Touch pennies.) *41, 42, 43, 44, 45.* (Repeat until firm.)
- How many cents is this row of coins worth? (Touch.) *45.*

b. Let's do it again, the fast way.
- (Point to 4 dimes.) Count for the dimes. (Touch.) *10, 20, 30, 40.* Keep counting for the pennies. (Touch pennies.) *41, 42, 43, 44, 45.*
- How many cents is this row of coins worth? (Touch.) *45.* (Repeat step b until firm.)

c. (Point to dimes in 2nd row.) You'll count for the coins in this row.
- Count for the dimes. Get ready. (Touch dimes.) *10, 20.* (Keep touching last dime.)
- How many cents do you have? (Signal.) *20.*
- Get 20 going. *Twentyyy.* Count for each penny. (Touch pennies.) *21, 22, 23, 24, 25.* (Repeat until firm.)
- How many cents is this row of coins worth? (Touch.) *25.*

d. Let's do it again, the fast way.
- (Point to 2 dimes.) Count for the dimes. (Touch.) *10, 20.* Keep counting for the pennies. (Touch pennies.) *21, 22, 23, 24, 25.*
- How many cents is this row of coins worth? (Touch.) *25.* (Repeat step d until firm.)

EXERCISE 4: NUMBER FAMILIES
PLUS/MINUS DISCRIMINATION (SMALL NUMBER 1, 2)

a. (Display:) [46:4A]

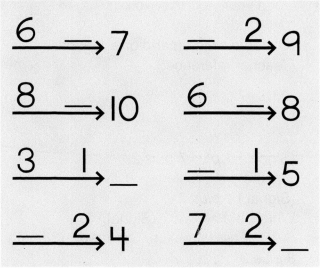

These families have a small number of 2 or a small number of 1. A number is missing in each family. You're going to say the problems for the missing numbers.
- (Point to $\overset{6}{\longrightarrow}$7.) Say the problem for the missing number. Get ready. (Signal.) *7 minus 6.*
- What's the missing number? (Signal.) *1.*
- Say the fact. (Signal.) *7 – 6 = 1.*

b. (Point to $\overset{8}{\longrightarrow}$10.) Say the problem for the missing number. Get ready. (Signal.) *10 minus 8.*
- What's the missing number? (Signal.) *2.*
- Say the fact. (Signal.) *10 – 8 = 2.*

c. (Repeat the following tasks for remaining problems:)

(Point to __.) Say the problem for the missing number.	What's the missing number?	Say the fact.
$\overset{3}{\longrightarrow}$_	3 + 1	4	3 + 1 = 4
$\overset{2}{\longrightarrow}$4	4 – 2	2	4 – 2 = 2
$\overset{2}{\longrightarrow}$9	9 – 2	7	9 – 2 = 7
$\overset{6}{\longrightarrow}$8	8 – 6	2	8 – 6 = 2
$\overset{1}{\longrightarrow}$5	5 – 1	4	5 – 1 = 4
$\overset{7\ 2}{\longrightarrow}$_	7 + 2	9	7 + 2 = 9

(Repeat for families that were not firm.)

EXERCISE 5: ADDITION
3 ADDENDS

a. (Distribute unopened workbooks to students.)
- Open your workbook to Lesson 46 and find part 1.
  (Observe children and give feedback.)
  (Teacher reference:)

  a. 1 + 7 + 2   c. 1 + 9 + 3
  b. 2 + 8 + 5

  Touch and read problem A. Get ready.
  (Signal.) *1 plus 7 plus 2.*
- Say the first problem you'll work. Get ready.
  (Signal.) *1 plus 7.*
- What's the answer? (Signal.) *8.*
- Say the next problem. Get ready. (Signal.)
  8 plus 2.
- What's the answer? (Signal.) *10.*
- Complete the equation. ✔
- Touch and read equation A. Get ready.
  (Signal.) *1 + 7 + 2 = 10.*
b. Touch and read problem B. Get ready.
  (Signal.) *2 plus 8 plus 5.*
- Say the first problem you'll work. Get ready.
  (Signal.) *2 plus 8.*
- What's the answer? (Signal.) *10.*
- Say the next problem. (Signal.) *10 plus 5.*
- What's the answer? (Signal.) *15.*
c. Touch and read problem C. Get ready.
  (Signal.) *1 plus 9 plus 3.*
- Say the first problem you'll work. Get ready.
  (Signal.) *1 plus 9.*
- What's the answer? (Signal.) *10.*
- Say the next problem. Get ready. (Signal.)
  10 plus 3.
  What's the answer? (Signal.) *13.*
  (Repeat steps b and c until firm.)
d. Work problems B and C. Put your pencil down when you've completed the equations for part 1.
  (Observe children and give feedback.)
e. Check your work.
- Touch and read equation B. Get ready.
  (Signal.) *2 + 8 + 5 = 15.*
- Touch and read equation C. Get ready.
  (Signal.) *1 + 9 + 3 = 13.*

EXERCISE 6: WORD PROBLEMS
WRITING IN COLUMNS

a. Find part 2 on worksheet 46. ✔
  (Teacher reference:)

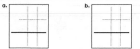

You're going to write the symbols for word problems in columns and work them. The equals bars are already shown.
- Touch where you'll write the symbols for problem A. ✔
  Listen to problem A: There were 54 mice in a field. Then 13 of those mice ran out of the field. How many mice ended up in the field?
- Listen again: There were 54 mice in a field. What number do you write first? (Signal.) *54.*
- Listen to the next part: Then 13 of those mice ran out of the field.
- Tell me the sign and the number you write for: Then 13 of those mice ran out of the field. Get ready. (Signal.) *Minus 13.*
b. Listen again and tell me the symbols for the whole problem: There were 54 mice in a field. Then 13 of those mice ran out of the field.
- Tell me the symbols. Get ready. (Signal.)
  54 minus 13.
  (Repeat step b until firm.)
c. Touch where you'll write 54. ✔
- Touch where you'll write minus. ✔
- Touch where you'll write 13. ✔
  (Repeat step c until firm.)
d. Do 54 and 13 start with hundreds, tens, or ones? (Signal.) *Tens.*
- Write 54 minus 13.
  (Observe children and give feedback.)
e. Work problem A. Put your pencil down when you know how many mice ended up in the field.
  (Observe children and give feedback.)
- Read the problem and the answer you wrote for A. Get ready. (Signal.) *54 – 13 = 41.*
- How many mice ended up in the field?
  (Signal.) *41.*

f. Touch where you'll write the symbols for problem B. ✔
   Listen to problem B: A crate had 162 grapes in it. Mr. Brown put 215 more grapes in the crate. How many grapes ended up in the crate?
   • Listen again: A crate had 162 grapes in it. What number do you write first? (Signal.) *162.*
   • Listen to the next part: Mr. Brown put 215 more grapes in the crate.
   • Tell me the sign and the number you write for: Mr. Brown put 215 more grapes in the crate. Get ready. (Signal.) *Plus 215.*
g. Listen again and tell me the symbols for the whole problem: A crate had 162 grapes in it. Mr. Brown put 215 more grapes in the crate.
   • Tell me the symbols. Get ready. (Signal.) *162 plus 215.*
   (Repeat step g until firm.)
h. Touch where you'll write 162. ✔
   • Touch where you'll write plus. ✔
   • Touch where you'll write 215. ✔
   (Repeat step h until firm.)
i. Do 162 and 215 start with hundreds, tens, or ones? (Signal.) *Hundreds.*
   • Write 162 plus 215.
   (Observe children and give feedback.)
j. Work problem B. Put your pencil down when you know how many grapes ended up in the crate.
   (Observe children and give feedback.)
   (Answer key:)

   • Read the problem and the answer you wrote for B. Get ready. (Signal.) *162 + 215 = 377.*
   • How many grapes ended up in the crate? (Signal.) *377.*

EXERCISE 7: NUMBER FAMILIES
SMALL NUMBER OF 1, 2

a. Find part 3 on worksheet 46. ✔
   (Teacher reference:)

   a. 3 2, ___
   b. ___ 2, 8
   c. 10 ⟶ 11
   d. 2 ⟶ 4
   e. ___ 1, 2
   f. 9 2, ___
   g. 8 1, ___
   h. ___ 2, 9

   Some of these families have a small number of 1. Some have a small number of 2.

• Write the missing number in each family. Raise your hand when you are finished.
   (Observe children and give feedback.)
b. Check your work.
• Family A. The numbers shown are 3 and 2. What's the missing number? (Signal.) *5.*
• Say the problem for the numbers that are shown. (Signal.) *3 plus 2.*
• Say the fact. (Signal.) *3 + 2 = 5.*
c. Family B. The numbers shown are 2 and 8. What's the missing number? (Signal.) *6.*
• Say the problem for the numbers that are shown. (Signal.) *8 minus 2.*
• Say the fact. (Signal.) *8 − 2 = 6.*
d. (Repeat the following tasks for remaining families:)

Family __. The numbers shown are __ and __. What's the missing number?			Say the problem for the numbers that are shown.	Say the fact.
C	10, 11	1	11 − 10	11 − 10 = 1
D	2, 4	2	4 − 2	4 − 2 = 2
E	1, 2	1	2 − 1	2 − 1 = 1
F	9, 2	11	9 + 2	9 + 2 = 11
G	8, 1	9	8 + 1	8 + 1 = 9
H	2, 9	7	9 − 2	9 − 2 = 7

(Repeat families that were not firm.)

EXERCISE 8: PLACE-VALUE ADDITION

a. Find part 4 on worksheet 46. ✔
   (Teacher reference:)

   a. ___ + ___ + ___ = 306
   b. ___ + ___ + ___ = 219
   c. ___ + ___ + ___ = 231
   d. ___ + ___ + ___ = 213
   e. ___ + ___ = 78

   You'll work these problems, but be careful. One of them is a two-digit number.
• Complete each equation.
   (Observe children and give feedback.)
   (Answer key:)

   a. 300 + 0 + 6 = 306
   b. 200 + 10 + 9 = 219
   c. 200 + 30 + 1 = 231
   d. 200 + 10 + 3 = 213
   e. 70 + 8 = 78

b. Check your work.
- Say the place-value addition for problem A.
  (Signal.) *300 + 0 + 6 = 306.*
- Say the place-value addition for problem B.
  (Signal.) *200 + 10 + 9 = 219.*
- Say the place-value addition for problem C.
  (Signal.) *200 + 30 + 1 = 231.*
- Say the place-value addition for problem D.
  (Signal.) *200 + 10 + 3 = 213.*
- Say the place-value addition for problem E.
  (Signal.) *70 + 8 = 78.*

EXERCISE 9: FACTS
SUBTRACTION (SMALL NUMBER OF 1 OR 2)

a. Turn to the other side of worksheet 46 and find
   part 5. ✔
   (Teacher reference:)

   a. $10-8$ d. $8-2$ g. $7-5$
   b. $10-1$ e. $7-6$ h. $9-2$
   c. $12-10$ f. $5-2$

   Some of these problems are from families with
   a small number of 1. Some are from families
   with a small number of 2.
- All these problems minus. So are all the
  answers big numbers or small numbers?
  (Signal.) *Small numbers.*
- Touch and read problem A. (Signal.)
  10 minus 8.
- What numbers are shown? (Signal.)
  10 (and) 8.
- So what's the missing small number?
  (Signal.) *2.*
- Say the fact for 10 minus 8.
  (Signal.) *10 – 8 = 2.*
b. Touch and read problem B. (Signal.) *10 minus 1.*
- What numbers are shown? (Signal.) *10 (and) 1.*
- So what's the missing small number?
  (Signal.) *9.*
- Say the fact for 10 minus 1.
  (Signal.) *10 – 1 = 9.*

c. (Repeat the following tasks for problems
   C through H:)

Touch and read problem __.	What numbers are shown?	So what's the missing small number?	Say the fact for __.
C	12 (and) 10	2	$12 - 10 = 2$
D	8 (and) 2	6	$8 - 2 = 6$
E	7 (and) 6	1	$7 - 6 = 1$
F	5 (and) 2	3	$5 - 2 = 3$
G	7 (and) 5	2	$7 - 5 = 2$
H	9 (and) 2	7	$9 - 2 = 7$

d. Complete the equations for the problems in
   part 5.
   (Observe children and give feedback.)
   (Answer key:)

   a. $10-8=2$ d. $8-2=6$ g. $7-5=2$
   b. $10-1=9$ e. $7-6=1$ h. $9-2=7$
   c. $12-10=2$ f. $5-2=3$

EXERCISE 10: INDEPENDENT WORK

a. Find part 6 on worksheet 46. ✔
   (Teacher reference:)

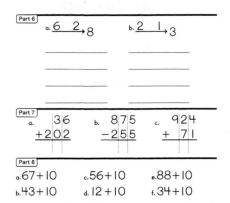

   You'll write four facts below each family.
b. Find part 7. ✔
   You'll work the column problems in part 7.
   Then you'll complete the equations in part 8.
c. Your turn: Complete worksheet 46.
   (Observe children and mark incorrect responses
   on children's worksheets as you give feedback.)

Lesson 47

EXERCISE 1: NUMBER FAMILIES
PLUS/MINUS DISCRIMINATION (SMALL NUMBERS 1, 2)

a. (Display:) [47:1A]

1 + 7	10 − 9	2 + 7
5 − 2	2 + 4	5 − 4
8 − 7	11 − 9	1 + 10

Some of these problems are from families with a small number of 1. Some are from families with a small number of 2.

- (Point to **1 + 7.**) This is 1 plus 7. Is that from a family with a small number of 1 or 2? (Signal.) *1.*
- (Point to **5 − 2.**) This is 5 minus 2. Is that from a family with a small number of 1 or 2? (Signal.) *2.*

b. (Point to **8 − 7.**) Is 8 minus 7 from a family with a small number of 1 or 2? (Signal.) *1.*

c. (Repeat the following task for remaining problems:)

- (Point to __.) Is __ from a family with a small number of 1 or 2?

d. This time, you'll say the facts.

- (Point to **1 + 7.**) What does 1 plus 7 equal? (Signal.) *8.*
- Say the fact. (Signal.) *1 + 7 = 8.*

e. (Point to **5 − 2.**) What does 5 minus 2 equal? (Signal.) *3.*

- Say the fact. (Signal.) *5 − 2 = 3.*

f. (Repeat the following tasks for remaining problems:)

- (Point to __.) What does __ equal?
- Say the fact.

EXERCISE 2: MIXED COUNTING

a. Count by hundreds to 1000. (Signal.) *100, 200, 300, 400, 500, 600, 700, 800, 900, 1000.*

- What's 100 plus 100? (Signal.) *200.*
- What's 400 plus 100? (Signal.) *500.*
- What's 400 plus 200? (Signal.) *600.*

(To Correct:)

- Listen: 4 plus 2. What's 4 plus 2? (Signal.) *6.*
- So what's 4 **hundred** plus 2 **hundred**? (Signal.) *600.*

- What's 800 plus 100? (Signal.) *900.*
- What's 900 plus 100? (Signal.) *1000.*

b. Your turn: Start with 73 and plus tens to 133. Get 73 going. *Seventy-threee.* Plus tens. (Tap.) *83, 93, 103, 113, 123, 133.*

- Start with 96 and plus tens to 136. Get 96 going. *Ninety-siiix.* Plus tens. (Tap.) *106, 116, 126, 136.*
- Start with 89 and plus tens to 139. Get 89 going. *Eighty-niiine.* Plus tens. (Tap.) *99, 109, 119, 129, 139.*

c. You have 45. When you count by ones, what's the next number? (Signal.) *46.*

- You have 45. When you count by tens, what's the next number? (Signal.) *55.*
- You have 75. When you count by ones, what's the next number? (Signal.) *76.*
- You have 75. When you count by tens, what's the next number? (Signal.) *85.*

EXERCISE 3: NUMBER FAMILIES
SMALL NUMBER OF 10   `REMEDY`

a. (Display:) [47:3A]

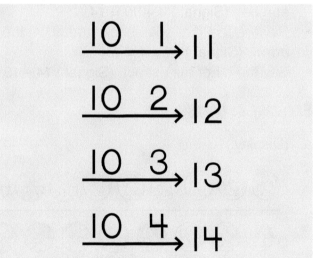

All of these families have a small number of 10.

- (Point to 10 __.) What are the small numbers in this family? (Signal.) *10 (and) 1.*
- What's the big number? (Signal.) *11.*
- Say the fact that starts with the first small number. (Signal.) *10 + 1 = 11.*
- Say the fact that starts with the other small number. (Signal.) *1 + 10 = 11.*
- Say the fact that goes backward down the arrow. (Signal.) *11 − 1 = 10.*
- Say the other minus fact. (Signal.) *11 − 10 = 1.*

b. (Point to 10 2→.) What are the small numbers in this family? (Signal.) *10 (and) 2.*
• What's the big number? (Signal.) *12.*
• Say the fact that starts with the first small number. (Signal.) *10 + 2 = 12.*
• Say the fact that starts with the other small number. (Signal.) *2 + 10 = 12.*
• Say the fact that goes backward down the arrow. (Signal.) *12 – 2 = 10.*
• Say the other minus fact. (Signal.) *12 – 10 = 2.*
c. (Point to 10 3→.) What are the small numbers in this family? (Signal.) *10 (and) 3.*
• What's the big number? (Signal.) *13.*
• Say the fact that starts with the first small number. (Signal.) *10 + 3 = 13.*
• Say the fact that starts with the other small number. (Signal.) *3 + 10 = 13.*
• Say the fact that goes backward down the arrow. (Signal.) *13 – 3 = 10.*
• Say the other minus fact. (Signal.) *13 – 10 = 3.*
d. (Point to 10 4→.) What are the small numbers in this family? (Signal.) *10 (and) 4.*
• What's the big number? (Signal.) *14.*
• Say the fact that starts with the first small number. (Signal.) *10 + 4 = 14.*
• Say the fact that starts with the other small number. (Signal.) *4 + 10 = 14.*
• Say the fact that goes backward down the arrow. (Signal.) *14 – 4 = 10.*
• Say the other minus fact. (Signal.) *14 – 10 = 4.*

EXERCISE 4: COINS

a. (Display:)   [47:4A]

(Point to display.) Each of these rows has dimes and pennies in it.
• What do you count by for each penny? (Touch.) *One.*
• What do you count by for each dime? (Touch.) *Ten.*
• (Point to 1st row.) Count for the dimes. Get ready. (Touch dimes.) *10, 20, 30, 40, 50, 60.* (Keep touching last dime.)
• Get 60 going. *Sixtyyy.* Count for each penny. (Touch pennies.) *61, 62, 63.*
• How many cents is this row worth? (Touch.) *63.*

b. Let's do that again the fast way. Count for the dimes. (Touch dimes.) *10, 20, 30, 40 50, 60.* Count for pennies. (Touch pennies.) *61, 62, 63.*
• How many cents is this row worth? (Touch.) *63.*
c. (Point to 2nd row.) Count for the dimes. Get ready. (Touch dimes.) *10, 20, 30.* (Keep touching last dime.)
• Get 30 going. *Thirtyyy.* Count for each penny. (Touch pennies.) *31, 32, 33, 34, 35.*
• How many cents is this row worth? (Touch.) *35.*
d. Let's do that again the fast way. Count for the dimes. (Touch dimes.) *10, 20, 30.* Count for pennies. (Touch pennies.) *31, 32, 33, 34, 35.*
• How many cents is this row worth? (Touch.) *35.*
(Repeat step c and d until firm.)

EXERCISE 5: FACTS
PLUS/MINUS MIX   REMEDY

a. (Distribute unopened workbooks to students.)
• Open your workbook to Lesson 47 and find part 1.
(Observe children and give feedback.)
(Teacher reference:)   R Part T

a. $9 - 8$	e. $8 - 7$	i. $2 + 7$
b. $6 - 1$	f. $3 - 2$	j. $6 - 4$
c. $5 - 3$	g. $1 + 10$	
d. $1 + 4$	h. $7 - 1$	

You're going to read each problem and say the answer. Then you'll go back and work them.
• Touch and read problem A. (Signal.) *9 minus 8.*
• What's the answer? (Signal.) *1.*
b. Touch and read problem B. (Signal.) *6 minus 1.*
• What's the answer? (Signal.) *5.*
c. (Repeat the following tasks for problems C though J:)
• Touch and read problem __.
• What's the answer?
d. Write answers to all the problems.
(Observe children and give feedback.)
e. Check your work. You'll touch and read each fact.
• Fact A. (Signal.) *9 – 8 = 1.*
• (Repeat for remaining facts:) *B, 6 – 1 = 5; C, 5 – 3 = 2; D, 1 + 4 = 5; E, 8 – 7 = 1; F, 3 – 2 = 1; G, 1 + 10 = 11; H, 7 – 1 = 6; I, 2 + 7 = 9; J, 6 – 4 = 2.*

EXERCISE 6: PLACE-VALUE ADDITION

a. Find part 2 on worksheet 47. ✔
(Teacher reference:)

a. ___ + ___ + __ = 735
b. ___ + ___ + __ = 412
c. ___ + ___ + __ = 208
d. ___ + ___ + __ = 691

• Complete each place-value equation.
(Observe children and give feedback.)
(Answer key:)

a. 700 + 30 + 5 = 735
b. 400 + 10 + 2 = 412
c. 200 + 0 + 8 = 208
d. 600 + 90 + 1 = 691

b. Check your work.
• Say the place-value addition for problem A.
(Signal.) *700 + 30 + 5 = 735.*
• Say the place-value addition for problem B.
(Signal.) *400 + 10 + 2 = 412.*
• Say the place-value addition for problem C.
(Signal.) *200 + 0 + 8 = 208.*
• Say the place-value addition for problem D.
(Signal.) *600 + 90 + 1 = 691.*

EXERCISE 7: NUMBER FAMILIES
SMALL NUMBER OF 10

REMEDY

a. Find part 3 on worksheet 47. ✔
(Teacher reference:)   **R Part O**

a. 10 4, __
b. 10 8, __
c. 10 6, __
d. 10 9, __
e. 10 3, __
f. 10 5, __

These families have the big number missing.
I'll say the small numbers. You'll tell me the big
number.
• Touch family A. ✔
• The small numbers are 10 and 4. What's the
big number? (Signal.) *14.*
b. Touch family B. ✔
• The small numbers are 10 and 8. What's the
big number? (Signal.) *18.*
c. Touch family C. ✔
• The small numbers are 10 and 6. What's the
big number? (Signal.) *16.*

d. Touch family D. ✔
• The small numbers are 10 and 9. What's the
big number? (Signal.) *19.*
e. Touch family E. ✔
• The small numbers are 10 and 3. What's the
big number? (Signal.) *13.*
f. Touch family F. ✔
• The small numbers are 10 and 5. What's the
big number? (Signal.) *15.*
(Repeat for families that were not firm.)

g. Write the missing number in each family. Put
your pencil down when the families in part 3
are complete.
(Observe children and give feedback.)
(Answer key:)

a. 10 4, 14
b. 10 8, 18
c. 10 6, 16
d. 10 9, 19
e. 10 3, 13
f. 10 5, 15

EXERCISE 8: WORD PROBLEMS
WRITING IN COLUMNS

REMEDY

a. Find part 4. ✔
(Teacher reference:)   **R Part D**

You're going to write the symbols for word
problems in columns and work them. The
equals bars are already shown.
• Touch where you'll write the symbols for
problem A. ✔
Listen to problem A: 13 fish were in a tank.
Jim put 42 more fish in the tank. How many
fish ended up in the tank?
• Listen again: 13 fish were in a tank. What
number do you write first? (Signal.) *13.*
• Listen to the next part: Jim put 42 more fish in
the tank. Tell me the sign and the number. Get
ready. (Signal.) *Plus 42.*
b. Listen again and tell me the symbols for the
whole problem: 13 fish were in a tank. Jim put
42 more fish in a tank.
• Tell me the symbols. Get ready. (Signal.)
13 plus 42.
(Repeat step b until firm.)

c. Touch where you'll write 13. ✔
• Touch where you'll write plus. ✔
• Touch where you'll write 42. ✔
   (Repeat step c until firm.)
d. Do 13 and 42 start with hundreds, tens, or
   ones? (Signal.) *Tens.*
• Write 13 plus 42.
   (Observe children and give feedback.)
e. Work problem A. Put your pencil down when
   you know how many fish ended up in the tank.
   (Observe children and give feedback.)
• Read the problem and the answer you wrote
   for A. Get ready. (Signal.) *13 + 42 = 55.*
• How many fish ended up in the tank?
   (Signal.) *55.*
f. Touch where you'll write the symbols for
   problem B. ✔
   Listen to problem B: A shop had 276 tires in it.
   Then people bought 166 of those tires. How
   many tires did the shop end up with?
• Listen again: A shop had 276 tires in it. What
   number do you write first? (Signal.) *276.*
• Listen to the next part: Then people bought
   166 of those tires. Tell me the sign and the
   number. Get ready. (Signal.) *Minus 166.*
g. Listen again and tell me the symbols for the
   whole problem: A shop had 276 tires in it.
   Then people bought 166 of those tires.
• Tell me the symbols. Get ready. (Signal.)
   276 minus 166.
   (Repeat step g until firm.)
h. Touch where you'll write 276. ✔
• Touch where you'll write minus. ✔
• Touch where you'll write 166. ✔
   (Repeat step h until firm.)
• Write 276 minus 166.
   (Observe children and give feedback.)
i. Work problem B. Put your pencil down when
   you know how many tires ended up in the shop.
   (Observe children and give feedback.)
• Read the problem and the answer you wrote
   for B. Get ready. (Signal.) *276 – 166 = 110.*
• How many tires ended up in the shop?
   (Signal.) *110.*

j. Touch where you'll write the symbols for
   problem C. ✔
   Listen to problem C: Jan had 39 dollars in her
   bank account. Jan took 8 of those dollars out
   of her bank account. How many dollars ended
   up in Jan's bank account?
• Listen again: Jan had 39 dollars in her bank
   account. What number do you write first?
   (Signal.) *39.*
• Listen to the next part: Jan took 8 of those
   dollars out of her bank account. Tell me the
   sign and the number. Get ready. (Signal.)
   Minus 8.
k. Listen again and tell me the symbols for the
   whole problem: Jan had 39 dollars in her bank
   account. Jan took 8 of those dollars out of her
   bank account.
• Tell me the symbols. Get ready. (Signal.)
   39 minus 8.
   (Repeat step k until firm.)
l. Touch where you'll write 39. ✔
• Touch where you'll write minus. ✔
• Touch where you'll write 8. ✔
   (Repeat step l until firm.)
m. Does 39 start with hundreds, tens, or ones?
   (Signal.) *Tens.*
• Does 8 start with hundreds, tens, or ones?
   (Signal.) *Ones.*
• Write 39 minus 8.
   (Observe children and give feedback.)
n. Work problem C. Put your pencil down when
   you know how many dollars ended up in Jan's
   bank account.
   (Observe children and give feedback.)
   (Answer key:)

• Read the problem and the answer you wrote
   for C. Get ready. (Signal.) *39 – 8 = 31.*
• How many dollars ended up in Jan's bank
   account? (Signal.) *31.*

EXERCISE 9: ADDITION
3 ADDENDS

a. Turn to the other side of worksheet 47 and find part 5. ✔
(Teacher reference:)

a. 1 + 6 + 2   c. 1 + 9 + 4
b. 8 + 2 + 6   d. 3 + 2 + 1

• Touch and read problem A. Get ready.
(Signal.) *1 plus 6 plus 2.*
• Say the first problem you'll work. Get ready.
(Signal.) *1 plus 6.*
• What's the answer? (Signal.) *7.*
• Say the next problem. Get ready. (Signal.)
7 plus 2.
• What's the answer? (Signal) *9.*
b. Touch and read problem B. Get ready.
(Signal.) *8 plus 2 plus 6.*
• Say the first problem you'll work. Get ready.
(Signal.) *8 plus 2.*
• What's the answer? (Signal.) *10.*
• Say the next problem. Get ready. (Signal.)
10 plus 6.
• What's the answer? (Signal.) *16.*
(Repeat until firm.)
c. Work problems A and B. Put your pencil down when you've completed the equations for problems A and B.
(Observe children and give feedback.)
d. Check your work.
• Touch and read equation A. Get ready.
(Signal.) *1 + 6 + 2 = 9.*
• Touch and read equation B. Get ready.
(Signal.) *8 + 2 + 6 = 16.*
e. Touch and read problem C. (Signal.) *1 plus 9 plus 4.*
• Work problem C. Put your pencil down when you're finished.
(Observe children and give feedback.)
f. Check your work for problem C.
• Say the first problem you worked. Get ready.
(Signal.) *1 plus 9.*
• What's the answer? (Signal.) *10.*
• Say the next problem you worked. Get ready.
(Signal.) *10 plus 4.*
• What's the answer? (Signal.) *14.*
• Touch and read equation C. Get ready.
(Signal.) *1 + 9 + 4 = 14.*

g. Touch and read problem D. Get ready.
(Signal.) *3 plus 2 plus 1.*
• Work problem D. Put your pencil down when you're finished.
(Observe children and give feedback.)
h. Check your work for problem D.
• Say the first problem you worked. Get ready.
(Signal.) *3 plus 2.*
• What's the answer? (Signal.) *5.*
• Say the next problem you worked. Get ready.
(Signal.) *5 plus 1.*
• What's the answer? (Signal.) *6.*
• Touch and read equation D. Get ready.
(Signal.) *3 + 2 + 1 = 6.*

EXERCISE 10: INDEPENDENT WORK

a. Find part 6 on worksheet 47. ✔
(Teacher reference:)

Part 6
a. 3 ⟶ 4   d. 6 1 ⟶ _

b. 2 2 ⟶ _   e. _ 2 ⟶ 7

c. _ 1 ⟶ 9   f. 3 ⟶ 5

Part 7
a. 758   b. 463   c. 892
   −107   +210   − 82

Below each family you'll write the problem and the answer to find the missing number. Then you'll write the missing number in each family. Start plus problems with the first small number.

b. Find part 7. ✔
You'll work the column problems.
c. Your turn: Complete worksheet 47.
(Observe children and mark incorrect responses on children's worksheets as you give feedback.)

Lesson 48

EXERCISE 1: NUMBER FAMILIES
SMALL NUMBER OF 10

REMEDY

a. (Display:) [48:1A]

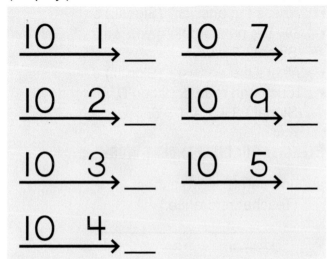

These number families have a small number of 10. The big number is missing.
- (Point to 10→1.) What are the small numbers in this family? (Signal.) *10 (and) 1.*
- What's the big number? (Signal.) *11.*

b. (Point to 10→2.) What are the small numbers in this family? (Signal.) *10 (and) 2.*
- What's the big number? (Signal.) *12.*

c. (Point to 10→3.) What are the small numbers in this family? (Signal.) *10 (and) 3.*
- What's the big number? (Signal.) *13.*
- Say the fact that starts with the first small number. (Signal.) *10 + 3 = 13.*

d. (Repeat the following tasks for remaining families:)
- (Point to __.) What are the small numbers in this family?
- What's the big number?
- Say the fact that starts with the first small number.

EXERCISE 2: MIXED COUNTING

a. Count by hundreds to 1000. (Tap.) *100, 200, 300, 400, 500, 600, 700, 800, 900, 1000.*
- Listen: 6 plus 2. What's 6 plus 2? (Signal.) *8.*
- So what's 600 plus 200? (Signal.) *800.*
- What's 200 plus 200? (Signal.) *400.*
- What's 800 plus 100? (Signal.) *900.*

b. Your turn: Start with 75 and plus tens to 135. Get 75 going. *Seventy-fiiive.* Plus tens. (Tap.) *85, 95, 105, 115, 125, 135.*

- Start with 91 and plus tens to 131. Get 91 going. *Ninety-wuuun.* Plus tens. (Tap.) *101, 111, 121, 131.*

c. I'll start with 4 and plus tens to 34. Fouuur, 14, 24, 34.
- Your turn: Start with 4 and plus tens to 34. Get 4 going. *Fouuur.* Plus tens. (Tap.) *14, 24, 34.*
- Start with 7 and plus tens to 37. Get 7 going. *Sevennn.* Plus tens. (Tap.) *17, 27, 37.*

d. You have 85. When you plus ones, what's the next number? (Signal.) *86.*
- You have 85. When you plus tens, what's the next number? (Signal.) *95.*
- You have 85. When you minus ones, what's the next number? (Signal.) *84.*

EXERCISE 3: NUMBER FAMILIES
PLUS/ MINUS DISCRIMINATION (SMALL NUMBERS 1, 2)

a. (Display:) [48:3A]

7 – 2	8 + 2
1 + 6	10 – 1
8 – 6	7 – 5
4 – 1	9 – 8

Some of these problems are from families with a small number of 1. Some are from families with a small number of 2.
- (Point to **7 – 2.**) This is 7 minus 2. Is that from a family with a small number of 1 or 2? (Signal.) *2.*
- (Point to **1 + 6.**) This is 1 plus 6. Is that from a family with a small number of 1 or 2? (Signal.) *1.*
- (Point to **8 – 6.**) This is 8 minus 6. Is that from a family with a small number of 1 or 2? (Signal.) *2.*

b. (Repeat the following task for remaining problems:)
- (Point to __.) This is __. Is that from a family with a small number of 1 or 2?

c. This time, you'll say the facts.
- (Point to **7 – 2.**) What does 7 minus 2 equal? (Signal.) *5.*
- Say the fact. (Signal.) *7 – 2 = 5.*

Connecting Math Concepts

d. (Point to **1 + 6.**) What does 1 plus 6 equal? (Signal.) *7.*
- Say the fact. (Signal.) *1 + 6 = 7.*
e. (Repeat the following tasks for remaining problems:)
- (Point to __.) What does __ equal?
- Say the fact.

EXERCISE 4: COUNTING BY FIVES

a. (Display:)   [48:4A]

5	10
15	20
25	30
35	40
45	50

(Point to **10.**) The numbers in this column are numbers for counting by 10.
- Read the numbers. Get ready. (Touch tens numbers.) *10, 20, 30, 40, 50.*
b. (Point to **5.**) If I say all of the numbers, I count by 5. My turn to count by fives to 20: 5, 10, 15, 20.
- Your turn: Say the first four numbers with me. Get ready. (Touch numbers as you and children say:) *5, 10, 15, 20.*
(Repeat step b until firm.)
c. (Point to **5.**) Your turn: Say those numbers. Get ready. (Touch.) *5, 10, 15, 20.*
(Repeat until firm.)
- Say the first four numbers without looking. Get ready. (Signal.) *5, 10, 15, 20.*
d. (Point to **25.**) Say the next two numbers with me. Get ready. (Touch numbers.) *25, 30.*
e. (Point to **25.**) Your turn: Say those numbers. Get ready. (Touch numbers.) *25, 30.*
f. (Point to **5.**) Say the numbers for counting by fives to 30. Get ready. (Touch numbers.) *5, 10, 15, 20, 25, 30.*
g. (Do not show display.) My turn to count by fives to 30: 5, 10, 15, 20, 25, 30.
Your turn to count by fives to 30: Get ready. (Tap.) *5, 10, 15, 20, 25, 30.*
(Repeat step g until firm.)

EXERCISE 5: ADDITION
3 ADDENDS

a. (Distribute unopened workbooks to students.)
- Open your workbook to Lesson 48 and find part 1.
(Observe children and give feedback.)
(Teacher reference:)

a. $2 + 6 + 1$ c. $10 + 0 + 6$

b. $1 + 9 + 3$ d. $1 + 5 + 2$

- Touch and read problem A. Get ready. (Signal.) *2 plus 6 plus 1.*
- Say the first problem you'll work. Get ready. (Signal.) *2 plus 6.*
- What's the answer? (Signal.) *8.*
- Say the next problem. Get ready. (Signal.) *8 plus 1.*
- What's the answer? (Signal.) *9.*
- Complete the equation. ✔
- Touch and read equation A. Get ready. (Signal.) *2 + 6 + 1 = 9.*
b. Touch and read problem B. Get ready. (Signal.) *1 plus 9 plus 3.*
- Say the first problem you'll work. Get ready. (Signal.) *1 plus 9.*
- What's the answer? (Signal.) *10.*
- Say the next problem. Get ready. (Signal.) *10 plus 3.*
- What's the answer? (Signal.) *13.*
c. Touch and read problem C. Get ready. (Signal.) *10 plus zero plus 6.*
- Say the first problem you'll work. Get ready. (Signal.) *10 plus zero.*
- What's the answer? (Signal.) *10.*
- Say the next problem. Get ready. (Signal.) *10 plus 6.*
- What's the answer? (Signal.) *16.*
d. Touch and read problem D. Get ready. (Signal.) *1 plus 5 plus 2.*
- Say the first problem you'll work. Get ready. (Signal.) *1 plus 5.*
- What's the answer? (Signal.) *6.*
- Say the next problem. Get ready. (Signal.) *6 plus 2.*
- What's the answer? (Signal.) *8.*
(Repeat steps b through d that were not firm.)
e. Work problems B, C, and D. Put your pencil down when you've completed the equations for part 1.
(Observe children and give feedback.)

f. Check your work.
- Touch and read equation B. Get ready. (Signal.) *1 + 9 + 3 = 13.*
- Touch and read equation C. Get ready. (Signal.) *10 + 0 + 6 = 16.*
- Touch and read equation D. Get ready. (Signal.) *1 + 5 + 2 = 8.*

EXERCISE 6: FACTS

SUBTRACTION (SMALL NUMBER OF 1 OR 2)

a. Find part 2 on worksheet 48. ✔
(Teacher reference:)

a. 9 – 1   d. 3 – 2   g. 7 – 5

b. 7 – 6   e. 10 – 2   h. 6 – 1

c. 9 – 7   f. 12 – 2

Some of these problems are from families with a small number of 2. Some of them are from families with a small number of 1.

- All of these problems minus, so is the answer a small number or the big number? (Signal.) *A small number.*
- Touch and read problem A. (Signal.) *9 minus 1.*
- What's the answer? (Signal.) *8.*

b. Touch and read problem B. (Signal.) *7 minus 6.*
- What's the answer? (Signal.) *1.*

c. (Repeat the following tasks for problems C through H:)

- Touch and read problem __.
- What's the answer?
(Repeat for problems that were not firm.)

d. Complete the facts for part 2. Put your pencil down when you're finished.
(Observe children and give feedback.)
(Answer key:)

a. 9 – 1 = 8   d. 3 – 2 = 1   g. 7 – 5 = 2

b. 7 – 6 = 1   e. 10 – 2 = 8   h. 6 – 1 = 5

c. 9 – 7 = 2   f. 12 – 2 = 10

e. Check your work. You'll touch and read each fact.
- Fact A. (Signal.) *9 – 1 = 8.*
- Fact B. (Signal.) *7 – 6 = 1.*
- (Repeat for:) C, *9 – 7 = 2;* D, *3 – 2 = 1;* E, *10 – 2 = 8;* F, *12 – 2 = 10;* G, *7 – 5 = 2;* H, *6 – 1 = 5.*

EXERCISE 7: COINS

a. (Display:)   REMEDY [48:7A]

- (Point to dimes.) This row has dimes and pennies. My turn to count for the dimes, then count for each penny. (Touch dimes.) 10, 20, 30. (Keep your finger on last dime.) Now I count for each penny. Thirtyyy, (Touch pennies.) 31, 32, 33, 34.
- (Point to dimes.) Your turn: Count for the dimes. Then count for each penny. Get ready. (Touch dimes.) *10, 20, 30.* (Keep your finger on last dime.)
- How many cents do you have? (Signal.) *30.*
- Get 30 going and count for each penny. Get it going. *Thirtyyy.* (Touch pennies.) *31, 32, 33, 34.*

b. Find part 3 on worksheet 48. ✔
(Teacher reference:)   R Part I

I'll count for the coins in row A. Touch each coin as I count.
- Finger over the first coin in row A. ✔
- (Children touch dimes as you count.) 10, 20, thirtyyy. (Children touch pennies as you count.) 31, 32, 33, 34.

c. Now you'll touch and count the coins in row A.
- Finger over the first coin in row A. ✔
- Count for the dimes. (Tap 3.) *10, 20, thirtyyy.* Count for the pennies. *(Tap 4.) 31, 32, 33, 34.* (Repeat until firm.)
- How many cents is row A worth? (Signal.) *34.*

d. I'll count for the coins in row B. Touch each coin as I count.
- Finger over the first coin in row B. ✔
- (Children touch dimes as you count.) 10, twentyyy. (Children touch pennies as you count.) 21, 22, 23.

e. Now you'll touch and count the coins in row B.
- Finger over the first coin in row B. ✔
- Count for the dimes. (Tap 2.) *10, twentyyy.* Count for the pennies. *(Tap 3.) 21, 22, 23.* (Repeat until firm.)
- How many cents is row B worth? (Signal.) *23.* Later, you'll count these rows of coins to yourself and write the number of cents after the equals signs.

EXERCISE 8: WORD PROBLEMS
WRITING IN COLUMNS

a. Find part 4. ✔
   (Teacher reference:)

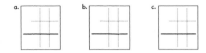

You're going to write the symbols for word problems in columns and work them. The equals bars are already shown.

• Touch where you'll write the symbols for problem A. ✔
  Listen to problem A: A truck had 140 packages on it. Then Ms. Taylor put 29 more packages on it. How many packages ended up on the truck?

• Listen again: A truck had 140 packages on it. What number do you write first? (Signal.) *140.*

• Listen to the next part: Then Ms. Taylor put 29 more packages on it. Tell me the sign and the number you write for that part. Get ready. (Signal.) *Plus 29.*

b. Listen again and tell me the symbols for the whole problem: A truck had 140 packages on it. Then Ms. Taylor put 29 more packages on it.
• Tell me the symbols. Get ready. (Signal.) *140 plus 29.*
  (Repeat step b until firm.)

c. Touch where you'll write 140. ✔
• Touch where you'll write plus. ✔
• Touch where you'll write 29. ✔
  (Repeat step c until firm.)

d. Write 140 plus 29.
   (Observe children and give feedback.)

e. Work problem A. Put your pencil down when you know how many packages ended up on the truck.
   (Observe children and give feedback.)

• Read the problem and the answer you wrote for A. Get ready. (Signal.) *140 + 29 = 169.*

• How many packages ended up on the truck? (Signal.) *169.*

f. Touch where you'll write the symbols for problem B. ✔
  Listen to problem B: Jill rode 7 kilometers in the morning. Jill rode 32 more kilometers in the afternoon. How many kilometers did Jill end up riding?

• Listen again: Jill rode 7 kilometers in the morning. What number do you write first? (Signal.) *7.*

• Listen to the next part: Jill rode 32 more kilometers in the afternoon. Tell me the sign and the number you write for that part. Get ready. (Signal.) *Plus 32.*

g. Listen again and tell me the symbols for the whole problem: Jill rode 7 kilometers in the morning. Jill rode 32 more kilometers in the afternoon.
• Tell me the symbols. Get ready. (Signal.) *7 plus 32.*
  (Repeat step g until firm.)

h. Touch where you'll write 7. ✔
• Touch where you'll write plus. ✔
• Touch where you'll write 32. ✔
  (Repeat step h until firm.)

i. Write 7 plus 32.
   (Observe children and give feedback.)

j. Work problem B. Put your pencil down when you know how many kilometers Jill ended up riding.
   (Observe children and give feedback.)

• Read the problem and the answer you wrote for B. Get ready. (Signal.) *7 + 32 = 39.*

• How many kilometers did Jill end up riding? (Signal.) *39.*

k. Touch where you'll write the symbols for problem C. ✔
  Listen to problem C: There were 258 ducks on a pond. 17 of those ducks flew away. How many ducks ended up on the pond?

• Listen again: There were 258 ducks on a pond. What number do you write first? (Signal.) *258.*

• Listen to the next part: 17 of those ducks flew away. Tell me the sign and the number you write for that part. Get ready. (Signal.) *Minus 17.*

l. Listen again and tell me the symbols for the whole problem: There were 258 ducks on a pond. 17 of those ducks flew away.
• Tell me the symbols. Get ready. (Signal.) *258 minus 17.*
  (Repeat step l until firm.)

m. Touch where you'll write 258. ✔
• Touch where you'll write minus. ✔
• Touch where you'll write 17. ✔
  (Repeat step m until firm.)

n. Write 258 minus 17.
   (Observe children and give feedback.)

o. Work problem C. Put your pencil down when you know how many ducks ended up on the pond.
(Observe children and give feedback.)
(Answer key:)

- Read the problem and the answer you wrote for C. Get ready. (Signal.) *258 – 17 = 241.*
- How many ducks ended up on the pond? (Signal.) *241.*

EXERCISE 9: NUMBER FAMILIES
SMALL NUMBER OF 10   REMEDY

a. Turn to the other side of worksheet 48 and find part 5. ✔
(Teacher reference:)   R Part P

a. 10 3, __   c. 10 5, __   e. 10 6, __
b. 10 7, __   d. 10 2, __   f. 10 1, __

These families have the big number missing. I'll say small numbers. You'll tell me the big number.
- Touch family A. ✔
- The small numbers are 10 and 3. What's the big number? (Signal.) *13.*
b. Touch family B. ✔
- The small numbers are 10 and 7. What's the big number? (Signal.) *17.*
c. Touch family C. ✔
- The small numbers are 10 and 5. What's the big number? (Signal.) *15.*
d. Touch family D. ✔
- The small numbers are 10 and 2. What's the big number? (Signal.) *12.*
e. Touch family E. ✔
- The small numbers are 10 and 6. What's the big number? (Signal.) *16.*
f. Touch family F. ✔
- The small numbers are 10 and 1. What's the big number? (Signal.) *11.*
(Repeat for families that were not firm.)

g. Write the missing big number in each family. Put your pencil down when the families in part 5 are complete.
(Observe children and give feedback.)
(Answer key:)

a. 10 3, 13   c. 10 5, 15   e. 10 6, 16
b. 10 7, 17   d. 10 2, 12   f. 10 1, 11

EXERCISE 10: FACTS
SUBTRACTION (SMALL NUMBER OF 1 OR 2)

a. Find part 6 on worksheet 48. ✔
(Teacher reference:)

            c. 8 – 6   f. 4 – 1
a. 7 – 2   d. 9 – 8   g. 7 – 5
b. 6 – 1   e. 10 – 2   h. 6 – 2

Some of these problems are from families with a small number of 1. Some are from families with a small number of 2.
- All these problems minus. So are all the answers big numbers or small numbers? (Signal.) *Small numbers.*
- Touch and read problem A. (Signal.) *7 minus 2.*
- What numbers are shown? (Signal.) *7 (and) 2.*
- So what's the missing small number? (Signal.) *5.*
- Say the fact for 7 minus 2. (Signal.) *7 – 2 = 5.*
b. Touch and read problem B. (Signal.) *6 minus 1.*
- What numbers are shown? (Signal.) *6 (and) 1.*
- So what's the missing small number? (Signal.) *5.*
- Say the fact for 6 minus 1. (Signal.) *6 – 1 = 5.*
c. (Repeat the following tasks for problems C through H:)
- Touch and read problem __.
- What numbers are shown?
- So what's the missing small number?
- Say the fact for __.
d. Complete the equations for the problems in part 6.
(Observe children and give feedback.)
(Answer key:)

              c. 8 – 6 = 2   f. 4 – 1 = 3
a. 7 – 2 = 5   d. 9 – 8 = 1   g. 7 – 5 = 2
b. 6 – 1 = 5   e. 10 – 2 = 8 h. 6 – 2 = 4

EXERCISE 11: INDEPENDENT WORK

a. Find part 7 on worksheet 48. ✔
   (Teacher reference:)

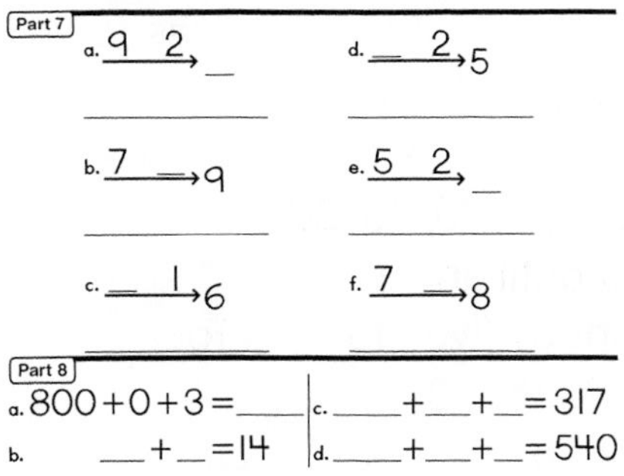

Part 7

a. $\underline{9\quad 2}\longrightarrow\underline{\quad}$ d. $\underline{\quad 2}\longrightarrow 5$

b. $\underline{7}\longrightarrow 9$ e. $\underline{5\quad 2}\longrightarrow\underline{\quad}$

c. $\underline{\quad 1}\longrightarrow 6$ f. $\underline{7}\longrightarrow 8$

Part 8

a. $800 + 0 + 3 =$ ____   c. ____ + __ + _ = 317

b. __ + _ = 14   d. ____ + __ + _ = 540

You'll write the fact for finding the missing number below each family. You'll write the missing number in each family. Start each plus fact with the first small number.

b. Find part 8. ✔
   You'll complete the place-value addition for each problem.

c. Turn to the other side of your worksheet and find part 2. ✔
   You'll count for the coins to yourself and write the number of cents after the equals.

d. Your turn: Complete worksheet 48.
   (Observe children and mark incorrect responses on children's worksheets as you give feedback.)

Lesson 49

EXERCISE 1: FACTS

SUBTRACTION (SMALL NUMBER OF 1 OR 2)   `REMEDY`

a. (Display:)   [49:1A]

4 – 2	6 – 5
6 – 4	11 – 9
5 – 2	10 – 1
8 – 6	7 – 2

Some of these problems are from families with a small number of 1. Some are from families with a small number of 2.

• All of these problems minus. So are all of the answers the big number or a small number? (Signal.) *A small number.*
• (Point to **4 – 2.**) Read the problem. (Touch.) *4 minus 2.*
• What's the answer? (Signal.) *2.*
• Say the fact for 4 minus 2. (Signal.) *4 – 2 = 2.*

b. (Point to **6 – 4.**) Read the problem. (Touch.) *6 minus 4.*
• What's the answer? (Signal.) *2.*
• Say the fact for 6 minus 4. (Signal.) *6 – 4 = 2.*

c. (Repeat the following tasks for the remaining problems:)
• (Point to __.) Read the problem.
• What's the answer?
• Say the fact for __.

EXERCISE 2: COUNTING BY FIVES

a. (Display:)   [49:2A]

5	10
15	20
25	30
35	40
45	50

• (Point to **10.**) Say the numbers for counting by tens to 50. Get ready. (Touch tens numbers.) *10, 20, 30, 40, 50.*

b. (Point to **5.**) My turn to count by fives to 20. (Touch numbers.) 5, 10, 15, 20.
• Your turn: Count by fives to 20. Get ready. (Touch numbers.) *5, 10, 15, 20.*
  (Repeat step b until firm.)

c. My turn to count by fives to 30. (Touch numbers.) 5, 10, 15, 20, 25, 30.
• Your turn: Count by fives to 30. Get ready. (Touch numbers.) *5, 10, 15, 20, 25, 30.*
  (Repeat step c until firm.)

d. (Display:)   [49:2B]

5	—
15	—
25	—
35	—
45	—

The tens numbers are missing.
• (Point to **5.**) Your turn to do it without looking at all the numbers. Count by fives to 30. Get ready. (Touch.) *5, 10, 15, 20, 25, 30.*
  (Repeat step d until firm.)

EXERCISE 3: MIXED COUNTING

a. Start with 5 and plus tens to 35. Get 5 going. *Fiiive.* Plus tens. (Tap.) *15, 25, 35.*
• Start with 8 and plus tens to 38. Get 8 going. *Eieieight.* Plus tens. (Tap.) *18, 28, 38.*
• Start with 73 and plus tens to 133. Get 73 going. *Seventy-threee.* Plus tens. (Tap.) *83, 93, 103, 113, 123, 133.*
• Start with 104 and plus tens to 134. Get 104 going. *One hundred fouuur.* Plus tens. (Tap.) *114, 124, 134.*

b. Count by hundreds to 1000. (Signal.) *100, 200, 300, 400, 500, 600, 700, 800, 900, 1000.*
• What's 600 plus 200? (Signal.) *800.*
• What's 200 plus 300? (Signal.) *500.*
• What's 700 plus 100? (Signal.) *800.*
• What's 500 plus 200? (Signal.) *700.*

c. You have 35. When you count by ones, what's the next number? (Signal.) *36.*
• You have 76. When you count by tens, what's the next number? (Signal.) *86.*
d. Count backward by ones from 30 to 20. Get 30 going. *Thirtyyy.* Count backward. (Tap.) *29, 28, 27, 26, 25, 24, 23, 22, 21, 20.*
• Start with 50 and count backward by ones to 40. Get 50 going. *Fiftyyy.* Count backward. (Tap.) *49, 48, 47, 46, 45, 44, 43, 42, 41, 40.*
• Start with 90 and count backward by tens. Get 90 going. *Ninetyyy.* Count backward. (Tap.) *80, 70, 60, 50, 40, 30, 20, 10.*

EXERCISE 4: NUMBER FAMILIES
SMALL NUMBER OF 10

a. (Display:)   [49:4A]

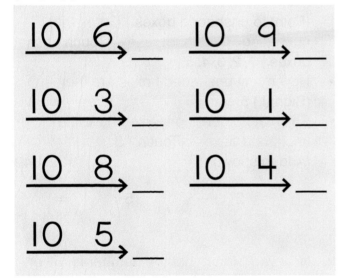

These are families that have a small number of 10.
• (Point to 10 6→.) What are the small numbers in this family? (Signal.) *10 (and) 6.*
• What's the big number? (Signal.) *16.*
b. (Point to 10 3→.) What are the small numbers in this family? (Signal.) *10 (and) 3.*
• What's the big number? (Signal.) *13.*
c. (Repeat the following tasks for remaining families:)
• (Point to __.) What are the small numbers in this family?
• What's the big number?

d. (Point to 10 6→.) This time you'll tell me the facts.
• Say the fact that starts with 10. (Signal.) *10 + 6 = 16.*
• Say the fact that starts with the other small number. (Signal.) *6 + 10 = 16.*
• Say the fact that goes backward down the arrow. (Signal.) *16 − 6 = 10.*
• Say the other minus fact. (Signal.) *16 − 10 = 6.*
e. (Point to 10 3→.) Say the fact that starts with 10. (Signal.) *10 + 3 = 13.*
• Say the fact that starts with the other small number. (Signal.) *3 + 10 = 13.*
• Say the fact that goes backward down the arrow. (Signal.) *13 − 3 = 10.*
• Say the other minus fact. (Signal.) *13 − 10 = 3.*
f. For the rest of the families, you'll say the fact that starts with the first small number and the fact that goes backward down the arrow.
• (Point to 10 8→.) Say the fact that starts with the first small number. (Signal.) *10 + 8 = 18.*
• Say the fact that goes backward down the arrow. (Signal.) *18 − 8 = 10.*
g. (Point to 10 5→.) Say the fact that starts with the first small number. (Signal.) *10 + 5 = 15.*
• Say the fact that goes backward down the arrow. (Signal.) *15 − 5 = 10.*
h. (Repeat the following tasks for remaining families:)
• Say the fact that starts with the first small number.
• Say the fact that goes backward down the arrow.

EXERCISE 5: COUNT ON
PART/WHOLE

a. You're going to get numbers going and count.
• You'll start with 4 and count to 11. What number will you start with? (Signal.) *4.*
• What number will you count to? (Signal.) *11.*
• Get 4 going. *Fouuur.* Count. (Tap.) *5, 6, 7, 8, 9, 10, 11.*
(Repeat until firm.)
b. Now you'll start with 4 and count to 7. What number will you start with? (Signal.) *4.*
• What number will you count to? (Signal.) *7.*
• Get 4 going. *Fouuur.* Count. (Signal.) *5, 6, 7.*
(Repeat step b until firm.)
c. Now you'll start with 2 and count to 6. What number will you start with? (Signal.) *2.*
• What number will you count to? (Signal.) *6.*
• Get 2 going. *Twooo.* Count. (Signal.) *3, 4, 5, 6.*
(Repeat step c until firm.)

d. Now you'll start with 52 and count to 56. What number will you start with? (Signal.) *52.*
• What number will you count to? (Signal.) *56.*
• Get 52 going. *Fifty-twooo.* Count. (Signal.) *53, 54, 55, 56.*
  (Repeat step d until firm.)

e. (Display:) W   [49:5A]

(Point to first shaded box.) You're going to count the shaded boxes and then count the unshaded boxes.
• Count the shaded boxes. Get ready. (Touch.) *1, 2, 3, 4.*
• How many shaded boxes? (Touch.) *4.*
  (Add to show:)   [49:5B]

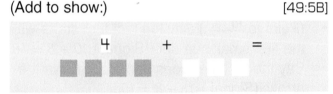

f. (Point to unshaded boxes.) Raise your hand when you know the number for the unshaded boxes. ✔
• How many unshaded boxes are there? (Signal.) *3.*
• So what number do I write above the unshaded boxes? (Touch.) *3.*
  (Add to show:)   [49:5C]

g. Now you'll figure out the number for all of the boxes — the shaded boxes and the unshaded boxes.
• (Point to **4.**) What's the number for the shaded boxes? (Touch.) *4.*
• My turn to get 4 going and count for the unshaded boxes. (Touch 4.) Fouuur. (Touch unshaded boxes.) *5, 6, 7.*
• Your turn: Get 4 going and count for the unshaded boxes. *Fouuur.* Count. (Tap 3.) *5, 6, 7.*
  (Repeat until firm.)
  So I write 7 after the equals.
  (Add to show:)   [49:5D]

h. (Point to **4.**) Read the fact. Get ready. (Touch.) *4 + 3 = 7.*
  (Repeat until firm.)
  Yes, 4 shaded boxes (touch) plus 3 unshaded boxes (touch) equals 7 boxes (touch 7).

i. (Display:) W   [49:5E]

• Raise your hand when you know the number for the shaded boxes. ✔
• How many shaded boxes? (Touch.) *2.*
  (Add to show:)   [49:5F]

j. (Point to unshaded boxes.) Count the unshaded boxes. Get ready. (Touch unshaded boxes.) *1, 2, 3, 4, 5.*
• How many unshaded boxes are there? (Signal.) *5.*
• So what number do I write above the unshaded boxes? (Touch.) *5.*
  (Add to show:)   [49:5G]

k. Now you'll figure out the number for all of the boxes.
• (Point to **2.**) What's the number for the shaded boxes? (Touch.) *2.*
• You'll get 2 going and count for the unshaded boxes. (Touch 2.) Get 2 going. *Twooo.* Count. (Touch unshaded boxes.) *3, 4, 5, 6, 7.*
  (Repeat until firm.)
• How many boxes are there altogether? (Signal.) *7.*
• Where do I write 7? (Signal.) *After the equals.*
  (Add to show:)   [49:5H]

l. (Point to **2.**) Read the fact. Get ready. (Touch.) *2 + 5 = 7.*
  (Repeat until firm.)
  Yes, 2 shaded boxes (touch) plus 5 unshaded boxes (touch) equals 7 boxes (touch 7).

EXERCISE 6: NUMBER FAMILIES

SMALL NUMBER OF 1, 2

a. (Distribute unopened workbooks to students.)
- Open your workbook to Lesson 49 and find part 1.
  (Observe children and give feedback.)
  (Teacher reference:)

a. $2+9$ c. $5-3$ f. $10-9$
b. $7+2$ d. $8+1$ g. $1+7$
             e. $7-2$ h. $6-2$

Some of these problems are from families with a small number of 1. Some are from families with a small number of 2.

- Touch and read problem A. (Signal.) *2 plus 9.*
- Is the answer the big number or a small number? (Signal.) *The big number.*
- What's 2 plus 9? (Signal.) *11.*

b. Touch and read problem B. (Signal.) *7 plus 2.*
- Is the answer the big number or a small number? (Signal.) *The big number.*
- What's 7 plus 2? (Signal.) *9.*

c. (Repeat the following tasks for problems C through H:)
- Touch and read problem __.
- Is the answer the big number or a small number?
- What's __?

d. Complete the equations in part 1. Put your pencil down when you're finished.
  (Observe children and give feedback.)
  (Answer key:)

a. $2+9=11$ c. $5-3=2$ f. $10-9=1$
b. $7+2=9$ d. $8+1=9$ g. $1+7=8$
                e. $7-2=5$ h. $6-2=4$

e. Check your work. You'll touch and read each fact.
- Fact A. (Signal.) *2 + 9 = 11.*
- (Repeat for:) B, *7 + 2 = 9;* C, *5 − 3 = 2;* D, *8 + 1 = 9;* E, *7 − 2 = 5;* F, *10 − 9 = 1;* G, *1 + 7 = 8;* H, *6 − 2 = 4.*

EXERCISE 7: COINS   REMEDY

a. (Display:)   [49:7A]

- (Point to dimes.) This row has dimes and pennies. My turn to count for the dimes and then count for the pennies. (Touch dimes.) 10, 20, 30. (Keep your finger on last dime.) Now I count for each penny. Thirtyyy, (touch pennies) 31, 32.
- (Point to dimes.) Your turn: Count for the dimes. Then count for each penny. Get ready. (Touch dimes.) *10, 20, 30.* (Keep your finger on last dime.)
- How many cents do you have? (Signal.) *30.*
- Get 30 going. *Thirtyyy.* Count for each penny. (Touch pennies.) *31, 32.*

b. Find part 2 on worksheet 49. ✔
  (Teacher reference:)   R Part J

I'll count for the coins in row A. Touch each coin as I count.
- Finger over the first coin in row A. ✔
- (Children touch dimes as you count.) 10, 20, 30, fortyyy. (Children touch pennies as you count.) 41, 42.

c. Now you'll touch and count the coins in row A.
- Finger over the first coin in row A. ✔
- Count for the dimes. (Tap 4.) *10, 20, 30, fortyyy.* Count for the pennies. (Tap 2.) *41, 42.*
  (Repeat until firm.)
- How many cents is row A worth? (Signal.) *42.*

d. I'll count for the coins in row B. Touch each coin as I count.
- Finger over the first coin in row B. ✔
- (Children touch dimes as you count.) 10, twentyyy. (Children touch pennies as you count.) 21, 22, 23, 24, 25.
  (Repeat until firm.)

e. Now you'll touch and count for the coins in row B.
- Finger over the first coin in row B. ✔
- Count for the dimes. (Tap 2.) *10, twentyyy.* Count for the pennies. (Tap 5.) *21, 22, 23, 24, 25.*
  (Repeat until firm.)
- How many cents is row B worth? (Signal.) *25.*
  Later, you'll count for these rows of coins to yourself and write the number of cents after the equals signs.

EXERCISE 8: 3 ADDENDS IN COLUMNS

a. (Display:) W [49:8A]

```
    6          5
    2          2
  + 1        + 2
  ___        ___
```

(Point to **6**.) My turn to read these problems.
(Touch.) 6 plus 2 plus 1.
(Point to **5**.) (Touch.) 5 plus 2 plus 2.
• (Point to **6**.) Your turn: Read this problem.
(Touch.) *6 plus 2 plus 1.*
• (Point to **5**.) Read this problem. (Touch.) *5 plus 2 plus 2.*

b. (Point to **6**.) Read this problem, again. (Touch.)
6 plus 2 plus 1.
• Say the first problem you'll work. (Signal.)
6 plus 2.
• What's the answer? (Signal.) *8.*
• Say the next problem. (Signal.) *8 plus 1.*
• What's the answer? (Signal.) *9.*
(Add to show:) [49:8B]

```
    6          5
    2          2
  + 1        + 2
  ___        ___
    9
```

c. (Point to **5**.) Touch and read this problem. Get
ready. (Signal.) *5 plus 2 plus 2.*
• Say the first problem you'll work. (Signal.)
5 plus 2.
• What's the answer? (Signal.) *7.*
• Say the next problem. (Signal.) *7 plus 2.*
• What's the answer? (Signal.) *9.*
(Add to show:) [49:8C]

```
    6          5
    2          2
  + 1        + 2
  ___        ___
    9          9
```

d. Find part 3 on worksheet 49. ✔
(Teacher reference:)

```
a.    2
      4
   + 1 0
   _____

b.    1
      9
   +  3
   _____

c.    2
      1
   +  2
   _____
```

These problems are like the problems you just
worked.
• Touch and read problem A. (Signal.) *2 plus
4 plus 10.*
• Say the problem for the first part. (Signal.)
2 plus 4.
• What's the answer? (Signal.) *6.*
• Say the next problem. (Signal.) *6 plus 10.*
• What's the answer? (Signal.) *16.*
• Write the answer. ✔
• Touch and read equation A. Get ready.
(Signal.) *2 + 4 + 10 = 16.*

e. Work the rest of the problems in part 3. When
you figure out the answer just write it.
(Observe children and give feedback.)

f. Check your work.
• Touch and read equation B. Get ready.
(Signal.) *1 + 9 + 3 = 13.*
• Touch and read equation C. Get ready.
(Signal.) *2 + 1 + 2 = 5.*

EXERCISE 9: WORD PROBLEMS
WRITING IN COLUMNS

REMEDY

a. Find part 4. ✔
    (Teacher reference:)

R Part E

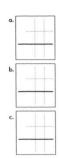

You're going to write the symbols for word problems in columns and work them. The equals bars are already shown.

• Touch where you'll write the symbols for problem A. ✔
    Listen to problem A: Greg had 142 goldfish. Greg bought 6 more goldfish. How many goldfish did Greg end up with?

• Listen again: Greg had 142 goldfish. What number do you write first? (Signal.) *142.*

• Listen to the next part: Greg bought 6 more goldfish. Tell me the sign and the number you write for that part. Get ready. (Signal.) *Plus 6.*

b. Listen again and tell me the symbols for the whole problem: Greg had 142 goldfish. Greg bought 6 more goldfish.

• Tell me the symbols. Get ready. (Signal.) *142 plus 6.*
    (Repeat step b until firm.)

c. Write 142 plus 6.
    (Observe children and give feedback.)

d. Work problem A. Put your pencil down when you know how many goldfish Greg ended up with.
    (Observe children and give feedback.)

• Read the problem and the answer you wrote for A. Get ready. (Signal.) *142 + 6 = 148.*

• How many goldfish did Greg end up with? (Signal.) *148.*

e. Touch where you'll write the symbols for problem B. ✔
    Listen to problem B: There were 435 ants in an anthill. 33 of those ants left the anthill. How many ants were still in the anthill?

• Listen again: There were 435 ants in an anthill. What number do you write first? (Signal.) *435.*

• Listen to the next part: 33 of those ants left the anthill. Tell me the sign and the number you write for that part. Get ready. (Signal.) *Minus 33.*

f. Listen again and tell me the symbols for the whole problem: There were 435 ants in an anthill. 33 of those ants left the anthill.

• Tell me the symbols. Get ready. (Signal.) *435 minus 33.*
    (Repeat step f until firm.)

g. Write 435 minus 33.
    (Observe children and give feedback.)

h. Work problem B. Put your pencil down when you know how many ants were still in the anthill.
    (Observe children and give feedback.)

• Read the problem and the answer you wrote for B. Get ready. (Signal.) *435 – 33 = 402.*

• How many ants were still in the anthill? (Signal.) *402.*

i. Touch where you'll write the symbols for problem C. ✔
    Listen to problem C: There were 28 children in a park. 6 of those children left the park. How many children ended up in the park?

• Listen again: There were 28 children in a park. What number do you write first? (Signal.) *28.*

• Listen to the next part: 6 of those children left the park. Tell me the sign and the number you write for that part. Get ready. (Signal.) *Minus 6.*

j. Listen again and tell me the symbols for the whole problem: There were 28 children in a park. 6 of those children left the park.

• Tell me the symbols. Get ready. (Signal.) *28 minus 6.*
    (Repeat step j until firm.)

k. Write 28 minus 6.
    (Observe children and give feedback.)

l. Work problem C. Put your pencil down when you know how many children ended up in the park.
(Observe children and give feedback.)
(Answer key:)

- Read the problem and the answer you wrote for C. Get ready. (Signal.) *28 – 6 = 22.*
- How many children ended up in the park? (Signal.) *22.*

EXERCISE 10: INDEPENDENT WORK

a. Find part 5. ✔
(Teacher reference:)

a. 9 ⟶ 11

b. 6 2, ⟶ _

c. _ 7 ⟶ 10

d. 4 ⟶ 6

The number families in part 5 have a missing number. Below each family you'll write the problem for finding the missing number and complete the equation. Then you'll write the missing number in the family.

b. Turn to the other side of worksheet 49 and find part 6. ✔
(Teacher reference:)

Part 6

a.	b.	c.	d.
61	398	516	207
+427	−227	− 14	+251

Part 7

a. 800 + 10 + 5   d.   __ + __ = 78

b. __ + __ + __ = 625   e. __ + __ + __ = 450

c. __ + __ + __ = 703   f.   60 + 0

Part 8

a. 17 + 10	d. 54 + 10	g. 65 + 10
b. 85 + 10	e. 23 + 10	h. 45 + 10
c. 73 + 10	f. 12 + 10	i. 19 + 10

Part 9

a. 4 − 4	d. 2 − 2	g. 9 − 0
b. 5 − 0	e. 7 − 1	h. 9 − 9
c. 6 − 1	f. 4 − 2	i. 5 − 1

You'll work the column problems in part 6. You'll complete the place-value addition equations for part 7. You'll complete the equations for parts 8 and 9.

c. Turn to the other side of your worksheet and find part 2. ✔
You'll count for the coins to yourself and write the number of cents after the equals.

d. Complete worksheet 49.
(Observe children and mark incorrect responses on children's worksheets as you give feedback.)

Lesson 50

EXERCISE 1: FACTS
SUBTRACTION

a. (Display:)   [50:1A]

6 – 2	8 – 6
8 – 1	6 – 5
5 – 3	10 – 5

Some of these problems are from families with a small number of 1. Some of them are from families with a small number of 2.
- All of these problems minus. So is the answer the big number or a small number? (Signal.) *A small number.*
- (Point to **6 – 2.**) Read the problem. (Touch.) *6 minus 2.*
- What's 6 minus 2? (Touch.) *4.*
- Say the fact. (Touch.) *6 – 2 = 4.*

b. (Point to **8 – 1.**) Read the problem. (Touch.) *8 minus 1.*
- What's 8 minus 1? (Touch.) *7.*
- Say the fact. (Touch.) *8 – 1 = 7.*

c. (Repeat the following tasks for the remaining problems:)
- (Point to __.) Read the problem.
- What's __?
- Say the fact.

EXERCISE 2: MIXED COUNTING

a. Count by tens to 100. Get ready. (Tap.) *10, 20, 30, 40, 50, 60, 70, 80, 90, 100.*
- Start with 100 and count backward by tens. Get 100 going. *One huuundred.* Count backward. (Tap.) *90, 80, 70, 60, 50, 40, 30, 20, 10.*

b. Start with 10 and count to 20. Get 10 going. *Tennn.* Count. (Tap.) *11, 12, 13, 14, 15, 16, 17, 18, 19, 20.*
- Start with 20 and count backward to 10. Get 20 going. *Twentyyy.* Count backward. (Tap.) *19, 18, 17, 16, 15, 14, 13, 12, 11, 10.*
- Start with 50 and count backward to 40. Get 50 going. *Fiftyyy.* Count backward. (Tap.) *49, 48, 47, 46, 45, 44, 43, 42, 41, 40.*
- (Repeat until firm.)

c. Start at 100 and count by hundreds to 1000. Get 100 going. *One huuundred.* Count. (Tap.) *200, 300, 400, 500, 600, 700, 800, 900, 1000.*
- Start at 900 and count backward by hundreds. Get 900 going. *Nine huuundred.* Count backward. (Tap.) *800, 700, 600, 500, 400, 300, 200, 100.*
- (Repeat step c until firm.)

EXERCISE 3: NUMBER FAMILIES
SMALL NUMBER OF 10

a. (Display:)   [50:3A]

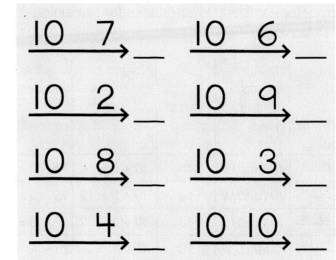

These are families that have a small number of 10.
- (Point to $\xrightarrow{10 \quad 7}$.) What are the small numbers in this family? (Signal.) *10 (and) 7.*
- What's the big number? (Signal.) *17.*
- Say the fact that starts with the first small number. (Signal.) *10 + 7 = 17.*
- Say the fact that starts with the other small number. (Signal.) *7 + 10 = 17.*
- Say the fact that goes backward down the arrow. (Signal.) *17 – 7 = 10.*
- Say the other minus fact. (Signal.) *17 – 10 = 7.*

b. (Point to $\xrightarrow{10 \quad 2}$.) What are the small numbers in this family? (Signal.) *10 (and) 2.*
- What's the big number? (Signal.) *12.*
- Say the fact that starts with the first small number. (Signal.) *10 + 2 = 12.*
- Say the fact that starts with the other small number. (Signal.) *2 + 10 = 12.*
- Say the fact that goes backward down the arrow. (Signal.) *12 – 2 = 10.*
- Say the other minus fact. (Signal.) *12 – 10 = 2.*

c. For the rest of the families, you'll say the fact that starts with the first small number and the fact that goes backward down the arrow.

- (Point to $\overset{10\quad 8}{\longrightarrow}$.) What are the small numbers in this family? (Signal.) *10 (and) 8.*
- What's the big number? (Signal.) *18.*
- Say the fact that starts with the first small number. (Signal.) *10 + 8 = 18.*
- Say the fact that goes backward down the arrow. (Signal.) *18 − 8 = 10.*

d. (Point to $\overset{10\quad 4}{\longrightarrow}$.) What are the small numbers in this family? (Signal.) *10 (and) 4.*

- What's the big number? (Signal.) *14.*
- Say the fact that starts with the first small number. (Signal.) *10 + 4 = 14.*
- Say the fact that goes backward down the arrow. (Signal.) *14 − 4 = 10.*

e. (Repeat the following tasks for remaining families:)

(Point to __.)	What are the small numbers in this family?	What's the big number?	Say the fact that starts with the first small number.	Say the fact that goes backward down the arrow.
$\overset{10\quad 6}{\longrightarrow}$	10 (and) 6	16	10 + 6 = 16	16 − 6 = 10
$\overset{10\quad 9}{\longrightarrow}$	10 (and) 9	19	10 + 9 = 19	19 − 9 = 10
$\overset{10\quad 3}{\longrightarrow}$	10 (and) 3	13	10 + 3 = 13	13 − 3 = 10
$\overset{10\quad 10}{\longrightarrow}$	10 (and) 10	20	10 + 10 = 20	20 − 10 = 10

EXERCISE 4: COUNTING BY FIVES

a. (Display:)   [50:4A]

5	10
15	20
25	30
35	40
45	50

My turn to count by fives to 50. (Touch numbers.) 5, 10, 15, 20, 25, 30, 35, 40, 45, 50.
- Do it with me. (Touch numbers.) *5, 10, 15, 20, 25, 30, 35, 40, 45, 50.*

b. Your turn: Count by fives to 50. Get ready. (Touch numbers.) *5, 10, 15, 20, 25, 30, 35, 40, 45, 50.*
(Repeat step b until firm.)

c. (Display:)   [50:4B]

5	—
15	—
25	—
35	—
45	—

The tens numbers are missing.
- I'll touch. You count by fives to 50. Get ready. (Touch.) *5, 10, 15, 20, 25, 30, 35, 40, 45, 50.*
(Repeat until firm.)

d. (Point to **25.**) When you count by fives what number comes after 25? (Touch.) *30.*
- (Point to **35.**) What number comes after 35? (Touch.) *40.*

e. Start with 25 and count by fives to 50. Get 25 going. *Twenty-fiiive.* Count. (Touch.) *30, 35, 40, 45, 50.*
(Repeat step e until firm.)

EXERCISE 5: MIXED COUNTING
PLUS TENS AND HUNDREDS

a. You have 35. When you count by ones, what's the next number? (Signal.) *36.*
- You have 35. When you plus tens, what's the next number? (Signal.) *45.*
- You have 4. When you count by ones, what's the next number? (Signal.) *5.*
- You have 4. When you plus tens, what's the next number? (Signal.) *14.*
- You have 11. When you count by ones, what's the next number? (Signal.) *12.*
- You have 11. When you plus tens, what's the next number? (Signal.) *21.*

b. What's 200 plus 300? (Signal.) *500.*
- What's 500 plus 200? (Signal.) *700.*

c. My turn: What's 500 minus 100? 400.
- Your turn: What's 500 minus 100? (Signal.) *400.*
(Repeat until firm.)

d. What's 700 minus 100? (Signal.) *600.*
- What's 300 minus 100? (Signal.) *200.*

EXERCISE 6: COUNT ON

PART/WHOLE

REMEDY

a. (Display:) W [50:6A]

You're going to count the shaded boxes, then count the unshaded boxes.
- (Point to first shaded box.) Count the shaded boxes. (Touch.) *1, 2, 3, 4.*
- How many shaded boxes? (Signal.) *4.*
(Add to show:) [50:6B]

b. (Point to unshaded boxes.) Raise your hand when you know the number for the unshaded boxes. ✔
- How many unshaded boxes are there? (Signal.) *2.*
- So what number do I write above the unshaded boxes? (Touch.) *2.*
(Add to show:) [50:6C]

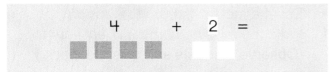

c. Now you'll figure out the number for all of the boxes.
- (Point to **4.**) What's the number for the shaded boxes? (Touch.) *4.*
- My turn to get 4 going and count for the unshaded boxes. Fouuur. (Touch unshaded boxes.) *5, 6.*
- Your turn: Get 4 going and count for the unshaded boxes. *Fouuur.* Count. (Tap 2.) *5, 6.*
(Repeat until firm.)
- How many boxes are there altogether? (Signal.) *6.*
So I write 6 after the equals.
(Add to show:) [50:6D]

d. (Point to **4.**) Read the fact. Get ready. (Touch.) *4 + 2 = 6.*
(Repeat until firm.)
Yes, 4 shaded boxes (touch) plus 2 unshaded boxes (touch) equals 6 boxes (touch 6).

e. (Display:) W [50:6E]

(Point to big circles.) Here's a group of big circles and a group of small circles. Raise your hand when you know the number for the big circles. ✔
- How many big circles are there? (Touch.) *2.*
(Add to show:) [50:6F]

f. (Point to small circles.) Count the small circles. Get ready. (Touch small circles.) *1, 2, 3, 4, 5, 6.*
- How many small circles? (Signal.) *6.*
(Add to show:) [50:6G]

g. Now you'll figure out the number for all of the circles.
- (Point to **2.**) What's the number for the big circles? (Touch.) *2.*
- Get it going. *Twooo.* Count. (Touch small circles.) *3, 4, 5, 6, 7, 8.*
(Repeat until firm.)
- How many circles are there altogether? (Signal.) *8.*
- Where do I write 8? (Signal.) *After the equals.*
(Add to show:) [50:6H]

h. (Point to **2.**) Read the fact. Get ready. (Touch.) *2 + 6 = 8.*
(Repeat until firm.)
Yes, 2 big circles (touch) plus 6 small circles (touch) equals 8 circles (touch 8).

EXERCISE 7: NUMBER FAMILIES
SMALL NUMBER OF 5

a. (Display:)   [50:7A]

$$\underset{\longrightarrow}{\underline{5 \quad 3}} \; \underline{\quad}$$

Here's a number family.
- What are the small numbers? (Signal.)
  5 (and) 3.
- What's the big number? (Signal.) *8.*
- Say the fact that starts with the first small number. (Signal.) *5 + 3 = 8.*
- Say the other plus fact. (Signal.) *3 + 5 = 8.*
- Say the fact that goes backward along the arrow. (Signal.) *8 – 3 = 5.*
- Say the other minus fact. (Signal.) *8 – 5 = 3.*
  (Repeat until firm.)

b. (Display:)   [50:7B]

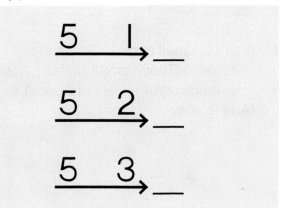

Here are families with a small number of 5. The big number is missing in all of the families.
- (Point to $\overset{5 \quad 1}{\longrightarrow}$.) Say the fact that starts with 5. (Signal.) *5 + 1 = 6.*
- (Point to $\overset{5 \quad 2}{\longrightarrow}$.) Say the fact that starts with 5. (Signal.) *5 + 2 = 7.*
- (Point to $\overset{5 \quad 3}{\longrightarrow}$.) Say the fact that starts with 5. (Signal.) *5 + 3 = 8.*
  (Repeat until firm.)

c. (Distribute unopened workbooks to students.)
- Open your workbook to Lesson 50 and find part 1.
  (Observe children and give feedback.)
  (Teacher reference:)

a. $\underline{\quad 3}_{\longrightarrow}8$ c. $\underline{5 \quad 3}_{\longrightarrow}\underline{\quad}$ e. $\underline{5 \quad}_{\longrightarrow}8$
b. $\underline{5 \quad}_{\longrightarrow}6$ d. $\underline{\quad 2}_{\longrightarrow}8$ f. $\underline{5 \quad 2}_{\longrightarrow}\underline{\quad}$

These are families that have a number missing. Some of them are the new family — small numbers of 5 and 3, big number of 8.
- Touch family A. ✔
- Say the problem for the missing number. (Signal.) *8 minus 3.*
- What's the answer? (Signal.) *5.*
- Say the fact for family A. (Signal.) *8 – 3 = 5.*
d. Touch family B. ✔
- Say the problem for the missing number. (Signal.) *6 minus 5.*
- What's the answer? (Signal.) *1.*
- Say the fact for family B. (Signal.) *6 – 5 = 1.*
e. (Repeat the following tasks for families C through F:)
- Touch family __.
- Say the problem for the missing number.
- What's the answer?
- Say the fact for family __.
f. Write the missing number in each family.
  (Observe students and give feedback.)
  (Answer key:)

a. $\underline{5 \quad 3}_{\longrightarrow}8$ c. $\underline{5 \quad 3}_{\longrightarrow}\underline{8}$ e. $\underline{5 \quad 3}_{\longrightarrow}8$
b. $\underline{5 \quad 1}_{\longrightarrow}6$ d. $\underline{6 \quad 2}_{\longrightarrow}8$ f. $\underline{5 \quad 2}_{\longrightarrow}\underline{7}$

EXERCISE 8: WORD PROBLEMS
WRITING IN COLUMNS

a. Find part 2 on worksheet 50. ✔
(Teacher reference:)

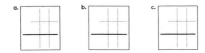

You're going to write the symbols for word problems in columns and work them. The equals bars are already shown.

• Touch where you'll write the symbols for problem A. ✔
Listen to problem A: There were 146 people on a plane. Then 32 of those people got off of the plane. How many people were still on the plane?

• Listen again: There were 146 people on a plane. What number do you write first? (Signal.) *146.*

• Listen to the next part: Then 32 of those people got off of the plane. Tell me the sign and the number you write for that part. Get ready. (Signal.) *Minus 32.*

b. Listen again and tell me the symbols for the whole problem: There were 146 people on a plane. Then 32 of those people got off the plane.

• Tell me the symbols. Get ready. (Signal.) *146 minus 32.*
(Repeat until firm.)

• Write 146 minus 32.
(Observe children and give feedback.)

c. Work problem A. Put your pencil down when you know how many people ended up on the plane.
(Observe children and give feedback.)

• Read the problem and the answer you wrote for A. Get ready. (Signal.) *146 – 32 = 114.*

• How many people were still on the plane? (Signal.) *114.*

d. Touch where you'll write the symbols for problem B. ✔
Listen to problem B: A truck had 107 packages on it. Then somebody put 31 more packages on it. How many packages ended up on the truck?

• Listen again: A truck had 107 packages on it. What number do you write first? (Signal.) *107.*

• Listen to the next part: Then somebody put 31 more packages on it. Tell me the sign and the number you write for that part. Get ready. (Signal.) *Plus 31.*

e. Listen again and tell me the symbols for the whole problem: A truck had 107 packages on it. Then somebody put 31 more packages on it.

• Tell me the symbols. Get ready. (Signal.) *107 plus 31.*
(Repeat until firm.)

• Write 107 plus 31.
(Observe children and give feedback.)

f. Work problem B. Put your pencil down when you know how many packages ended up on the truck.
(Observe children and give feedback.)

• Read the problem and the answer you wrote for B. Get ready. (Signal.) *107 + 31 = 138.*

• How many packages ended up on the truck? (Signal.) *138.*

g. Touch where you'll write the symbols for problem C. ✔
Listen to problem C: Donna had 98 cents. Then she gave 7 cents to her brother. How many cents did she still have left?

• Listen again: Donna had 98 cents. What number do you write first? (Signal.) *98.*

• Listen to the next part: Then she gave 7 cents to her brother. Tell me the sign and the number you write for that part. Get ready. (Signal.) *Minus 7.*

h. Listen again and tell me the symbols for the whole problem: Donna had 98 cents. Then she gave 7 cents to her brother.

• Tell me the symbols. Get ready. (Signal.) *98 minus 7.*
(Repeat until firm.)

• Write 98 minus 7.
(Observe children and give feedback.)

i. Work problem C. Put your pencil down when you know how many cents Donna ended up with.
(Observe children and give feedback.)
(Answer key:)

- Read the problem and the answer you wrote for C. Get ready. (Signal.) *98 – 7 = 91.*
- How many cents did Donna end up with? (Signal.) *91.*

EXERCISE 9: FACTS
PLUS/MINUS MIX

a. Find part 3 on worksheet 50. ✔
(Teacher reference:)

a. 10 – 2   e. 1 + 7
b. 4 – 2   f. 9 – 7
c. 2 + 6   g. 6 – 4
d. 5 – 2   h. 2 + 8

Some of the problems in part 3 plus and some of them minus. You're going to tell me if the big number or a small number is missing.
- Touch and read problem A. Get ready. (Signal.) *10 minus 2.*
- Is the answer the big number or a small number? (Signal.) *A small number.*
- What's 10 minus 2? (Signal.) *8.*
b. Touch and read problem B. Get ready. (Signal.) *4 minus 2.*
- Is the answer the big number or a small number? (Signal.) *A small number.*
- What's 4 minus 2? (Signal.) *2.*
c. (Repeat the following tasks for problems C through H:)

Touch and read problem __. Get ready.	Is the answer the big number or a small number?	What's __?	
C	2 + 6	*The big number.*	2 + 6	8
D	5 – 2	*A small number.*	5 – 2	3
E	1 + 7	*The big number.*	1 + 7	8
F	9 – 7	*A small number.*	9 – 7	2
G	6 – 4	*A small number.*	6 – 4	2
H	2 + 8	*The big number.*	2 + 8	10

d. Complete the equations in part 3. Put your pencil down when you're finished.
(Observe children and give feedback.)

e. Check your work. You'll touch and read each fact.
- Fact A. (Signal.) *10 – 2 = 8.*
- Fact B. (Signal.) *4 – 2 = 2.*
- (Repeat for remaining facts.)

EXERCISE 10: 3 ADDENDS IN COLUMNS

a. Find part 4 on worksheet 50. ✔
(Teacher reference:)

a. 5   c. 7
    2   1
  + 1   +   1
  ___   ___

b. 1   d. 2
    4   3
  + 2   +   2
  ___   ___

- Touch and read problem A. (Signal.) *5 plus 2 plus 1.*
- Say the problem for the first part. (Signal.) *5 plus 2.*
- What's the answer? (Signal.) *7.*
- Say the next problem. Get ready. (Signal.) *7 plus 1.*
- What's the answer? (Signal.) *8.*
- Write the answer. ✔
- Touch and read equation A. Get ready. (Signal.) *5 + 2 + 1 = 8.*
b. Work the rest of the problems in part 4.
(Observe children and give feedback.)
(Answer key:)

a. 5   c. 7
    2   1
  + 1   +   1
  ___   ___
    8   9

b. 1   d. 2
    4   3
  + 2   +   2
  ___   ___
    7   7

c. Check your work.
- Touch and read equation B. Get ready. (Signal.) *1 + 4 + 2 = 7.*
- Touch and read equation C. Get ready. (Signal.) *7 + 1 + 1 = 9.*
- Touch and read equation D. Get ready. (Signal.) *2 + 3 + 2 = 7.*

Connecting Math Concepts

EXERCISE 11: INDEPENDENT WORK

a. Find part 5 on worksheet 50. ✔
   (Teacher reference:)

   $\underline{7 \quad 2}_{,9}$

   _____
   _____
   _____
   _____

   You'll write both plus facts and both minus
   facts on the lines below the family in part 5.

b. Turn to the other side of worksheet 50 and find
   part 6. ✔
   (Teacher reference:)

   Part 6
   a. [coins] =   b. [coins] =

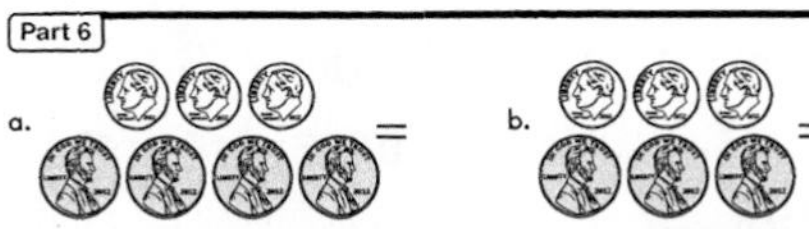

   Part 7
   a.   10+3   d.   __+__=72
   b. ___+__+__=705 e. ___+__+__=190
   c. ___+__+__=284 f. 600+30+0

   Part 8
   a. 238   b. 6 1   c. 375   d. 794
      − 31   +527   + 20   −524

   Part 9
   a. $\underline{6}{\longrightarrow}7$ c. $\underline{1}{\longrightarrow}2$ e. $\underline{5 \quad 2}_{,}\underline{\quad}$

   _____   _____   _____

   b. $\underline{8 \quad 2}_{,}\underline{\quad}$ d. $\underline{3}{\longrightarrow}5$ f. $\underline{\quad 2}_{,8}$

   _____   _____   _____

   You'll count for each group of coins and write
   the number to show how many cents it equals.

c. Find part 7. ✔
   You'll complete the place-value addition
   equations for part 7.
   • Find part 8. ✔
     You'll work the column problems for part 8.
   • Find part 9. ✔
     The number families in part 9 have a missing
     number. Below each family you'll write the
     problem for finding the missing number and
     complete the equation. Then you'll write the
     missing number in the family.

d. Complete worksheet 50.
   (Observe children and mark incorrect responses
   on children's worksheets as you give feedback.)

Mastery Test ⑤

Teacher Presentation

a. Find Test 5 in your test booklet. ✔
• Touch part 1. ✔
(Teacher reference:)

You're going to write numbers in a column.
• Look at the spaces where you'll write the numbers for part 1. ✔
These columns don't have letters for hundreds, tens, and ones.
• Touch the space for the hundreds digit in row A. ✔
• Touch the space for the tens digit in row A. ✔
• Touch the space for the ones digit in row A. ✔
(Repeat until firm.)

b. Listen: You're going to write 514 in row A. What number? (Signal.) *514.*
• Write 514 in row A.
(Observe children.)

c. You're going to write 3 in row B. What number? (Signal.) *3.*
• Write 3 in row B.
(Observe children.)

d. You're going to write 72 in row C. What number? (Signal.) *72.*
• Write 72.
(Observe children.)

e. You're going to write 835 in row D. What number? (Signal.) *835.*
• Write 835.
(Observe children.)

f. You're going to write 18 in row E. What number? (Signal.) *18.*
• Write 18.
(Observe children.)

g. Touch part 2 on Test 5. ✔
(Teacher reference:)

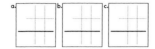

You're going to write the symbols for word problems in columns and work them. The equals bars are already shown.
• Touch where you'll write the symbols for problem A. ✔
Listen to problem A: A shop had 276 tires in it. People bought 166 of those tires. How many tires did the shop end up with?
• Listen again: A shop had 276 tires in it. What number do you write first? (Signal.) *276.*
• Listen to the next part: Then people bought 166 of those tires. Tell me the sign and the number. Get ready. (Signal.) *Minus 166.*
• Listen again and write the symbols for the whole problem: A shop had 276 tires in it. People bought 166 of those tires.
(Observe children.)
• Work problem A. Put your pencil down when you know how many tires ended up in the shop.
(Observe children.)

h. Touch where you'll write the symbols for problem B. ✔
Listen to problem B: A man started with 62 nails. The man found 14 nails. How many nails did the man end up with?
• Listen again: A man had 62 nails. What number do you write first? (Signal.) *62.*
• Listen to the next part: The man found 14 nails. Tell me the sign and the number. Get ready. (Signal.) *Plus 14.*
• Listen again and write the symbols for the whole problem: A man had 62 nails. The man found 14 nails.
(Observe children.)
• Work problem B. Put your pencil down when you know how many nails the man ended up with.
(Observe children.)

i. Touch where you'll write the symbols for problem C. ✔
    Listen to problem C: There were 435 ants in an anthill. 33 of those ants left the anthill. How many ants were still in the anthill?

•   Listen again: There were 435 ants in an anthill. What number do you write first? (Signal.) *435.*

•   Listen to the next part: 33 of those ants left the anthill. Tell me the sign and the number. Get ready. (Signal.) *Minus 33.*

•   Listen again and write the symbols for the whole problem: There were 435 ants in an anthill. 33 of those ants left the anthill.
    (Observe children.)

•   Work problem C. Put your pencil down when you know how many ants ended up in the anthill.
    (Observe children.)

j. Let's go over the problems you'll work for the rest of the test.

•   Touch part 3. ✔
    (Teacher reference:)

    a. 8 ⟶ 9

    b. 9 2,

    c. ___ 2, 9

    d. 6 ⟶ 8

    e. 5 1,

    In part 3, you'll write the fact for finding the missing number in each number family.
    For each family, say the problem for the missing number to yourself. Then write the equation on the space below. Remember to write the missing number in the family after you've completed the equation below.

•   Touch part 4. ✔
    (Teacher reference:)

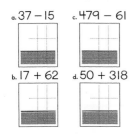

    In part 4, you'll write each problem in a column. Don't work the problem, just write the problem so the digits and the sign are in the right places.

•   Touch part 5. ✔
    (Teacher reference:)

    a. ⓟⓟⓟⓟⓟ
    b. ⓟⓟⓝⓝⓝⓝⓝ
    c. ⓟⓟⓟⓟⓝⓝ
    d. ⓟⓟⓟⓟⓝⓝⓝ

    In part 5, you'll count the cents for each group of coins. Then you'll write an equals and the number to show how many cents each group is worth.

k. Turn to the other side of Test 5 and touch part 6. ✔
    (Teacher reference:)

    a. ___ + ___ + ___ = 209
    b. ___ + ___ + ___ = 813
    c. ___ + ___ + ___ = 500
    d. ___ + ___ + ___ = 406
    e. ___ + ___ + ___ = 620

    You'll complete the place-value equation for each problem in part 6.

•   Touch part 7. ✔
    (Teacher reference:)

    e. 10 5, ___
    a. 10 7, ___   f. 10 2, ___
    b. 10 1, ___   g. 10 9, ___
    c. 10 8, ___   h. 10 6, ___
    d. 10 3, ___   i. 10 4, ___

    The families in part 7 have the big number missing. You'll write the missing big number in each family.

•   Touch part 8. ✔
    (Teacher reference:)

    a. 1 + 7 + 2   c. 1 + 4 + 2
    b. 9 + 1 + 6   d. 1 + 9 + 5

    In part 8 you'll complete equations that plus three numbers.

- Touch part 9. ✔
  (Teacher reference:)

  a. $5+2$

  b. $9-7$

  c. $6-2$

  d. $2+6$

  e. $7-5$

  f. $8-2$

  g. $9+2$

  h. $5-3$

  i. $2+7$

  All of these problems are from number families
  with a small number of 2. You'll complete
  equations in part 9.
- Your turn: Work the rest of the problems on
  Test 5. ✔
  (Direct students where to put their
  assessment books when they are finished.)

Scoring Notes

a. Collect test booklets. Use the Answer Key and
   Passing Criteria Table to score the tests.

Part	Score	Possible Score	Passing Score
Passing Criteria Table — Mastery Test 5			
1	2 for each number	10	10
2	3 for each problem	9	6
3	3 for each problem	15	12
4	2 for each problem	8	6
5	2 for each problem	8	6
6	3 for each equation	15	12
7	1 for each problem	9	8
8	2 for each equation	8	6
9	2 for each equation	18	14
	Total	100	

b. Complete the Mastery Test 5 Remedy
   Summary Sheet to determine whether group
   remedies are needed. Reproducible Remedy
   Summary Sheets are at the back of the
   Answer Key and at the back of the *Teacher's
   Guide.*
- If ¼ or more of the students did not pass a test
  part, present the remedy for that part before
  beginning Lesson 51. The Remedy Table
  follows and is also at the end of the Mastery
  Test 5 Answer Key. Remedies worksheets
  follow Mastery Test 5 in the *Student
  Assessment Book.*

Part	Test Items	Remedy Lesson	Remedy Exercise	Student Material Remedies Worksheet
Remedy Table — Mastery Test 5				
1	Writing Numbers in Columns	35	2	—
		37	7	Part A
		41	10	Part B
2	Word Problems	44	7	Part C
		47	8	Part D
		49	9	Part E
3	Number Families (Writing Facts to Find Missing Numbers)	37	3	—
		42	4	—
		45	8	Part F
4	Writing Column Problems	42	9	Part G
		43	7	Part H
5	Coins	42	5	—
		44	4	—
		48	7	Part I
		49	7	Part J
6	Place-Value Equations	37	6	Part K
		39	8	Part L
		40	7	Part M
		41	7	Part N
7	Number Family (Small Number of 10)	47	3	—
		47	7	Part O
		48	1	—
		48	9	Part P
8	3 Addends	37	2	—
		42	10	Part Q
		43	4	Part R
9	Facts (Mix) (Small Number of 2)	43	1	—
		43	8	Part S
		46	1	—
		47	5	Part T

Retest

Retest individual students on any part failed.

Lessons 51-55 Planning Page

	Lesson 51	Lesson 52	Lesson 53	Lesson 54	Lesson 55
Student Learning Objectives	**Exercises** 1. **Say addition and subtraction facts** 2. Count forward or backward from a given number 3. **Find missing numbers in number families with a small number of 10** 4. Solve problems with three addends in columns 5. Solve addition and subtraction facts 6. Count a part in a whole and count on for the whole 7. Find and write missing numbers in number families 8. Write the symbols for word problems in columns and solve 9. **Learn about the number family table** 10. Complete work independently	**Exercises** 1. Say addition and subtraction facts 2. Count forward or backward from a given number 3. **Say addition facts with 5s and 10s** 4. Solve problems with three addends in columns 5. **Say and write addition and subtraction facts to find missing numbers in number families** 6. Count a part in a whole and count on for the whole 7. Write the symbols for word problems in columns and solve 8. **Solve addition facts with 2s and 3s** 9. Complete work independently	**Exercises** 1. Count by fives 2. Say addition and subtraction facts 3. Count forward or backward from a given number 4. Say addition facts with 5s and 10s 5. Count a part in a whole and count on for the whole 6. Solve addition and subtraction facts 7. **Say and write place-value column equations for 3-digit numbers** 8. Write the symbols for word problems in columns and solve 9. Solve problems with three addends in columns 10. Complete work independently	**Exercises** 1. Say addition and subtraction facts 2. Count forward or backward from a given number 3. **Say two addition and two subtraction facts with a small number of 6** 4. **Recognize that zero is not used as the first digit in a number** 5. Say and write place-value column equations for 3-digit numbers 6. **Solve addition and subtraction facts with 5s and 10s** 7. Count a part in a whole and count on for the whole 8. Solve addition and subtraction facts 9. Write the symbols for word problems in columns and solve 10. Complete work independently	**Exercises** 1. Say addition and subtraction facts 2. Count forward or backward from a given number 3. **Say addition and subtraction facts with a small number of 10** 4. Recognize that zero is not used as the first digit in a number 5. **Find missing numbers in number families with a small number of 6** 6. **Solve 2- and 3-digit subtraction problems when the answer begins with zero** 7. Count a part in a whole and count on for the whole 8. Solve addition and subtraction facts 9. **Identify and count the value of nickels; count and write the value of mixed groups of coins** 10. Say and write place-value column equations for 3-digit numbers 11. Complete work independently

Common Core State Standards for Mathematics					
1.OA 3	✔	✔	✔	✔	✔
1.OA 5-6	✔	✔	✔	✔	✔
1.OA 8	✔	✔	✔	✔	✔
1.NBT 1-2, 4	✔	✔	✔	✔	✔
1.NBT 5		✔	✔	✔	✔
1.NBT 6				✔	
Teacher Materials	Presentation Book 2, Board Displays CD or chalkboard				
Student Materials	Workbook 1, Pencil				
Additional Practice	• Student Practice Software: Block 2: Activity 1 (1.NBT.1), Activity 2 (1.NBT.1 and 1.NBT.2), Activity 3 (1.NBT.1), Activity 4 (1.NBT.4 and 1.NBT.5), Activity 5 (1.NBT.2 and 1.NBT.4), Activity 6 (1.NBT.2) • Math Fact Worksheets 17–18 (After Lesson 51), 19 (After Lesson 53), 20–24 (After Lesson 54)				
Mastery Test					

Lesson 51

EXERCISE 1: FACTS
ADDITION/SUBTRACTION

a. (Display:)   [51:1A]

8 – 2	9 – 7	11 – 2
7 + 2	8 + 1	9 – 8
6 – 4	4 – 3	6 + 2

Some of these problems are from families with a small number of 1. Some of them are from families with a small number of 2.

- (Point to **8 – 2.**) Read the problem. (Touch.) *8 minus 2.*
- What's 8 minus 2? (Signal.) *6.*
- Say the fact. (Signal.) *8 – 2 = 6.*

b. (Point to **7 + 2.**) Read the problem. (Touch.) *7 plus 2.*
- What's 7 plus 2? (Signal.) *9.*
- Say the fact. (Signal.) *7 + 2 = 9.*

c. (Repeat the following tasks for remaining problems:)
- (Point to __.) Read the problem.
- What's __?
- Say the fact.

EXERCISE 2: MIXED COUNTING

a. Count by tens to 50. Get ready. (Signal.) *10, 20, 30, 40, 50.*
(Display:)   [51:2A]

5	—
15	—
25	—
35	—
45	—

Here are some of the numbers for counting by fives.
- (Point to **5.**) Your turn: Count by fives to 50. (Touch.) *5, 10, 15, 20, 25, 30, 35, 40, 45, 50.* (Repeat until firm.)

(Display:)   [51:2B]

—	—
=	=
—	—
=	=
—	—

b. (Point to 1st line.) You can't see any numbers.
- Your turn: Count by fives to 50. Get ready. (Touch.) *5, 10, 15, 20, 25, 30, 35, 40, 45, 50.* (Repeat until firm.)

c. Start with 100 and count backward by tens. Get 100 going. *One huuundred.* Count backward. (Tap.) *90, 80, 70, 60, 50, 40, 30, 20, 10.*

d. Start with 60 and count backward by tens. Get 60 going. *Sixtyyy.* Count backward. (Tap.) *50, 40, 30, 20, 10.*

e. Start with 1000 and count backward by hundreds. Get 1000 going. *One thouuusand.* (Tap.) *900, 800, 700, 600, 500, 400, 300, 200, 100.*

EXERCISE 3: NUMBER FAMILIES
SMALL NUMBER OF 10   [REMEDY]

a. (Display:)   [51:3A]

$$\underset{=}{\overset{4}{\longrightarrow}}14 \qquad \underset{=}{\overset{3}{\longrightarrow}}13$$

$$10 \longrightarrow 12 \qquad 10\ 9 \longrightarrow \_$$

$$10\ 6 \longrightarrow \_ \qquad \overset{7}{\longrightarrow}17$$

$$10 \longrightarrow 17 \qquad 10 \longrightarrow 15$$

All of these families have a small number of 10.
- (Point to ⟹14.) Is the big number or a small number missing? (Signal.) *A small number.*
- Say the problem for the missing number. (Signal.) *14 minus 4.*
- What's 14 minus 4? (Signal.) *10.*

b. (Point to $\overset{10}{\longrightarrow}$12.) Is the big number or a small number missing? (Signal.) *A small number.*
- Say the problem for the missing number. (Signal.) *12 minus 10.*
- What's 12 minus 10? (Signal.) *2.*

c. (Repeat the following tasks for the remaining families:)

(Point to __.) Is the big number or a small number missing?	Say the problem for the missing number.	What's __?	
$\overset{10\ \ 6}{\longrightarrow}\_$	The big number.	10 + 6	10 + 6	16
$\overset{10}{\longrightarrow}17$	A small number.	17 − 10	17 − 10	7
$\overset{3}{\longrightarrow}13$	A small number.	13 − 3	13 − 3	10
$\overset{10\ \ 9}{\longrightarrow}\_$	The big number.	10 + 9	10 + 9	19
$\overset{7}{\longrightarrow}17$	A small number.	17 − 7	17 − 7	10
$\overset{10}{\longrightarrow}15$	A small number.	15 − 10	15 − 10	5

EXERCISE 4: 3 ADDENDS IN COLUMNS

a. (Distribute unopened workbooks to students.)
- Open your workbook to Lesson 51 and find part 1.
(Observe children and give feedback.)
(Teacher reference:)

$$
\begin{array}{ll}
\text{a.} & \begin{array}{r} 2 \\ 8 \\ +\ 4 \\ \hline \end{array} \qquad
\text{c.} \ \begin{array}{r} 3 \\ 2 \\ +\ 1 \\ \hline \end{array} \\[2em]
\text{b.} & \begin{array}{r} 1 \\ 6 \\ +\ 1 \\ \hline \end{array} \qquad
\text{d.} \ \begin{array}{r} 1 \\ 9 \\ +\ 3 \\ \hline \end{array}
\end{array}
$$

- Touch and read problem A. (Signal.) *2 plus 8 plus 4.*
- Say the problem for the first part. (Signal.) *2 plus 8.*
- What's the answer? (Signal.) *10.*
- Say the next problem. Get ready. (Signal.) *10 plus 4.*
- What's the answer? (Signal.) *14.*
- Write the answer. ✔
- Touch and read equation A. Get ready. (Signal.) *2 + 8 + 4 = 14.*

b. Work the rest of the problems in part 1. When you figure out the answer just write it. (Observe children and give feedback.) (Answer key:)

$$
\begin{array}{ll}
\text{a.} & \begin{array}{r} 2 \\ 8 \\ +\ 4 \\ \hline 14 \end{array} \qquad
\text{c.} \ \begin{array}{r} 3 \\ 2 \\ +\ 1 \\ \hline 6 \end{array} \\[2em]
\text{b.} & \begin{array}{r} 1 \\ 6 \\ +\ 1 \\ \hline 8 \end{array} \qquad
\text{d.} \ \begin{array}{r} 1 \\ 9 \\ +\ 3 \\ \hline 13 \end{array}
\end{array}
$$

c. Check your work.
- Touch and read equation B. Get ready. (Signal.) *1 + 6 + 1 = 8.*
- Touch and read equation C. Get ready. (Signal.) *3 + 2 + 1 = 6.*
- Touch and read equation D. Get ready. (Signal.) *1 + 9 + 3 = 13.*

EXERCISE 5: FACTS
PLUS/MINUS MIX

a. Find part 2 on worksheet 51. ✔
(Teacher reference:)

$$
\begin{array}{ll}
\text{a.}\ 8-2 & \text{e.}\ 10-8 \\
\text{b.}\ 6+1 & \text{f.}\ 2+6 \\
\text{c.}\ 11-2 & \text{g.}\ 4-2 \\
\text{d.}\ 4-3 & \text{h.}\ 2+8
\end{array}
$$

These problems are from families that have a small number of 1 or a small number of 2. You're going to tell me if the big number or a small number is missing. Then you'll tell me the answer.
- Touch and read problem A. Get ready. (Touch.) *8 minus 2.*
- Is the answer the big number or a small number? (Signal.) *A small number.*
- What's 8 minus 2? (Signal.) *6.*

b. Touch and read problem B. Get ready. (Touch.) *6 plus 1.*
- Is the answer the big number or a small number? (Signal.) *The big number.*
- What's 6 plus 1? (Signal.) *7.*

c. (Repeat the following tasks for problems C through H:)

Touch and read problem __.		Is the answer the big number or a small number?	What's __?	
C	11 – 2	A small number.	11 – 2	9
D	4 – 3	A small number.	4 – 3	1
E	10 – 8	A small number.	10 – 8	2
F	2 + 6	The big number.	2 + 6	8
G	4 – 2	A small number.	4 – 2	2
H	2 + 8	The big number.	2 + 8	10

d. Complete the equations in part 2. Put your pencil down when you're finished.
(Observe children and give feedback.)
(Answer key:)

a. 8–2=6   e. 10–8=2
b. 6+1=7   f. 2+6=8
c. 11–2=9   g. 4–2=2
d. 4–3=1   h. 2+8=10

e. Check your work. You'll touch and read each fact.
• Fact A. (Signal.) 8 – 2 = 6.
• (Repeat for:) B, 6 + 1 = 7; C, 11 – 2 = 9; D, 4 – 3 = 1; E, 10 – 8 = 2; F, 2 + 6 = 8; G, 4 – 2 = 2; H, 2 + 8 = 10.

EXERCISE 6: COUNT ON
PART/WHOLE

a. Find part 3 on worksheet 51. ✔
(Teacher reference:)

a.
□□ + ⊞ =

These are big boxes and small boxes. You're going to write the number of big boxes, the number of small boxes, and the number for all of the boxes.
• Touch where you'll write the number for the big boxes. ✔
• Touch where you'll write the number for the small boxes. ✔
• Touch where you'll write the number for all of the boxes. ✔
(Repeat until firm.)

b. Write the number for the big boxes and the small boxes. Then stop.
(Observe children and give feedback.)
• What's the number for the big boxes? (Signal.) 2.

• What's the number for the small boxes? (Signal.) 4.
Now you're going to count all the boxes.
• Touch the number you'll get going. ✔
• What number are you touching? (Signal.) 2.
• Get it going. (Signal.) Twooo.
• Touch and count. (Tap.) 3, 4, 5, 6.
(Repeat until firm.)

c. How many boxes are there altogether? (Signal.) 6.
• Touch where you will write 6. ✔
• Do it.
(Observe children and give feedback.)
• Everybody, start with 2. Touch and read the fact. Get ready. (Tap as children say:) 2 + 4 = 6.
(Repeat until firm.)

EXERCISE 7: NUMBER FAMILIES

a. (Display:)   [51:7A]

$$\overset{5\quad\quad 2}{\underset{\longrightarrow}{\rule{3cm}{0.4pt}}}\underline{}$$

$$\overset{5\quad\quad 3}{\underset{\longrightarrow}{\rule{3cm}{0.4pt}}}\underline{}$$

Here are families with a small number of 5. The big number is missing in both of the families.
• (Point to 5—3→.) Here's the family you learned last time.
• (Point to 5—2→.) You know the facts for this number family. What are the small numbers in this family? (Signal.) 5 (and) 2.
• What's the big number? (Signal.) 7.
• (Point to 5—3→.) What are the small numbers in this family? (Signal.) 5 (and) 3.
• What's the big number? (Signal.) 8.
• Say the fact that starts with the first small number. (Signal.) 5 + 3 = 8.
• Say the other plus fact. (Signal.) 3 + 5 = 8.
• Say the fact that goes backward along the arrow. (Signal.) 8 – 3 = 5.
• Say the other minus fact. (Signal.) 8 – 5 = 3.
(Repeat until firm.)

b. Find part 4 on worksheet 51. ✔
(Teacher reference:)

a. $\underline{7} \rightarrow 8$ c. $\underline{\quad 2} \rightarrow 8$ e. $\underline{5 \quad} \rightarrow 6$
b. $\underline{\quad 3} \rightarrow 8$ d. $\underline{5 \quad 3} \rightarrow$ f. $\underline{5 \quad} \rightarrow 8$

These are families that have a number missing. Some of them are the new family — small numbers of 5 and 3, big number of 8.

- Touch family A. ✔
- Say the problem for the missing number. (Signal.) *8 minus 7.*
- What's the answer? (Signal.) *1.*
- Say the fact for family A. (Signal.) *8 – 7 = 1.*

c. (Repeat the following tasks for families B through F:)

- Touch family __.
- Say the problem for the missing number.
- What's the answer?
- Say the fact for family __.

d. Write the missing number in each family. (Observe students and give feedback.)
(Answer key:)

a. $\underline{7 \quad 1} \rightarrow 8$ c. $\underline{6 \quad 2} \rightarrow 8$ e. $\underline{5 \quad 1} \rightarrow 6$
b. $\underline{5 \quad 3} \rightarrow 8$ d. $\underline{5 \quad 3} \rightarrow 8$ f. $\underline{5 \quad 3} \rightarrow 8$

EXERCISE 8: WORD PROBLEMS (COLUMNS)

a. Find part 5 on your worksheet. ✔
(Teacher reference:)

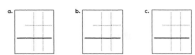

You're going to write the symbols for word problems in columns and work them. The equals bars are already shown.

- Touch where you'll write the symbols for problem A. ✔
Listen to problem A: 148 cows were in a field. Then 41 of those cows left the field. How many cows were still in the field?
- Listen again: 148 cows were in a field. What number do you write first? (Signal.) *148.*
- Listen to the next part: Then 41 of those cows left the field. Tell me the sign and the number you write for that part. Get ready. (Signal.) *Minus 41.*

b. Listen again and tell me the symbols for the whole problem: 148 cows were in a field. Then 41 of those cows left the field.
- Tell me the symbols. Get ready. (Signal.) *148 minus 41.*
(Repeat step b until firm.)

c. Write 148 minus 41 and work problem A. Put your pencil down when you know how many cows were still in the field.
(Observe children and give feedback.)
- Read the problem and the answer you wrote for A. Get ready. (Signal.) *148 – 41 = 107.*
- How many cows were still in the field? (Signal.) *107.*

d. Touch where you'll write the symbols for problem B. ✔
Listen to problem B: There were 224 children in the school. Then 52 more children came to the school. How many children ended up in the school?
- Listen again: There were 224 children in the school. What number do you write first? (Signal.) *224.*
- Listen to the next part: Then 52 more children came to the school. Tell me the sign and the number you write for that part. Get ready. (Signal.) *Plus 52.*

e. Listen again and tell me the symbols for the whole problem: There were 224 children in the school. Then 52 more children came to the school.
- Tell me the symbols. Get ready. (Signal.) *224 plus 52.*
(Repeat step e until firm.)

f. Write 224 plus 52 and work problem B. Put your pencil down when you know how many children ended up in the school.
(Observe children and give feedback.)
- Read the problem and the answer you wrote for B. Get ready. (Signal.) *224 + 52 = 276.*
- How many children ended up in the school? (Signal.) *276.*

g. Touch where you'll write the symbols for problem C. ✔
Listen to problem C: Mr. Jones sold 15 frying pans in the morning. He sold 42 more frying pans in the afternoon. How many frying pans did Mr. Jones end up selling?
- Listen again: Mr. Jones sold 15 frying pans in the morning. What number do you write first? (Signal.) *15.*
- Listen to the next part: He sold 42 more frying pans in the afternoon. Tell me the sign and the number you write for that part. Get ready. (Signal.) *Plus 42.*

h. Listen again and tell me the symbols for the whole problem: Mr. Jones sold 15 frying pans in the morning. He sold 42 more frying pans in the afternoon.

- Tell me the symbols. Get ready. (Signal.)
  15 plus 42.
  (Repeat step h until firm.)

i. Write 15 plus 42 and work problem C. Put your pencil down when you know how many frying pans Mr. Jones ended up selling.
  (Observe children and give feedback.)
  (Answer key:)

- Read the problem and the answer you wrote for C. Get ready. (Signal.) *15 + 42 = 57.*
- How many frying pans did Mr. Jones end up selling? (Signal.) *57.*

EXERCISE 9: NUMBER FAMILY TABLE

a. Open your workbook to the inside front cover and touch the number families. ✔
  (Teacher reference:)

Number Family Table

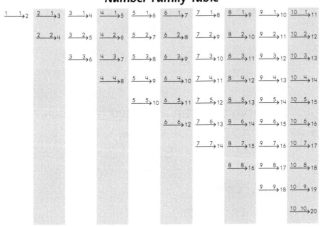

(Praise children who find the right page quickly. Assist children who need help.)
This page shows the number family tables. All the families you're learning are in this table.

- Look at the top row. ✔
  ALL of the families in this row have a small number of 1.
- Touch the family with the small numbers of 1 and 1. ✔
- Touch the family with the small numbers of 2 and 1. ✔
- Touch the family with the small numbers of 3 and 1. ✔

- What are the small numbers in the next family? (Signal.) *4 and 1.*
- What are the small numbers in the next family? (Signal.) *5 and 1.*
- Touch the last family in the row. ✔
- What are the small numbers in that family? (Signal.) *10 and 1.*

b. Go back to the beginning of the next row. ✔
- The small numbers in the family are 2 and 2. All of the families in this row have a small number of 2.
- Touch the family with the small numbers of 3 and 2. ✔
- What are the small numbers in the next family? (Signal.) *4 and 2.*
- What are the small numbers in the next family? (Signal.) *5 and 2.*
- Touch the last family in the row. ✔
- What are the small numbers in that family? (Signal.) *10 and 2.*

c. All of the families below that family have a small number of 10.
  The next family down has small number of 10 and 3.
- What are the small numbers in the next family down? (Signal.) *10 and 4.*
- Touch the number family that is all the way down at the bottom. ✔
- What are the small numbers in that family? (Signal.) *10 and 10.*

EXERCISE 10: INDEPENDENT WORK

a. Turn back to Lesson 51 in your workbook and find part 6. ✔
  (Teacher reference:)

(Praise children who find the page quickly. Assist children who need help.)
- Touch a dime. ✔
- What do you count by for each dime? (Signal.) *10.*
- Touch a penny. ✔
- What do you count by for each penny? (Signal.) *1.*
  Later, you're going to count the cents and write the number to show how much each group of coins is worth.

b. Turn to the other side of worksheet 51 and find part 7. ✔

(Teacher reference:)

Turn to the other side of worksheet 51 and find part 7.

Part 7

a. $\underline{9 \quad 2} \rightarrow 11$　　b. $\underline{10 \quad 7} \rightarrow 17$

_____　　　_____

_____　　　_____

_____　　　_____

Part 8

a.
$$\begin{array}{r} 6\,7\,3 \\ -2\,5\,2 \\ \hline \end{array}$$
b.
$$\begin{array}{r} 8\,3 \\ +4\,1\,2 \\ \hline \end{array}$$
c.
$$\begin{array}{r} 2\,8\,5 \\ -1\,8\,3 \\ \hline \end{array}$$

Part 9

a. ___ + ___ + ___ = 735

b. ___ + ___ + ___ = 806

c. 60 + 9

d. 300 + 80 + 4

e. ___ + ___ = 30

f. ___ + ___ + ___ = 512

Part 10

a.
$$\begin{array}{r} 9 \\ +2 \\ \hline \end{array}$$
b.
$$\begin{array}{r} 7 \\ +2 \\ \hline \end{array}$$
c.
$$\begin{array}{r} 7 \\ -2 \\ \hline \end{array}$$
d.
$$\begin{array}{r} 2 \\ +5 \\ \hline \end{array}$$
e.
$$\begin{array}{r} 8 \\ -6 \\ \hline \end{array}$$
f.
$$\begin{array}{r} 8 \\ +2 \\ \hline \end{array}$$
g.
$$\begin{array}{r} 11 \\ +2 \\ \hline \end{array}$$

You're going to write the two plus facts and the two minus facts for each family. Then you'll work the column problems in part 8, complete the place-value addition equations in part 9, and write answers to the facts in part 10.

c. Complete worksheet 51.

(Observe children and mark incorrect responses on children's worksheets as you give feedback.)

Lesson 52

EXERCISE 1: FACTS
ADDITION/SUBTRACTION

a. (Display:)   [52:1A]

1 + 10	17 − 10	10 + 9
18 − 10	13 − 3	14 − 10
5 + 10	17 − 7	8 + 10

All of these problems are from families with a small number of 10. Remember, if the problem minuses, it starts with the big number.

- (Point to **1 + 10.**) Read the problem. (Touch.) *1 plus 10.*
- Is the answer the big number or a small number? (Signal.) *The big number.*
- Say the fact. (Signal.) *1 + 10 = 11.*

b. (Point to **18 − 10.**) Read the problem. (Touch.) *18 minus 10.*
- Is the answer the big number or a small number? (Signal.) *A small number.*
- Say the fact. (Signal.) *18 − 10 = 8.*

c. (Repeat the following tasks for remaining problems:)

(Point to ___.) Read the problem.	Is the answer the big number or a small number?	Say the fact.
5 + 10	The big number.	5 + 10 = 15
17 − 10	A small number.	17 − 10 = 7
13 − 3	A small number.	13 − 3 = 10
17 − 7	A small number.	17 − 7 = 10
10 + 9	The big number.	10 + 9 = 19
14 − 10	A small number.	14 − 10 = 4
8 + 10	The big number.	8 + 10 = 18

EXERCISE 2: MIXED COUNTING

a. Count by fives to 50. Get ready. (Tap.) *5, 10, 15, 20, 25, 30, 35, 40, 45, 50.*
b. Now you'll plus tens.
- Start with 88 and plus tens to 128. Get 88 going. *Eighty-eieieight.* Plus tens. (Tap.) *98, 108, 118, 128.*
- Start with 83 and plus tens to 123. Get 83 going. *Eighty-threee.* Plus tens. (Tap.) *93, 103, 113, 123.*
- Start with 83 and count by ones to 93. Get 83 going. *Eighty-threee.* Count by ones. (Tap.) *84, 85, 86, 87, 88, 89, 90, 91, 92, 93.*
c. Start with 100 and count backward by tens. Get 100 going. *One huuundred.* Count backward. (Tap.) *90, 80, 70, 60, 50, 40, 30, 20, 10.*
d. You have 24. When you plus tens, what's the next number? (Signal.) *34.*
- You have 104. When you plus tens, what's the next number? (Signal.) *114.*
- You have 104. When you count backward by ones, what's the next number? (Signal.) *103.*

EXERCISE 3: FACTS
PLUS 5, 10

a. (Display:)   [52:3A]

30 + 10	25 + 5
45 + 10	15 + 5
45 + 5	10 + 5
20 + 5	5 + 10

Some of these problems plus 5 and some plus 10.

- (Point to **30 + 10.**) Read the problem. (Touch.) *30 plus 10.*
- What's 30 plus 10? (Signal.) *40.*
- Say the fact. (Signal.) *30 + 10 = 40.*
b. (Point to **45 + 10.**) Read the problem. (Touch.) *45 plus 10.*
- What's 45 plus 10? (Signal.) *55.*
- Say the fact. (Signal.) *45 + 10 = 55.*

c. (Repeat the following tasks for remaining problems:)

(Point to __.) Read the problem.	What's __?		Say the fact.
45 + 5	45 + 5	50	45 + 5 = 50
20 + 5	20 + 5	25	20 + 5 = 25
25 + 5	25 + 5	30	25 + 5 = 30
15 + 5	15 + 5	20	15 + 5 = 20
10 + 5	10 + 5	15	10 + 5 = 15
5 + 10	5 + 10	15	5 + 10 = 15

EXERCISE 4: 3 ADDENDS IN COLUMNS

a. (Distribute unopened workbooks to students.)
• Open your workbook to Lesson 52 and find part 1.
(Observe children and give feedback.)
(Teacher reference:)

a. 7   b. 3   c. 1   d. 1
   2   5   2   9
+ 2   + 2   + 3   + 7

• Touch and read problem A. (Signal.) *7 plus 2 plus 2.*
• Say the problem for the first part. (Signal.) *7 plus 2.*
• What's the answer? (Signal.) *9.*
• Say the next problem. Get ready. (Signal.) *9 plus 2.*
• What's the answer? (Signal.) *11.*
• Write the answer for problem A. ✔
• Touch and read equation A. Get ready. (Signal.) *7 + 2 + 2 = 11.*
b. Work the rest of the problems in part 1.
(Observe children and give feedback.)
(Answer key:)

a. 7   b. 3   c. 1   d. 1
   2   5   2   9
+ 2   + 2   + 3   + 7
  11   10   6   17

c. Check your work.
• Touch and read equation B. Get ready. (Signal.) *3 + 5 + 2 = 10.*
• Touch and read equation C. Get ready. (Signal.) *1 + 2 + 3 = 6.*
• Touch and read equation D. Get ready. (Signal.) *1 + 9 + 7 = 17.*

EXERCISE 5: NUMBER FAMILIES

a. (Display:) [52:5A]

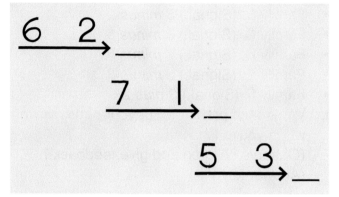

You're going to tell me the big number for each family.
• (Point to ⁶—²→.) What are the small numbers? (Signal.) *6 (and) 2.*
• What's the big number? (Signal.) *8.*
b. (Point to ⁷—¹→.) What are the small numbers? (Signal.) *7 (and) 1.*
• What's the big number? (Signal.) *8.*
c. (Point to ⁵—³→.) What are the small numbers? (Signal.) *5 (and) 3.*
• What's the big number? (Signal.) *8.*
d. Now you're going to say a plus fact and a minus fact for each family.
• (Point to ⁶—²→.) Say the fact that starts with 6. (Signal.) *6 + 2 = 8.*
• Say the fact that goes backward along the arrow. (Signal.) *8 – 2 = 6.*
e. (Point to ⁷—¹→.) Say the fact that starts with 7. (Signal.) *7 + 1 = 8.*
• Say the fact that goes backward along the arrow. (Signal.) *8 – 1 = 7.*
f. (Point to ⁵—³→.) Say the fact that starts with 5. (Signal.) *5 + 3 = 8.*
• Say the fact that goes backward along the arrow. (Signal.) *8 – 3 = 5.*
g. Find part 2 on worksheet 52. ✔
(Teacher reference:)

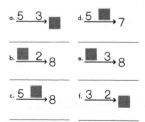

These number families have a missing number.
• On the line below each family write the problem for the missing number. If a big number is missing, write the plus problem that starts with the first small number.
(Observe children and give feedback.)

h. Check your work. You'll touch and read the problem you wrote for each family.
- Family A. (Signal.) *5 plus 3.*
- Family B. (Signal.) *8 minus 2.*
- Family C. (Signal.) *8 minus 5.*
- Family D. (Signal.) *7 minus 5.*
- Family E. (Signal.) *8 minus 3.*
- Family F. (Signal.) *3 plus 2.*

i. Work each problem. Put your pencil down when you're finished.
(Observe children and give feedback.)
(Answer key:)

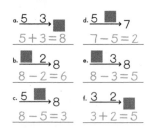

j. Check your work. You'll touch and read the fact for each family.
- Family A. (Signal.) *5 + 3 = 8.*
- Family B. (Signal.) *8 – 2 = 6.*
- Family C. (Signal.) *8 – 5 = 3.*
- Family D. (Signal.) *7 – 5 = 2.*
- Family E. (Signal.) *8 – 3 = 5.*
- Family F. (Signal.) *3 + 2 = 5.*

EXERCISE 6: COUNT ON
PART/WHOLE

a. Find part 3 on worksheet 52. ✔
(Teacher reference:)

You're going to complete the equation to show the number of big boxes, small boxes, and all the boxes.
- Touch where you'll write the number for the big boxes. ✔
- Touch where you'll write the number for the small boxes. ✔
- Touch where you'll write the number for all the boxes. ✔

b. Write the number for the big boxes and the small boxes. Then stop.
(Observe children and give feedback.)
- What's the number for the big boxes? (Signal.) *4.*
- What's the number for the small boxes? (Signal.) *3.*
Now you're going to count all the boxes.

- Touch the number you'll get going. ✔
Then you'll count the small boxes.
- What number are you touching? (Signal.) *4.*
- Get it going. (Signal.) *Fouuur.*
- Touch and count. (Signal.) *5, 6, 7.*
(Repeat until firm.)

c. How many boxes are there altogether? (Signal.) *7.*
- Touch where you will write 7. ✔
- Do it.
(Observe children and give feedback.)
- Touch and read the fact. Get ready. (Tap as children say:) *4 + 3 = 7.*
(Repeat until firm.)

EXERCISE 7: WORD PROBLEMS (COLUMNS)

a. Find part 4 on worksheet 52. ✔
(Teacher reference:)

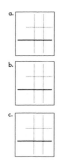

You're going to write the symbols for word problems in columns and work them. The equals bars are already shown.
- Touch where you'll write the symbols for problem A. ✔
Listen to problem A: 217 boys were swimming in the lake. Then 105 of those boys got out of the lake. How many boys were still in the lake?
- Listen again: 217 boys were swimming in the lake. What number do you write first? (Signal.) *217.*
- Listen to the next part: Then 105 of those boys got out of the lake. Tell me the sign and the number you write for that part. Get ready. (Signal.) *Minus 105.*

b. Listen again and tell me the symbols for the whole problem: 217 boys were swimming in the lake. Then 105 of those boys got out of the lake.
- Tell me the symbols. Get ready. (Signal.) *217 minus 105.*
(Repeat step b until firm.)

c. Write 217 minus 105 and work problem A. Put your pencil down when you know how many boys were still in the lake.
(Observe children and give feedback.)
• Read the problem and the answer you wrote for A. Get ready. (Signal.) *217 − 105 = 112.*
• How many boys were still in the lake? (Signal.) *112.*
d. Touch where you'll write the symbols for problem B. ✔
Listen to problem B: There were 358 children in a school. Then 56 of those children left the school. How many children were still in the school?
• Listen again: There were 358 children in a school. What number do you write first? (Signal.) *358.*
• Listen to the next part: Then 56 of those children left the school. Tell me the sign and the number you write for that part. Get ready. (Signal.) *Minus 56.*
e. Listen again and tell me the symbols for the whole problem: There were 358 children in a school. Then 56 of those children left the school.
• Tell me the symbols. Get ready. (Signal.) *358 minus 56.*
(Repeat step e until firm.)
f. Write 358 minus 56 and work problem B. Put your pencil down when you know how many children were still in the school.
(Observe children and give feedback.)
• Read the problem and the answer you wrote for B. Get ready. (Signal.) *358 − 56 = 302.*
• How many children were still in the school? (Signal.) *302.*
g. Touch where you'll write the symbols for problem C. ✔
Listen to problem C: 222 people were on a train. Then 543 more people got on the train. How many people ended up on the train?
• Listen again: 222 people were on a train. What number do you write first? (Signal.) *222.*
• Listen to the next part: Then 543 more people got on the train. Tell me the sign and the number you write for that part. Get ready. (Signal.) *Plus 543.*

h. Listen again and tell me the symbols for the whole problem: 222 people were on a train. Then 543 more people got on the train.
• Tell me the symbols. Get ready. (Signal.) *222 plus 543.*
(Repeat step h until firm.)
i. Write 222 plus 543 and work problem C. Put your pencil down when you know how many people ended up on the train.
(Observe children and give feedback.)
(Answer key:)
(*The problems are not in the configuration they appear on worksheet.*)

• Read the problem and the answer you wrote for C. Get ready. (Signal.) *222 + 543 = 765.*
• How many people ended up on the train? (Signal.) *765.*

EXERCISE 8: FACTS
ADDITION (SMALL NUMBER OF 2 OR 3)   REMEDY

a. Find part 5 on worksheet 52. ✔
(Teacher reference:)   R Part F

a. 2 + 5	d. 6 + 2	g. 3 + 2
b. 2 + 7	e. 8 + 2	h. 5 + 3
c. 3 + 5	f. 3 + 10	i. 2 + 10

Some of these problems are from families with a small number of 2. Some problems are from families with a small number of 3.
• All of these problems plus, so is the answer the big number or a small number? (Signal.) *The big number.*
• Complete the equations.
(Observe children and give feedback.)
b. You'll touch and read each fact.
• Fact A. (Signal.) *2 + 5 = 7.*
• Fact B. (Signal.) *2 + 7 = 9.*
• Fact C. (Signal.) *3 + 5 = 8.*
• Fact D. (Signal.) *6 + 2 = 8.*
• Fact E. (Signal.) *8 + 2 = 10.*
• Fact F. (Signal.) *3 + 10 = 13.*
• Fact G. (Signal.) *3 + 2 = 5.*
• Fact H. (Signal.) *5 + 3 = 8.*
• Fact I. (Signal.) *2 + 10 = 12.*

a. Turn to the other side of worksheet 52 and find part 6. ✔
(Teacher reference:)

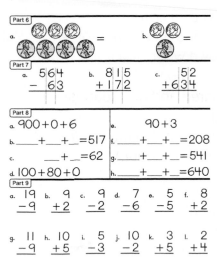

• Touch a dime. ✔
• What do you count by for each dime? (Signal.) *10.*
• Touch a penny. ✔
• What do you count by for each penny? (Signal.) *1.*
You'll count for each group of coins and write the cents after the equals to show how much the group is worth. Then you'll work the problems in parts 7, 8, and 9.

b. Complete worksheet 52.
(Observe children and mark incorrect responses on children's worksheets as you give feedback.)

Lesson 53

EXERCISE 1: COUNT BY FIVES

a. Everybody, count by fives to 50. Get ready.
   (Tap.) *5, 10, 15, 20, 25, 30, 35, 40, 45, 50.*
   (Repeat step a until firm.)
b. This time, you'll tell me the next number for counting by fives.
- Listen: 40. What's the next number for counting by fives? (Signal.) *45.*
- So what's 40 plus 5? (Signal.) *45.*
- Say the fact for 40 plus 5. (Signal.) *40 + 5 = 45.*
c. Listen: 25. What's the next number for counting by fives? (Signal.) *30.*
- So what's 25 plus 5? (Signal.) *30.*
- Say the fact for 25 plus 5. (Signal.) *25 + 5 = 30.*
d. (Repeat the following tasks for 20 and 5:)

Listen: __.	What's the next number for counting by fives?	So what's __?		Say the fact for __.
20	25	20 + 5	25	20 + 5 = 25
5	10	5 + 5	10	5 + 5 = 10

EXERCISE 2: FACTS
ADDITION/SUBTRACTION

a. (Display:)   [53:2A]

12 − 2	7 + 2
2 + 5	6 − 1
8 − 5	5 + 3
5 − 2	9 − 2

Some of these problems are from families with small numbers of 5 and 3. For each problem, you're going to tell me if the big number or a small number is missing. Then you're going to tell me the missing number and say the fact.
- (Point to **12 − 2.**) Read the problem. (Touch.) *12 minus 2.*
- Is the big number or a small number missing? (Signal.) *A small number.*
- What's 12 minus 2? (Signal.) *10.*
- Say the fact for 12 minus 2. (Signal.) *12 − 2 = 10.*

b. (Point to **2 + 5.**) Read the problem. (Touch.) *2 plus 5.*
- Is the big number or a small number missing? (Signal.) *The big number.*
- What's 2 plus 5? (Signal.) *7.*
- Say the fact for 2 plus 5. (Signal.) *2 + 5 = 7.*
c. (Repeat the following tasks for remaining problems:)

(Point to __.) Read the problem.	Is the big number or a small number missing?	What's __?		Say the fact.
8 − 5	A small number.	8 − 5	3	8 − 5 = 3
5 − 2	A small number.	5 − 2	3	5 − 2 = 3
7 + 2	The big number.	7 + 2	9	7 + 2 = 9
6 − 1	A small number.	6 − 1	5	6 − 1 = 5
5 + 3	The big number.	5 + 3	8	5 + 3 = 8
9 − 2	A small number.	9 − 2	7	9 − 2 = 7

(Repeat problems that were not firm.)

EXERCISE 3: MIXED COUNTING

a. Count by fives to 50. Get ready. (Tap.) *5, 10, 15, 20, 25, 30, 35, 40, 45, 50.*
b. Start with 1000 and count backward by hundreds. Get 1000 going. *One thouuusand.* Count backward. (Tap.) *900, 800, 700, 600, 500, 400, 300, 200, 100.*
c. Listen: You have 27. When you plus tens, what's the next number? (Signal.) *37.*
- You have 327. When you plus tens, what's the next number? (Signal.) *337.*
- You have 327. When you count backward by ones, what's the next number? (Signal.) *326.*
- You have 45. When you count by fives, what's the next number? (Signal.) *50.*

EXERCISE 4: FACTS
PLUS 5, 10

a. (Display:)   [53:4A]

20 + 10	5 + 5
20 + 5	45 + 10
10 + 10	45 + 5
35 + 5	60 + 10

Some of these problems plus 5 and some plus 10.

- (Point to **20 + 10.**) Read the problem. (Touch.) *20 plus 10.*
- What's 20 plus 10? (Signal.) *30.*
- Say the fact. (Signal.) *20 + 10 = 30.*

b. (Point to **20 + 5.**) Read the problem. (Touch.) *20 plus 5.*
- What's 20 plus 5? (Signal.) *25.*
- Say the fact. (Signal.) *20 + 5 = 25.*

c. (Repeat the following tasks for remaining problems:)

(Point to __.) Read the problem.	What's __?		Say the fact.
10 + 10	10 + 10	20	10 + 10 = 20
35 + 5	35 + 5	40	35 + 5 = 40
5 + 5	5 + 5	10	5 + 5 = 10
45 + 10	45 + 10	55	45 + 10 = 55
45 + 5	45 + 5	50	45 + 5 = 50
60 + 10	60 + 10	70	60 + 10 = 70

EXERCISE 5: COUNT ON
PART/WHOLE

a. (Distribute unopened workbooks to students.)
- Open your workbook to Lesson 53 and find part 1.
  (Observe children and give feedback.)
  (Teacher reference:)

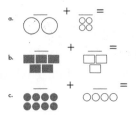

- For problem A, you're going to write an equation for all the circles.
- Write the number of big circles and small circles. Then stop.
  (Observe children and give feedback.)
- What's the number for the big circles? (Signal.) *2.*
- What's the number for the small circles? (Signal.) *4.*

b. Now you're going to count all the circles. You'll get 2 going. Then you'll touch and count for the small circles.
- Touch the number 2. ✔
- Get it going. (Signal.) *Twooo.* Touch and count. (Tap 4 as children touch small circles and count.) *3, 4, 5, 6.*
  (Repeat until firm.)

c. How many circles are there in problem A? (Signal.) *6.*
- Write 6 after the equals. ✔
- Everybody, start with 2. Touch the symbols and read the fact. Get ready. (Signal.) *2 + 4 = 6.*
  (Repeat until firm.)

d. Touch problem B. ✔
- Write the number for the shaded boxes and the unshaded boxes. Then stop.
  (Observe children and give feedback.)
- What's the number for the shaded boxes? (Signal.) *5.*
- What's the number for the unshaded boxes? (Signal.) *3.*

e. Touch the number you'll get going. ✔
- What number are you touching? (Signal.) *5.*
- Get it going. (Signal.) *Fiiive.* Touch and count. (Tap 3.) *6, 7, 8.*
  (Repeat step e until firm.)

f. How many boxes are there altogether? (Signal.) *8.*
- Where will you write 8? (Signal.) *After the equals.*
- Write 8. ✔
- Touch the symbols and say the fact. Get ready. (Signal.) *5 + 3 = 8.*

g. Touch problem C.
- Write the number for the shaded circles and the unshaded circles. Then stop.
  (Observe children and give feedback.)
- What's the number for the shaded circles? (Signal.) *8.*
- What's the number for the unshaded circles? (Signal.) *4.*
- Touch the number you'll get going. ✔
- What number are you touching? (Signal.) *8.*
- Get it going. (Signal.) *Eieieight.* Touch and count. (Tap 4.) *9, 10, 11, 12.*
  (Repeat until firm.)
h. How many circles are there altogether? (Signal.) *12.*
- Where will you write 12? (Signal.) *After the equals.*
- Write 12. ✔
- Touch the symbols and say the fact. Get ready. (Signal.) *8 + 4 = 12.*

EXERCISE 6: FACTS
PLUS/MINUS MIX

a. Find part 2 on worksheet 53. ✔
   (Teacher reference:)

a. 7 – 5 =   f. 11 – 2 =
b. 8 + 10 =   g. 6 + 2 =
c. 2 + 7 =   h. 5 + 10 =
d. 5 – 3 =   i. 8 – 6 =
e. 14 – 4 =   j. 17 – 7 =

These problems are from families that have a small number of 2 or a small number of 10. You're going to tell me if the big number or a small number is missing. Then you'll tell me the answer.
- Touch and read problem A. Get ready. (Signal.) *7 minus 5.*
- Is the answer the big number or a small number? (Signal.) *A small number.*
- What's 7 minus 5? (Signal.) *2.*
b. Touch and read problem B. Get ready. (Signal.) *8 plus 10.*
- Is the answer the big number or a small number? (Signal.) *The big number.*
- What's 8 plus 10? (Signal.) *18.*

c. (Repeat the following tasks for problems C through J:)

Touch and read problem __.		Is the answer the big number or a small number?	What's __?	
C	2 + 7	*The big number.*	2 + 7	9
D	5 – 3	*A small number.*	5 – 3	2
E	14 – 4	*A small number.*	14 – 4	10
F	11 – 2	*A small number.*	11 – 2	9
G	6 + 2	*The big number.*	6 + 2	8
H	5 + 10	*The big number.*	5 + 10	15
I	8 – 6	*A small number.*	8 – 6	2
J	17 – 7	*A small number.*	17 – 7	10

d. Complete the equations in part 2. Put your pencil down when you're finished.
   (Observe children and give feedback.)
   (Answer key:)

a. 7 – 5 = 2   f. 11 – 2 = 9
b. 8 + 10 = 18   g. 6 + 2 = 8
c. 2 + 7 = 9   h. 5 + 10 = 15
d. 5 – 3 = 2   i. 8 – 6 = 2
e. 14 – 4 = 10   j. 17 – 7 = 10

e. Check your work. You'll touch and read each fact.
- Fact A. (Signal.) *7 – 5 = 2.*
- (Repeat for:) B, *8 + 10 = 18;* C, *2 + 7 = 9;* D, *5 – 3 = 2;* E, *14 – 4 = 10;* F, *11 – 2 = 9;* G, *6 + 2 = 8;* H, *5 + 10 = 15;* I, *8 – 6 = 2;* J, *17 – 7 = 10.*

EXERCISE 7: PLACE-VALUE ADDITION (COLUMNS)

REMEDY

a. Find part 3 on worksheet 53. ✔
   (Teacher reference:)

R Part K

You're going to complete the place-value equations in columns.
- Touch the number shown in problem A. ✔
- What number? (Signal.) *594.*
- Say the place-value addition for 594. (Signal.) *500 + 90 + 4 = 594.*
  (Repeat until firm.)

b. Complete the equation. Write 500 on the top line. Write 90 on the next line. Write 4 on the bottom line. Remember to start each number in the correct column.
(Observe children and give feedback.)
• Touch and read the place-value equation for problem A. Get ready. (Signal.)
$500 + 90 + 4 = 594$.
c. Touch the number shown in problem B. ✔
• What number? (Signal.) *382*.
• Say the place-value addition for 382. (Signal.)
$300 + 80 + 2 = 382$.
(Repeat until firm.)
d. Complete the equation for B.
(Observe children and give feedback.)
• Touch and read the place-value equation for problem B. Get ready. (Signal.)
$300 + 80 + 2 = 382$.
e. Tell me the number shown in problem C. Get ready. (Signal.) *107*.
• Say the place-value equation for 107. (Signal.)
$100 + 0 + 7 = 107$.
(Display:) W   [53:7A]

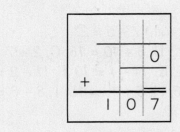

• (Point to **0**.) Here's where you write zero.
f. Complete the equation for C. Remember, zero starts in the ones column.
(Observe children and give feedback.)
• Touch and read the place-value equation for problem C. (Signal.) $100 + 0 + 7 = 107$.

EXERCISE 8: WORD PROBLEMS (COLUMNS)

a. Find part 4 on worksheet 53. ✔
(Teacher reference:)

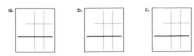

You're going to write the symbols for word problems in columns and work them. The equals bars are already shown.
• Touch where you'll write the symbols for problem A. ✔
Listen to problem A: There were 19 red bugs in the yard. There were 120 white bugs in the yard. How many bugs were there altogether?

• Listen again: There were 19 red bugs in the yard. What number do you write first?
(Signal.) *19*.
• Listen to the next part: There were 120 white bugs in the yard. Tell me the sign and the number you write for that part. Get ready.
(Signal.) *Plus 120*.
b. Listen again and tell me the symbols for the whole problem: There were 19 red bugs in the yard. There were 120 white bugs in the yard.
• Tell me the symbols. Get ready. (Signal.)
19 plus 120.
(Repeat step b until firm.)
c. Write 19 plus 120 and work problem A. Put your pencil down when you know how many bugs were there altogether.
(Observe children and give feedback.)
• Read the problem and the answer you wrote for A. Get ready. (Signal.) $19 + 120 = 139$.
• How many bugs were there altogether?
(Signal.) *139*.
d. Touch where you'll write the symbols for problem B. ✔
Listen to problem B: A truck had 125 boxes on it. Then workers put 51 more boxes on the truck. How many boxes ended up on the truck?
• Listen again: A truck had 125 boxes on it. What number do you write first? (Signal.) *125*.
• Listen to the next part: Then workers put 51 more boxes on the truck. Tell me the sign and the number you write for that part. Get ready. (Signal.) *Plus 51*.
e. Listen again and tell me the symbols for the whole problem: A truck had 125 boxes on it. Then workers put 51 more boxes on the truck.
• Tell me the symbols. Get ready. (Signal.)
125 plus 51.
(Repeat step e until firm.)
f. Write 125 plus 51 and work problem B. Put your pencil down when you know how many boxes ended up on the truck.
(Observe children and give feedback.)
• Read the problem and the answer you wrote for B. Get ready. (Signal.) $125 + 51 = 176$.
• How many boxes ended up on the truck?
(Signal.) *176*.

g. Touch where you'll write the symbols for problem C. ✔
Listen to problem C: There were 397 birds on a wire. Then 92 of those birds flew away. How many of those birds were still on the wire?

• Listen again: There were 397 birds on a wire. What number do you write first? (Signal.) *397.*

• Listen to the next part: Then 92 of those birds flew away. Tell me the sign and the number you write for that part. Get ready. (Signal.) *Minus 92.*

h. Listen again and tell me the symbols for the whole problem: There were 397 birds on a wire. Then 92 of those birds flew away.

• Tell me the symbols. Get ready. (Signal.) *397 minus 92.*
(Repeat step h until firm.)

i. Write 397 minus 92 and work problem C. Put your pencil down when you know how many birds were still on the wire.
(Observe children and give feedback.)
(Answer key:)

• Read the problem and the answer you wrote for C. Get ready. (Signal.) *397 – 92 = 305.*

• How many birds were still on the wire? (Signal.) *305.*

EXERCISE 9: 3 ADDENDS IN COLUMNS

a. Turn to the other side of worksheet 53 and find part 5. ✔
(Teacher reference:)

a. 1 b. 8 c. 1
   7   2   9
  +2   +2   +5

• Touch and read problem A. (Signal.) *1 plus 7 plus 2.*

• Say the problem for the first part. (Signal.) *1 plus 7.*

• What's the answer? (Signal.) *8.*

• Say the next problem. Get ready. (Signal.) *8 plus 2.*

• What's the answer? (Signal.) *10.*

• Write the answer for problem A. ✔

• Touch and read equation A. Get ready. (Signal.) *1 + 7 + 2 = 10.*

b. Work the rest of the problems in part 5. (Observe children and give feedback.)
(Answer key:)

a. 1   b. 8   c. 1
   7   2   9
  +2   +2   +5
  ──   ──   ──
  10   12   15

c. Check your work.

• Touch and read equation B. Get ready. (Signal.) *8 + 2 + 2 = 12.*

• Touch and read equation C. Get ready. (Signal.) *1 + 9 + 5 = 15.*

EXERCISE 10: INDEPENDENT WORK

a. Find part 6 on worksheet 53. ✔
(Teacher reference:)

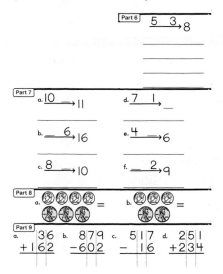

You'll write four facts below the number family.

b. Find part 7. ✔
The families in part 7 have a missing number. Below each family, you'll write the problem for the missing number and complete the fact. Then you'll write the missing number in the family.

c. Find part 8. ✔
You'll count for each group of coins and write the cents after the equals to show how much it's worth. Then you'll work the column problems in part 9.

d. Complete worksheet 53.
(Observe children and mark incorrect responses on children's worksheets as you give feedback.)

Lesson 54

EXERCISE 1: FACTS
ADDITION/SUBTRACTION

a. (Display:)   [54:1A]

11 − 2	5 + 2
3 + 5	6 − 1
7 − 5	8 − 5
4 − 2	2 + 7

Some of these problems are from the family with small numbers 5 and 2. For each problem, you're going to tell me if the big number or a small number is missing. Then you're going to tell me the missing number and say the fact.

• (Point to **11 − 2.**) Read the problem. (Touch.) *11 minus 2.*
• Is the big number or a small number missing? (Signal.) *A small number.*
• What's 11 minus 2? (Signal.) *9.*
• Say the fact for 11 minus 2. (Signal.) *11 − 2 = 9.*

b. (Point to **3 + 5.**) Read the problem. (Touch.) *3 plus 5.*
• Is the big number or a small number missing? (Signal.) *The big number.*
• What's 3 plus 5? (Signal.) *8.*
• Say the fact for 3 plus 5. (Signal.) *3 + 5 = 8.*

c. (Repeat the following tasks for the remaining problems:)

(Point to __.) Read the problem.	Is the big number or a small number missing?	What's __?		Say the fact for __.
7 − 5	A small number.	7 − 5	2	7 − 5 = 2
4 − 2	A small number.	4 − 2	2	4 − 2 = 2
5 + 2	The big number.	5 + 2	7	5 + 2 = 7
6 − 1	A small number.	6 − 1	5	6 − 1 = 5
8 − 5	A small number.	8 − 5	3	8 − 5 = 3
2 + 7	The big number.	2 + 7	9	2 + 7 = 9

(Repeat problems that were not firm.)

EXERCISE 2: MIXED COUNTING

a. Start with 146 and plus tens to 206. Get 146 going. *One hundred forty-siiix.* Count. (Tap.) *156, 166, 176, 186, 196, 206.*
• Start with 149 and plus tens to 209. Get 149 going. *One hundred forty-niiine.* Count. (Tap.) *159, 169, 179, 189, 199, 209.*

b. Start with 1000 and count backward by hundreds. Get 1000 going. *One thouuusand.* Count backward. (Tap.) *900, 800, 700, 600, 500, 400, 300, 200, 100.*

c. Listen: You have 600. When you count backward by hundreds, what's the next number? (Signal.) *500.*
• So 600 minus 100 equals 500. What's 600 minus 100? (Signal.) *500.*
• Listen: You have 400. When you count backward by hundreds, what's the next number? (Signal.) *300.*
• So what's 400 minus 100? (Signal.) *300.*

d. Start with 100 and count backward by tens. Get 100 going. *One huuundred.* Count backward. (Tap.) *90, 80, 70, 60, 50, 40, 30, 20, 10.*
- Listen: You have 100. When you count backward by tens, what's the next number? (Signal.) *90.*
- So what's 100 minus 10? (Signal.) *90.*
- Say the fact. (Signal.) *100 – 10 = 90.*
e. What's 90 minus 10? (Signal.) *80.*
- What's 80 minus 10? (Signal.) *70.*
- What's 30 minus 10? (Signal.) *20.*
  (Repeat step e until firm.)
f. Count by fives to 50. (Tap.) *5, 10, 15, 20, 25, 30, 35, 40, 45, 50.*
- Count by tens to 50. (Tap.) *10, 20, 30, 40, 50.*
g. Listen: You have 30. When you count by tens, what's the next number? (Signal.) *40.*
- So what's 30 plus 10? (Signal.) *40.*
h. Listen: You have 30. When you count by fives, what's the next number? (Signal.) *35.*
- So what's 30 plus 5? (Signal.) *35.*
i. Listen: 45 plus 5. What's 45 plus 5? (Signal.) *50.*
- Listen: 15 plus 10. What's 15 plus 10? (Signal.) *25.*
- Listen: 15 plus 5. What's 15 plus 5? (Signal.) *20.*
- Listen: 15 plus 1. What's 15 plus 1? (Signal.) *16.*

EXERCISE 3: NUMBER FAMILIES
SMALL NUMBER OF 6

REMEDY

a. (Display:) [54:3A]

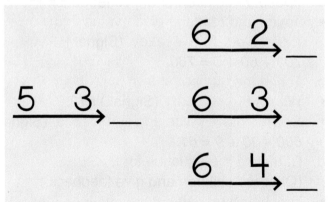

These families have a small number of 5 or a small number of 6. The big number is missing in each family. Some of these families are new.
- (Point to 6 → 2.) What are the small numbers? (Touch.) *6 (and) 2.*
- What's the big number? (Signal.) *8.*
b. (Point to 6 → 3.) What are the small numbers? (Touch.) *6 (and) 3.*
- What's the big number? (Signal.) *9.*

c. (Point to 6 → 4.) What are the small numbers? (Touch.) *6 (and) 4.*
- What's the big number? (Signal.) *10.*
d. (Point to 6 → 2.) Say the fact that starts with the first small number. Get ready. (Signal.) *6 + 2 = 8.*
- Say the fact that starts with the other small number. Get ready. (Signal.) *2 + 6 = 8.*
e. (Point to 6 → 3.) What are the small numbers in this family? (Signal.) *6 (and) 3.*
- What's the big number? (Signal.) *9.*
- Say the fact that starts with the first small number. Get ready. (Signal.) *6 + 3 = 9.*
- Say the fact that starts with the other small number. Get ready. (Signal.) *3 + 6 = 9.*
f. (Point to 6 → 4.) What are the small numbers in this family? (Signal.) *6 (and) 4.*
- What's the big number? (Signal.) *10.*
- Say the fact that starts with the first small number. Get ready. (Signal.) *6 + 4 = 10.*
- Say the fact that starts with the other small number. Get ready. (Signal.) *4 + 6 = 10.*
  (Repeat step f until firm.)
g. (Point to 5 → 3.) Here's a number family you know.
- What are the small numbers? (Touch.) *5 (and) 3.*
- What's the big number? (Signal.) *8.*
- Say the fact that starts with the first small number. Get ready. (Signal.) *5 + 3 = 8.*
- Say the fact that starts with the other small number. Get ready. (Signal.) *3 + 5 = 8.*
h. (Point to 6 → 3.) Say the fact that starts with the first small number. Get ready. (Signal.) *6 + 3 = 9.*
- Say the fact that starts with the other small number. Get ready. (Signal.) *3 + 6 = 9.*
  (Repeat step h until firm.)
i. This time, you'll tell me about some minus facts.
- (Point to 5 → 3.) Say the fact that goes backward along the arrow. Get ready. (Signal.) *8 – 3 = 5.*
- Say the other minus fact. (Signal.) *8 – 5 = 3.*
j. (Point to 6 → 3.) Say the fact that goes backward along the arrow. Get ready. (Signal.) *9 – 3 = 6.*
- Say the other minus fact. (Signal.) *9 – 6 = 3.*
k. (Point to 6 → 4.) Say the fact that goes backward along the arrow. Get ready. (Signal.) *10 – 4 = 6.*
- Say the other minus fact. (Signal.) *10 – 6 = 4.*

EXERCISE 4: DIGITS
WHEN 1ST EQUALS ZERO

a. (Display:) W   [54:4A]

041	80	09	105	013

Zero can't be the digit a number starts with. So some of these numbers are wrong.
- (Point to **041**.) This number starts with zero. What does this number start with? (Touch.) *Zero.*
- So is this number right or wrong? (Signal.) *Wrong.*

b. (Point to **80**.) What does this number start with? (Touch.) *8.*
- So is this number right or wrong? (Signal.) *Right.*

c. (Repeat the following tasks for 09, 105, 013:)
- (Point to ___.) What does this number start with?
- So is this number right or wrong?

d. (Point to **041**.) Is this number right or wrong? (Touch.) *Wrong.*
- This should be 41. What are the digits of 41? (Signal.) *4 and 1.*
(Change to show:)   [54:4B]

41	80	09	105	013

- (Point to **41**.) Now is this number right or wrong? (Touch.) *Right.*
- What number? (Touch.) *41.*

e. (Point to **80**.) Is this number right or wrong? (Touch.) *Right.*
- What number is this? (Touch.) *80.*
- What are the digits of 80? (Signal.) *8 and zero.*

f. (Point to **09**.) Is this number right or wrong? (Touch.) *Wrong.*
- What number should this be? (Signal.) *9.*
(Change to show:)   [54:4C]

41	80	9	105	013

- (Point to **9**.) Is this 9 now? (Touch.) *Yes.*

g. (Point to **105**.) Is this number right or wrong? (Touch.) *Right.*
- What number is this? (Touch.) *105.*
- What are the digits of 105? (Signal.) *1, zero, 5.*

h. (Point to **013**.) Is this number right or wrong? (Touch.) *Wrong.*
- What number should this be? (Signal.) *13.*
(Change to show:)   [54:4D]

41	80	9	105	13

- (Point to **13**.) Is this 13 now? (Touch.) *Yes.*

EXERCISE 5: PLACE-VALUE ADDITION (COLUMNS)   REMEDY

a. (Distribute unopened workbooks to students.)
- Open your workbook to Lesson 54 and find part 1.
(Observe children and give feedback.)
(Teacher reference:)   R Part L

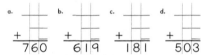

You're going to complete the place-value equations for these numbers.
- Touch the number shown in problem A. ✔
- What's the number? (Signal.) *760.*
- Say the place-value addition for 760. (Signal.) *700 + 60 + 0 = 760.*
(Repeat until firm.)

b. Complete the equation. Write 700 on the top line. Write 60 on the next line. Write zero on the bottom line. Remember to start each number in the correct column.
(Observe children and give feedback.)
- Touch and read the place-value equation for problem A. Get ready. (Signal.) *700 + 60 + 0 = 760.*

c. Touch the number shown in problem B. ✔
- What's the number? (Signal.) *619.*
- Say the place-value addition for 619. (Signal.) *600 + 10 + 9 = 619.*

d. Complete the equation for B.
(Observe children and give feedback.)
- Touch and read the place-value equation for problem B. Get ready. (Signal.) *600 + 10 + 9 = 619.*

e. Touch the number shown in problem C. ✔
- What's the number? (Signal.) *181.*
- Say the place-value addition for 181. (Signal.) *100 + 80 + 1 = 181.*

f. Complete the equation for C.
   (Observe children and give feedback.)
• Touch and read the place-value equation for problem C. Get ready. (Signal.)
   100 + 80 + 1 = 181.
g. Touch the number shown in problem D. ✔
• What's the number? (Signal.) *503.*
• Say the place-value addition for 503. (Signal.)
   500 + 0 + 3 = 503.
h. Complete the equation for D. Remember where to write zero.
   (Observe children and give feedback.)
• Touch and read the place-value equation for problem D. Get ready. (Signal.)
   500 + 0 + 3 = 503.

EXERCISE 6: FACTS
PLUS 5, 10

a. Find part 2 on worksheet 54. ✔
   (Teacher reference:)

   a. 20 + 5   e. 40 + 5
   b. 35 + 5   f. 15 + 5
   c. 15 + 10   g. 5 + 10
   d. 27 + 10 h. 10 + 5

   Some of these problems plus 10 and some plus 5.
• Read problem A. (Signal.) *20 plus 5.*
• What's 20 plus 5? (Signal.) *25.*
b. Read problem B. (Signal.) *35 plus 5.*
• What's 35 plus 5? (Signal.) *40.*
c. (Repeat the following tasks for problems C through H:)
• Read problem __.
• What's __?
d. Complete the facts in part 2.
   (Observe children and give feedback.)
   (Answer key:)

   a. 20 + 5 = 25 e. 40 + 5 = 45
   b. 35 + 5 = 40 f. 15 + 5 = 20
   c. 15 + 10 = 25 g. 5 + 10 = 15
   d. 27 + 10 = 37 h. 10 + 5 = 15

e. Check your work. Touch and read each fact.
• Fact A. (Signal.) *20 + 5 = 25.*
• Fact B. (Signal.) *35 + 5 = 40.*
• (Repeat for:) C, *15 + 10 = 25;* D, *27 + 10 = 37;* E, *40 + 5 = 45;* F, *15 + 5 = 20;* G, *5 + 10 = 15;* H, *10 + 5 = 15.*

EXERCISE 7: COUNT ON
PART/WHOLE
REMEDY

a. Find part 3 on worksheet 54. ✔
   (Teacher reference:)

R Part C

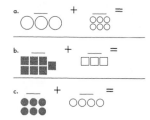

For problem A, you're going to write an equation for all the circles.
• Write the number of big circles and small circles. Then stop.
   (Observe children and give feedback.)
• What's the number for the big circles? (Signal.) *3.*
• What's the number for the small circles? (Signal.) *6.*
b. Now you're going to count all the circles. You'll get 3 going. Then you'll touch and count for the small circles.
• Touch the number 3. ✔
• Get it going. (Signal.) *Threee.* Touch and count. (Tap 6 as children touch and count.) *4, 5, 6, 7, 8, 9.*
   (Repeat until firm.)
c. How many circles are there in problem A? (Signal.) *9.*
• Write 9 after the equals. ✔
• Everybody, start with 3. Touch the symbols and read the fact. Get ready. (Signal.) *3 + 6 = 9.*
   (Repeat until firm.)
d. Touch problem B. ✔
• Write the number for the shaded boxes and the unshaded boxes. Then stop.
   (Observe children and give feedback.)
• What's the number for the shaded boxes? (Signal.) *7.*
• What's the number for the unshaded boxes? (Signal.) *3.*
e. Touch the number you'll get going. ✔
• What number are you touching? (Signal.) *7.*
• Get it going. *Sevennn.* Touch and count. (Tap 3.) *8, 9, 10.*
   (Repeat step e until firm.)

f. How many boxes are there altogether?
(Signal.) *10.*
- Where will you write 10? (Signal.) *After the equals.*
- Write 10. ✔
- Touch the symbols and read the fact. Get ready. (Signal.) *7 + 3 = 10.*

g. Touch problem C. ✔
- Write the number for the shaded circles and the unshaded circles. Then stop.
(Observe children and give feedback.)
- What's the number for the shaded circles? (Signal.) *6.*
- What's the number for the unshaded circles? (Signal.) *4.*

h. Touch the number you'll get going. ✔
- What number are you touching? (Signal.) *6.*
- Get it going. *Siiix.* Touch and count. (Tap 4.) *7, 8, 9, 10.*
(Repeat step h until firm.)

i. How many circles are there altogether? (Signal.) *10.*
- Write 10. ✔
- Touch the symbols and read the fact. Get ready. (Signal.) *6 + 4 = 10.*

EXERCISE 8: FACTS
PLUS/MINUS MIX

a. Find part 4 on worksheet 54. ✔
(Teacher reference:)

a. $14 - 10$
b. $1 + 8$
c. $2 + 7$
d. $5 - 3$
e. $17 - 7$
f. $11 - 2$
g. $6 + 2$
h. $7 - 6$
i. $10 - 8$
j. $3 + 10$

These problems are from families you know. You're going to tell me if the big number or a small number is missing. Then you'll tell me the answer.
- Touch and read problem A. Get ready. (Signal.) *14 minus 10.*
- Is the answer the big number or a small number? (Signal.) *A small number.*
- What's 14 minus 10? (Signal.) *4.*

b. Touch and read problem B. Get ready. (Signal.) *1 plus 8.*
- Is the answer the big number or a small number? (Signal.) *The big number.*
- What's 1 plus 8? (Signal.) *9.*

c. (Repeat the following tasks for problems C through J:)

Touch and read problem __.		Is the answer the big number or a small number?	What's __?	
C	2 + 7	The big number.	2 + 7	9
D	5 − 3	A small number.	5 − 3	2
E	17 − 7	A small number.	17 − 7	10
F	11 − 2	A small number.	11 − 2	9
G	6 + 2	The big number.	6 + 2	8
H	7 − 6	A small number.	7 − 6	1
I	10 − 8	A small number.	10 − 8	2
J	3 + 10	The big number.	3 + 10	13

(Repeat problems that were not firm.)

d. Complete the equations in part 4. Put your pencil down when you've finished.
(Observe children and give feedback.)
(Answer key:)

a. $14 - 10 = 4$
b. $1 + 8 = 9$
c. $2 + 7 = 9$
d. $5 - 3 = 2$
e. $17 - 7 = 10$
f. $11 - 2 = 9$
g. $6 + 2 = 8$
h. $7 - 6 = 1$
i. $10 - 8 = 2$
j. $3 + 10 = 13$

e. Check your work. You'll touch and read each fact.
- Fact A. (Signal.) *14 − 10 = 4.*
- (Repeat for:) B, *1 + 8 = 9;* C, *2 + 7 = 9;* D, *5 − 3 = 2;* E, *17 − 7 = 10;* F, *11 − 2 = 9;* G, *6 + 2 = 8;* H, *7 − 6 = 1;* I, *10 − 8 = 2;* J, *3 + 10 = 13.*

EXERCISE 9: WORD PROBLEMS (COLUMNS)

a. Turn to the other side of worksheet 54 and find part 5. ✔
    (Teacher reference:)

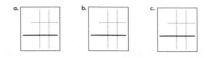

You're going to write the symbols for word problems in columns and work them. The equals bars are already shown.

- Touch where you'll write the symbols for problem A. ✔
  Listen to problem A: There were 247 children on the playground. 46 of those children left the playground. How many children ended up on the playground?

- Listen again: There were 247 children on the playground. What number do you write first? (Signal.) *247.*

- Listen to the next part: 46 of those children left the playground. Tell me the sign and the number. (Signal.) *Minus 46.*

b. Listen again and tell me the symbols for the whole problem: There were 247 children on the playground. 46 of those children left the playground.

- Tell me the symbols. Get ready. (Signal.)
  247 minus 46.
  (Repeat until firm.)

c. Write 247 minus 46 and work problem A. Put your pencil down when you know how many children ended up on the playground.
    (Observe children and give feedback.)

- Read the problem and the answer you wrote for A. Get ready. (Signal.) *247 – 46 = 201.*

- How many children ended up on the playground? (Signal.) *201.*

d. Touch where you'll write the symbols for problem B. ✔
    Listen to problem B: A table had 28 apples on it. Then Mrs. Jones put 31 more apples on the table. How many apples ended up on the table?

- Listen again: A table had 28 apples on it. What number do you write first? (Signal.) *28.*

- Listen to the next part: Then Mrs. Jones put 31 more apples on the table. Tell me the sign and the number. (Signal.) *Plus 31.*

e. Listen again and tell me the symbols for the whole problem: A table had 28 apples on it. Then Mrs. Jones put 31 more apples on the table.

- Tell me the symbols. Get ready. (Signal.)
  28 plus 31.
  (Repeat until firm.)

f. Write 28 plus 31 and work problem B. Put your pencil down when you know how many apples ended up on the table.
    (Observe children and give feedback.)

- Read the problem and the answer you wrote for B. Get ready. (Signal.) *28 + 31 = 59.*

- How many apples ended up on the table? (Signal.) *59.*

g. Touch where you'll write the symbols for problem C. ✔
    Listen to problem C: Jan had 365 cards. She gave 30 of those cards to her brother. How many cards does Jan still have?

- Listen again: Jan had 365 cards. What number do you write first? (Signal.) *365.*

- Listen to the next part: She gave 30 of those cards to her brother. Tell me the sign and the number. (Signal.) *Minus 30.*

h. Listen again and tell me the symbols for the whole problem: Jan had 365 cards. She gave 30 of those cards to her brother.

- Tell me the symbols. Get ready. (Signal.)
  365 minus 30.
  (Repeat until firm.)

i. Write 365 minus 30 and work problem C. Put your pencil down when you know how many cards Jan still has.
    (Observe children and give feedback.)
    (Answer key:)

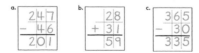

- Read the problem and the answer you wrote for C. Get ready. (Signal.) *365 – 30 = 335.*

- How many cards does Jan still have? (Signal.) *335.*

EXERCISE 10: INDEPENDENT WORK

a. Find part 6 on worksheet 54. ✔
(Teacher reference:)

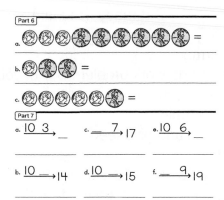

You'll count for each group of coins and write the cents for the group after the equals.

b. Find part 7. ✔
The families in part 7 have a missing number. Below each family, you'll write the problem for finding the missing number and complete the fact. Then you'll write the missing number in the family.

c. Complete worksheet 54.
(Observe children and mark incorrect responses on children's worksheets as you give feedback.)

Lesson 55

EXERCISE 1: FACTS
ADDITION/SUBTRACTION

a. (Display:)   [55:1A]

$10 - 8$	$11 - 9$
$2 + 6$	$3 - 2$
$5 - 2$	$3 + 5$
$2 + 7$	$6 - 4$

Some of these problems are from the family with small numbers of 2 and 3. For each problem, you're going to tell me if the big number or a small number is missing. Then you're going to tell me the missing number and say the fact.

- (Point to **10 − 8.**) Read the problem. (Touch.) *10 minus 8.*
- Is the big number or a small number missing? (Signal.) *A small number.*
- What's 10 minus 8? (Signal.) *2.*
- Say the fact for 10 minus 8. (Signal.) *10 − 8 = 2.*

b. (Point to **2 + 6.**) Read the problem. (Touch.) *2 plus 6.*
- Is the big number or a small number missing? (Signal.) *The big number.*
- What's 2 plus 6? (Signal.) *8.*
- Say the fact for 2 plus 6. (Signal.) *2 + 6 = 8.*

c. (Repeat the following tasks for remaining problems:)

(Point to __.) Read the problem.	Is the big number or a small number missing?	What's __?	Say the fact for __.	
$5 - 2$	A small number.	$5 - 2$ 3	$5 - 2$	$5 - 2 = 3$
$2 + 7$	The big number.	$2 + 7$ 9	$2 + 7$	$2 + 7 = 9$
$11 - 9$	A small number.	$11 - 9$ 2	$11 - 9$	$11 - 9 = 2$
$3 - 2$	A small number.	$3 - 2$ 1	$3 - 2$	$3 - 2 = 1$
$3 + 5$	The big number.	$3 + 5$ 8	$3 + 5$	$3 + 5 = 8$
$6 - 4$	A small number.	$6 - 4$ 2	$6 - 4$	$6 - 4 = 2$

(Repeat problems that were not firm.)

EXERCISE 2: MIXED COUNTING

a. Start with 140 and count backward to 130. Get 140 going. *One hundred fortyyy.* Count backward. (Tap.) *139, 138, 137, 136, 135, 134, 133, 132, 131, 130.*

b. Count by hundreds to 1000. Get ready. (Tap.) *100, 200, 300, 400, 500, 600, 700, 800, 900, 1000.*

c. Start with 100 and count by tens to 200. Get 100 going. *One huuundred.* Count. (Tap.) *110, 120, 130, 140, 150, 160, 170, 180, 190, 200.*

- Start with 100 and count backward by tens. Get 100 going. *One huuundred.* Count backward. (Tap.) *90, 80, 70, 60, 50, 40, 30, 20, 10.*
(Repeat until firm.)

d. Start with 100 and count backward by ones to 90. Get 100 going. *One huuundred.* Count backward. (Tap.) *99, 98, 97, 96, 95, 94, 93, 92, 91, 90.*

e. Count by fives to 50. Get ready. (Tap.) *5, 10, 15, 20, 25, 30, 35, 40, 45, 50.*

- Start with 35 and count by fives to 80. Get 35 going. *Thirty-fiiive.* Count. (Tap.) *40, 45, 50, 55, 60, 65, 70, 75, 80.*
(Repeat step e until firm.)

EXERCISE 3: NUMBER FAMILIES
SMALL NUMBER OF 10
REMEDY

a. (Display:) [55:3A]

4 + 10	19 − 9
12 − 2	18 − 10
16 − 10	10 + 5
7 + 10	11 − 1
13 − 10	20 − 10

All of these problems are from families that have a small number of 10. You'll tell me if a small number or the big number is missing. Then you'll say the fact.

• (Point to **4 + 10.**) Is a small number or the big number missing? (Touch.) *The big number.*
• Say the fact for 4 plus 10. (Signal.)
  4 + 10 = 14.
b. (Point to **12 − 2.**) Is a small number or the big number missing? (Touch.) *A small number.*
• Say the fact for 12 minus 2. (Signal.)
  12 − 2 = 10.
c. (Repeat the following tasks for remaining problems:)

(Point to __.) Is a small number or the big number missing?		Say the fact for __.	
16 − 10	*A small number.*	16 − 10	*16 − 10 = 6*
7 + 10	*The big number.*	7 + 10	*7 + 10 = 17*
13 − 10	*A small number.*	13 − 10	*13 − 10 = 3*
19 − 9	*A small number.*	19 − 9	*19 − 9 = 10*
18 − 10	*A small number.*	18 − 10	*18 − 10 = 8*
10 + 5	*The big number.*	10 + 5	*10 + 5 = 15*
11 − 1	*A small number.*	11 − 1	*11 − 1 = 10*
20 − 10	*A small number.*	20 − 10	*20 − 10 = 10*

EXERCISE 4: DIGITS
WHEN 1ST EQUALS ZERO
REMEDY

a. (Display:) W [55:4A]

150	019	05	406	081

• One of the digits in each of these numbers is zero. Can a number start with zero? (Signal.) *No.*
• (Point to **150.**) What does this number start with? (Touch.) *1.*
• So is this number right or wrong? (Signal.) *Right.*
b. (Point to **019.**) What does this number start with? (Touch.) *Zero.*
• So is this number right or wrong? (Signal.) *Wrong.*
c. (Repeat the following tasks for 05, 406, 081:)
• (Point to __.) What does this number start with?
• So is this number right or wrong?
d. (Point to **150.**) Is this number right or wrong? (Touch.) *Right.*
• What number is this? (Touch.) *150.*
• What are the digits of 150? (Signal.) *1, 5, and zero.*
e. (Point to **019.**) Is this number right or wrong? (Touch.) *Wrong.*
• What number should it be? (Signal.) *19.*
• What are the digits of 19? (Signal.) *1 and 9.* (Change to show:) [55:4B]

150	19	05	406	081

• (Point to **19.**) Now is this number right or wrong? (Touch.) *Right.*
• What number? (Touch.) *19.*
f. (Point to **05.**) Is this number right or wrong? (Touch.) *Wrong.*
• What number should it be? (Signal.) *5.* (Change to show:) [55:4C]

150	19	5	406	081

• (Point to **5.**) Is this 5 now? (Touch.) *Yes.*
g. (Point to **406.**) Is this number right or wrong? (Touch.) *Right.*
• What number is this? (Signal.) *406.*
• What are the digits of 406? (Signal.) *4, zero, 6.*

h. (Point to **081**.) Is this number right or wrong? (Touch.) *Wrong.*
- What number should this be? (Signal.) *81.*
- What are the digits of 81? (Signal.) *8 and 1.* (Change to show:)   55:4D]

150	19	5	406	81

- (Point to **81**.) Is this 81 now? (Touch.) *Yes.*

EXERCISE 5: NUMBER FAMILIES
SMALL NUMBER OF 6

a. (Display:)   [55:5A]

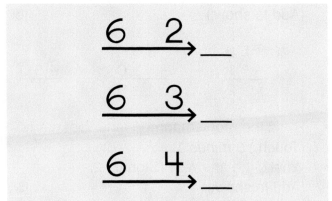

You worked with these families last time.
- (Point to 6 2 arrow.) What are the small numbers? (Touch.) *6 (and) 2.*
- What's the big number? (Signal.) *8.*
b. (Point to 6 3 arrow.) What are the small numbers? (Touch.) *6 (and) 3.*
- What's the big number? (Signal.) *9.*
c. (Point to 6 4 arrow.) What are the small numbers? (Touch.) *6 (and) 4.*
- What's the big number? (Signal.) *10.*
d. (Point to 6 2 arrow.) The small numbers are 6 and 2. What's the big number in this family? (Signal.) *8.*
- Say the fact that starts with 6. Get ready. (Signal.) *6 + 2 = 8.*
- Say the fact that starts with the other small number. (Signal.) *2 + 6 = 8.*
- Say the fact that goes backward along the arrow. (Signal.) *8 − 2 = 6.*
- Say the other minus fact. Get ready. (Signal.) *8 − 6 = 2.*

e. (Point to 6 3 arrow.) The small numbers are 6 and 3. What's the big number in this family? (Signal.) *9.*
- Say the fact that starts with 6. Get ready. (Signal.) *6 + 3 = 9.*
- Say the fact that starts with the other small number. (Signal.) *3 + 6 = 9.*
- Say the fact that goes backward along the arrow. (Signal.) *9 − 3 = 6.*
- Say the other minus fact. Get ready. (Signal.) *9 − 6 = 3.*
f. (Point to 6 4 arrow.) The small numbers are 6 and 4. What's the big number in this family? (Signal.) *10.*
- Say the fact that starts with 6. Get ready. (Signal.) *6 + 4 = 10.*
- Say the fact that starts with the other small number. (Signal.) *4 + 6 = 10.*
- Say the fact that goes backward along the arrow. (Signal.) *10 − 4 = 6.*
- Say the other minus fact. Get ready. (Signal.) *10 − 6 = 4.*
(Repeat step f until firm.)
g. (Display:)   [55:5B]

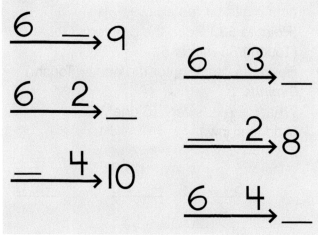

These are families with a small number of 6. Each family has a number missing. You'll say the problem and the missing number for each family.
- (Point to 6 arrow 9.) Is a small number or the big number missing in this family? (Signal.) *A small number.*
- Say the problem for the missing number. (Signal.) *9 minus 6.*
- What's the answer? (Signal.) *3.*
h. (Point to 6 2 arrow __.) Is a small number or the big number missing in this family? (Signal.) *The big number.*
- Say the problem for the missing number. (Signal.) *6 plus 2.*
- What's the answer? (Signal.) *8.*

i. (Repeat the following tasks for remaining families:)

(Point to __.)	Is a small number or the big number missing in this family?	Say the problem for the missing number.	What's the answer?
$\xrightarrow{4}$ 10	A small number.	10 – 4	6
6 3 $\rightarrow$ __	The big number.	6 + 3	9
$\xrightarrow{2}$ 8	A small number.	8 – 2	6
6 4 $\rightarrow$ __	The big number.	6 + 4	10

(Repeat families that were not firm.)

EXERCISE 6: COLUMN SUBTRACTION

WHEN 1ST EQUALS ZERO [REMEDY]

a. (Display:) W   [55:6A]

$$\begin{array}{r} 5\,3 \\ -\,5\,1 \\ \hline \end{array} \qquad \begin{array}{r} 4\,3\,9 \\ -\,4\,1\,0 \\ \hline \end{array} \qquad \begin{array}{r} 6\,4 \\ -\,6\,0 \\ \hline \end{array}$$

The beginning digit of the answer for these problems is zero. But we're going to write the right digits for the answer.

• (Point to **53.**) Read the problem. Get ready. (Touch.) *53 minus 51.*
• Read the problem for the ones. (Touch.) *3 minus 1.*
• What's the answer? (Signal.) *2.*
(Add to show:)   [55:6B]

$$\begin{array}{r} 5\,3 \\ -\,5\,1 \\ \hline 2 \end{array} \qquad \begin{array}{r} 4\,3\,9 \\ -\,4\,1\,0 \\ \hline \end{array} \qquad \begin{array}{r} 6\,4 \\ -\,6\,0 \\ \hline \end{array}$$

• Read the problem for the tens. Get ready. (Touch.) *5 minus 5.*
• What's the answer? (Signal.) *Zero.*
(Add to show:)   [55:6C]

$$\begin{array}{r} 5\,3 \\ -\,5\,1 \\ \hline 0\,2 \end{array} \qquad \begin{array}{r} 4\,3\,9 \\ -\,4\,1\,0 \\ \hline \end{array} \qquad \begin{array}{r} 6\,4 \\ -\,6\,0 \\ \hline \end{array}$$

(Point to **0.**) Is the number right or wrong? (Touch.) *Wrong.*

(Erase to show:)   [55:6D]

$$\begin{array}{r} 5\,3 \\ -\,5\,1 \\ \hline 2 \end{array} \qquad \begin{array}{r} 4\,3\,9 \\ -\,4\,1\,0 \\ \hline \end{array} \qquad \begin{array}{r} 6\,4 \\ -\,6\,0 \\ \hline \end{array}$$

Now the answer is right.
• Read the whole equation. (Touch.) *53 – 51 = 2.*
b. (Point to **439.**) Read the problem. Get ready. (Touch.) *439 minus 410.*
• Read the problem for the ones. (Touch.) *9 minus zero.*
• What's the answer? (Signal.) *9.*
(Add to show:)   [55:6E]

$$\begin{array}{r} 5\,3 \\ -\,5\,1 \\ \hline 2 \end{array} \qquad \begin{array}{r} 4\,3\,9 \\ -\,4\,1\,0 \\ \hline 9 \end{array} \qquad \begin{array}{r} 6\,4 \\ -\,6\,0 \\ \hline \end{array}$$

• Read the problem for the tens. Get ready. (Touch.) *3 minus 1.*
• What's the answer? (Signal.) *2.*
(Add to show:)   [55:6F]

$$\begin{array}{r} 5\,3 \\ -\,5\,1 \\ \hline 2 \end{array} \qquad \begin{array}{r} 4\,3\,9 \\ -\,4\,1\,0 \\ \hline 2\,9 \end{array} \qquad \begin{array}{r} 6\,4 \\ -\,6\,0 \\ \hline \end{array}$$

• Read the problem for the hundreds. Get ready. (Touch.) *4 minus 4.*
• What's the answer? (Signal.) *Zero.*
• Do I write zero? (Signal.) *No.*
• Read the whole equation. (Touch.) *439 – 410 = 29.*
c. (Point to **64.**) Read the problem. Get ready. (Touch.) *64 minus 60.*
• Read the problem for the ones. (Touch.) *4 minus zero.*
• What's the answer? (Signal.) *4.*
(Add to show:)   [55:6G]

$$\begin{array}{r} 5\,3 \\ -\,5\,1 \\ \hline 2 \end{array} \qquad \begin{array}{r} 4\,3\,9 \\ -\,4\,1\,0 \\ \hline 2\,9 \end{array} \qquad \begin{array}{r} 6\,4 \\ -\,6\,0 \\ \hline 4 \end{array}$$

• Read the problem in the tens. Get ready. (Touch.) *6 minus 6.*
• What's the answer? (Signal.) *Zero.*
• Do I write zero? (Signal.) *No.*
• Read the whole equation. (Touch.) *64 – 60 = 4.*
Remember, zero can't be the first digit of the answer.

EXERCISE 7: COUNT ON
PART/WHOLE

a. (Distribute unopened workbooks to students.)
- Open your workbook to Lesson 55 and find part 1.
  (Observe children and give feedback.)
  (Teacher reference:)

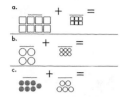

- Touch problem A. ✔
- Write the number for the big boxes and the small boxes. Then stop.
  (Observe children and give feedback.)
- What's the number for the big boxes? (Signal.) *8.*
- What's the number for the small boxes? (Signal.) *6.*
- Touch the number you'll get going. ✔
- What number are you touching? (Signal.) *8.*
- Get it going. (Signal.) *Eieieight.* Touch and count. (Tap 6 as children touch small boxes and count.) *9, 10, 11, 12, 13, 14.*
  (Repeat until firm.)

b. How many boxes are there altogether? (Signal.) *14.*
- Write 14. ✔
- Touch the symbols and read the fact for the boxes in problem A. Get ready. (Signal.) *8 + 6 = 14.*
  (Repeat until firm.)

c. Touch problem B. ✔
- Write the number for the big circles and the small circles. Then stop.
  (Observe children and give feedback.)
- What's the number for the big circles? (Signal.) *4.*
- What's the number for the small circles? (Signal.) *6.*
- Touch the number you'll get going. ✔
- What number are you touching? (Signal.) *4.*
- Get it going. (Signal.) *Fouuur.*
- Touch and count. (Tap 6 as children touch small circles and count.) *5, 6, 7, 8, 9, 10.*
  (Repeat until firm.)

d. How many circles are there altogether? (Signal.) *10.*
- Write 10. ✔
- Touch the numbers and read the fact for the circles in problem B. Get ready. (Signal.) *4 + 6 = 10.*

e. Touch problem C. ✔
- Write the number for the shaded circles and the unshaded circles. Then stop.
  (Observe children and give feedback.)
- What's the number for the shaded circles? (Signal.) *7.*
- What's the number for the unshaded circles? (Signal.) *5.*
- Touch the number you'll get going. ✔
- What number are you touching? (Signal.) *7.*
- Get it going. (Signal.) *Sevennn.*
- Touch and count. (Tap 5 as children touch unshaded circles and count.) *8, 9, 10, 11, 12.*
  (Repeat until firm.)

f. How many circles are there altogether? (Signal.) *12.*
- Write 12. ✔
- Touch the numbers and read the fact for the circles in problem C. Get ready. (Signal.) *7 + 5 = 12.*

EXERCISE 8: FACTS

PLUS/MINUS MIX

a. Find part 2 on worksheet 55. ✔
   (Teacher reference:)

   a. $19 - 9$
   b. $8 - 3$
   c. $4 + 10$
   d. $2 + 9$
   e. $16 - 10$
   f. $15 - 5$
   g. $2 + 6$
   h. $5 - 3$
   i. $5 + 3$

These problems are from families you know.
You're going to tell me if the big number or a
small number is missing. Then you'll tell me
the answer.

- Touch and read problem A. Get ready.
  (Signal.) *19 minus 9.*
- Is the answer the big number or a small
  number? (Signal.) *A small number.*
- What's 19 minus 9? (Signal.) *10.*

b. Touch and read problem B. Get ready.
  (Signal.) *8 minus 3.*
- Is the answer the big number or a small
  number? (Signal.) *A small number.*
- What's 8 minus 3? (Signal.) *5.*

c. (Repeat the following tasks for problems C
   through I:)

Touch and read problem __.		Is the answer the big number or a small number?	What's __?	
C	4 + 10	The big number.	4 + 10	14
D	2 + 9	The big number.	2 + 9	11
E	16 − 10	A small number.	16 − 10	6
F	15 − 5	A small number.	15 − 5	10
G	2 + 6	The big number.	2 + 6	8
H	5 − 3	A small number.	5 − 3	2
I	5 + 3	The big number.	5 + 3	8

(Repeat problems that were not firm.)

d. Complete the equations in part 2. Put your
   pencil down when you've finished.
   (Observe children and give feedback.)
   (Answer key:)
   (*The problems are not in the configuration
   they appear on worksheet.*)

   a. $19 - 9 = 10$ d. $2 + 9 = 11$ g. $2 + 6 = 8$
   b. $8 - 3 = 5$ e. $16 - 10 = 6$ h. $5 - 3 = 2$
   c. $4 + 10 = 14$ f. $15 - 5 = 10$ i. $5 + 3 = 8$

e. Check your work. You'll touch and read
   each fact.
- Fact A. (Signal.) *19 − 9 = 10.*
- (Repeat for:) B, *8 − 3 = 5*; C, *4 + 10 = 14*;
  D, *2 + 9 = 11*; E, *16 − 10 = 6*; F, *15 − 5 = 10*;
  G, *2 + 6 = 8*; H, *5 − 3 = 2*; I, *5 + 3 = 8.*

EXERCISE 9: COINS

NICKELS   `REMEDY`

a. (Display:)   [55:9A]

- (Point to nickels.) These coins are **nickels.**
  What are they? (Touch.) *Nickels.*
- A nickel is worth 5 cents. How much is a
  nickel worth? (Touch.) *5 cents.*
- So when you count nickels, what do you count
  by? (Signal.) *Fives.*
- (Point to 1st row.) Count by fives and figure
  out how much this row is worth. Get ready.
  (Touch.) *5, 10, 15, 20, 25, 30.*
- How much is this row worth? (Signal.)
  30 cents.

b. (Point to 2nd row.) You'll count and figure
   out how much this row is worth. Get ready.
   (Touch.) *5, 10, 15, 20, 25, 30, 35, 40.*
- How much is this row worth? (Signal.)
  40 cents.

c. Find part 3 on worksheet 55. ✔
   (Teacher reference:)   `R Part G`

   a. ⑤⑤⑤⑤⑤=
   b. ⑤⑤⑤⑤⑤=
   c. ⑤⑤⑤⑤
      ⑤⑤⑤ =

Row A has nickels. You'll touch and count to
figure out how much the row is worth.

- What will you count by for each nickel?
  (Signal.) *Five.*
- Fingers ready. ✔
- Touch and count. Get ready. (Tap.) *5, 10, 15,
  20, 25.*
- How much is row A worth? (Signal.) *25 cents.*
  (Repeat until firm.)

d. Touch row B. ✔
- Row B has dimes and pennies.
- What do you count by for each dime? (Signal.) *Ten.*
- What do you count by for each penny? (Signal.) *One.*

e. You'll touch and count the coins in row B.
- Finger over the first dime. ✔
- Touch and count for the dimes. (Tap.) *10, 20.* Touch and count for the pennies. (Tap.) *21, 22, 23.* (Repeat until firm.)
- How much is row B worth? (Signal.) *23 cents.*

f. You'll touch and count for the coins in row C.
- Finger over the first nickel. ✔
- Touch and count for the nickels. (Tap.) *5, 10, 15, 20, 25, 30, 35.* (Repeat until firm.)
- How much is row C worth? (Signal.) *35 cents.* Later, you'll count the cents for each row and write the number of cents after the equals.

EXERCISE 10: PLACE-VALUE ADDITION (COLUMNS)

a. Find part 4 on worksheet 55. ✔ (Teacher reference:)

You're going to complete the place-value equations in part 4.
- Touch the number shown in problem A. ✔
- What's the number? (Signal.) *607.*
- Say the place-value addition for 607. (Signal.) *600 + 0 + 7 = 607.* (Repeat until firm.)
- Complete equation A. (Observe children and give feedback.)
- Everybody, touch and read the place-value equation for 607. Get ready. (Signal.) *600 + 0 + 7 = 607.*

b. Complete the rest of the place-value equations in part 4. (Observe children and give feedback.)

c. Check your work.
- Touch and read the place-value equation for problem B. Get ready. (Signal.) *400 + 10 + 8 = 418.*
- Touch and read the place-value equation for problem C. Get ready. (Signal.) *300 + 90 + 2 = 392.*

EXERCISE 11: INDEPENDENT WORK

a. Find part 5 on worksheet 55. ✔ (Teacher reference:)

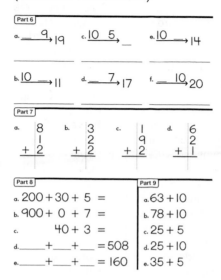

You'll work these column problems later.

b. Turn to the other side of worksheet 55 and find part 6. ✔ (Teacher reference:)

The families in part 6 have a missing number. Below each family you'll write the problem for finding the missing number and complete the equation. Then you'll write the missing number in the family.

c. Find parts 7, 8, and 9. ✔ You'll work the problems in parts 7, 8, and 9.

d. Turn to the other side of your worksheet and find part 3. ✔ You'll count for the coins to yourself and write the number of cents after the equals.

e. Your turn: Complete worksheet 55. (Observe children and mark incorrect responses on children's worksheets as you give feedback.)

Lessons 56-60 Planning Page

	Lesson 56	Lesson 57	Lesson 58	Lesson 59	Lesson 60
Student Learning Objectives	**Exercises** 1. Say two addition and two subtraction facts for number families; find missing numbers in number families 2. Count forward or backward from a given number 3. **Use ordinal numbers through fifth** 4. **Solve addition and subtraction facts in columns** 5. Count and write the value of mixed groups of coins 6. Recognize that zero is not used as the first digit in a number 7. Solve addition and subtraction facts 8. Count a part in a whole and count on for the whole 9. Write the symbols for word problems in columns and solve 10. Complete work independently	**Exercises** 1. **Count by 25s** 2. Say two addition and two subtraction facts for number families; find missing numbers in number families 3. Count forward from a given number 4. Use ordinal numbers through fifth 5. **Solve addition facts** 6. Count and write the value of mixed groups of coins 7. Recognize that zero is not used as the first digit in a number 8. **Count on for the whole** 9. Solve addition and subtraction facts 10. Write the symbols for word problems in columns and solve 11. Complete work independently	**Exercises** 1. Count by 25s 2. Say two addition and two subtraction facts for number families; find missing numbers in number families 3. Count forward or backward from a given number 4. Use ordinal numbers through fifth 5. **Solve subtraction facts** 6. Count on for the whole 7. Solve 2-digit/3-digit addition and subtraction problems 8. Count and write the value of mixed groups of coins 9. Solve addition and subtraction facts 10. Complete work independently	**Exercises** 1. Count by 25s 2. **Identify and count the value of bills** 3. Use ordinal numbers through fifth 4. Count forward or backward from a given number 5. Solve addition and subtraction facts 6. Count and write the value of a mixed group of coins 7. Solve subtraction facts 8. Write the symbols for word problems in columns and solve 9. Count on for the whole 10. Solve 2-digit/3-digit addition and subtraction problems 11. Complete work independently	**Exercises** 1. Find missing numbers in number families 2. Count forward from a given number 3. **Learn to measure inches with a ruler** 4. Identify and count the value of bills 5. Use ordinal numbers through fifth 6. Solve addition facts 7. Write the symbols for word problems in columns and solve 8. Learn to measure inches with a ruler 9. Solve subtraction facts 10. Solve 2-digit/3-digit addition and subtraction problems 11. Complete work independently
Common Core State Standards for Mathematics					
1.OA 3	✔	✔	✔		
1.OA 5	✔	✔	✔	✔	
1.OA 6	✔	✔	✔	✔	✔
1.OA 8	✔	✔	✔	✔	✔
1.NBT 1	✔	✔	✔	✔	✔
1.NBT 2	✔	✔	✔	✔	✔
1.NBT 4	✔	✔	✔	✔	✔
Teacher Materials	Presentation Book 2, Board Displays CD or chalkboard				
Student Materials	Workbook 1, Pencil, ruler				
Additional Practice	Student Practice Software: Block 2: Activity 1 (1.NBT.1), Activity 2 (1.NBT.1 and 1.NBT.2), Activity 3 (1.NBT.1), Activity 4 (1.NBT.4 and 1.NBT.5), Activity 5 (1.NBT.2 and 1.NBT.4), Activity 6 (1.NBT.2)				
Mastery Test					Student Assessment Book (Present Mastery Test 6 following Lesson 60.)

Lesson 56

EXERCISE 1: NUMBER FAMILIES

a. (Display:)   [56:1A]

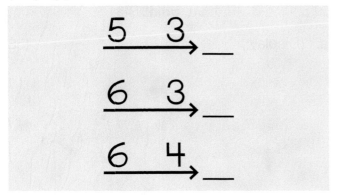

You worked with these number families before.
- (Point to 5 →3.) What are the small numbers in this family? (Touch.) *5 (and) 3.*
- What's the big number? (Signal.) *8.*
b. (Point to 6 →3.) What are the small numbers in this family? (Touch.) *6 (and) 3.*
- What's the big number? (Signal.) *9.*
c. (Point to 6 →4.) What are the small numbers in this family? (Touch.) *6 (and) 4.*
- What's the big number? (Signal.) *10.*
d. (Point to 5 →3.) The small numbers are 5 and 3. What's the big number in this family? (Signal.) *8.*
- Say the fact that starts with 5. Get ready. (Signal.) *5 + 3 = 8.*
- Say the fact that starts with the other small number. (Signal.) *3 + 5 = 8.*
- Say the fact that goes backward along the arrow. (Signal.) *8 − 3 = 5.*
- Say the other minus fact. Get ready. (Signal.) *8 − 5 = 3.*
e. (Point to 6 →3.) The small numbers are 6 and 3. What's the big number in this family? (Signal.) *9.*
- Say the fact that starts with 6. Get ready. (Signal.) *6 + 3 = 9.*
- Say the fact that starts with the other small number. (Signal.) *3 + 6 = 9.*
- Say the fact that goes backward along the arrow. (Signal.) *9 − 3 = 6.*
- Say the other minus fact. Get ready. (Signal.) *9 − 6 = 3.*

f. (Point to 6 →4.) The small numbers are 6 and 4. What's the big number in this family? (Signal.) *10.*
- Say the fact that starts with 6. Get ready. (Signal.) *6 + 4 = 10.*
- Say the fact that starts with the other small number. (Signal.) *4 + 6 = 10.*
- Say the fact that goes backward along the arrow. (Signal.) *10 − 4 = 6.*
- Say the other minus fact. Get ready. (Signal.) *10 − 6 = 4.*
(Repeat step f until firm.)

g. (Display:)   [56:1B]

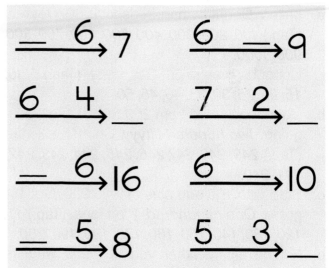

Each family has a number missing. You'll say the problem and the missing number for each family.
- (Point to →6 7.) Is a small number or the big number missing from this family? (Signal.) *A small number.*
- Say the problem for the missing number. Get ready. (Signal.) *7 minus 6.*
- What's the answer? (Signal.) *1.*
h. (Point to 6 →4 __.) Is a small number or the big number missing from this family? (Signal.) *The big number.*
- Say the problem for the missing number. Get ready. (Signal.) *6 plus 4.*
- What's the answer? (Signal.) *10.*

i. (Repeat the following tasks for remaining families:)

(Point to __.)	Is a small number or the big number missing from this family?	Say the problem for the missing number.	What's the answer?
$\xrightarrow{6}_{16}$	A small number.	16 – 6	10
$\xrightarrow{5}_{8}$	A small number.	8 – 5	3
$\xrightarrow{6}_{9}$	A small number.	9 – 6	3
$\xrightarrow{7 \quad 2}$	The big number.	7 + 2	9
$\xrightarrow{6}_{10}$	A small number.	10 – 6	4
$\xrightarrow{5 \quad 3}$	The big number.	5 + 3	8

EXERCISE 2: MIXED COUNTING

a. Listen: Count by hundreds to 1000. Get ready. (Tap.) *100, 200, 300, 400, 500, 600, 700, 800, 900, 1000.*
• Count by fives to 50. Get ready. (Tap.) *5, 10, 15, 20, 25, 30, 35, 40, 45, 50.*
b. Count backward from 250 to 240. Get 250 going. *Two hundred fiftyyy.* Count backward. (Tap.) *249, 248, 247, 246, 245, 244, 243, 242, 241, 240.*
c. Start with 100 and plus tens to 200. Get 100 going. *One huuundred.* Plus tens. (Tap.) *110, 120, 130, 140, 150, 160, 170, 180, 190, 200.*
d. You have 340. When you plus tens, what's the next number? (Signal.) *350.*
• You have 340. When you count by ones, what's the next number? (Signal.) *341.*
• You have 340. When you count by fives, what's the next number? (Signal.) *345.*
e. You have 161. When you count by ones, what's the next number? (Signal.) *162.*
• You have 161. When you minus ones, what's the next number? (Signal.) *160.*

f. You have 325. When you plus tens, what's the next number? (Signal.) *335.*
• You have 325. When you count by fives, what's the next number? (Signal.) *330.*
• You have 325. When you count by ones, what's the next number? (Signal.) *326.*

EXERCISE 3: ORDINAL NUMBERS

a. (Display:)   [56:3A]

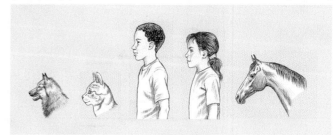

• (Point to dog.) Count the things in this row. Get ready. (Touch.) *1, 2, 3, 4, 5.*
• How many things are in this row? (Signal.) *5.*
• (Point to dog.) This is the first thing. What is the first thing? (Touch.) *(A) dog.*
• Say **first.** (Signal.) *First.*
b. (Point to cat.) This is the second thing. What is the second thing? (Touch.) *(A) cat.*
• Say **second.** (Signal.) *Second.*
c. (Point to boy.) This is the third thing. What is the third thing? (Touch.) *(A) boy.*
• Say **third.** (Signal.) *Third.*
d. (Point to dog.) My turn: (Touch dog, cat, boy.) First, second, third.
• Say that with me. (Touch.) ***First, second, third.***
 (Repeat until firm.)
e. What's the first thing in the row? (Signal.) *(A) dog.*
• What's the second thing in the row? (Signal.) *(A) cat.*
• What's the third thing in the row? (Signal.) *(A) boy.*
 (Repeat step e until firm.)

EXERCISE 4: FACTS
PLUS/MINUS MIX (COLUMNS)

a. (Distribute unopened workbooks to students.)
• Open your workbook to Lesson 56 and find part 1.
   (Observe children and give feedback.)
   (Teacher reference:)

a. 2
  +6

d. 9
  −2

g. 2
  +9

b. 10
  − 8

e. 4
  −3

h. 2
  +8

c. 8
  −5

f. 7
  − 1

i. 3
  +5

All these problems are from number families you know.
You're going to read each problem and say the answer. Then you'll go back and write the answers.
• Touch and read problem A. Get ready. (Signal.) *2 plus 6.*
• What's 2 plus 6? (Signal.) *8.*
b. Touch and read problem B. Get ready. (Signal.) *10 minus 8.*
• What's 10 minus 8? (Signal.) *2.*
c. (Repeat the following tasks for problems C through I:)

Touch and read problem __.		What's the answer?
C	8 − 5	3
D	9 − 2	7
E	4 − 3	1
F	7 − 1	6
G	2 + 9	11
H	2 + 8	10
I	3 + 5	8

d. Complete the equations for all the problems in part 1.
   (Observe children and give feedback.)
e. Check your work. You'll touch and read each fact.
• Fact A. (Signal.) *2 + 6 = 8.*
• (Repeat for fact:) B, *10 − 8 = 2;* C, *8 − 5 = 3;* D, *9 − 2 = 7;* E, *4 − 3 = 1;* F, *7 − 1 = 6;* G, *2 + 9 = 11;* H, *2 + 8 = 10;* I, *3 + 5 = 8.*

EXERCISE 5: COINS

a. Find part 2 on worksheet 56. ✔
   (Teacher reference:)

Group A has nickels and pennies.
• How much is each nickel worth? (Signal.) *5 cents.*
• So what do you count by for each nickel? (Signal.) *Five.*
• What do you count by for each penny? (Signal.) *One.*
b. I'll count for the coins in group A. You'll touch the coins as I count.
• Finger over the first nickel. ✔
• Get ready. (Count as children touch nickels.) 5, 10, 15, 20, twenty-fiiive. (Count as children touch pennies.) 26, 27, 28.
   (Repeat step b until firm.)
c. Now you'll count the nickels and then keep counting for the pennies.
• Fingers ready. (Children hold fingers over 1st nickel.) ✔
• Touch and count for the nickels. (Tap 5.) 5, 10, 15, 20, twenty-fiiive. Count for the pennies. (Tap 3.) 26, 27, 28.
   (Repeat until firm.)
• How many cents is group A worth? (Signal.) *28.*
d. Group B has dimes and pennies. You'll touch and count for the coins.
• Fingers ready. ✔
• Count for the dimes. (Tap.) 10, 20, thirtyyy. Count for the pennies. (Tap.) 31, 32, 33.
   (Repeat until firm.)
• How many cents is group B worth? (Signal.) *33.*
e. Row C has dimes and pennies. You'll touch and count for the coins.
• Fingers ready. ✔
• Count for the dimes. (Tap.) 10, 20, 30, fortyyy. Count for the pennies. (Tap.) 41, 42.
   (Repeat until firm.)
• How many cents is group C worth? (Signal.) *42.*

f. Group D has nickels and pennies. You'll touch and count for the coins.
- • Fingers ready. ✔
- • Count for the nickels. (Tap.) *5, 10, fifteeen.*
  Count for the pennies. (Tap.) *16, 17, 18, 19.*
  (Repeat until firm.)
- • How many cents is group D worth? (Signal.) *19.*

g. Count for each group to yourself and write the answer after the equals. Put your pencil down when you've completed the equations for part 2. (Observe children and give feedback.)

EXERCISE 6: DIGITS
WHEN 1ST EQUALS ZERO (COLUMNS)

a. (Display:) W [56:6A]

$$\begin{array}{r} 269 \\ -219 \\ \hline \end{array}$$

The beginning digit of the answer for this problem is zero. But we're going to write the right digit for the answer.
- • (Point to **269**.) Read the problem. Get ready. (Touch.) *269 minus 219.*
- • Read the problem in the ones. (Touch.) *9 minus 9.*
- • What's the answer? (Signal.) *Zero.*
- • Do we write the zero? (Signal.) *Yes.*
  (Add to show:) [56:6B]

$$\begin{array}{r} 269 \\ -219 \\ \hline 0 \end{array}$$

b. Read the problem in the tens. Get ready. (Touch.) *6 minus 1.*
- • What's the answer? (Signal.) *5.*
  (Add to show:) [56:6C]

$$\begin{array}{r} 269 \\ -219 \\ \hline 50 \end{array}$$

- • Read the problem in the hundreds. Get ready. (Touch.) *2 minus 2.*
- • What's the answer? (Signal.) *Zero.*
- • Do we write the zero? (Signal.) *No.*
- • Read the whole equation. (Touch.)
  269 – 219 = 50.

c. Find part 3 on worksheet 56. ✔
  (Teacher reference:)

a. $\begin{array}{r} 516 \\ -506 \\ \hline \end{array}$ b. $\begin{array}{r} 269 \\ -219 \\ \hline \end{array}$

- • Touch and read problem A. (Signal.)
  516 minus 506.
- • Write the digits for the answer in the right places.
  Remember, we do not write zero if it is the beginning digit.
  (Observe children and give feedback.)
- • Everybody, touch and read the equation for A. (Signal.) *516 – 506 = 10.*

d. Touch and read problem B. (Signal.)
  269 minus 219.
- • Write the digits for the answer in the right places.
  (Observe children and give feedback.)
- • Everybody, touch and read the equation for B. (Signal.) *269 – 219 = 50.*

EXERCISE 7: FACTS
PLUS/MINUS MIX

a. Find part 4. ✔
  (Teacher reference:)

a. $10+6$ e. $2+3$ h. $19-10$
b. $2+9$ f. $5-3$ i. $2+5$
c. $20-10$ g. $2+7$ j. $8-6$
d. $17-7$

These problems are from families you know. You're going to tell me if the big number or a small number is missing. Then you'll tell me the answer.
- • Touch and read problem A. Get ready. (Signal.) *10 plus 6.*
- • Is the answer the big number or a small number? (Signal.) *The big number.*
- • What's 10 plus 6? (Signal.) *16.*

b. Touch and read problem B. Get ready. (Signal.) *2 plus 9.*
- • Is the answer the big number or a small number? (Signal.) *The big number.*
- • What's 2 plus 9? (Signal.) *11.*

c. (Repeat the following tasks for problems C through J:)

Touch and read problem __.		Is the answer the big number or a small number?	What's __?	
C	20 – 10	A small number.	20 – 10	10
D	17 – 7	A small number.	17 – 7	10
E	2 + 3	The big number.	2 + 3	5
F	5 – 3	A small number.	5 – 3	2
G	2 + 7	The big number.	2 + 7	9
H	19 – 10	A small number.	19 – 10	9
I	2 + 5	The big number.	2 + 5	7
J	8 – 6	A small number.	8 – 6	2

(Repeat problems that were not firm.)

d. Complete the equations in part 4. Put your pencil down when you've finished.
(Observe children and give feedback.)
(Answer key:)

a. $10 + 6 = 16$ e. $2 + 3 = 5$ h. $19 - 10 = 9$
b. $2 + 9 = 11$ f. $5 - 3 = 2$ i. $2 + 5 = 7$
c. $20 - 10 = 10$ g. $2 + 7 = 9$ j. $8 - 6 = 2$
d. $17 - 7 = 10$

e. Check your work. You'll touch and read each fact.
• Fact A. (Signal.) $10 + 6 = 16$.
• (Repeat for:) B, $2 + 9 = 11$; C, $20 - 10 = 10$; D, $17 - 7 = 10$; E, $2 + 3 = 5$; F, $5 - 3 = 2$; G, $2 + 7 = 9$; H, $19 - 10 = 9$; I, $2 + 5 = 7$; J, $8 - 6 = 2$.

EXERCISE 8: COUNT ON
PART/WHOLE

a. Turn to the other side of worksheet 56 and find part 5. ✔
(Teacher reference:)

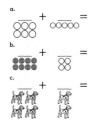

For problem A, you're going to write an equation for all the circles.
• Write the number for the big circles and the small circles. Then stop.
(Observe children and give feedback.)

• What's the number for the big circles? (Signal.) 6.
• What's the number for the small circles? (Signal.) 5.

b. Now you're going to count for all the circles.
• You'll get 6 going. Then you'll touch and count for the small circles.
• Touch the number 6. ✔
• Get it going. (Signal.) Siiix. Touch and count. (Tap 5 as children touch small circles and count.) 7, 8, 9, 10, 11.
(Repeat until firm.)

c. How many circles are there in problem A? (Signal.) 11.
• Write 11 after the equals. ✔
• Everybody, start with 6. Touch the symbols and read the fact. Get ready. (Tap as children say:) $6 + 5 = 11$.
(Repeat until firm.)

d. Touch problem B. ✔
• Write the number for the shaded circles and the unshaded circles. Then stop.
(Observe children and give feedback.)
• What's the number for the shaded circles? (Signal.) 8.
• What's the number for the unshaded circles? (Signal.) 4.

e. Touch the number you'll get going. ✔
• What number are you touching? (Signal.) 8.
• Get it going. Eieieight. Touch and count. (Tap 4 as children touch unshaded circles and count.) 9, 10, 11, 12.
(Repeat until firm.)

f. How many circles are there altogether? (Signal.) 12.
• Write 12. ✔
• Touch the symbols and say the fact. Get ready. (Signal.) $8 + 4 = 12$.

g. Touch problem C. ✔
• Write the number for the shaded dogs and the unshaded dogs. Then stop.
(Observe children and give feedback.)
• What's the number for the shaded dogs? (Signal.) 4.
• What's the number for the unshaded dogs? (Signal.) 2.

h. Touch the number you'll get going. ✔
- What number are you touching? (Signal.) *4.*
- Get it going. *Fouuur.* Touch and count. (Tap 2.) *5, 6.*
(Repeat step h until firm.)

i. How many dogs are there altogether? (Signal.) *6.*
- Write 6. ✔
- Touch the symbols and say the fact. Get ready. (Signal.) *4 + 2 = 6.*

EXERCISE 9: WORD PROBLEMS (COLUMNS)

a. Find part 6 on worksheet 56. ✔
(Teacher reference:)

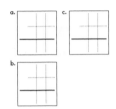

You're going to write the symbols for word problems in columns and work them. The equals bars are already shown.
- Touch where you'll write the symbols for problem A. ✔
Listen to problem A: Henry had 27 cents. Jan had 42 cents. How much did the children have altogether?
- Listen again: Henry had 27 cents. What number do you write first? (Signal.) *27.*
- Listen to the next part: Jan had 42 cents. Tell me the sign and the number you write for that part. Get ready. (Signal.) *Plus 42.*

b. Listen again and tell me the symbols for the whole problem: Henry had 27 cents. Jan had 42 cents.
- Tell me the symbols. Get ready. (Signal.) *27 plus 42.*
(Repeat step b until firm.)

c. Write 27 plus 42 and work problem A. Put your pencil down when you know how much the children had altogether.
(Observe children and give feedback.)
- Read the problem and the answer you wrote for A. Get ready. (Signal.) *27 + 42 = 69.*
- How much did the children have altogether? (Signal.) *69 cents.*

d. Touch where you'll write the symbols for problem B. ✔
Listen to problem B: There were 376 ducks on the lake. Then 126 of those ducks flew away. How many ducks were still on the lake?
- Listen again: There were 376 ducks on the lake. What number do you write first? (Signal.) *376.*
- Listen to the next part: Then 126 of those ducks flew away. Tell me the sign and the number you write for that part. Get ready. (Signal.) *Minus 126.*

e. Listen again and tell me the symbols for the whole problem: There were 376 ducks on the lake. Then 126 of those ducks flew away.
- Tell me the symbols. Get ready. (Signal.) *376 minus 126.*
(Repeat step e until firm.)

f. Write 376 minus 126 and work problem B. Put your pencil down when you know how many ducks were still on the lake.
(Observe children and give feedback.)
- Read the problem and the answer you wrote for B. Get ready. (Signal.) *376 – 126 = 250.*
- How many ducks were still on the lake? (Signal.) *250.*

g. Touch where you'll write the symbols for problem C. ✔
Listen to problem C: A brown dog had 66 fleas. A spotted dog had 202 fleas. How many fleas did both dogs have altogether?
- Listen again: A brown dog had 66 fleas. What number do you write first? (Signal.) *66.*
- Listen to the next part: A spotted dog had 202 fleas. Tell me the sign and the number you write for that part. Get ready. (Signal.) *Plus 202.*

h. Listen again and tell me the symbols for the whole problem: A brown dog had 66 fleas. A spotted dog had 202 fleas.
- Tell me the symbols. Get ready. (Signal.) *66 plus 202.*
(Repeat step h until firm.)

i. Write 66 plus 202 and work problem C. Put your pencil down when you know how many fleas both dogs had altogether.
(Observe children and give feedback.)
(Answer key:)

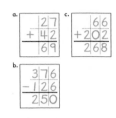

• Read the problem and the answer you wrote for C. Get ready. (Signal.) *66 + 202 = 268.*
• How many fleas did both dogs have altogether? (Signal.) *268.*

EXERCISE 10: INDEPENDENT WORK

a. Find part 7 on worksheet 56. ✔
(Teacher reference:)

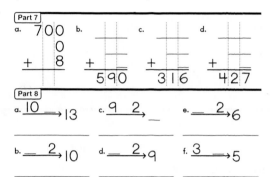

These are place-value addition problems in columns. You'll complete the place-value addition for each problem.

b. Find part 8. ✔
The families in part 8 have a number missing. Below each family you'll write the problem for finding the missing number and complete the fact. Then you'll write the missing number in each family.

c. Complete worksheet 56.
(Observe children and mark incorrect responses on children's worksheets as you give feedback.)

Lesson 57

EXERCISE 1: COUNTING BY 25s

a. (Display:)   [57:1A]

```
25        50        75        100
●─────────●─────────●─────────────▶
```

These are the numbers for counting by 25.
- (Point to number line.) My turn to say the numbers. (Touch and count.) 25, 50, 75, 100.
- Your turn: Say the numbers for counting by 25. Get ready. (Touch as children count.) 25, 50, 75, 100.
  (Repeat until firm.)
b. What's the first number you say when you count by 25? (Signal.) 25.
- What's the next number? (Signal.) 50.
- (Repeat for 75, 100.)
c. (Do not show display.)
  See if you can count by 25s to 100 without looking at the numbers.
- Count by 25s to 100. Get ready. (Tap.) 25, 50, 75, 100.
  (Repeat until firm.)
d. When you count by 25, what's the first number? (Signal.) 25.
- What's the next number? (Signal.) 50.
- (Repeat for 75, 100.)
  (Repeat step d until firm.)

EXERCISE 2: NUMBER FAMILIES

a. (Display:)   [57:2A]

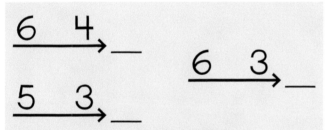

You worked with these families last time.
- (Point to 6 ⟶ 4.) What are the small numbers in this family? (Touch.) 6 (and) 4.
- What's the big number? (Signal.) 10.

b. (Point to 6 ⟶ 3.) What are the small numbers in this family? (Touch.) 6 (and) 3.
- What's the big number? (Signal.) 9.
c. (Point to 5 ⟶ 3.) What are the small numbers in this family? (Touch.) 5 (and) 3.
- What's the big number? (Signal.) 8.
d. (Point to 6 ⟶ 4.) The small numbers are 6 and 4. What's the big number in this family? (Signal.) 10.
- Say the fact that starts with 6. Get ready. (Signal.) 6 + 4 = 10.
- Say the fact that starts with the other small number. (Signal.) 4 + 6 = 10.
- Say the fact that goes backward along the arrow. (Signal.) 10 – 4 = 6.
- Say the other minus fact. (Signal.) 10 – 6 = 4.
e. (Point to 6 ⟶ 3.) The small numbers are 6 and 3. What's the big number in this family? (Signal.) 9.
- Say the fact that starts with 6. Get ready. (Signal.) 6 + 3 = 9.
- Say the fact that starts with the other small number. (Signal.) 3 + 6 = 9.
- Say the fact that goes backward along the arrow. (Signal.) 9 – 3 = 6.
- Say the other minus fact. (Signal.) 9 – 6 = 3.
f. (Point to 5 ⟶ 3.) The small numbers are 5 and 3. What's the big number in this family? (Signal.) 8.
- Say the fact that starts with 5. Get ready. (Signal.) 5 + 3 = 8.
- Say the fact that starts with the other small number. (Signal.) 3 + 5 = 8.
- Say the fact that goes backward along the arrow. (Signal.) 8 – 3 = 5.
- Say the other minus fact. Get ready. (Signal.) 8 – 5 = 3.
  (Repeat step f until firm.)

g. (Display:) [57:2B]

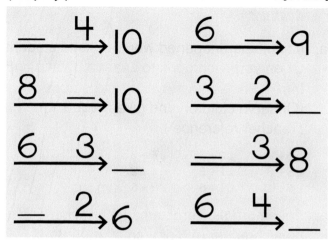

Each of these families has a number missing. You'll say the problem and the missing number for each family.

- (Point to ═4→10.) Is a small number or the big number missing from this family? (Signal.) *A small number.*
- Say the problem for the missing number. Get ready. (Signal.) *10 minus 4.*
- What's the answer? (Signal.) *6.*

h. (Point to 8═→10.) Is a small number or the big number missing from this family? (Signal.) *A small number.*
- Say the problem for the missing number. (Signal.) *10 minus 8.*
- What's the answer? (Signal.) *2.*

i. (Repeat the following tasks for remaining families:)

(Point to __.)	Is a small number or the big number missing from this family?	Say the problem for the missing number.	What's the answer?
6 3→_	The big number.	6 + 3	9
═2→6	A small number.	6 − 2	4
6 →9	A small number.	9 − 6	3
3 2→_	The big number.	3 + 2	5
═3→8	A small number.	8 − 3	5
6 4→_	The big number.	6 + 4	10

EXERCISE 3: MIXED COUNTING

a. Count by fives to 50. Get ready. (Tap.) *5, 10, 15, 20, 25, 30, 35, 40, 45, 50.*

b. My turn to start with 45 and count by fives to 100. Forty-fiiive, 50, 55, 60, 65, 70, 75, 80, 85, 90, 95, 100.
   Let's take turns saying the numbers for counting by fives from 45 to 100. The first number is 45. What's the next number? (Signal.) *50.*
- The next number is 55. What's the next number? (Signal.) *60.*
- The next number is 65. What's the next number? (Signal.) *70.*
- The next number is 75. What's the next number? (Signal.) *80.*
- The next number is 85. What's the next number? (Signal.) *90.*
- The next number is 95. What's the next number? (Signal.) *100.*
- Your turn: Start with 45 and count by fives to 100. Get 45 going. *Forty-fiiive.* Count. (Tap.) *50, 55, 60, 65, 70, 75, 80, 85, 90, 95, 100.* (Repeat until firm.)

c. Now you're going to count by tens. Start with 100 and count by tens to 200. Get 100 going. *One huuundred.* Count. (Tap.) *110, 120, 130, 140, 150, 160, 170, 180, 190, 200.*

d. Start with 100 and count by hundreds to 1000. Get 100 going. *One huuundred.* Count. (Tap.) *200, 300, 400, 500, 600, 700, 800, 900, 1000.* (Repeat step d until firm.)

e. Start with 45 and plus tens to 105. Get 45 going. *Forty-fiiive.* Plus tens. (Tap.) *55, 65, 75, 85, 95, 105.*

f. Start with 100 and count backward by tens. Get 100 going. *One huuundred.* Count backward. (Tap.) *90, 80, 70, 60, 50, 40, 30, 20, 10.*
   (Repeat steps e and f that were not firm.)

EXERCISE 4: ORDINAL NUMBERS

a. (Display:) [57:4A]

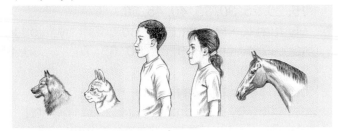

- (Point to dog.) Count the things in this row. Get ready. (Touch.) *1, 2, 3, 4, 5.*
- How many things are in this row? (Signal.) *5.*
b. (Point to dog.) My turn: What's the first thing in this row? (Touch.) A dog.
- Your turn: What's the first thing in this row? (Touch.) *(A) dog.*
- (Point to cat.) What's the second thing in this row? (Touch.) *(A) cat.*
- (Point to boy.) What's the third thing in this row? (Touch.) *(A) boy.*
c. Tell me first, second, third.
- (Point to cat.) Is this the first, second, or third thing? (Touch.) *Second (thing).*
- (Point to dog.) Is this the first, second, or third thing? (Touch.) *First (thing).*
- (Point to boy.) Is this the first, second, or third thing? (Touch.) *Third (thing).*
- (Repeat step c until firm.)
d. (Point to girl.) This is the fourth thing. Say **fourth.** (Touch.) *Fourth.*
- What's the fourth thing? (Touch.) *(A) girl.*
e. (Point to horse.) This is the fifth thing. Say **fifth.** (Touch.) *Fifth.*
- What's the fifth thing? (Touch.) *(A) horse.*
f. (Point to row.) What's the first thing in the row? (Signal.) *(A) dog.*
- What's the second thing in the row? (Signal.) *(A) cat.*
- What's the third thing in the row? (Signal.) *(A) boy.*
- What's the fourth thing in the row? (Signal.) *(A) girl.*
- What's the fifth thing in the row? (Signal.) *(A) horse.*
g. (Point to dog.) My turn to touch the things that are first, second, third, fourth, and fifth. (Touch things as you say:) First, second, third, fourth, and fifth.
- (Point to dog.) Your turn: (Touch things as children say:) *First, second, third, fourth, and fifth.*
- (Repeat step g until firm.)

EXERCISE 5: FACTS
ADDITION

a. (Distribute unopened workbooks to students.)
- Open your workbook to Lesson 57 and find part 1.
  (Observe children and give feedback.)
  (Teacher reference:)

a. 6 + 4	d. 3 + 6	g. 6 + 3
b. 5 + 3	e. 2 + 5	h. 5 + 2
c. 5 + 10	f. 1 + 6	i. 4 + 6

These are plus problems for families with a small number of 5 or a small number of 6. You're going to read the problems and tell me the answers. Then you'll go back and work them.
- Touch and read problem A. Get ready. (Signal.) *6 plus 4.*
- What's 6 plus 4? (Signal.) *10.*
b. Problem B. (Signal.) *5 plus 3.*
- What's 5 plus 3? (Signal.) *8.*
c. (Repeat the following tasks for problems C through I:)

Problem __.		What's __?	
C	5 + 10	5 + 10	15
D	3 + 6	3 + 6	9
E	2 + 5	2 + 5	7
F	1 + 6	1 + 6	7
G	6 + 3	6 + 3	9
H	5 + 2	5 + 2	7
I	4 + 6	4 + 6	10

d. Complete the equations for all of the problems in part 1.
  (Observe children and give feedback.)
  (Answer key:)

a. 6 + 4 = 10	d. 3 + 6 = 9	g. 6 + 3 = 9
b. 5 + 3 = 8	e. 2 + 5 = 7	h. 5 + 2 = 7
c. 5 + 10 = 15	f. 1 + 6 = 7	i. 4 + 6 = 10

e. Check your work. You'll touch and read each fact.
- Fact A. (Signal.) *6 + 4 = 10.*
- (Repeat for:) B, *5 + 3 = 8;* C, *5 + 10 = 15;* D, *3 + 6 = 9;* E, *2 + 5 = 7;* F, *1 + 6 = 7;* G, *6 + 3 = 9;* H, *5 + 2 = 7;* I, *4 + 6 = 10.*

Connecting Math Concepts

EXERCISE 6: COINS

a. Find part 2 on worksheet 57. ✔
(Teacher reference:)

Group A has dimes and nickels.
- What do you count by for each dime?
(Signal.) *Ten.*
- What do you count by for each nickel?
(Signal.) *Five.*

b. I'll count. You'll touch the coins.
- Fingers ready. ✔
- (Children touch dimes.) 10, twentyyyy.
(Children touch nickels.) 25, 30, 35, 40.

c. Your turn to touch and count.
- Fingers ready. ✔
- Touch and count for the dimes. (Signal.)
10, twentyyy. Count for the nickels. (Signal.)
25, 30, 35, 40.
(Repeat until firm.)
- How many cents is group A worth? (Signal.) *40.*

d. Group B has nickels and pennies.
- What do you count by for each nickel?
(Signal.) *Five.*
- What do you count by for each penny?
(Signal.) *One.*

e. You'll touch and count for the nickels. Then
you'll touch and count for the pennies.
- Fingers ready. ✔
- Count for the nickels. (Tap.) *5, 10, 15, 20,
twenty-fiiive.* Count for the pennies. (Tap.)
26, 27, 28, 29, 30.
(Repeat until firm.)
- How many cents is group B worth?
(Signal.) *30.*

f. Group C has dimes and nickels.
- What do you count by for each dime?
(Signal.) *Ten.*
- What do you count by for each nickel?
(Signal.) *Five.*

g. You'll touch and count for the dimes. Then
you'll touch and count for the nickels.
- Fingers ready. ✔
- Count for the dimes. (Tap.) *10, 20, 30 fortyyy.*
Count for the nickels. (Tap.) *45, 50, 55.*
(Repeat until firm.)

- How many cents is group C worth?
(Signal.) *55.*
Later, you'll count for these rows to yourself
and write the number of cents after the equals.

EXERCISE 7: DIGITS

WHEN 1ST EQUALS ZERO (COLUMNS)   REMEDY

a. Find part 3 on worksheet 57. ✔
(Teacher reference:)   R Part M

$$\begin{array}{r} 58 \\ -57 \\ \hline \end{array} \quad \begin{array}{r} 364 \\ -343 \\ \hline \end{array} \quad \begin{array}{r} 496 \\ -416 \\ \hline \end{array}$$

a.   b.   c.

The beginning digit of the answer for these
problems is zero. But we're going to write the
correct digits for the answer.
- Touch and read problem A. Get ready.
(Signal.) *58 minus 57.*
- Read the problem for the ones. (Signal.)
8 minus 7.
- What's the answer? (Signal.) *1.*
- Write 1. ✔

b. Read the problem for the tens. Get ready.
(Signal.) *5 minus 5.*
- What's the answer? (Signal.) *Zero.*
- Do we write zero if it is the beginning digit?
(Signal.) *No.*
- Touch and read the whole equation. (Signal.)
58 – 57 = 1.

c. Touch and read problem B. Get ready.
(Signal.) *364 minus 343.*
- Read the problem for the ones. (Signal.)
4 minus 3.
- What's the answer? (Signal.) *1.*
- Write 1. ✔

d. Read the problem for the tens. Get ready.
(Signal.) *6 minus 4.*
- What's the answer? (Signal.) *2.*
- Write 2. ✔

e. Read the problem for the hundreds. Get ready.
(Signal.) *3 minus 3.*
- What's the answer? (Signal.) *Zero.*
- Do we write zero? (Signal.) *No.*
- Touch and read the whole equation. (Signal.)
364 – 343 = 21.

f. Touch and read problem C. Get ready.
(Signal.) *496 minus 416.*
- Read the problem for the ones. (Signal.)
6 minus 6.
- What's the answer? (Signal.) *Zero.*
- Do we write zero? (Signal.) *Yes.*
- Write zero. ✔

g. Read the problem for the tens. Get ready.
(Signal.) *9 minus 1.*
- What's the answer? (Signal.) *8.*
- Write 8. ✔
h. Read the problem for the hundreds. Get ready.
(Signal.) *4 minus 4.*
- What's the answer? (Signal.) *Zero.*
- Do we write zero? (Signal.) *No.*
- Touch and read the whole equation. (Signal.)
496 – 416 = 80.

EXERCISE 8: COUNTING OBJECTS   REMEDY

a. Find part 4. ✔
(Teacher reference:)   R Part D

a. 18 ___ =
     ⌐◯   oooo ,
_____
b. 37 ___ =
     ⌐□   ⊞

This is a new kind of problem. You can see the
big circles for the first number, but you can't
count them. The number is written above them.
- Touch the number for the big circles. ✔
- How many big circles are there? (Signal.) *18.*
- Count the small circles and write the number
above them. Then write the plus sign between
the numbers.
(Observe children and give feedback.)
- Everybody, touch and read the problem for A.
(Signal.) *18 plus 4 (equals).*
b. You'll get the number for the big circles
going. Then you'll touch and count for the
small circles.
- Touch the number you'll get going. ✔
- Get it going. *Eighteeen.* Touch and count.
(Tap.) *19, 20, 21, 22.*
(Repeat until firm.)
- How many circles are there in all? (Signal.) *22.*
- Write 22. ✔
- Touch and read fact A for all the circles. Get
ready. (Signal.) *18 + 4 = 22.*
c. Touch problem B. ✔
- Problem B shows big boxes and small boxes.
- Touch the number for the big boxes. ✔
- How many big boxes are there? (Signal.) *37.*
- Write the plus sign and the number for the
small boxes.
(Observe children and give feedback.)
- Everybody, touch and read the problem for B.
(Signal.) *37 plus 6 (equals).*

d. You'll get the number for the big boxes going.
Then you'll touch and count for the small boxes.
- Touch the number you'll get going. ✔
- Get it going. *Thirty-sevennn.* Touch and count.
(Tap.) *38, 39, 40, 41, 42, 43.*
(Repeat until firm.)
- How many boxes are there in all? (Signal.) *43.*
- Write 43. ✔
- Touch and read the fact for all the boxes. Get
ready. (Signal.) *37 + 6 = 43.*

EXERCISE 9: FACTS
PLUS/MINUS MIX

a. Turn to the other side of worksheet 57 and
find part 5. ✔
(Teacher reference:)

a. 8 – 2   f. 2 + 6
b. 8 + 10   g. 4 – 2
c. 11 – 2   h. 9 + 2
d. 10 – 2   i. 17 – 7
e. 13 – 3   j. 6 – 4

These problems are from families you know.
You're going to touch and read each problem.
Then you'll tell me the answer.
- Touch and read problem A. Get ready.
(Signal.) *8 minus 2.*
- What's 8 minus 2? (Signal.) *6.*
b. Touch and read problem B. (Signal.) *8 plus 10.*
- What's 8 plus 10? (Signal.) *18.*
c. (Repeat the following tasks for problems C
through J:)

Problem __.		What's __?	
C	11 – 2	11 – 2	9
D	10 – 2	10 – 2	8
E	13 – 3	13 – 3	10
F	2 + 6	2 + 6	8
G	4 – 2	4 – 2	2
H	9 + 2	9 + 2	11
I	17 – 7	17 – 7	10
J	6 – 4	6 – 4	2

(Repeat problems that were not firm.)

d. Complete the equations in part 5. Put your pencil down when you've finished.
(Observe children and give feedback.)
(Answer key:)

a. $8 - 2 = 6$ f. $2 + 6 = 8$
b. $8 + 10 = 18$ g. $4 - 2 = 2$
c. $11 - 2 = 9$ h. $9 + 2 = 11$
d. $10 - 2 = 8$ i. $17 - 7 = 10$
e. $13 - 3 = 10$ j. $6 - 4 = 2$

e. Check your work. You'll touch and read each fact.
• Fact A. (Signal.) $8 - 2 = 6$.
• (Repeat for:) B, $8 + 10 = 18$; C, $11 - 2 = 9$; D, $10 - 2 = 8$; E, $13 - 3 = 10$; F, $2 + 6 = 8$; G, $4 - 2 = 2$; H, $9 + 2 = 11$; I, $17 - 7 = 10$; J, $6 - 4 = 2$.

EXERCISE 10: WORD PROBLEMS (COLUMNS) REMEDY

a. Find part 6 on worksheet 57. ✔
(Teacher reference:)   R Part A

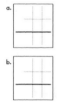

You're going to write the symbols for word problems in columns and work them. The equals bars are already shown.
• Touch where you'll write the symbols for problem A. ✔
Listen to problem A: 765 students were in school. 562 of those students left the school. How many students ended up in the school?
• Listen again and write the number for the first part: 765 students were in school. ✔
• Write the sign and the number for the next part: 562 of those students left the school. ✔
• Touch and read the symbols you wrote for problem A. Get ready. (Signal.) 765 minus 562 (equals).
b. Work problem A and figure out how many students ended up in the school.
(Observe children and give feedback.)
• For problem A, you wrote 765 minus 562. What's the answer? (Signal.) 203.
• So how many students ended up in the school? (Signal.) 203.

c. Touch where you'll write the symbols for problem B. ✔
Listen to problem B: Workers picked 83 apples in the morning. Workers picked 215 more apples in the afternoon. How many apples did the workers end up picking?
• Listen again and write the number for the first part: Workers picked 83 apples in the morning. ✔
• Write the sign and the number for the next part: Workers picked 215 more apples in the afternoon. ✔
• Touch and read the symbols you wrote for problem B. Get ready. (Signal.) 83 plus 215 (equals).
d. Work problem B and figure out how many apples the workers ended up picking.
(Observe children and give feedback.)
(Answer key:)
(The problems are not in the configuration they appear on worksheet.)

a. $\begin{array}{r} 765 \\ -562 \\ \hline 203 \end{array}$ b. $\begin{array}{r} 83 \\ +215 \\ \hline 298 \end{array}$

• For problem B, you wrote 83 plus 215. What's the answer? (Signal.) 298.
• So how many apples did the workers end up picking? (Signal.) 298.

EXERCISE 11: INDEPENDENT WORK

a. Find part 7 on worksheet 57. ✔
(Teacher reference:)

Part 7
a. $200 + 0 + 5$

b. $\_\_ + \_\_ + \_\_ = 184$
c. $\_\_ + \_\_ = 16$
d. $\begin{array}{r} + \\ \hline 721 \end{array}$
e. $\begin{array}{r} 40 \\ + \quad 9 \\ \hline \end{array}$

Part 8
a. $\begin{array}{r} 265 \\ +213 \\ \hline \end{array}$ b. $\begin{array}{r} 857 \\ -255 \\ \hline \end{array}$ c. $\begin{array}{r} 416 \\ + 83 \\ \hline \end{array}$

You'll complete the place-value addition equation for each problem. Some of the problems in part 7 are written in rows. Some are written in columns. Then, you'll work the column problems in part 8.
b. Find part 2 on the other side of worksheet 57. ✔
• Count for each group of coins and write the cents to show how much each group is worth. Then complete the other side of worksheet 57.
(Observe children and mark incorrect responses on children's worksheets as you give feedback.)

Lesson 58

EXERCISE 1: COUNTING BY 25S

a. (Display:)   [58:1A]

(Point to number line.) These are the numbers for counting by 25.

- What's the first number? (Signal.) *25.*
- What's the next number? (Signal.) *50.*
- What's the next number? (Signal.) *75.*
- What's the next number? (Signal.) *100.*
- Count by 25s to 100. (Touch as children count.) *25, 50, 75, 100.*
  (Repeat until firm.)

b. (Do not show display.) See if you can do it without looking at the numbers.
- Count by 25s to 100. Get ready. (Tap.) *25, 50, 75, 100.*
- When you count by 25, what's the first number? (Signal.) *25.*
- What's the next number? (Signal.) *50.*
- What's the next number? (Signal.) *75.*
- What's the next number? (Signal.) *100.*

c. Listen: When you count by 25s what comes after 75? (Signal.) *100.*
- What comes after 50? (Signal.) *75.*
- Start with 50 and count by 25s to 100. Get 50 going. *Fiftyyyy.* Count. (Tap.) *75, 100.*
- Start with 25 and count by 25s to 100. Get 25 going. *Twenty-fiiive.* Count. (Tap.) *50, 75, 100.*
  (Repeat step c until firm.)

EXERCISE 2: NUMBER FAMILIES   `REMEDY`

a. (Display:)   [58:2A]

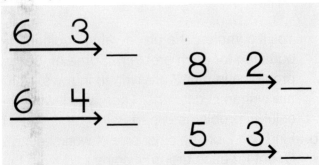

You worked with these families before.

- (Point to 6 → 3.) What are the small numbers in this family? (Touch.) *6 (and) 3.*
- What's the big number? (Signal.) *9.*

b. (Point to 8 → 2.) What are the small numbers in this family? (Touch.) *8 (and) 2.*
- What's the big number? (Signal.) *10.*

c. (Point to 6 → 4.) What are the small numbers in this family? (Touch.) *6 (and) 4.*
- What's the big number? (Signal.) *10.*

d. (Point to 5 → 3.) What are the small numbers in this family? (Touch.) *5 (and) 3.*
- What's the big number? (Signal.) *8.*

e. (Point to 6 → 3.) The small numbers are 6 and 3. What's the big number in this family? (Signal.) *9.*
- Say the fact that starts with 6. Get ready. (Signal.) *6 + 3 = 9.*
- Say the fact that starts with the other small number. (Signal.) *3 + 6 = 9.*
- Say the fact that goes backward along the arrow. (Signal.) *9 − 3 = 6.*
- Say the other minus fact. (Signal.) *9 − 6 = 3.*

f. (Point to 8 → 2.) The small numbers are 8 and 2. What's the big number in this family? (Signal.) *10.*
- Say the fact that starts with 8. Get ready. (Signal.) *8 + 2 = 10.*
- Say the fact that starts with the other small number. (Signal.) *2 + 8 = 10.*
- Say the fact that goes backward along the arrow. (Signal.) *10 − 2 = 8.*
- Say the other minus fact. (Signal.) *10 − 8 = 2.*

g. (Point to 6 → 4.) The small numbers are 6 and 4. What's the big number in this family? (Signal.) *10.*
- Say the fact that starts with 6. Get ready. (Signal.) *6 + 4 = 10.*
- Say the fact that starts with the other small number. (Signal.) *4 + 6 = 10.*
- Say the fact that goes backward along the arrow. Get ready. (Signal.) *10 − 4 = 6.*
- Say the other minus fact. (Signal.) *10 − 6 = 4.*

h. (Point to $\xrightarrow{5\quad3}$.) The small numbers are 5 and 3. What's the big number in this family? (Signal.) *8.*

- Say the fact that starts with 5. Get ready. (Signal.) *5 + 3 = 8.*
- Say the fact that starts with the other small number. (Signal.) *3 + 5 = 8.*
- Say the fact that goes backward along the arrow. (Signal.) *8 − 3 = 5.*
- Say the other minus fact. (Signal.) *8 − 5 = 3.*
  (Repeat step h until firm.)

i. (Display:) [58:2B]

Each of these families has a number missing. You'll say the problem and the missing number for each family.

- (Point to $\xrightarrow{2}$10.) Is a small number or the big number missing in this family? (Signal.) *A small number.*
- Say the problem for the missing number. (Signal.) *10 minus 2.*
- What's the answer? (Signal.) *8.*

j. (Point to $\xrightarrow{6}$10.) Is a small number or the big number missing in this family? (Signal.) *A small number.*

- Say the problem for the missing number. (Signal.) *10 minus 6.*
- What's the answer? (Signal.) *4.*

k. (Repeat the following tasks for remaining families:)

(Point to __.)	Is a small number or the big number missing in this family?	Say the problem for the missing number.	What's the answer?
$\xrightarrow{4}$14	A small number.	14 − 4	10
$\xrightarrow{5\quad3}$_	The big number.	5 + 3	8
$\xrightarrow{4}$6	A small number.	6 − 4	2
$\xrightarrow{3}$9	A small number.	9 − 3	6
$\xrightarrow{5}$7	A small number.	7 − 5	2
$\xrightarrow{6\quad4}$_	The big number.	6 + 4	10

EXERCISE 3: MIXED COUNTING

a. Your turn: Count by fives to 50. (Tap.) *5, 10, 15, 20, 25, 30, 35, 40, 45, 50.*

b. Start with 45 and count by fives to 100. Get 45 going. *Forty-fiiive.* Count. (Tap.) *50, 55, 60, 65, 70, 75, 80, 85, 90, 95, 100.*
   (Repeat step b until firm.)

c. Count by ones from 210 to 220. Get 210 going. *Two hundred tennn.* Count. (Signal.) *211, 212, 213, 214, 215, 216, 217, 218, 219, 220.*

d. Start with 17 and plus tens to 107. Get 17 going. *Seventeeen.* Plus tens. (Signal.) *27, 37, 47, 57, 67, 77, 87, 97, 107.*

e. Listen: You have 25. When you plus tens, what's the next number? (Signal.) *35.*

- You have 25. When you count by 25s, what's the next number? (Signal.) *50.*
- You have 25. When you count by ones, what's the next number? (Signal.) *26.*
- You have 25. When you count by fives, what's the next number? (Signal.) *30.*
  (Repeat step e until firm.)

f. Listen: You have 40. When you plus tens, what's the next number? (Signal.) *50.*

- You have 40. When you count by fives, what's the next number? (Signal.) *45.*
- You have 40. When you count by ones, what's the next number? (Signal.) *41.*
  (Repeat step f until firm.)

EXERCISE 4: ORDINAL NUMBERS

a. (Display:)   [58:4A]

(Point to **first** row.) There are 5 things in this row.

- (Point to **dog**.) I'll touch each thing in this row. Tell me first, second, third, fourth, or fifth for each thing I touch. (Touch things as children say:) *First, second, third, fourth, fifth.* (Repeat until firm.)
- (Point to **dog**.) What's the first thing in this row? (Touch.) *(A) dog.*

b. (Keep touching dog.) Look at the fifth thing in this row. ✔
- What's the fifth thing in this row? (Signal.) *(A) horse.*

c. (Keep touching dog.) Look at the fourth thing in this row. ✔
- What's the fourth thing in this row? (Signal.) *(A) girl.*

d. (Keep touching dog.) Look at the second thing in this row. ✔
- What's the second thing in this row? (Signal.) *(A) cat.*

e. (Keep touching dog.) Look at the third thing in this row. ✔
- What's the third thing in this row? (Signal.) *(A) boy.*

f. (Point to **second** row.) Look at this row. ✔
- How many things are in this row? (Touch.) *5.*

g. (Touch the girl.) Look at the first thing in this row. ✔
  What's the first thing in this row? (Signal.) *(A) girl.*
- (Keep touching girl.) Look at the fifth thing in this row. ✔
  What's the fifth thing in this row? (Signal.) *(A) boy.*

- (Keep touching girl.) Look at the fourth thing in this row. ✔
  What's the fourth thing in this row? (Signal.) *(A) horse.*

h. (Keep touching girl.) Tell me if I touch the second thing in this row.
- (Point to **dog**.) Is this the second thing? (Touch.) *No.*
- (Point to **cat**.) Is this the second thing? (Touch.) *Yes.*
- (Keep touching cat.) The cat is the second thing. Say that sentence. (Signal.) *The cat is the second thing.*

i. (Point to **girl**.) Look at the first thing in this row. ✔
- What's the first thing in this row? (Signal.) *(A) girl.*
- Yes, the girl is the first thing. Say the sentence about the girl. (Signal.) *The girl is the first thing.*

j. (Keep touching girl.) Look at the fifth thing in this row. ✔
- What's the fifth thing in this row? (Signal.) *(A) boy.*
- Say the sentence about the boy. (Signal.) *The boy is the fifth thing.*

k. (Keep touching girl.) Look at the fourth thing in this row. ✔
- What's the fourth thing in this row? (Signal.) *(A) horse.*
- Say the sentence about the horse. (Signal.) *The horse is the fourth thing.*

EXERCISE 5: FACTS
SUBTRACTION

a. (Distribute unopened workbooks to students.)
- Open your workbook to Lesson 58 and find part 1.
  (Observe children and give feedback.)
  (Teacher reference:)

a.	b.	c.	d.	e.	f.	g.	h.
10	8	7	9	15	10	8	6
− 6	− 3	− 2	− 6	− 10	− 8	− 1	− 4

These problems are from families you know.
- All of these problems minus, so is the answer the big number or a small number? (Signal.) *A small number.*

b. Touch and read problem A. Get ready. (Signal.) *10 minus 6.*
- What's 10 minus 6? (Signal.) *4.*

c. (Repeat the following tasks for problems B through H:)
- Problem __.
- What's __?

d. Complete the facts in part 1. Put your pencil down when you're finished.
(Observe children and give feedback.)
(Answer key:)

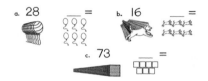

e. Check your work. You'll touch and read each fact.
- Fact A. (Signal.) *10 – 6 = 4.*
- (Repeat for:) B, *8 – 3 = 5;* C, *7 – 2 = 5;* D, *9 – 6 = 3;* E, *15 – 10 = 5;* F, *10 – 8 = 2;* G, *8 – 1 = 7;* H, *6 – 4 = 2.*

EXERCISE 6: COUNTING OBJECTS

a. Find part 2 on worksheet 58. ✔
(Teacher reference:)

a. 28 ___ =
b. 16 ___ =
c. 73 ___ =

You're going to complete each equation.
- Touch problem A. ✔
It shows striped balloons and white balloons.
- What's the number for the striped balloons? (Signal.) *28.*
- Write the plus sign and the number for the white balloons.
(Observe children and give feedback.)

b. Everybody, touch and read the problem. (Signal.) *28 plus 6 (equals).*
You'll get the number for the striped balloons going. Then you'll touch and count for the white balloons.
- Touch the number you'll get going. ✔
- Get it going. (Signal.) *Twenty-eieieight.* Touch and count. (Tap.) *29, 30, 31, 32, 33, 34.*
(Repeat until firm.)
- How many balloons are there altogether? (Signal.) *34.*
- Write 34. ✔
- Touch and read the fact for the balloons. Get ready. (Signal.) *28 + 6 = 34.*

c. Touch problem B. ✔
It shows big dogs and small dogs.
- What's the number for the big dogs? (Signal.) *16.*
- Write the plus sign and the number for the small dogs.
(Observe children and give feedback.)

d. Everybody, touch and read the problem. (Signal.) *16 plus 6 (equals).*
You'll get the number for the big dogs going. Then you'll touch and count for the small dogs.
- Touch the number you'll get going. ✔
- Get it going. (Signal.) *Sixteeen.* Touch and count. (Tap.) *17, 18, 19, 20, 21, 22.*
(Repeat until firm.)
- How many dogs are there altogether? (Signal.) *22.*
- Write 22. ✔
- Touch and read the fact for the dogs. Get ready. (Signal.) *16 + 6 = 22.*

e. Touch problem C. ✔
It shows shaded boxes and unshaded boxes.
- What's the number for the shaded boxes? (Signal.) *73.*
- Write the plus sign and the number for the unshaded boxes.
(Observe children and give feedback.)

f. Everybody, touch and read the problem. (Signal.) *73 plus 7 (equals).*
You'll get the number for the shaded boxes going. Then you'll touch and count for the unshaded boxes.
- Touch the number you'll get going. ✔
- Get it going. (Signal.) *Seventy-threee.* Touch and count. (Tap.) *74, 75, 76, 77, 78, 79, 80.*
(Repeat until firm.)
- How many boxes are there altogether? (Signal.) *80.*
- Write 80. ✔
- Touch and read the fact for the boxes. Get ready. (Signal.) *73 + 7 = 80.*

EXERCISE 7: COLUMN PROBLEMS

a. Find part 3 on worksheet 58. ✔
    (Teacher reference:)

a. 725   b. 53   c. 96   d. 364
   −705   +821   − 91   − 54

For some of these problems the beginning
digit of the answer is zero.

- Touch and read problem A. (Signal.)
  725 minus 705.
- Touch and read the problem in the ones
  column. (Signal.) *5 minus 5.*
- What's the answer? (Signal.) *Zero.*
- Touch and read the problem in the tens
  column. (Signal.) *2 minus zero.*
- What's the answer? (Signal.) *2.*
- Touch and read the problem in the hundreds
  column. (Signal.) *7 minus 7.*
- What's the answer? (Signal.) *Zero.*
- Do we write zero if it's the beginning digit?
  (Signal.) *No.*
  (Repeat until firm.)
- Write the answer to problem A.
  (Observe children and give feedback.)
- Everybody, touch and read the equation for A.
  (Signal.) *725 − 705 = 20.*

b. Touch and read problem B. (Signal.)
    53 plus 821.
- Touch and read the problem in the ones
  column. (Signal.) *3 plus 1.*
- What's the answer? (Signal.) *4.*
- Touch and read the problem in the tens
  column. (Signal.) *5 plus 2.*
- What's the answer? (Signal.) *7.*
- What do you write in the answer for the
  hundreds? (Signal.) *8.*
- Write the answer to problem B.
  (Observe children and give feedback.)
- Everybody, touch and read the equation for B.
  (Signal.) *53 + 821 = 874.*

c. Touch and read problem C. (Signal.)
    96 minus 91.
- Touch and read the problem in the ones
  column. (Signal.) *6 minus 1.*
- What's the answer? (Signal.) *5.*
- Touch and read the problem in the tens
  column. (Signal.) *9 minus 9.*
- What's the answer? (Signal.) *Zero.*
- Do we write zero if it's the beginning digit?
  (Signal.) *No.*
  (Repeat until firm.)

- Write the answer to problem C.
  (Observe children and give feedback.)
- Everybody, touch and read the equation for C.
  (Signal.) *96 − 91 = 5.*

d. Touch and read problem D. (Signal.)
    364 minus 54.
- Touch and read the problem in the ones
  column. (Signal.) *4 minus 4.*
- What's the answer? (Signal.) *Zero.*
- Touch and read the problem in the tens
  column. (Signal.) *6 minus 5.*
- What's the answer? (Signal.) *1.*
- What do you write in the hundreds column?
  (Signal.) *3.*
- Write the answer to problem D.
  (Observe children and give feedback.)
- Everybody, touch and read the equation for D.
  (Signal.) *364 − 54 = 310.*

EXERCISE 8: COINS

DIMES, NICKELS, PENNIES

a. Find part 4. ✔
    (Teacher reference:) R Part H

You're going to count the cents for each group
of coins. Group A has dimes, nickels, and
pennies.

- Touch a dime in group A. ✔
  What do you count by for each dime?
  (Signal.) *Ten.*
- Touch a nickel in group A. ✔
  What do you count by for each nickel?
  (Signal.) *Five.*
- Touch a penny in group A. ✔
  What do you count by for each penny?
  (Signal.) *One.*

b. Now you'll touch and count the coins in group A.
- Touch and count for the dimes. Get ready.
  (Tap 3.) *10, 20, 30.* Count for the nickels. (Tap 3.)
  35, 40, 45. Count for the pennies. (Tap 3.)
  46, 47, 48.
  (Repeat until firm.)
- How much is group A worth? (Signal.)
  48 cents.

c. Now you'll touch and count for the coins in group B.

- Touch and count for the dimes. Get ready. (Tap 2.) *10, 20.* Count for the nickels. (Tap 2.) *25, 30.* Count for the pennies. (Tap 2.) *31, 32.* (Repeat until firm.)
- How much is group B worth? (Signal.) *32 cents.*

Later, you'll count these groups to yourself and write the number of cents after the equals.

EXERCISE 9: FACTS
PLUS/MINUS MIX

a. Find part 5 on worksheet 58. ✔
(Teacher reference:)

a. $10-8$ d. $7-2$ g. $10-6$
b. $8-5$ e. $9-3$ h. $3+5$
c. $10+2$ f. $11-9$

These problems are from families you know. You're going to tell me if the big number or a small number is missing. Then you'll tell me the answer.

- Touch and read problem A. Get ready. (Signal.) *10 minus 8.*
- Is the answer the big number or a small number? (Signal.) *A small number.*
- What's 10 minus 8? (Signal.) *2.*

b. Touch and read problem B. (Signal.) *8 minus 5.*
- Is the answer the big number or a small number? (Signal.) *A small number.*
- What's 8 minus 5? (Signal.) *3.*

c. (Repeat the following tasks for problems C through H:)

Problem __.		Is the answer the big number or a small number?	What's __?	
C	10 + 2	The big number.	10 + 2	12
D	7 – 2	A small number.	7 – 2	5
E	9 – 3	A small number.	9 – 3	6
F	11 – 9	A small number.	11 – 9	2
G	10 – 6	A small number.	10 – 6	4
H	3 + 5	The big number.	3 + 5	8

(Repeat problems that were not firm.)

d. Complete the equations in part 5. Put your pencil down when you're finished.
(Observe children and give feedback.)
(Answer key:)

a. $10-8=2$ d. $7-2=5$ g. $10-6=4$
b. $8-5=3$ e. $9-3=6$ h. $3+5=8$
c. $10+2=12$ f. $11-9=2$

e. Check your work. You'll touch and read each fact.

- Fact A. (Signal.) *10 – 8 = 2.*
- (Repeat for facts:) B, *8 – 5 = 3;* C, *10 + 2 = 12;* D, *7 – 2 = 5;* E, *9 – 3 = 6;* F, *11 – 9 = 2;* G, *10 – 6 = 4;* H, *3 + 5 = 8.*

EXERCISE 10: INDEPENDENT WORK

a. Turn to the other side of worksheet 58 and find part 6. ✔
(Teacher reference:)

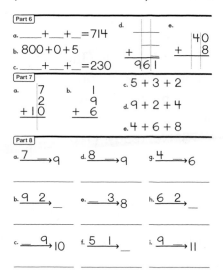

Some of the problems are written in rows and some of them are written in columns. You'll complete the place-value addition for the problems in part 6.

b. Find part 7. ✔
The problems in part 7 add three numbers. Some of them are in rows and some are in columns.

c. Find part 8. ✔
The families in part 8 have a number missing. Below each family you'll write the problem for finding the missing number and complete the fact. Then you'll write the missing number in each family.

d. Turn to the other side of your worksheet and find part 4. ✔

- Count for each group of coins and write the cents to show how much each group is worth. Then complete the other side of worksheet 58.
(Observe children and mark incorrect responses on children's worksheets as you give feedback.)

Lesson 59

EXERCISE 1: COUNTING BY 25s

a. You're going to count by 25s to 100.
- What's the first number you'll say? (Signal.) *25.*
- Count by 25s to 100. Get ready. (Tap.) *25, 50, 75, 100.*
  (Repeat until firm.)

b. This time you're going to count by 25s and keep counting by ones.
- Count by 25s to 50. Get ready. (Tap.) *25, 50.*
- Now keep counting from 50 by ones. Get ready. (Tap.) *51, 52, 53, 54, 55.* Stop.
  (Repeat until firm.)

c. Count by 25s to 75. Get ready. (Tap.) *25, 50, 75.*
- Now keep counting from 75 by ones. Get ready. (Tap.) *76, 77, 78, 79, 80.* Stop.
  (Repeat step c until firm.)

EXERCISE 2: COUNTING DOLLARS REMEDY

a. (Display:) [59:2A]

- (Point to **5** in first row.) These are dollar bills. The number tells how many dollars each bill is worth.
- How many dollars is this bill worth? (Touch.) *5.*
- (Point to next bill.) How many dollars is this bill worth? (Touch.) *1.*
- (Point to another 1.) How many dollars is this bill worth? (Touch.) *1.*

b. (Point to **5**.) I'm going to count how many dollars this row of bills is worth. I'll get 5 going and then count for the ones. (Touch 5.) Fiiive. (Touch 1s.) *6, 7, 8, 9.*
- Your turn: (Touch 5.) Get it going. *Fiiive.* (Touch 1s.) *6, 7, 8, 9.*
  (Repeat until firm.)
- How many dollars is this row worth? (Touch.) *9.*

c. (Point to **20**.) The number tells how many dollars this bill is worth. How many dollars? (Touch.) *20.*

- (Point to next bill.) How many dollars is this bill worth? (Touch.) *1.*
- Raise your hand when you know how many bills in this row are worth 1 dollar. ✔
- How many bills are worth 1 dollar? (Signal.) *3.*

d. (Point to **20**.) My turn to get this number going and count. (Touch 20.) Twentyyy. (Touch 1s.) *21, 22, 23.*
- Your turn: (Touch 20.) Get it going. *Twentyyy.* (Touch 1s.) *21, 22, 23.*
  (Repeat until firm.)
- How much is this row worth? (Touch.) *23 dollars.*

e. (Point to **50**.) How many dollars is this bill worth? (Touch.) *50.*
- Raise your hand when you know how many bills in this row are worth 1 dollar. ✔
- How many bills are worth 1 dollar? (Touch.) *4.*
- (Point to **50**.) Your turn to get this number going and count. (Touch 50.) Get it going. *Fiftyyy.* (Touch 1s.) *51, 52, 53, 54.*
  (Repeat until firm.)
- How much is this row worth? (Touch.) *54 dollars.*

EXERCISE 3: ORDINAL NUMBERS

a. (Display:) [59:3A]

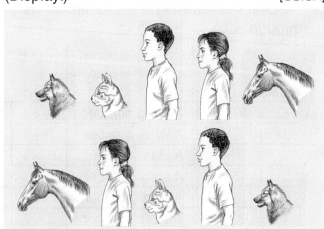

- (Point to horse in bottom row.) Count the things in this row. Get ready. (Touch.) *1, 2, 3, 4, 5.*

b. Now I'll point to one of the things in the bottom row. You'll tell me if it's first, second, third, fourth, or fifth.
- (Point to girl.) Tell me. (Touch.) *Second.*
- (Point to horse.) Tell me. (Touch.) *First.*
- (Point to dog.) Tell me. (Touch.) *Fifth.*

c. Look at the fifth thing in the bottom row. ✔
- What's the fifth thing in the bottom row? (Signal.) *(A) dog.*
d. Look at the top row. ✔
- How many things are in the top row? (Signal.) *5.*
- What's the first thing in the top row? (Signal.) *(A) dog.*
- What's the second thing in the top row? (Signal.) *(A) cat.*
- Look at the fourth thing in the top row. ✔
- What's the fourth thing? (Signal.) *(A) girl.*
- Look at the fifth thing in the top row. ✔
- What's the fifth thing? (Signal.) *(A) horse.*
  (Repeat step d until firm.)

EXERCISE 4: MIXED COUNTING

a. Listen: You have 35. When you plus tens, what's the next number? (Signal.) *45.*
- You have 35. When you count by ones, what's the next number? (Signal.) *36.*
- You have 35. When you count by fives, what's the next number? (Signal.) *40.*
- You have 35. When you minus ones, what's the next number? (Signal.) *34.*
  (Repeat step a until firm.)
b. Listen: You have 50. When you plus tens, what's the next number? (Signal.) *60.*
- You have 50. When you count by fives, what's the next number? (Signal.) *55.*
- You have 50. When you count by ones, what's the next number? (Signal.) *51.*
- You have 50. When you count by 25s, what's the next number? (Signal.) *75.*
- You have 50. When you minus ones, what's the next number? (Signal.) *49.*
  (Repeat step b until firm.)
c. Listen: You have 75. When you count by 25s, what's the next number? (Signal.) *100.*
- You have 75. When you plus tens, what's the next number? (Signal.) *85.*
- You have 75. When you count by fives, what's the next number? (Signal.) *80.*
- You have 75. When you count by ones, what's the next number? (Signal.) *76.*
- You have 75. When you minus ones, what's the next number? (Signal.) *74.*
  (Repeat step c until firm.)

EXERCISE 5: FACTS
PLUS/MINUS MIX

a. (Distribute unopened workbooks to students.)
- Open your workbook to Lesson 59 and find part 1.
  (Observe children and give feedback.)
  (Teacher reference:)

a. $5+3$ f. $3+6$
b. $8-2$ g. $2+4$
c. $11-9$ h. $9-3$
d. $6+4$ i. $9+10$
e. $13-3$ j. $10-6$

These problems are from families you know. You're going to touch and read each problem and tell me if the big number or a small number is missing. Then you'll tell me the answer.
- Problem A. (Signal.) *5 plus 3.*
- Is the answer the big number or a small number? (Signal.) *The big number.*
- What's 5 plus 3? (Signal.) *8.*
b. Problem B. (Signal.) *8 minus 2.*
- Is the answer the big number or a small number? (Signal.) *A small number.*
- What's 8 minus 2? (Signal.) *6.*
c. (Repeat the following tasks for problems C through J:)

Problem __.		Is the answer the big number or a small number?	What's __?	
C	$11 - 9$	A small number.	$11 - 9$	2
D	$6 + 4$	The big number.	$6 + 4$	10
E	$13 - 3$	A small number.	$13 - 3$	10
F	$3 + 6$	The big number.	$3 + 6$	9
G	$2 + 4$	The big number.	$2 + 4$	6
H	$9 - 3$	A small number.	$9 - 3$	6
I	$9 + 10$	The big number.	$9 + 10$	19
J	$10 - 6$	A small number.	$10 - 6$	4

(Repeat problems that were not firm.)

d. Complete the equations in part 1. Put your pencil down when you're finished.
  (Observe children and give feedback.)
  (Answer key:)

a. $5+3=8$ f. $3+6=9$
b. $8-2=6$ g. $2+4=6$
c. $11-9=2$ h. $9-3=6$
d. $6+4=10$ i. $9+10=19$
e. $13-3=10$ j. $10-6=4$

e. Check your work. You'll touch and read each fact.
- Fact A. (Signal.) *5 + 3 = 8.*
- (Repeat for:) B, *8 – 2 = 6;* C, *11 – 9 = 2;* D, *6 + 4 = 10;* E, *13 – 3 = 10;* F, *3 + 6 = 9;* G, *2 + 4 = 6;* H, *9 – 3 = 6;* I, *9 + 10 = 19;* J, *10 – 6 = 4.*

EXERCISE 6: COINS

DIMES, NICKELS, PENNIES

REMEDY

a. Find part 2 on worksheet 59. ✔
(Teacher reference:)

R Part I

You're going to count the cents for each row of coins. Group A has dimes, nickels, and pennies.
- Touch a dime in group A. ✔
What do you count by for each dime?
(Signal.) *Ten.*
- Touch a nickel in group A. ✔
What do you count by for each nickel?
(Signal.) *Five.*
- Touch a penny in group A. ✔
What do you count by for each penny?
(Signal.) *One.*

b. Now you'll touch and count for the coins in group A.
- Touch and count for the dimes. Get ready.
(Tap 2.) *10, 20.* Count for the nickels. (Tap 4.) *25, 30, 35, 40.* Count for the pennies. (Tap 2.) *41, 42.*
(Repeat until firm.)
- How much is group A worth? (Signal.) *42 cents.*

c. Now you'll touch and count for the coins in group B.
- Touch and count for the dimes. Get ready.
(Tap 4.) *10, 20, 30, 40.* Count for the nickels. (Tap 2.) *45, 50.* Count for the pennies. (Tap.) *51.*
(Repeat until firm.)
- How much is group B worth? (Signal.) *51 cents.*
Later, you'll count these groups to yourself and write the number of cents after the equals.

EXERCISE 7: FACTS

SUBTRACTION

a. Find part 3. ✔
(Teacher reference:)

a. 13 – 3   b. 6 – 4   c. 9 – 8   d. 9 – 6   e. 8 – 3
f. 12 – 10   g. 5 – 3   h. 8 – 2   i. 15 – 5   j. 10 – 4

All of these are minus problems, so they show the big number and a small number of a family. You're going to touch each problem as I read it. Then you'll tell me the answer.
- Problem A is 13 minus 3. What's the answer?
(Signal.) *10.*
b. Problem B is 6 minus 4. What's the answer?
(Signal.) *2.*
c. (Repeat the following tasks for problems C through J:)

Problem ___ is ___.		What's the answer?
C	9 – 8	1
D	9 – 6	3
E	8 – 3	5
F	12 – 10	2
G	5 – 3	2
H	8 – 2	6
I	15 – 5	10
J	10 – 4	6

(Repeat problems that were not firm.)
d. Complete the equations in part 3. Put your pencil down when you're finished.
(Observe children and give feedback.)
(Answer key:)

a. 13 – 3 = 10   b. 6 – 4 = 2   c. 9 – 8 = 1   d. 9 – 6 = 3   e. 8 – 3 = 5
f. 12 – 10 = 2   g. 5 – 3 = 2   h. 8 – 2 = 6   i. 15 – 5 = 10   j. 10 – 4 = 6

e. Check your work. You'll touch and read each fact.
- Fact A. (Signal.) *13 – 3 = 10.*
- (Repeat for:) B, *6 – 4 = 2;* C, *9 – 8 = 1;* D, *9 – 6 = 3;* E, *8 – 3 = 5;* F, *12 – 10 = 2;* G, *5 – 3 = 2;* H, *8 – 2 = 6;* I, *15 – 5 = 10;* J, *10 – 4 = 6.*

EXERCISE 8: WORD PROBLEMS (COLUMNS) [REMEDY]

a. Find part 4. ✔
   (Teacher reference:)   [R] **Part B**

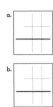

You're going to write the symbols for word problems in columns and work them. The equals bars are already shown.

- Touch where you'll write the symbols for problem A. ✔
  Listen to problem A: Denise had 56 cards. Donna had 132 cards. How many cards did the girls have altogether?
- Listen again and write the number for the first part: Denise had 56 cards. ✔
- Write the sign and the number for the next part: Donna had 132 cards. ✔
- Touch and read the symbols you wrote for problem A. Get ready. (Signal.) *56 plus 132.*

b. Work problem A and figure out how many cards the girls ended up with altogether.
- For problem A, you wrote 56 plus 132. What's the answer? (Signal.) *188.*
- How many cards did the girls end up with altogether? (Signal.) *188.*

c. Touch where you'll write the symbols for problem B. ✔
  Listen to problem B: A truck had 187 boxes on it. Then 127 of those boxes fell off the truck. How many boxes were still on the truck?
- Listen again and write the number for the first part: A truck had 187 boxes on it. ✔
- Write the sign and the number for the next part: Then 127 of those boxes fell off the truck. ✔
- Touch and read the symbols you wrote for problem B. Get ready. (Signal.) *187 minus 127.*

d. Work problem B and figure out how many boxes ended up on the truck.
   (Answer key:)

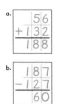

- For problem B, you wrote 187 minus 127. What's the answer? (Signal.) *60.*
- How many boxes ended up on the truck? (Signal.) *60.*

EXERCISE 9: COUNTING OBJECTS [REMEDY]

a. Find part 5 on worksheet 59. ✔
   (Teacher reference:)   [R] **Part E**

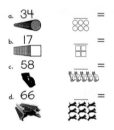

You're going to complete each equation.
- Touch problem A. ✔
  It shows shaded circles and unshaded circles.
- What's the number for the shaded circles? (Signal.) *34.*
- Write the plus sign and the number for the unshaded circles.
  (Observe children and give feedback.)

b. Everybody, touch and read the problem for A. (Signal.) *34 plus 6 (equals).*
  You'll get the number for the shaded circles going. Then you'll touch and count for the unshaded circles.
- Touch the number you'll get going. ✔
- Get it going. (Signal.) *Thirty-fouuur.*
- Touch and count. (Tap.) *35, 36, 37, 38, 39, 40.*
  (Repeat until firm.)
- How many circles are there altogether? (Signal.) *40.*
- Write 40. ✔
- Touch and read the fact for the circles. Get ready. (Signal.) *34 + 6 = 40.*

c. Touch problem B. ✔
It shows big boxes and small boxes.
- What's the number for the big boxes?
(Signal.) *17.*
- Write the plus sign and the number for the small boxes.
(Observe children and give feedback.)

d. Everybody, touch and read the problem for B. (Signal.) *17 plus 4 (equals).*
You'll get the number for the big boxes going. Then you'll touch and count for the small boxes.
- Touch the number you'll get going. ✔
- Get it going. (Signal.) *Seventeeen.*
- Touch and count. (Tap.) *18, 19, 20, 21.*
(Repeat until firm.)
- How many boxes are there altogether? (Signal.) *21.*
- Write 21. ✔
- Touch and read the fact for the boxes. Get ready. (Signal.) *17 + 4 = 21.*

e. Touch problem C. ✔
It shows black cats and white cats.
- What's the number for the black cats? (Signal.) *58.*
- Write the plus sign and the number for the white cats.
(Observe children and give feedback.)

f. Everybody, touch and read the problem for C. (Signal.) *58 plus 5 (equals).*
You'll get the number for the black cats going. Then you'll touch and count for the white cats.
- Touch the number you'll get going. ✔
- Get it going. (Signal.) *Fifty-eieieight.*
- Touch and count. (Tap.) *59, 60, 61, 62, 63.*
(Repeat until firm.)
- How many cats are there altogether? (Signal.) *63.*
- Write 63. ✔
- Touch and read the fact for the cats. Get ready. (Signal.) *58 + 5 = 63.*

g. Touch problem D. ✔
It shows big dogs and small dogs.
- What's the number for the big dogs? (Signal.) *66.*
- Write the plus sign and the number for the small dogs.
(Observe children and give feedback.)

h. Everybody, touch and read the problem for D. (Signal.) *66 plus 9 (equals).*
- Touch the number you'll get going. ✔
- Get it going. (Signal.) *Sixty-siiix.*
- Touch and count. (Tap.) *67, 68, 69, 70, 71, 72, 73, 74, 75.*
(Repeat until firm.)
- How many dogs are there altogether? (Signal.) *75.*
- Complete the equation for D. ✔
- Touch and read the fact for the dogs. Get ready. (Signal.) *66 + 9 = 75.*

EXERCISE 10: COLUMN PROBLEMS

a. Turn to the other side of worksheet 59 and find part 6. ✔
(Teacher reference:)

$$\text{a. } \begin{array}{r} 856 \\ -650 \\ \hline \end{array} \quad \text{b. } \begin{array}{r} 764 \\ -724 \\ \hline \end{array} \quad \text{c. } \begin{array}{r} 207 \\ +\ 51 \\ \hline \end{array} \quad \text{d. } \begin{array}{r} 69 \\ -\ 57 \\ \hline \end{array}$$

For one of these problems, the beginning digit of the answer is zero.
- Touch and read problem A. (Signal.) *856 minus 650.*
- Touch and read the problem in the ones column. (Signal.) *6 minus zero.*
- What's the answer? (Signal.) *6.*
- Touch and read the problem in the tens column. (Signal.) *5 minus 5.*
- What's the answer? (Signal.) *Zero.*
- Touch and read the problem in the hundreds column. (Signal.) *8 minus 6.*
- What's the answer? (Signal.) *2.*
(Repeat until firm.)
- Write the answer to problem A.
(Observe children and give feedback.)
- Everybody, touch and read the equation for A. (Signal.) *856 – 650 = 206.*

b. Touch and read problem B. (Signal.)
764 minus 724.
- Touch and read the problem in the ones column. (Signal.) *4 minus 4.*
- What's the answer? (Signal.) *Zero.*
- Touch and read the problem in the tens column. (Signal.) *6 minus 2.*
- What's the answer? (Signal.) *4.*
- Touch and read the problem in the hundreds column. (Signal.) *7 minus 7.*
- What's the answer? (Signal.) *Zero.*
- Do we write zero if it's the beginning digit? (Signal.) *No.*
- Write the answer to problem B.
(Observe children and give feedback.)
- Everybody, touch and read the equation for B. (Signal.) *764 – 724 = 40.*

c. Touch and read problem C. (Signal.)
207 plus 51.
- Touch and read the problem in the ones column. (Signal.) *7 plus 1.*
- What's the answer? (Signal.) *8.*
- Touch and read the problem in the tens column. (Signal.) *Zero plus 5.*
- What's the answer? (Signal.) *5.*
- What do you write in the answer for the hundreds? (Signal.) *2.*
(Repeat until firm.)

d. Touch and read problem D. (Signal.) *69 minus 57.*
- Touch and read the problem in the ones column. (Signal.) *9 minus 7.*
- What's the answer? (Signal.) *2.*
- Touch and read the problem in the tens column. (Signal.) *6 minus 5.*
- What's the answer? (Signal.) *1.*
- Write the answers to problems C and D.
(Observe children and give feedback.)

e. Check your work.
- Everybody, touch and read the equation for C. (Signal.) *207 + 51 = 258.*
- Touch and read the equation for D. (Signal.) *69 – 57 = 12.*

EXERCISE 11: INDEPENDENT WORK

a. Find part 7. ✔
(Teacher reference:)

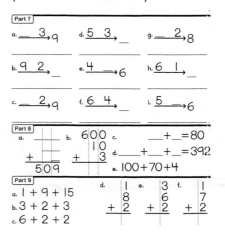

The families in part 7 have a number missing. Below each family you'll write the problem for finding the missing number and complete the fact. Then you'll write the missing number in each family.

b. The problems in parts 8 and 9 are written in rows and columns. The problems in part 8 are place-value addition problems.

c. Turn to the other side of your worksheet and find part 2. ✔
- Count for each group of coins and write the cents to show how much each group is worth. Then complete the other side of worksheet 59.
(Observe children and mark incorrect responses on children's worksheets as you give feedback.)

Lesson  60

EXERCISE 1: NUMBER FAMILIES
MISSING NUMBER IN FAMILY

a. (Display:)   [60:1A]

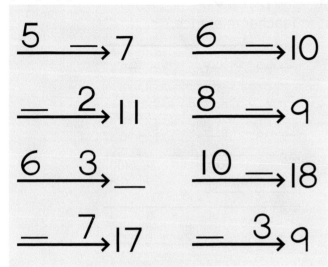

You're going to say the problem for the missing number in each family.

- (Point to $5 \rightarrow 7$.) Say the problem for the missing number. Get ready. (Touch.) *7 minus 5.*
- What's 7 minus 5? (Signal.) *2.*

b. (Point to $\rightarrow 2 \rightarrow 11$.) Say the problem for the missing number. (Touch.) *11 minus 2.*
- What's 11 minus 2? (Signal.) *9.*

c. (Repeat the following tasks for remaining families:)

(Point to __.)	Say the problem for the missing number.	What's __?	
$6 \quad 3 \rightarrow$	6 + 3	6 + 3	9
$\rightarrow 17$	17 – 7	17 – 7	10
$6 \rightarrow 10$	10 – 6	10 – 6	4
$8 \rightarrow 9$	9 – 8	9 – 8	1
$10 \rightarrow 18$	18 – 10	18 – 10	8
$\rightarrow 3 \rightarrow 9$	9 – 3	9 – 3	6

(Repeat for families that were not firm.)

EXERCISE 2: MIXED COUNTING

a. Listen: Count by 25s to 100. Get ready. (Tap.) *25, 50, 75, 100.*
- My turn to start with 100 and count by 25s to 200: One huuundred, 125, 150, 175, 200.
- Your turn: Start with 100 and count by 25s to 200. Get 100 going. *One huuundred.* Count. (Tap.) *125, 150, 175, 200.*

b. Listen: You have 50. When you count by 25s, what's the next number? (Signal.) *75.*
- You have 50. When you plus ones, what's the next number? (Signal.) *51.*
- You have 50. When you count by fives, what's the next number? (Signal.) *55.*
- You have 50. When you plus tens, what's the next number? (Signal.) *60.*
(Repeat step b until firm.)

c. Listen: You have 75. When you count by 25s, what's the next number? (Signal.) *100.*
- You have 75. When you plus tens, what's the next number? (Signal.) *85.*
- You have 75. When you count by fives, what's the next number? (Signal.) *80.*
- You have 75. When you plus ones, what's the next number? (Signal.) *76.*
(Repeat step c until firm.)

EXERCISE 3: RULER SKILLS   REMEDY

a. (Write a 1-inch line on the board:)
- This line is 1 inch long.
  How long is this line? (Signal.) *1 inch.*

b. (Display:)   [60:3A]

↓ 1 inch ↓

Here's a bigger picture. You can see that an inch is not as long as your thumb.
(Touch left end of line.) This is the beginning of the line.
(Touch right end of line.) This is the end of the line.

c. (Display:) W [60:3B]

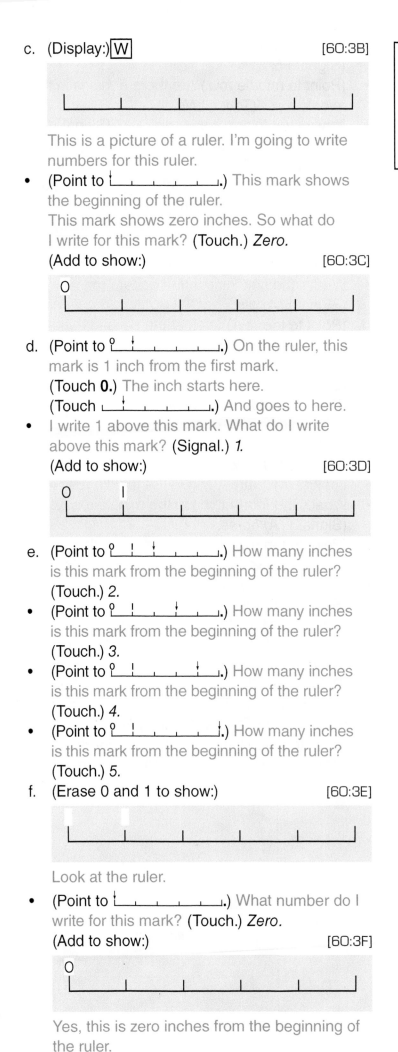

This is a picture of a ruler. I'm going to write numbers for this ruler.

- (Point to ⌊_____⌋.) This mark shows the beginning of the ruler.
  This mark shows zero inches. So what do I write for this mark? (Touch.) *Zero.*
  (Add to show:) [60:3C]

0 ⌊_____⌋

d. (Point to 0 ⌊_____⌋.) On the ruler, this mark is 1 inch from the first mark.
  (Touch **0**.) The inch starts here.
  (Touch ⌊_____⌋.) And goes to here.
- I write 1 above this mark. What do I write above this mark? (Signal.) *1.*
  (Add to show:) [60:3D]

0 1 ⌊_____⌋

e. (Point to 0 ⌊_____⌋.) How many inches is this mark from the beginning of the ruler? (Touch.) *2.*
- (Point to 0 ⌊_____⌋.) How many inches is this mark from the beginning of the ruler? (Touch.) *3.*
- (Point to 0 ⌊_____⌋.) How many inches is this mark from the beginning of the ruler? (Touch.) *4.*
- (Point to 0 ⌊_____⌋.) How many inches is this mark from the beginning of the ruler? (Touch.) *5.*

f. (Erase 0 and 1 to show:) [60:3E]

⌊_____⌋

Look at the ruler.
- (Point to ⌊_____⌋.) What number do I write for this mark? (Touch.) *Zero.*
  (Add to show:) [60:3F]

0 ⌊_____⌋

Yes, this is zero inches from the beginning of the ruler.

g. (Point to 0 ⌊_____⌋.) What number do I write for this mark? (Touch.) *1.*
- (Point to 0 ⌊_____⌋.) What number do I write for this mark? (Touch.) *2.*
- (Point to 0 ⌊_____⌋.) What number do I write for this mark? (Touch.) *3.*
  (Repeat step g until firm.)

h. (Display:) [60:3G]

0 ⌊__|__|__|__|__|__⌋
b d a c

(Point to **a**.) You're going to figure out how many inches this mark is from the beginning of the ruler.
- (Touch **0**.) Count the inches. Get ready. (Touch lines.) *1, 2, 3, 4, 5, 6.*
- (Keep touching **a**.) How many inches is this mark from the beginning of the ruler? (Tap.) *6.*

i. (Point to **b**.) You're going to figure out how many inches this mark is from the beginning of the ruler.
- (Touch **0**.) Count the inches. Get ready. (Touch lines.) *1, 2, 3, 4.*
- (Keep touching **b**.) How many inches is this mark from the beginning of the ruler? (Tap.) *4.*

EXERCISE 4: COUNTING DOLLARS   REMEDY

a. (Display:) [60:4A]

- (Point to **50**.) You're going to figure out how many dollars each row is worth.
- How much is this bill worth? (Touch.) *50 dollars.*
- (Point to **10**.) How much is this bill worth? (Touch.) *10 dollars.*
- (Point to each 10.) How many bills are worth 10 dollars? (Touch.) *2.*
- (Point to **1**.) How much is this bill worth? (Touch.) *1 dollar.*
- (Point to each 1.) How many bills are worth 1 dollar? (Touch.) *2.*

b. (Point to **50.**) You're going to get the number for this bill going. Then you'll count for the rest of the bills.
* (Touch 50.) Get it going. *Fiftyyy.* Count for the tens. (Touch 10s.) *60, seventyyyy.* Count for the ones. (Touch 1s.) *71, 72.*
  (Repeat until firm.)
* How many dollars is this row worth? (Touch.) *72.*
c. (Point to **20.**) How much is this bill worth? (Touch.) *20 dollars.*
* (Point to **10.**) How much is this bill worth? (Touch.) *10 dollars.*
* (Point to each 10.) How many bills are worth 10 dollars? (Touch.) *3.*
* (Point to **5.**) How much is this bill worth? (Touch.) *5 dollars.*
d. (Point to **20.**) You're going to get the number for this bill going. Then you'll count for the rest of the bills.
* (Touch **20.**) Get it going. *Twentyyy.* Count for the tens. (Touch 10s.) *30, 40, fiftyyy.* Count for the five. (Touch 5.) *55.*
  (Repeat until firm.)
* (Point to **20.**) How much is this row worth? (Touch.) *55 dollars.*

EXERCISE 5: ORDINAL NUMBERS

a. (Display:)   [60:5A]

* Raise your hand when you know how many things are in the top row. ✔

* How many things are in the top row? (Signal.) *5.*
* (Point to middle row.) Are there 5 things in the middle row? (Touch.) *No.*
* How many things are in this row? (Signal.) *4.*
b. (Point to bottom row.) Are there 5 things in the bottom row? (Touch.) *No.*
* How many things are in this row? (Signal.) *3.*
c. Look at the third thing in the bottom row. ✔
* What's the third thing in this row? (Signal.) *(A) boy.*
* Look at the first thing in that row. ✔
* What's the first thing in the bottom row? (Signal.) *(A) girl.*
d. (Point to top row.) Look at the second thing in the top row. ✔
* What's the second thing in the top row? (Signal.) *(A) cat.*
e. (Point to middle row.) Look at the second thing in the middle row. ✔
* What's the second thing in the middle row? (Signal.) *(A) boy.*
* Look at the fourth thing in the middle row. ✔
* What's the fourth thing in the middle row? (Signal.) *(A) horse.*
f. (Point to top row.) Look at the fourth thing in the top row. ✔
* What's the fourth thing in the top row? (Signal.) *(A) girl.*
* Look at the fifth thing in the top row. ✔
* What's the fifth thing in the top row? (Signal.) *(A) dog.*
g. (Point to middle row.) Tell me if I touch the first, second, third, or fourth thing.
* (Point to cat.) Tell me. (Touch.) *First.*
* (Point to horse.) Tell me. (Touch.) *Fourth.*
* (Point to boy.) Tell me. (Touch.) *Second.*

EXERCISE 6: FACTS

ADDITION

a. (Distribute unopened workbooks to students.)
- Open your workbook to Lesson 60 and find part 1.
  (Observe children and give feedback.)
  (Teacher reference:)   R Part J

a. 6+4   f. 5+10
b. 7+2   g. 2+9
c. 3+6   h. 4+6
d. 10+8   i. 5+2
e. 2+3   j. 3+5

These are plus problems for families you know. You're going to read each problem and tell me the answer. Then you'll go back and work all of the problems.
- Touch and read problem A. Get ready. (Signal.) *6 plus 4.*
- What's 6 plus 4? (Signal.) *10.*
b. Touch and read problem B. (Signal.) *7 plus 2.*
- What's 7 plus 2? (Signal.) *9.*
c. (Repeat the following tasks for problems C through J:)
- Touch and read problem __.
- What's __?
d. Complete the equations for all of the problems in part 1.
  (Observe children and give feedback.)
  (Answer key:)

a. 6+4= 10 f. 5+10= 15
b. 7+2= 9   g. 2+9 = 11
c. 3+6= 9   h. 4+6 = 10
d. 10+8= 18 i. 5+2 = 7
e. 2+3= 5   j. 3+5 = 8

e. Check your work. You'll touch and read each fact.
- Fact A. (Signal.) *6 + 4 = 10.*
- (Repeat for:) B, *7 + 2 = 9;* C, *3 + 6 = 9;* D, *10 + 8 = 18;* E, *2 + 3 = 5;* F, *5 + 10 = 15;* G, *2 + 9 = 11;* H, *4 + 6 = 10;* I, *5 + 2 = 7;* J, *3 + 5 = 8.*

EXERCISE 7: WORD PROBLEMS (COLUMNS)

a. Find part 2 on worksheet 60. ✔
   (Teacher reference:)

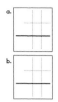

You're going to write the symbols for word problems in columns and work them. The equals bars are already shown.
- Touch where you'll write the symbols for problem A. ✔
  Listen to problem A: There were 176 biscuits on the table. People ate 62 of those biscuits. How many biscuits ended up on the table?
- Listen again and write the number for the first part: There were 176 biscuits on the table. ✔
- Write the sign and the number for the next part: People ate 62 of those biscuits. ✔
- Touch and read the symbols you write for problem A. Get ready. (Signal.) *176 minus 62 (equals).*
b. Work problem A and figure out how many biscuits ended up on the table.
  (Observe children and give feedback.)
- For problem A, you wrote 176 minus 62. What's the answer? (Signal.) *114.*
- So how many biscuits ended up on the table? (Signal.) *114.*
c. Touch where you'll write the symbols for problem B. ✔
  Listen to problem B: There were 316 people on a train. 43 more people got on the train. How many people ended up on the train?
- Listen again and write the number for the first part: There were 316 people on a train. ✔
- Write the sign and the number for the next part: 43 more people got on the train. ✔
- Touch and read the symbols you wrote for problem B. Get ready. (Signal.) *316 plus 43 (equals).*

d. Work problem B and figure out how many people ended up on the train.
(Observe children and give feedback.)
(Answer key:)
(The problems are not in the configuration they appear on worksheet.)

- For problem B, you wrote 316 plus 43. What's the answer? (Signal.) *359.*
- So how many people ended up on the train? (Signal.) *359.*

EXERCISE 8: RULER SKILLS

a. Find part 3 on worksheet 60. ✔
(Teacher reference:)

- Touch the star on the ruler. ✔
You're going to figure out how many inches the star is from the beginning of the ruler.
- Touch the beginning of the ruler and keep your finger there.
(Observe children and give feedback.)
- Touch and count the marks to the star. Get ready. (Tap 3.) *1, 2, 3.*
(Repeat until firm.)
- How many inches is it to the star? (Signal.) *3.*
- Write 3 above the star.
(Observe children and give feedback.)
b. This time, you'll figure out how many inches the heart is from the beginning of the ruler.
- Touch the beginning of the ruler. ✔
- Touch and count the marks to the heart. Get ready. (Tap 5.) *1, 2, 3, 4, 5.*
(Repeat until firm.)
- How many inches is it to the heart? (Signal.) *5.*
- Write 5 above the heart.
(Observe children and give feedback.)
c. This time, you'll figure out how many inches the circle is from the beginning of the ruler.
- Touch the beginning of the ruler. ✔
- Touch and count the marks to the circle. Get ready. (Tap 2.) *1, 2.*
(Repeat until firm.)
- How many inches is it to the circle? (Signal.) *2.*
- Write 2 above the circle.
(Observe children and give feedback.)

EXERCISE 9: FACTS
SUBTRACTION

a. Find part 4 on worksheet 60. ✔
(Teacher reference:)

a. $10-9$ d. $9-6$ g. $10-8$
b. $7-5$ e. $8-7$ h. $4-2$
c. $10-4$ f. $8-3$

All of these are minus problems, so they show the big number and a small number of a family. You're going to touch each problem as I read it. Then you'll tell me the answer.
- Problem A is 10 minus 9. What's the answer? (Signal.) *1.*
b. Problem B is 7 minus 5. What's the answer? (Signal.) *2.*
c. (Repeat the following tasks for problems C through H:)

Problem __ is __.		What's the answer?
C	$10 - 4$	6
D	$9 - 6$	3
E	$8 - 7$	1
F	$8 - 3$	5
G	$10 - 8$	2
H	$4 - 2$	2

d. Complete the equations in part 4. Put your pencil down when you're finished.
(Observe children and give feedback.)
(Answer key:)

a. $10-9=1$ d. $9-6=3$ g. $10-8=2$
b. $7-5=2$ e. $8-7=1$ h. $4-2=2$
c. $10-4=6$ f. $8-3=5$

e. Check your work. You'll touch and read each fact.
- Fact A. (Signal.) *10 – 9 = 1.*
- (Repeat for:) B, *7 – 5 = 2;* C, *10 – 4 = 6;* D, *9 – 6 = 3;* E, *8 – 7 = 1;* F, *8 – 3 = 5;* G, *10 – 8 = 2;* H, *4 – 2 = 2.*

EXERCISE 10: COLUMN PROBLEMS

a. Find part 5 on worksheet 60. ✔
(Teacher reference:)

a. $\begin{array}{r} 26 \\ +433 \\ \hline \end{array}$ b. $\begin{array}{r} 468 \\ -425 \\ \hline \end{array}$ c. $\begin{array}{r} 879 \\ -357 \\ \hline \end{array}$

For one of these problems the beginning digit of the answer is zero.

- Touch and read problem A. (Signal.) 26 plus 433.
- Touch and read the problem in the ones column. (Signal.) 6 plus 3.
- What's the answer? (Signal.) 9.
- Touch and read the problem in the tens column. (Signal.) 2 plus 3.
- What's the answer? (Signal.) 5.
- What do you write in the answer for the hundreds? (Signal.) 4.
- Write the answer to problem A. (Observe children and give feedback.)
- Everybody, touch and read the equation for A. (Signal.) 26 + 433 = 459. (Repeat until firm.)

b. Touch and read problem B. (Signal.) 468 minus 425.

- Touch and read the problem in the ones column. (Signal.) 8 minus 5.
- What's the answer? (Signal.) 3.
- Touch and read the problem in the tens column. (Signal.) 6 minus 2.
- What's the answer? (Signal.) 4.
- Touch and read the problem in the hundreds column. (Signal.) 4 minus 4.
- What's the answer? (Signal.) Zero.
- Do we write zero if it's the beginning digit? (Signal.) No.
- Write the answer to problem B. (Observe children and give feedback.)
- Everybody, touch and read the equation for B. (Signal.) 468 – 425 = 43.

c. Touch and read problem C. (Signal.) 879 minus 357.

- Touch and read the problem in the ones column. (Signal.) 9 minus 7.
- What's the answer? (Signal.) 2.

- Touch and read the problem in the tens column. (Signal.) 7 minus 5.
- What's the answer? (Signal.) 2.
- Touch and read the problem in the hundreds column. (Signal.) 8 minus 3.
- What's the answer? (Signal.) 5.
- Write the answer to problem C. (Observe children and give feedback.)
- Everybody, touch and read the equation for C. (Signal.) 879 – 357 = 522.

EXERCISE 11: INDEPENDENT WORK

a. Turn to the other side of worksheet 60 and find part 6. ✔
(Teacher reference:)

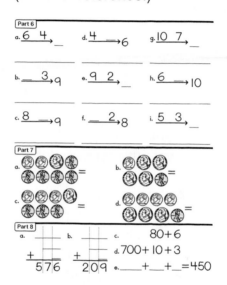

The families in part 6 have a number missing. Below each family you'll write the problem for finding the missing number and complete the fact. Then you'll write the missing number in each family.

b. Find part 7. ✔
You're going to count for the coins in each group and write the number of cents to show what it's worth.

c. Find part 8. ✔
You'll complete the place-value addition for the problems in rows and columns.

d. Complete worksheet 60.
(Observe children and mark incorrect responses on children's worksheets as you give feedback.)

Mastery Test (6)

Note: Mastery Tests are administered to all students in the group. Each student will need a pencil and a *Student Assessment Book.* Try to arrange students so they cannot look at other students' responses.

Teacher Presentation

a. Find Test 6 in your test booklet. ✔
- Touch part 1. ✔
  (Teacher reference:)

  You're going to write the symbols for word problems in columns and work them. The equals bars are already shown.
- Touch where you'll write the symbols for problem A. ✔
  Listen to problem A: A truck had 187 boxes on it. Then 127 of those boxes fell off the truck. How many boxes were still on the truck?
- Listen again and write the number for the first part: A truck had 187 boxes on it. ✔
- Write the sign and the number for the next part: Then 127 of those boxes fell off the truck. ✔
- Work problem A and figure out how many boxes were still on the truck.
  (Observe children.)
b. Touch where you'll write the symbols for problem B. ✔
  Listen to problem B: Workers picked 182 apples in the morning. Workers picked 215 more apples in the afternoon. How many apples did the workers end up picking?
- Listen again and write the number for the first part: Workers picked 182 apples in the morning. ✔
- Write the sign and the number for the next part: Workers picked 215 more apples in the afternoon. ✔
- Work problem B and figure out how many apples the workers ended up picking.
  (Observe children.)

c. Touch where you'll write the symbols for problem C. ✔
  Listen to problem C: 765 students were in school. 562 of those students left the school. How many students ended up in the school?
- Listen again and write the number for the first part: 765 students were in the school. ✔
- Write the sign and the number for the next part: 562 of those students left the school. ✔
- Work problem C and figure out how many students ended up in the school.
  (Observe children.)
d. Touch part 2 on Test 6. ✔
  (Teacher reference:)

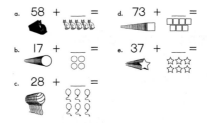

Each problem in part 2 shows two groups of objects. For each problem, you'll write the number for the second group. Then you'll get the number for the first group going and touch and count to yourself to figure out how many objects are in both groups.
- Write the number of objects in the second group for all of the problems in part 2. Put your pencil down when you've written the number for the second group in each problem.
  (Observe children.)
e. Touch the number you'll get going in problem A. ✔
- Get it going and touch and count for the white cats to yourself. Then complete equation A.
  (Observe children.)
f. Touch the number you'll get going in problem B. ✔
- Get it going and touch and count for the small circles to yourself. Then complete equation B.
  (Observe children.)
g. Touch the number you'll get going in problem C. ✔
- Get it going and touch and count for the white balloons to yourself. Then complete equation C.
  (Observe children.)

h. Touch the number you'll get going in
    problem D. ✔
•   Get it going and touch and count for the small
    boxes to yourself. Then complete equation D.
    (Observe children.)
i. Touch the number you'll get going in
    problem E. ✔
•   Get it going and touch and count for the small
    stars to yourself. Then complete equation E.
    (Observe children.)
j. Touch part 3. ✔
    (Teacher reference:)

	d. $11-2$	h. $2+6$
a. $6-4$	e. $8+10$	i. $17-7$
b. $9+2$	f. $8-2$	j. $4-2$
c. $13-3$	g. $10-8$	

These problems are from number families with
a small number of 2 or 10. You'll complete
equations in part 3.

k. Turn to the other side of Test 6 and touch
    part 4. ✔
    (Teacher reference:)

a. ⊛⊛⊛⊛⊛⊛⊛⊛
b. ⊛⊛⊛⊛⊛⊛⊛⊛⊛
c. ⊛⊛⊛⊛⊛⊛⊛⊛⊛⊛
d. ⊛⊛⊛⊛⊛⊛⊛

In part 4, you'll count the cents for each group
of coins. Then you'll write an equals and the
number to show how many cents each group
is worth.

l. Touch part 5. ✔
    (Teacher reference:)

	d. $9-3$	h. $3+6$
a. $8-3$	e. $6+4$	i. $10-4$
b. $10-6$	f. $9-6$	j. $5+3$
c. $3+5$	g. $8-5$	

These problems are from number families
with a small number of 5 or 6. You'll complete
equations in part 5.

m. Touch part 6. ✔
    (Teacher reference:)

a.	b.	c.	d.
$+$	$+$	$+$	$+$
382	107	619	760

You'll complete the place-value equations in
columns in part 6.

n. Touch part 7. ✔
    (Teacher reference:)

a. 265	b. 265	c. 468	d. 364
-210	$+210$	-426	$-\ \ 54$

In part 7, you'll work column problems.
•   Your turn: Work the rest of the problems on
    Test 6. ✔
    (Direct students where to put their
    assessment books when they are finished.)

Scoring Notes

a. Collect test booklets. Use the Answer Key and
    Passing Criteria Table to score the tests.

Passing Criteria Table — Mastery Test 6				
Part	Score	Possible Score	Passing Score
1	3 for each problem	9	6
2	3 for each problem	15	12
3	2 for each fact	20	18
4	3 for each group of coins	12	9
5	2 for each fact	20	18
6	3 for each equation	12	9
7	3 for each problem	12	9
	Total	100	

b. Complete the Mastery Test 6 Remedy
    Summary Sheet to determine whether group
    remedies are needed. Reproducible Remedy
    Summary Sheets are at the back of the
    Answer Key and at the back of the *Teacher's
    Guide.*
•   If ¼ or more of the students did not pass a test
    part, present the remedy for that part before
    beginning Lesson 61. The Remedy Table
    follows and is also at the end of the Mastery
    Test 6 Answer Key. Remedies worksheets
    follow Mastery Test 6 in the *Student
    Assessment Book.*

Part	Test Items	Remedy Lesson	Remedy Exercise	Student Material Remedies Worksheet
1	Word Problems	57	10	Part A
		59	8	Part B
2	Count-on	50	6	—
		54	7	Part C
		57	8	Part D
		59	9	Part E
3	Facts (Small Numbers of 2, 10)	49	1	—
		51	3	—
		52	8	Part F
		55	3	—
4	Coins	55	9	Part G
		58	8	Part H
		59	6	Part I
5	Facts (Small Numbers of 5, 6)	54	3	—
		58	2	—
		60	6	Part J
6	Place-Value Equations	53	7	Part K
		54	5	Part L
7	Column Problems	55	4	—
		55	6	—
		57	7	Part M
		58	7	Part N

Table title: **Remedy Table — Mastery Test 6**

Retest

Retest individual students on any part failed.

Cumulative Test 1

Note: Cumulative Tests are group-administered. Each student will need a pencil and the Student Assessment Book. Try to arrange students so they cannot look at other students' responses.

Cumulative Test 1 assesses the skills taught in Lessons 1–60 of CMC 1. Presenting Cumulative Test 1 to children who have not mastered these skills is not recommended.

Teacher Presentation

a. Find Cumulative Test 1 in your test booklet. ✔
- Touch part 1. ✔
  (Teacher reference:)

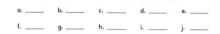

You're going to write numbers.
- Touch the space after A. ✔
- The number for A is 18. What number? (Signal.) *18.*
- Write 18 for A.
  (Observe children.)
- The number for B is 14. What number? (Signal.) *14.*
- Write 14 for B.
  (Observe children.)
- The number for C is 15. What number? (Signal.) *15.*
- Write 15 for C.
  (Observe children.)
- (Repeat for: D, 90; E, 80; F, 20; G, 42; H, 53; I, 27; J, 13.)

b. Touch part 2. ✔
  (Teacher reference:)

		i.	4 + 1	r.	7 + 2
a.	2 + 1	j.	14 + 1	s.	10 + 1
b.	12 + 1	k.	8 + 1	t.	10 + 2
c.	6 + 1	l.	8 + 2	u.	4 + 0
d.	16 + 1	m.	9 + 1	v.	1 + 0
e.	3 + 1	n.	9 + 2	w.	10 + 0
f.	13 + 1	o.	5 + 1	x.	38 + 0
g.	1 + 1	p.	5 + 2	y.	0 + 0
h.	11 + 1	q.	7 + 1		

- Complete these equations.
  (Observe children.)

c. Touch part 3. ✔
  (Teacher reference:)

$$\underline{90} + \underline{6} = 96$$

a. ___ + ___ = 37	d. ___ + ___ = 19
b. ___ + ___ = 81	e. ___ + ___ = 15
c. ___ + ___ = 90	f. ___ + ___ = 50

I'll read the place-value equation that's shown. 90 plus 6 equals 96.
- Complete the place-value equations in part 3.
  (Observe children.)

d. Touch part 4. ✔
  (Teacher reference:)

a.
b.
c.
d.
e.
f.
g.
h.

You're going to write numbers in a column.
- Look at the spaces where you'll write the numbers for part 4. ✔
  These columns don't have letters for hundreds, tens, and ones.
- Touch the space for the hundreds digit in row A. ✔
- Touch the space for the tens digit in row A. ✔
- Touch the space for the ones digit in row A. ✔
  (Repeat until firm.)

e. Listen: You're going to write 346 in row A. What number? (Signal.) *346.*
- Write 346 in row A.
  (Observe children.)

f. You're going to write 179 in row B. What number? (Signal.) *179.*
- Write 179 in row B.
  (Observe children.)

g. (Repeat for the following tasks for C, 805; D, 3; E, 514; F, 72; G, 835; H, 18.)
- You're going to write ___ in row ___. What number?
- Write ___ in row ___.
  (Observe children.)

h. Touch part 5. ✔
(Teacher reference:)

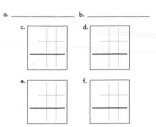

I'll tell you word problems. You'll write the symbols for each problem. Then you'll complete the equation and answer the question.

- Listen to problem A: There were 10 apples on a table. Then 1 of those apples dropped off the table. How many apples ended up on the table?
- Listen to the first part again: There were 10 apples on a table. What symbol will you write for that part? (Signal.) *10.*
- Write the symbol for the first part of problem A. ✔
- Listen to the next part: Then 1 of those apples dropped off the table. What symbols will you write for that part? (Signal.) *Minus 1.*
- Write the symbols and complete the equation. ✔

i. Listen to problem B: 30 leaves were on a plant. Then 6 more leaves grew on the plant. How many leaves ended up on the plant?
- Listen to the first part again: 30 leaves were on a plant. What symbol will you write for that part? (Signal.) *30.*
- Write the symbol for the first part of problem B. ✔
- Listen to the next part: Then 6 more leaves grew on the plant. What symbols will you write for that part? (Signal.) *Plus 6.*
- Write the symbols and complete the equation. ✔

j. Touch where you'll write the symbols for problem C. ✔
Listen to problem C: A shop had 276 tires in it. People bought 166 of those tires. How many tires did the shop end up with?
- Listen again: A shop had 276 tires in it. What number do you write first? (Signal.) *276.*
- Listen to the next part: Then people bought 166 of those tires. Tell me the sign and the number. Get ready. (Signal.) *Minus 166.*

- Listen again and write the symbols for the whole problem: A shop had 276 tires in it. People bought 166 of those tires. (Observe children.)
- Work problem C. Put your pencil down when you know how many tires ended up in the shop. (Observe children.)

k. Touch where you'll write the symbols for problem D. ✔
Listen to problem D: A man started with 62 nails. The man found 14 more nails. How many nails did the man end up with?
- Listen again: A man had 62 nails. What number do you write first? (Signal.) *62.*
- Listen to the next part: The man found 14 more nails. Tell me the sign and the number. Get ready. (Signal.) *Plus 14.*
- Listen again and write the symbols for the whole problem: A man started with 62 nails. The man found 14 more nails. (Observe children.)
- Work problem D. Put your pencil down when you know how many nails the man ended up with. (Observe children.)

l. Touch where you'll write the symbols for problem E. ✔
- Listen to problem E: Workers picked 182 apples in the morning. Workers picked 215 more apples in the afternoon. How many apples did the workers end up picking?
- Listen again and write the number for the first part: Workers picked 182 apples in the morning. ✔
- Write the sign and the number for the next part: Workers picked 215 more apples in the afternoon. ✔
- Work problem E and figure out how many apples the workers ended up picking. (Observe children.)

m. Touch where you'll write the symbols for problem F. ✔
- Listen to problem F: 765 students were in the school. 562 of those students left the school. How many students ended up in the school?
- Listen again and write the number for the first part: 765 students were in the school. ✔
- Write the sign and the number for the next part: 562 of those students left the school. ✔
- Work problem F and figure out how many students ended up in the school. (Observe children.)

n. Touch part 6. ✔
(Teacher reference:)

a. $300 + 40 + 8 =$ f. ___+___+__=620

b. $500 + 60 + 1 =$ g. | | | h. | | |

c. $100 + 70 + 6 =$

d. ___+___+__=209   +___ | | | +___ | | |

e. ___+___+__=813   6 1 9   7 6 0

- Complete the 3-digit place-value equations.
  Put your pencil down when you've completed
  the place-value equations in part 6.
  (Observe children.)

o. Touch part 7. ✔
(Teacher reference:)

a. $2 + 7 = 9$ b. $30 + 8 = 38$ c. $15 + 1 = 16$

_____   _____   _____

For each problem, I'll read the equation you'll
turn around. The turn-around is the other plus
equation that has the same three numbers.

- Equation A: 2 plus 7 equals 9.
- Equation B: 30 plus 8 equals 38.
- Equation C: 15 plus 1 equals 16.
  For each problem in part 7, write the turn-
  around equation below. That's the equation
  with the same three numbers.
  (Observe children.)

p. Touch part 8. ✔
(Teacher reference:)

a. $5 \underset{\longrightarrow 6}{1}$ b. $7 \underset{\longrightarrow 9}{2}$

In the spaces below each number family you'll
write the two plus facts and the two minus
facts the family tells about.

- Write the four facts for each family.
  (Observe children.)

q. Touch part 9. ✔
(Teacher reference:)

a. 46   b. 51   c. 75   d. 357
   –10   +23   +21   –51

e. 51   f. 946   g. 265   h. 265
  +628   –810   –210   +210

i. 468   j. 364
  –426   – 54

- Work the column problems. Put your pencil
  down when you're finished.
  (Observe children.)

r. Touch part 10. ✔
(Teacher reference:)

a. $8 \longrightarrow 9$ b. $9 \ 2 \longrightarrow$ c. $\longrightarrow 2 \longrightarrow 9$

d. $6 \longrightarrow 8$ e. $5 \ 1 \longrightarrow$

On the space below each family, write the
problem for finding the missing number. Then
complete the equation and write the missing
number in the family. Put your pencil down
when you've completed each family in part 10
and written the fact below.

s. Touch part 11. ✔
(Teacher reference:)

a. 58 + __ =   c. 28 + __ =

b. 17 + __ =   d. 73 + __ =

Each problem in part 11 shows two groups
of objects. For each problem, you'll write the
number for the second group. Then you'll get
the number for the first group going and touch
and count to yourself to figure out how many
objects are in both groups.

- Write the number of objects in the second
  group for all of the problems in part 11. Put
  your pencil down when you've written the
  number for the second group in each problem.
  (Observe children.)
- Touch the number you'll get going in
  problem A. ✔
- Get it going and touch and count for the white
  cats to yourself. Then complete equation A.
  (Observe children.)

t. Touch the number you'll get going in
  problem B. ✔
- Get it going and touch and count for the small
  circles to yourself. Then complete equation B.
  (Observe children.)

u. Touch the number you'll get going in
  problem C. ✔
- Get it going and touch and count for the
  white balloons to yourself. Then complete
  equation C.
  (Observe children.)

v. Touch the number you'll get going in
  problem D. ✔
- Get it going and touch and count for the small
  boxes to yourself. Then complete equation D.
  (Observe children.)

w. Touch part 12. ✔
(Teacher reference:)

	i. $10 - 9$	r. $2 + 7$
a. $7 - 1$	j. $5 + 2$	s. $13 - 3$
b. $1 + 7$	k. $9 - 7$	t. $11 - 1$
c. $8 - 7$	l. $6 - 2$	u. $8 + 10$
d. $1 + 8$	m. $2 + 6$	v. $10 - 8$
e. $4 - 3$	n. $7 - 5$	w. $2 + 10$
f. $4 + 1$	o. $8 - 2$	x. $17 - 7$
g. $8 - 1$	p. $9 + 2$	y. $4 - 2$
h. $1 + 5$	q. $5 - 3$	

• Complete these equations.
(Observe children.)

x. Touch part 13. ✔
(Teacher reference:)

a. $9 - 1$	f. $5 - 1$	k. $7 - 6$
b. $6 - 1$	g. $10 - 10$	l. $9 - 8$
c. $3 - 1$	h. $7 - 1$	m. $10 - 1$
d. $8 - 8$	i. $10 - 9$	n. $7 - 7$
e. $2 - 1$	j. $7 - 0$	o. $10 - 0$

• Complete these equations.
(Observe children.)

y. Touch part 14. ✔
(Teacher reference:)

In part 14, you'll count the cents for each group of coins. Then you'll write an equals and the number to show how many cents each group is worth.

• Touch part 15. ✔
(Teacher reference:)

a. $70 + 10$	e. $50 + 10$
b. $72 + 10$	f. $53 + 10$
c. $20 + 10$	g. $30 + 10$
d. $25 + 10$	h. $36 + 10$

• You'll complete equations in part 15.
• Touch part 16. ✔
(Teacher reference:)

a. $1 + 7 + 2$	c. $1 + 4 + 2$
b. $9 + 1 + 6$	d. $1 + 9 + 5$

• You'll complete equations that plus 3 numbers in part 16.
• Touch part 17. ✔
(Teacher reference:)

	d. $9 - 3$	h. $3 + 6$
a. $8 - 3$	e. $6 + 4$	i. $10 - 4$
b. $10 - 6$	f. $9 - 6$	j. $5 + 3$
c. $3 + 5$	g. $8 - 5$	

• You'll complete equations in part 17.
• Your turn: Work the rest of the problems on Cumulative Test 1.
(Direct students where to put their assessment books when they are finished.)

Scoring Notes

a. Collect test booklets. Use the Answer Key to score the tests. Children can earn 1 point for each problem in all parts. Record children's performance on the Summary Sheet. Reproducible Summary Sheets are at the back of the *Teacher's Guide*.

b. Use the Remedy Table for Cumulative Test 1, available online, to determine whether remedies are needed. Use the online Remedies Worksheets when presenting remedies for Cumulative Test 1.

Retest
Retest individual students on any part failed.

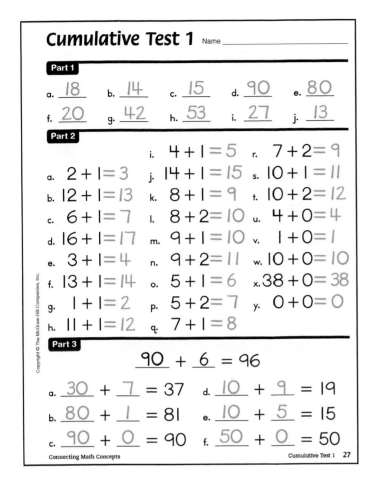

Cumulative Test 1 Name _____

Part 1

a. 18 b. 14 c. 15 d. 90 e. 80
f. 20 g. 42 h. 53 i. 27 j. 13

Part 2

i. $4+1=5$ r. $7+2=9$
a. $2+1=3$ j. $14+1=15$ s. $10+1=11$
b. $12+1=13$ k. $8+1=9$ t. $10+2=12$
c. $6+1=7$ l. $8+2=10$ u. $4+0=4$
d. $16+1=17$ m. $9+1=10$ v. $1+0=1$
e. $3+1=4$ n. $9+2=11$ w. $10+0=10$
f. $13+1=14$ o. $5+1=6$ x. $38+0=38$
g. $1+1=2$ p. $5+2=7$ y. $0+0=0$
h. $11+1=12$ q. $7+1=8$

Part 3

$90 + 6 = 96$

a. $30 + 7 = 37$ d. $10 + 9 = 19$
b. $80 + 1 = 81$ e. $10 + 5 = 15$
c. $90 + 0 = 90$ f. $50 + 0 = 50$

Connecting Math Concepts   Cumulative Test 1   27

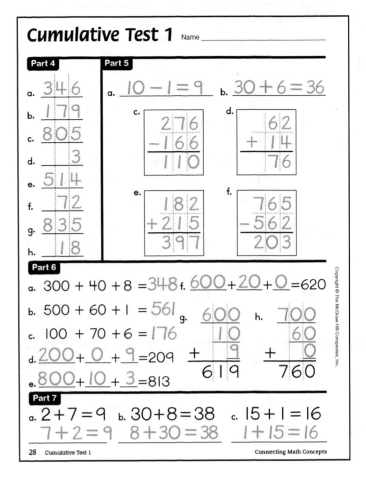

Cumulative Test 1 Name _____

Part 4

a. 346
b. 179
c. 805
d. 3
e. 514
f. 72
g. 835
h. 18

Part 5

a. $10 - 1 = 9$ b. $30 + 6 = 36$

c.
```
  276
- 166
  110
```
d.
```
  62
+ 14
  76
```
e.
```
  182
+ 215
  397
```
f.
```
  765
- 562
  203
```

Part 6

a. $300 + 40 + 8 = 348$ f. $600 + 20 + 0 = 620$
b. $500 + 60 + 1 = 561$
c. $100 + 70 + 6 = 176$
d. $200 + 0 + 9 = 209$
e. $800 + 10 + 3 = 813$

g.
```
  600
   10
+   9
  619
```
h.
```
  700
   60
+   0
  760
```

Part 7

a. $2+7=9$ b. $30+8=38$ c. $15+1=16$
   $7+2=9$ $8+30=38$ $1+15=16$

28   Cumulative Test 1   Connecting Math Concepts

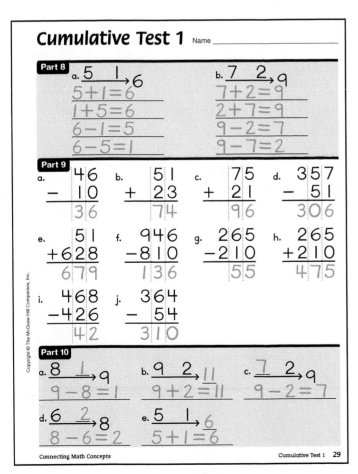

Cumulative Test 1 Name _____

Part 8

a. $5 \quad 1 \to 6$
$5+1=6$
$1+5=6$
$6-1=5$
$6-5=1$

b. $7 \quad 2 \to 9$
$7+2=9$
$2+7=9$
$9-2=7$
$9-7=2$

Part 9

a.
```
  46
- 10
  36
```
b.
```
  51
+ 23
  74
```
c.
```
  75
+ 21
  96
```
d.
```
  357
-  51
  306
```
e.
```
   51
+ 628
  679
```
f.
```
  946
- 810
  136
```
g.
```
  265
- 210
   55
```
h.
```
  265
+ 210
  475
```
i.
```
  468
- 426
   42
```
j.
```
  364
-  54
  310
```

Part 10

a. $8 \quad 1 \to 9$
$9-8=1$

b. $9 \quad 2 \to 11$
$9+2=11$

c. $7 \quad 2 \to 9$
$9-2=7$

d. $6 \quad 2 \to 8$
$8-6=2$

e. $5 \quad 1 \to 6$
$5+1=6$

Connecting Math Concepts   Cumulative Test 1   29

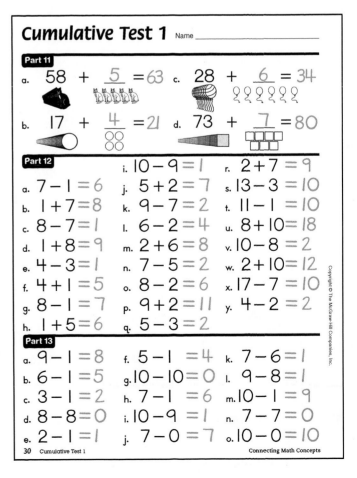

Cumulative Test 1 Name _____

Part 11

a. $58 + 5 = 63$ c. $28 + 6 = 34$
b. $17 + 4 = 21$ d. $73 + 7 = 80$

Part 12

i. $10-9=1$ r. $2+7=9$
a. $7-1=6$ j. $5+2=7$ s. $13-3=10$
b. $1+7=8$ k. $9-7=2$ t. $11-1=10$
c. $8-7=1$ l. $6-2=4$ u. $8+10=18$
d. $1+8=9$ m. $2+6=8$ v. $10-8=2$
e. $4-3=1$ n. $7-5=2$ w. $2+10=12$
f. $4+1=5$ o. $8-2=6$ x. $17-7=10$
g. $8-1=7$ p. $9+2=11$ y. $4-2=2$
h. $1+5=6$ q. $5-3=2$

Part 13

a. $9-1=8$ f. $5-1=4$ k. $7-6=1$
b. $6-1=5$ g. $10-10=0$ l. $9-8=1$
c. $3-1=2$ h. $7-1=6$ m. $10-1=9$
d. $8-8=0$ i. $10-9=1$ n. $7-7=0$
e. $2-1=1$ j. $7-0=7$ o. $10-0=10$

30   Cumulative Test 1   Connecting Math Concepts

Cumulative Test 1 Name _____

Copyright © The McGraw-Hill Companies, Inc.

Part 14

a.  = 50

b.  = 25

c.  = 42

d.  = 48

Part 15

a. $70 + 10 = 80$ e. $50 + 10 = 60$

b. $72 + 10 = 82$ f. $53 + 10 = 63$

c. $20 + 10 = 30$ g. $30 + 10 = 40$

d. $25 + 10 = 35$ h. $36 + 10 = 46$

Part 16

a. $1 + 7 + 2 = 10$ c. $1 + 4 + 2 = 7$

b. $9 + 1 + 6 = 16$ d. $1 + 9 + 5 = 15$

Part 17

d. $9 - 3 = 6$ h. $3 + 6 = 9$

a. $8 - 3 = 5$ e. $6 + 4 = 10$ i. $10 - 4 = 6$

b. $10 - 6 = 4$ f. $9 - 6 = 3$ j. $5 + 3 = 8$

c. $3 + 5 = 8$ g. $8 - 5 = 3$

Connecting Math Concepts   Cumulative Test 1   **31**

Lessons 61-65 Planning Page

	Lesson 61	Lesson 62	Lesson 63	Lesson 64	Lesson 65
Student Learning Objectives	**Exercises** 1. Say two addition and two subtraction facts for number families; find missing numbers in number families 2. **Identify 2-dimensional shapes (rectangle)** 3. Count forward or backward from a given number 4. **Count the value of mixed groups of bills** 5. Solve addition facts 6. **Identify and count the value of quarters; count and write the value of mixed groups of coins** 7. Solve subtraction facts 8. Learn to measure inches with a ruler 9. Solve 2-digit/ 3-digit addition and subtraction problems 10. Complete work independently	**Exercises** 1. **Use a centimeter ruler and count on for the whole to determine if an equation is true** 2. Solve 2-digit/ 3-digit subtraction problems when the answer begins with zero 3. Count forward or backward from a given number 4. **Identify 2-dimensional shapes (square)** 5. Solve subtraction facts 6. Count and write the value of mixed groups of coins 7. Write the symbols for word problems in columns and solve 8. **Count and write the value of mixed groups of bills** 9. Solve addition facts 10. Complete work independently	**Exercises** 1. Say two addition and two subtraction facts for number families; find missing numbers in number families 2. Count forward or backward from a given number 3. Identify 2-dimensional shapes 4. Solve subtraction facts 5. Count and write the value of mixed groups of coins 6. Solve addition facts 7. **Use a centimeter ruler and count on for the whole to solve equations** 8. Write the symbols for word problems in columns and solve 9. Count and write the value of mixed groups of bills 10. Complete work independently	**Exercises** 1. Say two addition and two subtraction facts for number families; find missing numbers in number families 2. **Identify 2-dimensional shapes (triangle)** 3. Count forward or backward from a given number 4. Solve subtraction facts 5. Count and write the value of mixed groups of coins 6. Use a centimeter ruler and count on for the whole to solve equations 7. Solve problems with three addends in columns 8. Count and write the value of mixed groups of bills 9. Write the symbols for word problems in columns and solve 10. Complete work independently	**Exercises** 1. **Say how many of each coin equals a dollar** 2. Say two addition and two subtraction facts for number families; find missing numbers in number families 3. Count forward from a given number 4. Identify 2-dimensional shapes 5. Say how many of each coin equals a dollar 6. Solve subtraction facts 7. Solve 2-digit/ 3-digit addition problems 8. **Use an inch ruler and count a part in a whole and count on for the whole to solve equations** 9. Write the symbols for word problems in columns and solve 10. Complete work independently
Common Core State Standards for Mathematics					
1.OA 3		✔	✔	✔	✔
1.OA 5		✔	✔	✔	✔
1.OA 6	✔	✔	✔	✔	✔
1.OA 7		✔	✔		
1.OA 8	✔	✔	✔	✔	✔
1.NBT 1	✔	✔	✔	✔	✔
1.NBT 2	✔			✔	✔
1.NBT 4	✔	✔	✔	✔	✔
1.G 1	✔	✔	✔	✔	✔
Teacher Materials	Presentation Book 2, Board Displays CD or chalkboard				
Student Materials	**Workbook 2,** Pencil, ruler				
Additional Practice	• Student Practice Software: Block 3: Activity 1 (1.NBT.2 and 1.NBT.4), Activity 2 (1.NBT.4 and 1.OA.6), Activity 3 (1.NBT.1), Activity 4 (1.NBT.1), Activity 5 (1.NBT.2), Activity 6 (1.MD.2), • Math Fact Worksheets 25–26 (After Lesson 61), 27 (After Lesson 63), 28–32 (After Lesson 64)				
Mastery Test					

Lesson 61

EXERCISE 1: NUMBER FAMILIES

a. (Display:) [61:1A]

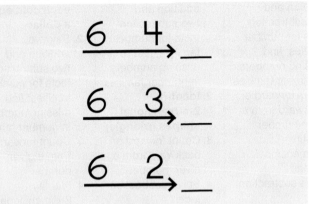

The big number is missing in these families.

- (Point to 6 4→.) What are the small numbers in this family? (Touch.) *6 and 4.*
- What's the big number? (Signal.) *10.*

b. (Point to 6 3→.) What are the small numbers in this family? (Touch.) *6 and 3.*
- What's the big number? (Signal.) *9.*

c. (Point to 6 2→.) What are the small numbers in this family? (Touch.) *6 and 2.*
- What's the big number? (Signal.) *8.*

d. (Point to 6 4→.) What are the small numbers in this family again? (Touch.) *6 and 4.*
- What's the big number? (Signal.) *10.*
- Say the fact that starts with the first small number. Get ready. (Signal.) *6 + 4 = 10.*
- Say the fact that starts with the other small number. Get ready. (Signal.) *4 + 6 = 10.*
- Say the fact that goes backward along the arrow. Get ready. (Signal.) *10 – 4 = 6.*
- Say the other minus fact. Get ready. (Signal.) *10 – 6 = 4.*

e. (Point to 6 3→.) What are the small numbers in this family? (Touch.) *6 and 3.*
- What's the big number? (Signal.) *9.*
- Say the fact that starts with the first small number. Get ready. (Signal.) *6 + 3 = 9.*
- Say the fact that starts with the other small number. Get ready. (Signal.) *3 + 6 = 9.*
- Say the fact that goes backward along the arrow. Get ready. (Signal.) *9 – 3 = 6.*
- Say the other minus fact. Get ready. (Signal.) *9 – 6 = 3.*

f. (Point to 6 2→.) What are the small numbers in this family? (Touch.) *6 and 2.*
- What's the big number? (Signal.) *8.*

- Say the fact that starts with the first small number. Get ready. (Signal.) *6 + 2 = 8.*
- Say the fact that starts with the other small number. Get ready. (Signal.) *2 + 6 = 8.*
- Say the fact that goes backward along the arrow. Get ready. (Signal.) *8 – 2 = 6.*
- Say the other minus fact. Get ready. (Signal.) *8 – 6 = 2.*

g. (Display:) [61:1B]

_ 4→10   6 3→ _

5 _→8   _ 2→10

9 2→ _   4 _→6

3 _→5   10 6→ _

You're going to say the problem for the missing number in each family.

- (Point to _ 4→10.) Say the problem for the missing number. Get ready. (Signal.) *10 minus 4.*
- What's 10 minus 4? (Signal.) *6.*

h. (Point to 5 _→8.) Say the problem for the missing number. (Signal.) *8 minus 5.*
- What's 8 minus 5? (Signal.) *3.*

i. (Repeat the following tasks for remaining families:)

(Point to __.)	Say the problem for the missing number.	What's __?	
9 2→ _	9 + 2	9 + 2	11
3 _→5	5 – 3	5 – 3	2
6 3→ _	6 + 3	6 + 3	9
_ 2→10	10 – 2	10 – 2	8
4 _→6	6 – 4	6 – 4	2
10 6→ _	10 + 6	10 + 6	16

(Repeat for families that were not firm.)

EXERCISE 2: SHAPES
INTRODUCTION OF RECTANGLES

a. (Display:) [61:2A]

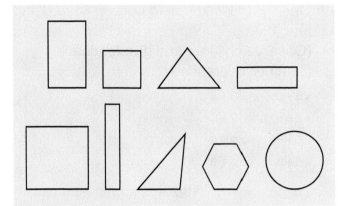

(Point to rectangles.) These are shapes. The shapes with 4 sides are rectangles.
* Say **rectangles.** (Signal.) *Rectangles.* (Repeat until firm.)
b. Listen: All rectangles have 4 sides.
* Say the rule about all rectangles. Get ready. (Signal.) *All rectangles have 4 sides.*
c. (Point to ⌷.) This shape has 4 sides. How many sides does this shape have? (Touch.) *4.*
* So is it a rectangle? (Signal.) *Yes.*
* (Point to ▭.) How many sides does this shape have? (Touch.) *4.*
* So is it a rectangle? (Signal.) *Yes.*
* (Point to △.) Does this shape have 4 sides? (Touch.) *No.*
* How many sides does this shape have? (Touch.) *3.*
* So is it a rectangle? (Signal.) *No.*
* (Point to ▭.) How many sides does this shape have? (Touch.) *4.*
* So is it a rectangle? (Signal.) *Yes.*
* (Point to ▢.) How many sides does this shape have? (Touch.) *4.*
* So is it a rectangle? (Signal.) *Yes.*
* (Point to ▯.) How many sides does this shape have? (Touch.) *4.*
* So is it a rectangle? (Signal.) *Yes.*
* (Point to ◿.) Does this shape have 4 sides? (Touch.) *No.*
* How many sides does this shape have? (Touch.) *3.*
* So is it a rectangle? (Signal.) *No.*

* (Point to ⬡.) Does this shape have 4 sides? (Touch.) *No.*
* So is it a rectangle? (Signal.) *No.*
* (Point to ◯.) Does this shape have 4 sides? (Touch.) *No.*
* So is it a rectangle? (Signal.) *No.*
d. Your turn to count the rectangles. Get ready. (Touch rectangles.) *1, 2, 3, 4, 5.*
* How many rectangles are there? (Signal.) *5.*

EXERCISE 3: MIXED COUNTING

a. Listen: Start with 50 and count by fives to 100.
* What number do you start with? (Signal.) *50.*
* Get 50 going. *Fiftyyy.* Count. (Tap.) *55, 60, 65, 70, 75, 80, 85, 90, 95, 100.*
b. Start with 100 and count backward by tens. Get 100 going. *One huuundred.* Count backward. (Tap.) *90, 80, 70, 60, 50, 40, 30, 20, 10.* (Repeat step b until firm.)
c. Start with 1000 and count backward by hundreds to 100. Get 1000 going. *One thouuusand.* Count backward. (Tap.) *900, 800, 700, 600, 500, 400, 300, 200, 100.*
d. Start with 21 and plus tens to 101. Get 21 going. *Twenty-wuuun.* Plus tens. (Tap.) *31, 41, 51, 61, 71, 81, 91, 101.*
e. Listen: Count by 25s to 100. Get ready. (Signal.) *25, 50, 75, 100.*
   My turn to start with 100 and count by 25s to 200: 100, 125, 150, 175, 200.
* Your turn: Start with 100 and count by 25s to 200. Get 100 going. *One huuundred.* Count. (Tap.) *125, 150, 175, 200.*
f. Listen: You have 25. When you plus 25s, what's the next number? (Signal.) *50.*
* You have 25. When you plus ones, what's the next number? (Signal.) *26.*
* You have 25. When you plus fives, what's the next number? (Signal.) *30.*
* You have 25. When you plus tens, what's the next number? (Signal.) *35.* (Repeat step f until firm.)

Exercise 4: Counting Dollars

a. (Display:) [61:4A]

(Point to .) All of these rows start with a 50 dollar bill.

- (Point to .) How much is this bill worth? (Touch.) *5 dollars.*

  Now you'll count for the bills in this row. You'll get 50 going and plus fives.
- (Touch .) Get it going. *Fiftyyy.* Count for the fives. (Touch 5s.) *55, 60, 65.*
  (Repeat until firm.)
- How many dollars is this row worth? (Touch.) *65 dollars.*

b. (Point to .) How much is this bill worth? (Touch.) *1 dollar.*

  You'll get 50 going and plus ones. (Touch .) Get it going. *Fiftyyy.* Count for the ones. (Touch 1s.) *51, 52, 53.*
  (Repeat until firm.)
- How many dollars is this row worth? (Touch.) *53 dollars.*

c. (Point to .) How much is this bill worth? (Touch.) *10 dollars.*

  You'll get 50 going and plus tens. (Touch .) Get it going. *Fiftyyy.* Count for the tens. (Touch 10s.) *60, 70, 80.*
  (Repeat until firm.)
- How many dollars is this row worth? (Touch.) *80 dollars.*

d. Let's do those again.
- (Touch in 1st row.) Get it going. *Fiftyyy.* Count. (Touch bills.) *55, 60, 65.*
- What's this row worth? (Touch.) *65 dollars.*

e. (Touch in 2nd row.) Get it going. *Fiftyyy.* Count. (Touch bills.) *51, 52, 53.*
- What's this row worth? (Touch.) *53 dollars.*

f. (Touch in 3rd row.) Get it going. *Fiftyyy.* Count. (Touch bills.) *60, 70, 80.*
- What's this row worth? (Touch.) *80 dollars.*

Exercise 5: Facts
Addition

a. (Distribute unopened workbooks to students.)
  Open your workbook to Lesson 61 and find part 1.
  (Observe children and give feedback.)
  (Teacher reference:)

  a. $3 + 5$ e. $3 + 6$
  b. $2 + 8$ f. $7 + 10$
  c. $6 + 4$ g. $2 + 3$
  d. $9 + 2$ h. $6 + 3$

  These are plus problems for families you know. You're going to read each problem and say the answer. Then you'll go back and work them.
- Touch and read problem A. Get ready.
  (Signal.) *3 plus 5.*
- (Repeat for:) B, $2 + 8$; C, $6 + 4$; D, $9 + 2$; E, $3 + 6$; F, $7 + 10$; G, $2 + 3$; H, $6 + 3$.

b. Complete the equations for all of the problems in part 1.
  (Observe children and give feedback.)
  (Answer key:)

  a. $3 + 5 = 8$ e. $3 + 6 = 9$
  b. $2 + 8 = 10$ f. $7 + 10 = 17$
  c. $6 + 4 = 10$ g. $2 + 3 = 5$
  d. $9 + 2 = 11$ h. $6 + 3 = 9$

c. Check your work. You'll touch and read each fact.
- Fact A. (Signal.) *3 + 5 = 8.*
- (Repeat for:) B, $2 + 8 = 10$; C, $6 + 4 = 10$; D, $9 + 2 = 11$; E, $3 + 6 = 9$; F, $7 + 10 = 17$; G, $2 + 3 = 5$; H, $6 + 3 = 9$.

Exercise 6: Coins
Quarters   REMEDY

a. (Display:) [61:6A]

- (Point to quarter.) This coin is a **quarter.** What coin is this? (Touch.) *(A) quarter.*
- A quarter is worth 25 cents. What's a quarter worth? (Signal.) *25 cents.*
  (Repeat until firm.)

b. Listen: Here are the numbers for counting the cents for quarters: 25, 50, 75, 100.

• Listen again: 25, 50, 75, 100. Say the numbers for counting the cents for quarters. Get ready. (Signal.) *25, 50, 75, 100.*

c. Find part 2 on worksheet 61. ✔
(Teacher reference:)

R **Part H**

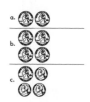

There are quarters in each of these groups of coins. We'll count for each group to figure out how much each group is worth.

• Touch the coins in group A. ✔
I'll count. You'll touch the coins as I count.

• Fingers over the first quarter. ✔

• Get ready. (Children touch quarters.) 25, 50.

d. Your turn to count the cents in group A. Get ready. (Tap 2.) 25, 50.

• How many cents is group A worth? (Signal.) *50.*

• Write an equals and the number of cents for group A. ✔

e. Touch the coins in group B. ✔
You'll touch and count for the coins in group B.

• Fingers over the first quarter. ✔

• Get ready. (Tap 4.) 25, 50, 75, 100.
(Repeat until firm.)

• How many cents is group B worth? (Signal.) *100.*

• Yes, 100 cents. 100 cents is a dollar. How many cents is a dollar? (Signal.) *100.*

• Write an equals and the number of cents for group B. ✔

f. Touch the coins in group C. ✔
This row has a quarter and nickels. I'll count. You'll touch the coins as I count.

• (Children touch quarters.) Twenty-fiiive.
(Children touch nickels.) 30, 35, 40.
(Repeat until firm.)

g. Your turn to touch and count for the coins in group C.

• Touch the quarter and get it going. *Twenty-fiiive.* Count for the nickels. (Tap 3.) 30, 35, 40.
(Repeat until firm.)

• How many cents is group C worth? (Signal.) *40.*

• Write an equals and the number of cents for group C. ✔

EXERCISE 7: FACTS
SUBTRACTION

a. Find part 3 on worksheet 61. ✔
(Teacher reference:)

a. $5 - 2$	e. $8 - 3$
b. $10 - 4$	f. $10 - 2$
c. $9 - 7$	g. $9 - 6$
d. $19 - 9$	h. $8 - 7$

These are minus problems for families you know. You're going to read the problems. Then you'll go back and work them.

• Touch and read problem A. Get ready. (Signal.) *5 minus 2.*

• What's the answer? (Signal.) *3.*

b. (Repeat the following tasks for the remaining problems:)

• Touch and read problem __.

• What's the answer?

c. Complete the equations for all of the problems in part 3.
(Observe children and give feedback.)
(Answer key:)

a. $5 - 2 = 3$	e. $8 - 3 = 5$
b. $10 - 4 = 6$	f. $10 - 2 = 8$
c. $9 - 7 = 2$	g. $9 - 6 = 3$
d. $19 - 9 = 10$	h. $8 - 7 = 1$

d. Check your work. You'll touch and read each fact.

• Fact A. (Signal.) *5 – 2 = 3.*

• (Repeat for:) B, *10 – 4 = 6;* C, *9 – 7 = 2;* D, *19 – 9 = 10;* E, *8 – 3 = 5;* F, *10 – 2 = 8;* G, *9 – 6 = 3;* H, *8 – 7 = 1.*

EXERCISE 8: RULER SKILLS

a. Find part 4 on worksheet 61. ✔
(Teacher reference:)

You're going to figure out how many inches the star, the circle, and the heart are from the beginning of the ruler. You'll start at the beginning of the ruler and touch and count for each inch.

- Touch the star. ✔
- Touch the beginning of the ruler. ✔
- Count and say stop when you get to the star. Get ready. (Tap 5.) *1, 2, 3, 4, 5, stop.*
(Repeat until firm.)
- How many inches is it to the star? (Signal.) *5.*
- Write 5 above the star. ✔

b. Touch the circle. ✔
- Touch the beginning of the ruler. ✔
- Count and say stop when you get to the circle. Get ready. (Tap 2.) *1, 2, stop.*
(Repeat until firm.)
- How many inches is it to the circle? (Signal.) *2.*
- Write 2 above the circle. ✔

c. Touch the heart. ✔
- Touch the beginning of the ruler. ✔
- Count and say stop when you get to the heart. Get ready. (Tap 4.) *1, 2, 3, 4, stop.*
(Repeat until firm.)
- How many inches is it to the heart? (Signal.) *4.*
- Write 4 above the heart. ✔

d. I'm going to tell you how much from the beginning of the ruler. You're going to touch the star, the circle, or the heart.
- Listen: Touch the object that is 2 inches from the beginning of the ruler. ✔
- What object did you touch? (Signal.) *(The) circle.*

e. Listen: Touch the object that is 5 inches from the beginning of the ruler. ✔
- What object did you touch? (Signal.) *(The) star.*

f. Listen: Touch the object that is 4 inches from the beginning of the ruler. ✔
- What object did you touch? (Signal.) *(The) heart.*
(Repeat steps d through f until firm.)

EXERCISE 9: COLUMN PROBLEMS

a. Find part 5 on worksheet 61. ✔
(Teacher reference:)

$$\begin{array}{r} a. \quad 487 \\ -456 \\ \hline \end{array}$$

$$\begin{array}{r} b. \quad 93 \\ +206 \\ \hline \end{array}$$

$$\begin{array}{r} c. \quad 827 \\ -317 \\ \hline \end{array}$$

For some of these problems the beginning digit of the answer is zero.
- Touch and read problem A. (Signal.) *487 minus 456.*
- Touch and read problem B. (Signal.) *93 plus 206.*
- Touch and read problem C. (Signal.) *827 minus 317.*

b. Work the problems in part 5. Put your pencil down when you're finished.
(Observe children and give feedback.)
(Answer key:)

$$\begin{array}{r} a. \quad 487 \\ -456 \\ \hline 31 \end{array}$$

$$\begin{array}{r} b. \quad 93 \\ +206 \\ \hline 299 \end{array}$$

$$\begin{array}{r} c. \quad 827 \\ -317 \\ \hline 510 \end{array}$$

c. Check your work.
- Touch and read equation A. (Signal.) *487 – 456 = 31.*
- Touch and read equation B. (Signal.) *93 + 206 = 299.*
- Touch and read equation C. (Signal.) *827 – 317 = 510.*

EXERCISE 10: INDEPENDENT WORK

a. Turn to the other side of worksheet 61 and find part 6. ✔

(Teacher reference:)

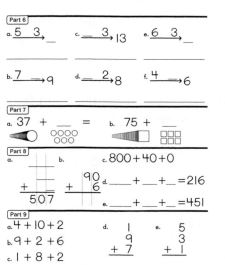

Below each family you'll write the problem and the answer for the missing number. Then you'll write the missing number in the family.

b. Find part 7. ✔
You'll complete the equation for the circles and the equation for the boxes. Then you'll complete the equations for parts 8 and 9.

c. Your turn: Complete worksheet 61.
(Observe children and mark incorrect responses on children's worksheets as you give feedback.)

Lesson

EXERCISE 1: RULER SKILLS

a. (Display:) [62:1A]

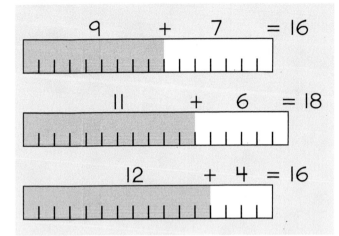

(Point to 1st ruler.) These rulers show centimeters. Each ruler has a shaded part (touch [shaded part]) and an unshaded part (touch [unshaded part].). Some of the numbers that show the total centimeters for these rulers are wrong.

• (Point to [ruler].) Is this the shaded part or the unshaded part? (Touch.) *(The) shaded part.*
• (Point to [ruler].) What part is this? (Touch.) *(The) unshaded part.*
 (Repeat until firm.)

b. You can see the numbers for the shaded and the unshaded parts.
• (Point to **9**.) How many centimeters is the shaded part? (Touch.) *9.*
• (Point to **7**.) How many centimeters is the unshaded part? (Touch.) *7.*
• (Point to **16**.) This is supposed to tell you how many centimeters there are in the shaded and the unshaded parts. How many centimeters altogether? (Touch.) *16.*

c. We're going to see if that's the right number. I touch the end of the shaded part. (Touch [ruler].) Then I get the number for the shaded part going and touch and count for the unshaded part.
• What number do I get going? (Signal.) *9.* Watch. Niiine. (Touch lines.) *10, 11, 12, 13, 14, 15, 16.*
• (Touch [ruler] ¹⁶ and keep touching.) Your turn to count for both parts.
• Get it going. *Niiine.* (Touch lines.) *10, 11, 12, 13, 14, 15, 16.* (Keep finger on end of ruler.)
• How many centimeters are there altogether? (Touch.) *16.*
• So are the numbers for this ruler right? (Signal.) *Yes.*

d. (Point to 2nd ruler.) Here's another ruler with a shaded part and an unshaded part.
• (Point to [ruler].) What part is this? (Touch.) *(The) shaded part.*
• How many centimeters is the shaded part? Get ready. (Signal.) *11.*
• (Point to [ruler].) How many centimeters is the unshaded part? (Signal.) *6.*
• (Point to **18**.) This is supposed to tell you how many centimeters there are in the shaded and unshaded parts. How many centimeters altogether? (Touch.) *18.*

e. We're going to see if that's the right number. I touch the end of the shaded part. (Touch and keep touching.) Then I get the number going and touch and count for the centimeters in the unshaded part.
• What number do I get going? (Signal.) *11.* Elevennn. (Touch lines.) *12, 13, 14, 15, 16, 17.* Your turn to count for both parts.
• (Touch [ruler].) Get it going. *Elevennn.* (Touch lines.) *12, 13, 14, 15, 16, 17.* (Repeat until firm.)
• How many centimeters are there altogether? (Touch.) *17.*
• (Point to **18**.) This number is wrong. What should this number be? (Touch.) *17.*

f. (Point to 3rd ruler.) What's the number for the shaded part? (Touch.) *12.*

- What's the number for the unshaded part? (Touch.) *4.*
- How many centimeters does it say there are altogether? (Touch.) *16.*
- I touch the end of the shaded part. (Touch ▭.) Do I touch here? (Touch.) *No.*
- (Point to ▭.) Do I touch here? (Touch.) *Yes.*

  I'll touch. You'll count to see if both parts are 16 centimeters. What number do you get going? (Signal.) *12.*

  (Touch ▭.) Get it going. *Tweeelve.* (Touch lines.) *13, 14, 15, 16.* (Repeat until firm.)
- How many centimeters are there altogether? (Signal.) *16.*
- (Point to **16.**) Is this the right number? (Touch.) *Yes.*

  Remember, to figure out how many there are altogether, touch the end of the shaded part. Get that number going and count.

EXERCISE 2: COLUMN SUBTRACTION   REMEDY

a. (Display:) W   [62:2A]

$$
\begin{array}{r}
6\ 5\ 8 \\
-\ 6\ 5\ 1 \\
\hline
\end{array}
$$

The beginning digit of the answer for this problem is zero. But we're going to write the right digits for the answer.

- (Point to **658.**) Read the problem. Get ready. (Touch.) *658 minus 651.*
- (Point to $-\overset{\downarrow}{6}5\overset{}{1}$.) Read the problem for the ones. (Touch.) *8 minus 1.*
- What's the answer? (Signal.) *7.* (Add to show:)   [62:2B]

$$
\begin{array}{r}
6\ 5\ 8 \\
-\ 6\ 5\ 1 \\
\hline
7
\end{array}
$$

- (Point to $-\overset{\downarrow}{6}\overset{}{5}\,1$.) Read the problem for the tens. (Touch.) *5 minus 5.*

- What's the answer? (Signal.) *Zero.* (Add to show:)   [62:2C]

$$
\begin{array}{r}
6\ 5\ 8 \\
-\ 6\ 5\ 1 \\
\hline
0\ 7
\end{array}
$$

- (Point to $\overset{\downarrow}{6}58 \atop -651 \atop 07$.) Read the problem for the hundreds. (Touch.) *6 minus 6.*
- What's the answer? (Signal.) *Zero.*
- Do we write zero if it's the beginning digit? (Signal.) *No.*
- (Point to **07.**) Look. Is zero the beginning digit? (Touch.) *Yes.*
- So is the number wrong? (Touch.) *Yes.*
- What should I do to fix the answer? (Call on a child. Idea: *Erase the zero.*) (Change to show:)   [62:2D]

$$
\begin{array}{r}
6\ 5\ 8 \\
-\ 6\ 5\ 1 \\
\hline
7
\end{array}
$$

This answer is right.

- (Point to **658.**) Read the whole equation. (Touch.) *658 − 651 = 7.*

  When you work problems, remember to look at your answers and make sure the first digit isn't zero.

EXERCISE 3: MIXED COUNTING

a. Listen: Start with 40 and count by fives to 100. Get 40 going. *Fortyyy.* Count. (Tap.) *45, 50, 55, 60, 65, 70, 75, 80, 85, 90, 95, 100.*

- Start with 100 and count by hundreds to 1000. Get 100 going. *One huuundred.* Count. (Tap.) *200, 300, 400, 500, 600, 700, 800, 900, 1000.*

b. Listen: You're going to start with 13 and plus tens to 103.

- What number will you start with? (Signal.) *13.*
- Start with 13 and plus tens to 103. Get 13 going. *Thirteeen.* Plus tens. (Tap.) *23, 33, 43, 53, 63, 73, 83, 93, 103.* (Repeat until firm.)

c. Start with 100 and count backward by tens. Get 100 going. *One huuundred.* Count backward. (Tap.) *90, 80, 70, 60, 50, 40, 30, 20, 10.*

- Start with 20 and count backward by ones to 10. Get 20 going. *Twentyyy.* Count backward. (Tap.) *19, 18, 17, 16, 15, 14, 13, 12, 11, 10.*

d. You have 85. When you plus tens, what's the next number? (Signal.) *95.*
- You have 85. When you plus fives, what's the next number? (Signal.) *90.*
- You have 85. When you plus ones, what's the next number? (Signal.) *86.*
- You have 85. When you minus 1, what's the next number? (Signal.) *84.*

EXERCISE 4: SHAPES
INTRODUCTION OF SQUARES

a. (Display:) [62:4A]

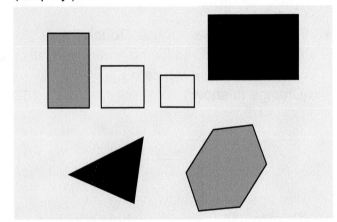

- Some of these shapes are rectangles. What are some of these shapes? (Signal.) *Rectangles.*
- Does a rectangle have 3 sides? (Signal.) *No.*
- How many sides does a rectangle have? (Signal.) *4.*
- Yes, rectangles have 4 sides. Say the rule. (Signal.) *Rectangles have 4 sides.*
b. (Point to shapes.) The shapes with 4 sides are rectangles.
- (Point to ▮.) This shape has 4 sides. How many sides does this shape have? (Touch.) *4.*
- So is it a rectangle? (Signal.) *Yes.*
c. (Point to ▢.) How many sides does this shape have? (Touch.) *4.*
- So is it a rectangle? (Signal.) *Yes.*
d. (Point to ▫.) How many sides does this shape have? (Touch.) *4.*
- So is it a rectangle? (Signal.) *Yes.*

e. (Point to ■.) How many sides does this shape have? (Touch.) *4.*
- So is it a rectangle? (Signal.) *Yes.*
f. (Point to ◀.) Does this shape have 4 sides? (Touch.) *No.*
- So is it a rectangle? (Signal.) *No.*
- (Point to ⬡.) Does this shape have 4 sides? (Touch.) *No.*
- So is it a rectangle? (Signal.) *No.*
g. You'll count the rectangles. Get ready. (Touch rectangles.) *1, 2, 3, 4.*
- How many rectangles are there? (Signal.) *4.*
h. Listen: Some of these rectangles are **squares.**
- What are some of these rectangles? (Signal.) *Squares.*
  (Repeat step h until firm.)
i. Listen: All sides of a square are the same size.
- (Point to top and left side of ▯.) Are these sides the same size? (Touch.) *No.*
- So is this rectangle a square? (Signal.) *No.*
- (Point to top and left side of ▢.) Are these sides the same size? (Touch.) *Yes.*
- So is this rectangle a square? (Signal.) *Yes.*
- (Point to top and left side of ▫.) Are these sides the same size? (Touch.) *Yes.*
- So is this rectangle a square? (Signal.) *Yes.*
- (Point to top and left side of ▬.) Are these sides the same size? (Touch.) *No.*
- So is this rectangle a square? (Signal.) *No.*
j. Raise your hand when you know how many of these rectangles are squares. ✔
- How many are squares? (Signal.) *2.*
k. Raise your hand when you know how many of these rectangles are not squares. ✔
- How many are not squares? (Signal.) *2.*
  (Repeat steps j and k until firm.)
l. Raise your hand when you know how many rectangles there are in all. ✔
- How many rectangles? (Signal.) *4.* Yes, 4 rectangles.

EXERCISE 5: FACTS

SUBTRACTION

a. (Distribute unopened workbooks to students.)
- Open your workbook to Lesson 62 and find part 1.
  (Observe children and give feedback.)
  (Teacher reference:)

	d. 10−4	h. 7−6
a. 13−10	e. 11−9	i. 8−3
b. 9−6	f. 8−5	j. 9−2
c. 9−7	g. 10−8	

These are minus problems for families you know. You're going to touch and read each problem and say the answer. Then you'll go back and work them.
- Problem A. (Signal.) *13 minus 10.*
- What's the answer? (Signal.) *3.*
b. (Repeat the following tasks for problems B through J:)
- Problem __.
- What's the answer?
c. Complete the equations for all of the problems in part 1.
  (Observe children and give feedback.)
  (Answer key:)

	d. 10−4=6	h. 7−6=1
a. 13−10=3	e. 11−9=2	i. 8−3=5
b. 9−6 =3	f. 8−5=3	j. 9−2=7
c. 9−7 =2	g. 10−8=2	

d. Check your work. You'll touch and read each fact.
- Fact A. (Signal.) *13 – 10 = 3.*
- (Repeat for:) B, *9 – 6 = 3;* C, *9 – 7 = 2;* D, *10 – 4 = 6;* E, *11 – 9 = 2;* F, *8 – 5 = 3;* G, *10 – 8 = 2;* H, *7 – 6 = 1;* I, *8 – 3 = 5;* J, *9 – 2 = 7.*

EXERCISE 6: COINS   REMEDY

a. Count by 25s to 100. Get ready. (Tap.) *25, 50, 75, 100.*
  (Repeat step a until firm.)
b. Find part 2 on worksheet 62. ✔
  (Teacher reference:)   R Part I

Each row has quarters.
- How much is a quarter worth? (Signal.) *25 cents.*
- Group A has quarters and nickels. I'll count the cents in group A. Touch the coins as I count. Get ready. 25, 50, seventy-fiiive. (Children touch quarters.) 80, 85, 90. (Children touch nickels.)
c. Your turn: Touch and count the cents for group A.
- Fingers ready. ✔
- Touch and count for the quarters. (Tap 3.) *25, 50, seventy-fiiive.* Count for the nickels. (Tap 3.) *80, 85, 90.*
  (Repeat until firm.)
- How much is group A worth? (Signal.) *90 cents.*
d. Group B has quarters and dimes. You'll touch and count for the quarters and then count for the dimes.
- What do you plus for each dime? (Signal.) *10.*
- Fingers ready. ✔
- Touch and count for the quarters. (Tap 2.) *25, fiftyyy.* Count for the dimes. (Tap 3.) *60, 70, 80.*
  (Repeat until firm.)
- How much is group B worth? (Signal.) *80 cents.*
e. Group C has a quarter and dimes. You'll touch the quarter and get it going. Then you'll count for the dimes.
- Touch the quarter. ✔
- Get it going. *Twenty-fiiive.* Count for the dimes. (Tap 3.) *35, 45, 55.*
  (Repeat until firm.)
- How many cents is group C worth? (Signal.) *55.*
f. Later, you'll count these groups to yourself and write an equals and the number of cents to show what each group is worth.

EXERCISE 7: WORD PROBLEMS (COLUMNS)

a. Find part 3 on worksheet 62. ✔
   (Teacher reference:)

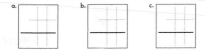

You're going to write the symbols for word problems in columns and work them. The equals bars are already shown.

• Touch where you'll write the symbols for problem A. ✔
   Listen to problem A: A train had 178 passengers on it. Then 27 of those passengers got off of the train. How many passengers were still on the train?

• Listen again and write the number for the first part: A train had 178 passengers on it. ✔

• Write the sign and the number for the next part: Then 27 of those passengers got off of the train. ✔

• Touch and read the symbols you wrote for problem A. Get ready. (Signal.) *178 minus 27.*

b. Work problem A and figure out how many passengers were still on the train.
   (Observe children and give feedback.)

• For problem A, you wrote 178 minus 27. What's the answer? (Signal.) *151.*

• So how many passengers were still on the train? (Signal.) *151.*

c. Touch where you'll write the symbols for problem B. ✔
   Listen to problem B: Bob had 13 marbles. Jan had 12 marbles. How many marbles did Bob and Jan have altogether?

• Listen again and write the number for the first part: Bob had 13 marbles. ✔

• Write the sign and the number for the next part: Jan had 12 marbles. ✔

• Touch and read the symbols you wrote for problem B. Get ready. (Signal.) *13 plus 12.*

d. Work problem B and figure out how many marbles Bob and Jan had altogether.
   (Observe children and give feedback.)

• For problem B, you wrote 13 plus 12. What's the answer? (Signal.) *25.*

• How many marbles did Bob and Jan have altogether? (Signal.) *25.*

e. Touch where you'll write the symbols for problem C. ✔
   Listen to problem C: Ken picked 37 apples. The children ate 5 of those apples. How many apples did Ken end up with?

• Listen again and write the number for the first part: Ken picked 37 apples. ✔

• Write the sign and the number for the next part: The children ate 5 of those apples. ✔

• Touch and read the symbols you wrote for problem C. Get ready. (Signal.) *37 minus 5.*

f. Work problem C and figure out how many apples Ken ended up with.
   (Observe children and give feedback.)
   (Answer key:)

• For problem C, you wrote 37 minus 5. What's the answer? (Signal.) *32.*

• How many apples did Ken end up with? (Signal.) *32.*

EXERCISE 8: COUNTING DOLLARS   REMEDY

a. Find part 4 on worksheet 62. ✔
   (Teacher reference:)   R Part D

You're going to figure out how many dollars each row is worth.

• Touch the first bill in row A. ✔

• How much is that bill worth? (Signal.) *20 dollars.*

• How much is the next bill worth? (Signal.) *10 dollars.*
   You'll touch the first bill and get 20 going. Then you'll touch and count for the tens.

• Touch the first bill. ✔

• Get it going. *Twentyyy.* Count for the tens. (Tap 3.) *30, 40, 50.*
   (Repeat until firm.)

• How much is row A worth? (Signal.) *50 dollars.*

b. Touch the first bill in row B. ✔
- How much is that bill worth? (Signal.) *50 dollars.*
- How much is the next bill worth? (Signal.) *5 dollars.*

  You'll get the number going for the first bill and touch and count for the fives.
- Touch the first bill and get it going. *Fiftyyy.* Count for the fives. (Tap 3.) *55, 60, 65.* (Repeat until firm.)
- How much is row B worth? (Signal.) *65 dollars.*
c. Row C starts with 20 dollars.
- What's the next bill? (Signal.) *5 dollars.*
- What's the next bill worth? (Signal.) *5 dollars.*
- What's the next bill worth? (Signal.) *1 dollar.*

  You'll get the number going for the first bill and touch and count for the fives and ones.
- Touch the first bill and get it going. *Twentyyy.* Count for the fives. (Tap 2.) *25, 30.* Count for the ones. (Tap 2.) *31, 32.* (Repeat until firm.)
- How much is row C worth? (Signal.) *32 dollars.*
d. Your turn: Go back to row A and figure out how much it is worth. Raise your hand when you know the answer.

  (Observe children and give feedback.)
- Everybody, how much is row A worth? (Signal.) *50 dollars.*
- So write equals 50 at the end of row A. ✔
e. Figure out how much row B is worth and write the equals and the number at the end of the row.

  (Observe children and give feedback.)
- Everybody, how much is row B worth? (Signal.) *65 dollars.*
f. Figure out how much row C is worth and write the equals and the number at the end of the row.

  (Observe children and give feedback.)
- Everybody, how much is row C worth? (Signal.) *32 dollars.*

EXERCISE 9: FACTS
ADDITION

a. Turn to the other side of worksheet 62 and find part 5.

  (Observe children and give feedback.)
  (Teacher reference:)

a. $2 + 8$	f. $2 + 4$
b. $3 + 6$	g. $6 + 4$
c. $2 + 9$	h. $3 + 5$
d. $4 + 6$	i. $6 + 3$
e. $7 + 10$	j. $2 + 2$

These are plus problems for families you know. You're going to read each problem and tell me the answer. Then you'll go back and work all of the problems.
- Touch and read problem A. Get ready. (Signal.) *2 plus 8.*
- What's 2 plus 8? (Signal.) *10.*
b. Touch and read problem B. Get ready. (Signal.) *3 plus 6.*
- What's 3 plus 6? (Signal.) *9.*
c. (Repeat the following tasks for remaining problems:)
- Touch and read problem __.
- What's __?
d. Complete the equations for all of the problems in part 5.

  (Observe children and give feedback.)
  (Answer key:)

a. $2 + 8 = 10$	f. $2 + 4 = 6$
b. $3 + 6 = 9$	g. $6 + 4 = 10$
c. $2 + 9 = 11$	h. $3 + 5 = 8$
d. $4 + 6 = 10$	i. $6 + 3 = 9$
e. $7 + 10 = 17$	j. $2 + 2 = 4$

e. Check your work. You'll touch and read each fact.
- Fact A. (Signal.) *2 + 8 = 10.*
- (Repeat for:) B, *3 + 6 = 9;* C, *2 + 9 = 11;* D, *4 + 6 = 10;* E, *7 + 10 = 17;* F, *2 + 4 = 6;* G, *6 + 4 = 10;* H, *3 + 5 = 8;* I, *6 + 3 = 9;* J, *2 + 2 = 4.*

a. Find part 6 on worksheet 62. ✔
    (Teacher reference:)

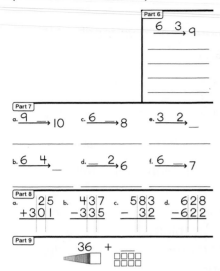

Below the family you'll write four facts for the family, two plus facts and two minus facts.

b. Find part 7. ✔
    Below each family you'll write the problem and the answer for the missing number. Then you'll write the missing number in the family. You'll work column problems in part 8.

c. Find part 9. ✔
    You'll complete the equation for the boxes.

d. Turn to the other side of your worksheet and find part 2. ✔
    Count for each group of coins and write an equals and the number of cents to show how much each group is worth. Then complete the other side of worksheet 62.

    (Observe children and mark incorrect responses on children's worksheets as you give feedback.)

Lesson  63

EXERCISE 1: NUMBER FAMILIES

a. (Display:)   [63:1A]

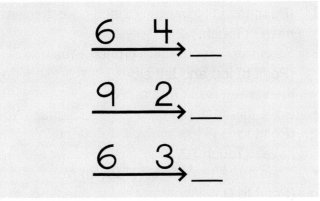

The big number is missing in these families.

- (Point to $\overset{6\quad 4}{\longrightarrow}$.) What are the small numbers in this family? (Touch.) *6 and 4.*
- What's the big number? (Signal.) *10.*
b. (Point to $\overset{9\quad 2}{\longrightarrow}$.) What are the small numbers in this family? (Touch.) *9 and 2.*
- What's the big number? (Signal.) *11.*
c. (Point to $\overset{6\quad 3}{\longrightarrow}$.) What are the small numbers in this family? (Touch.) *6 and 3.*
- What's the big number? (Signal.) *9.*
d. (Point to $\overset{6\quad 4}{\longrightarrow}$.) What are the small numbers in this family again? (Touch.) *6 and 4.*
- What's the big number? (Signal.) *10.*
- Say the fact that starts with the first small number. Get ready. (Signal.) *6 + 4 = 10.*
- Say the fact that starts with the other small number. Get ready. (Signal.) *4 + 6 = 10.*
- Say the fact that goes backward along the arrow. Get ready. (Signal.) *10 – 4 = 6.*
- Say the other minus fact. Get ready. (Signal.) *10 – 6 = 4.*
e. (Point to $\overset{9\quad 2}{\longrightarrow}$.) What are the small numbers in this family? (Touch.) *9 and 2.*
- What's the big number? (Signal.) *11.*
- Say the fact that starts with the first small number. Get ready. (Signal.) *9 + 2 = 11.*
- Say the fact that starts with the other small number. Get ready. (Signal.) *2 + 9 = 11.*
- Say the fact that goes backward along the arrow. Get ready. (Signal.) *11 – 2 = 9.*
- Say the other minus fact. Get ready. (Signal.) *11 – 9 = 2.*

f. (Point to $\overset{6\quad 3}{\longrightarrow}$.) What are the small numbers in this family? (Touch.) *6 and 3.*
- What's the big number? (Signal.) *9.*
- Say the fact that starts with the first small number. Get ready. (Signal.) *6 + 3 = 9.*
- Say the fact that starts with the other small number. Get ready. (Signal.) *3 + 6 = 9.*
- Say the fact that goes backward along the arrow. Get ready. (Signal.) *9 – 3 = 6.*
- Say the other minus fact. Get ready. (Signal.) *9 – 6 = 3.*
g. (Display:)   [63:1B]

$$\overset{5}{\underline{\quad}}\longrightarrow 8 \qquad \overset{\quad 3}{=\!\!=}\longrightarrow 9$$
$$\overset{\quad 2}{=\!\!=}\longrightarrow 11 \qquad \overset{6\quad 4}{\longrightarrow}\underline{\quad}$$
$$\overset{10\quad 4}{\longrightarrow}\underline{\quad} \qquad \overset{6}{\underline{\quad}}\longrightarrow 8$$
$$\overset{8}{\underline{\quad}}\longrightarrow 10 \qquad \overset{\quad 2}{=\!\!=}\longrightarrow 6$$

You're going to say the problem for the missing number in each family.

- (Point to $\overset{5}{\longrightarrow}$8.) Say the problem for the missing number. (Signal.) *8 minus 5.*
- What's 8 minus 5? (Signal.) *3.*
h. (Point to $\overset{\quad 2}{\longrightarrow}$11.) Say the problem for the missing number. (Signal.) *11 minus 2.*
- What's 11 minus 2? (Signal.) *9.*
i. (Repeat the following tasks for remaining families:)

(Point to __.)	Say the problem for the missing number.	What's __?	
$\overset{10\ 4}{\longrightarrow}\underline{\ }$	10 + 4	10 + 4	14
$\overset{8}{\longrightarrow}$10	10 – 8	10 – 8	2
$\overset{\ 3}{\longrightarrow}$9	9 – 3	9 – 3	6
$\overset{6\ 4}{\longrightarrow}\underline{\ }$	6 + 4	6 + 4	10
$\overset{6}{\longrightarrow}$8	8 – 6	8 – 6	2
$\overset{\ 2}{\longrightarrow}$6	6 – 2	6 – 2	4

(Repeat for families that were not firm.)

EXERCISE 2: MIXED COUNTING

a. You're going to start with 11 and plus tens to 101.
- What do you start with? (Signal.) *11.*
- Start with 11 and plus tens and stop at 101. Get 11 going. *Elevennn.* Plus tens. (Tap.) *21, 31, 41, 51, 61, 71, 81, 91, 101.*
- Start with 100 and count backward by tens. Get 100 going. *One huuundred.* Count backward. (Tap.) *90, 80, 70, 60, 50, 40, 30, 20, 10.*

b. Start at 75 and count by fives to 100. Get 75 going. *Seventy-fiiive.* Count. (Tap.) *80, 85, 90, 95, 100.*

c. You have 75. When you plus tens, what's the next number? (Signal.) *85.*
- You have 75. When you plus ones, what's the next number? (Signal.) *76.*
- You have 75. When you plus 25s, what's the next number? (Signal.) *100.*
- You have 75. When you minus ones, what's the next number? (Signal.) *74.*

EXERCISE 3: SHAPES

a. (Display:)   [63:3A]

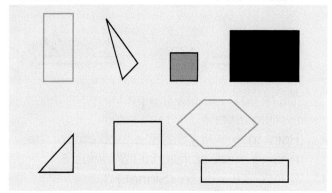

You learned about rectangles. The shapes with 4 sides in this picture are rectangles. Some rectangles are squares.
- How many sides does a rectangle have? (Signal.) *4.*
  Yes, all rectangles have 4 sides.
- Say the sentence. (Signal.) *All rectangles have 4 sides.*

b. (Point to shapes.) Raise your hand when you know how many rectangles are in this picture. ✔
- How many rectangles? (Signal.) *5.*

c. (Point to ◺.) Is this a rectangle? (Touch.) *No.*
- How do you know? (Call on a student.) *It doesn't have 4 sides.*
- (Point to ▪.) Is this a rectangle? (Touch.) *Yes.*
- How do you know? (Signal.) *It has 4 sides.* (Repeat step c until firm.)

d. Listen: Some rectangles are squares. What are some of these rectangles? (Signal.) *Squares.*

e. All sides of a square are the same size.
- (Point to ▪.) These sides are the same size, so this is a square.
- Is this rectangle a square? (Touch.) *Yes.*

f. (Point to ⬜.) How many sides does this shape have? (Touch.) *4.*
- So is this a rectangle? (Touch.) *Yes.*
- (Point to top and left sides.) Are these sides the same size? (Touch.) *No.*
- So is this rectangle a square? (Signal.) *No.*

g. (Point to ◹.) How many sides does this shape have? (Touch.) *3.*
- So is this a rectangle? (Touch.) *No.*

h. (Point to ▪.) How many sides does this shape have? (Touch.) *4.*
- So is this a rectangle? (Touch.) *Yes.*
- (Point to top and left sides.) Are these sides the same size? (Touch.) *Yes.*
- So is this rectangle a square? (Touch.) *Yes.*

i. (Point to ■.) How many sides does this shape have? (Touch.) *4.*
- So is this a rectangle? (Touch.) *Yes.*
- (Point to top and left sides.) Are these sides the same size? (Touch.) *No.*
- So is this rectangle a square? (Touch.) *No.*

j. (Point to ◺.) How many sides does this shape have? (Touch.) *3.*
- So is this a rectangle? (Touch.) *No.*

k. (Point to ⬜.) How many sides does this shape have? (Touch.) *4.*
- So is this a rectangle? (Touch.) *Yes.*
- (Point to top and left sides.) Are these sides the same size? (Touch.) *Yes.*
- So is this rectangle a square? (Touch.) *Yes.*

l. (Point to ⬡.) Raise your hand when you know how many sides this shape has. ✔
- How many sides does this shape have? (Touch.) *6.*
  (To correct:)
  - (Point to ⬡.) I'll touch each side of this shape. Count the sides to yourself. (Touch each side.)
  - Everybody, how many sides in this shape? (Touch.) *6.*
- So is this a rectangle? (Touch.) *No.*

m. (Point to ▭.) How many sides does this shape have? (Touch.) *4.*
- So is this a rectangle? (Touch.) *Yes.*
- (Point to top and left sides.) Are these sides the same size? (Touch.) *No.*
- So is this rectangle a square? (Signal.) *No.*

n. Look at all the shapes. Raise your hand when you know how many of these shapes are rectangles. ✔
- How many of these shapes are rectangles? (Signal.) *5.*
- Raise your hand when you know how many of these shapes are not rectangles. ✔
- How many of these shapes are not rectangles? (Signal.) *3.*

o. Raise your hand when you know how many of these rectangles are squares. ✔
- How many of these rectangles are squares? (Signal.) *2.*

EXERCISE 4: FACTS
SUBTRACTION

a. (Distribute unopened workbooks to students.)
- Open your workbook to Lesson 63 and find part 1.
  (Observe children and give feedback.)
  (Teacher reference:)

	d. $9-6$	h. $11-2$
a. $7-2$	e. $8-5$	i. $9-3$
b. $6-4$	f. $10-6$	j. $8-6$
c. $14-10$	g. $10-9$	

These are minus problems for families you know. You're going to read each problem and say the answer. Then you'll go back and work them.
- Touch and read problem A. Get ready. (Signal.) *7 minus 2.*
- What's the answer? (Signal.) *5.*

b. (Repeat the following tasks for the remaining problems:)
- Touch and read problem __.
- What's the answer?

c. Complete the equations for all of the problems in part 1.
  (Observe children and give feedback.)
  (Answer key:)

	d. $9-6=3$	h. $11-2=9$
a. $7-2=5$	e. $8-5=3$	i. $9-3=6$
b. $6-4=2$	f. $10-6=4$	j. $8-6=2$
c. $14-10=4$	g. $10-9=1$	

d. Check your work. You'll touch and read each fact.
- Fact A. (Signal.) $7-2=5.$
- (Repeat for:) B, $6-4=2$; C, $14-10=4$; D, $9-6=3$; E, $8-5=3$; F, $10-6=4$; G, $10-9=1$; H, $11-2=9$; I, $9-3=6$; J, $8-6=2.$

EXERCISE 5: COINS

a. Find part 2 on worksheet 63. ✔
  (Teacher reference:)

These rows have quarters in them.
- How much is a quarter worth? (Signal.) *25 cents.*
- Your turn to count by 25s to 100. Get ready. (Tap.) *25, 50, 75, 100.*
  (Repeat until firm.)

b. Row A has quarters and nickels. You'll touch and count for the coins.
- You count by 25 for the quarters. What do you count by for the nickels? (Signal.) *5.*
- Fingers ready. ✔
- Touch and count for the quarters. Get ready. (Tap 2.) *25, fiftyyy.* Count for the nickels. (Tap 4.) *55, 60, 65, 70.*
  (Repeat until firm.)
- How much is row A worth? (Signal.) *70 cents.*

c. Row B has a quarter, nickels, and dimes.
- You count by 25 for the quarter and by fives for the nickels. What do you count by for the dimes? (Signal.) *Tens.*
  You'll touch and count for the quarters. Then you'll count for the dimes.
- Fingers ready. ✔
- Touch the quarter and get it going. (Tap.) *Twenty-fiiive.* Count for the nickels. (Tap 2.) *30, thirty-fiiive.* Count for the dimes. (Tap 3.) *45, 55, 65.*
  (Repeat until firm.)
- How much is row B worth? (Signal.) *65 cents.*

d. Row C has a quarter, nickels, and dimes.
You'll touch and count for the coins.
  - Fingers ready. ✔
  - Touch the quarter and get it going. (Tap.)
    Twenty-fiiive. Count for the nickels. (Tap 3.) *30,
    35, fortyyy.* Count for the dimes. (Tap 2.) *50, 60.*
    (Repeat until firm.)
  - How much is row C worth? (Signal.) *60 cents.*
    Later, you'll count to yourself and write an
    equals and the number of cents for each row.

EXERCISE 6: FACTS
ADDITION

a. Find part 3 on worksheet 63. ✔
   (Teacher reference:)

	d. 2 + 3	h. 8 + 10
a. 6 + 3	e. 9 + 2	i. 6 + 4
b. 8 + 2	f. 5 + 3	j. 4 + 2
c. 10 + 6	g. 6 + 2	

   These are plus problems for families you
   know. You're going to read each problem
   and say the answer. Then you'll go back
   and work them.
   - Touch and read problem A. Get ready.
     (Signal.) *6 plus 3.*
   - What's 6 plus 3? (Signal.) *9.*
b. Touch and read problem B. (Signal.) *8 plus 2.*
   - What's 8 plus 2? (Signal.) *10.*
c. (Repeat the following tasks for remaining
   problems:)
   - Touch and read problem __.
   - What's __?
d. Complete the equations for all of the problems
   in part 3.
   (Observe children and give feedback.)
   (Answer key:)

	d. 2 + 3 = 5	h. 8 + 10 = 18
a. 6 + 3 = 9	e. 9 + 2 = 11	i. 6 + 4 = 10
b. 8 + 2 = 10	f. 5 + 3 = 8	j. 4 + 2 = 6
c. 10 + 6 = 16	g. 6 + 2 = 8	

e. Check your work. You'll touch and read
   each fact.
  - Fact A. (Signal.) *6 + 3 = 9.*
  - (Repeat for:) B, *8 + 2 = 10;* C, *10 + 6 = 16;*
    D, *2 + 3 = 5;* E, *9 + 2 = 11;* F, *5 + 3 = 8;*
    G, *6 + 2 = 8;* H, *8 + 10 = 18;* I, *6 + 4 = 10;*
    J, *4 + 2 = 6.*

EXERCISE 7: RULER SKILLS   [REMEDY]

a. (Display:)   [63:7A]

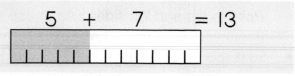

   This ruler shows centimeters. It has a shaded
   part and an unshaded part. We're going to
   check the centimeters for both parts and see if
   the number is right.
  - How many centimeters is the shaded part?
    (Signal.) *5.*
  - How many centimeters is the unshaded part?
    (Signal.) *7.*
  - How many centimeters are the shaded and
    the unshaded parts? (Signal.) *13.*
    (Repeat until firm.)
b. I'll touch the end of the shaded part. You'll
   get a number going and count as I touch the
   unshaded centimeters.
  - Which part will I touch the end of? (Signal.)
    The shaded part.
  - (Touch .) What's the number for that
    part? (Signal.) *5.*
  - Get it going. *Fiiive.* (Touch lines.) *6, 7, 8, 9, 10,
    11, 12.*
    (Repeat until firm.)
  - How many centimeters are there altogether?
    (Touch.) *12.*
  - (Point to **13.**) Is that the right number?
    (Touch.) *No.*
  - What is the right number? (Signal.) *12.*
    Yes, 5 plus 7 equals 12.
  - Say the correct fact. Get ready. (Signal.)
    5 + 7 = 12.
c. Find part 4 on worksheet 63. ✔
   (Teacher reference:)   [R |Part A]

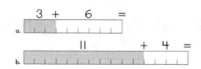

  - Touch ruler A. ✔
    It shows centimeters.
  - Touch the number of centimeters for the
    shaded part. ✔
  - How many centimeters? (Signal.) *3.*
  - Touch the number of centimeters for the
    unshaded part. ✔
  - How many centimeters? (Signal.) *6.*

d. You're going to figure out how many centimeters there are for both parts.
- Touch the end of the shaded part. ✔
- What number will you get going? (Signal.) *3.*
- Get it going. *Threee.* Touch and count the lines. (Tap.) *4, 5, 6, 7, 8, 9.*
  (Repeat until firm.)
- How many centimeters are there for both parts? (Signal.) *9.*
- Write 9 to complete the fact for both parts. ✔
- Everybody, touch and read the fact for ruler A. (Signal.) *3 + 6 = 9.*

e. Touch ruler B. ✔
- Touch the number of centimeters for the shaded part. ✔
- How many centimeters? (Signal.) *11.*
- Touch the number of centimeters for the unshaded part. ✔
- How many centimeters? (Signal.) *4.*

f. You're going to figure out how many centimeters there are for both parts.
- Touch the end of the shaded part. ✔
- What number will you get going? (Signal.) *11.*
- Get it going. *Elevennn.* Touch and count the lines. (Tap.) *12, 13, 14, 15.*
  (Repeat until firm.)
- How many centimeters are there for both parts? (Signal.) *15.*
- Write 15 to complete the fact for both parts. ✔
- Everybody, touch and read the fact for ruler B. (Signal.) *11 + 4 = 15.*

EXERCISE 8: WORD PROBLEMS (COLUMNS)

a. Turn to the other side of worksheet 63 and find part 5. ✔
(Teacher reference:)

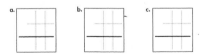

You're going to write the symbols for word problems in columns and work them. The equals bars are already shown.
- Touch where you'll write the symbols for problem A. ✔
  Listen to problem A: A store had 471 apples in it. The store got 216 more apples. How many apples did the store end up with?
- Listen again and write the number for the first part: A store had 471 apples. ✔

- Write the sign and the number for the next part: The store got 216 more apples. ✔
- Touch and read the symbols you wrote for problem A. Get ready. (Signal.) *471 plus 216.*

b. Work problem A and figure out how many apples the store ended up with.
(Observe children and give feedback.)
- For problem A, you wrote 471 plus 216. What's the answer? (Signal.) *687.*
- So how many apples did the store end up with? (Signal.) *687.*

c. Touch where you'll write the symbols for problem B. ✔
  Listen to problem B: There were 179 ducks on a lake. Then 53 of those ducks flew away. How many ducks were still on the lake?
- Listen again and write the number for the first part: There were 179 ducks on a lake. ✔
- Write the sign and the number for the next part: 53 of those ducks flew away. ✔
- Touch and read the symbols you wrote for problem B. Get ready. (Signal.) *179 minus 53.*

d. Work problem B and figure out how many ducks were still on the lake.
(Observe children and give feedback.)
- For problem B, you wrote 179 minus 53. What's the answer? (Signal.) *126.*
- So how many ducks were still on the lake? (Signal.) *126.*

e. Touch where you'll write the symbols for problem C. ✔
  Listen to problem C: A man had 85 dollars. His son had 113 dollars. How much did the man and his son have altogether?
- Listen again and write the number for the first part: A man had 85 dollars. ✔
- Write the sign and the number for the next part. His son had 113 dollars. ✔
- Touch and read the symbols you wrote for problem C. Get ready. (Signal.) *85 plus 113.*

f. Work problem C and figure out how many dollars the man and his son had altogether.
(Observe children and give feedback.)
(Answer key:)

- For problem C, you wrote 85 plus 113. What's the answer? (Signal.) *198.*
- So how many dollars did the man and his son have altogether? (Signal.) *198.*

EXERCISE 9: COUNTING DOLLARS

a. Find part 6. ✔
(Teacher reference:)

You're going to figure out how many dollars each row is worth.

- Touch the first bill in row A. ✔
- How much is that bill worth? (Signal.) *20 dollars.*
- How much is the next bill worth? (Signal.) *5 dollars.*
- How much is the last bill in that row worth? (Signal.) *1 dollar.*
  You'll touch the first bill and get 20 going. Then you'll touch and count for the 5s and the 1.
- Touch the 1st bill. ✔
- Get it going. *Twentyyy.* Count for the fives. (Tap 3.) *25, thirtyyy.* Count for the 1. (Tap.) *31.* (Repeat until firm.)
- How much is row A worth? (Signal.) *31 dollars.*

b. Touch the first bill in row B. ✔
- How much is that bill worth? (Signal.) *50 dollars.*
- How much is the next bill worth? (Signal.) *10 dollars.*
- How much is the last bill worth? (Signal.) *5 dollars.*
  You'll get the number going for the first bill and touch and count for the tens and the 5.
- Touch the 1st bill and get it going. *Fiftyyy.* Count for the tens. (Tap 2.) *60, seventyyy.* Count for the 5. (Tap.) *75.* (Repeat until firm.)
- How much is row B worth? (Signal.) *75 dollars.*

c. Your turn: Go back to row A and figure out how much it is worth. Raise your hand when you know the answer.
(Observe children and give feedback.)
- Everybody, how much is row A worth? (Signal.) *31 dollars.*
- So write equals 31 at the end of row A. ✔

d. Figure out how much row B is worth and write the equals and the number at the end of the row.
(Observe children and give feedback.)
- Everybody, how much is row B worth? (Signal.) *75 dollars.*

EXERCISE 10: INDEPENDENT WORK

a. Find part 7 on worksheet 63. ✔
(Teacher reference:)

Below each family you'll write the facts for finding the missing number. Then you'll write the missing number in the family.
You'll work the column problems in part 8.

b. Complete worksheet 63. Remember to write an equals sign and the cents for each row of coins in part 2.
(Observe children and mark incorrect responses on children's worksheets as you give feedback.)

Lesson 64

EXERCISE 1: NUMBER FAMILIES   [REMEDY]

a. (Display:)   [64:1A]

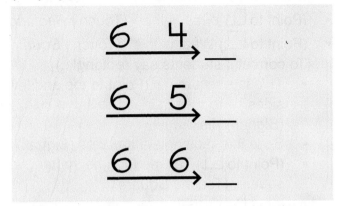

The big number is missing in all these families.
- (Point to 6 4→.) What are the small numbers in this family? (Touch.) *6 and 4.*
- What's the big number? (Signal.) *10.*
- (Point to 6 5→.) What are the small numbers in this family? (Touch.) *6 and 5.*
- What's the big number? (Signal.) *11.*
- (Point to 6 6→.) What are the small numbers in this family? (Touch.) *6 and 6.*
- What's the big number? (Signal.) *12.*

b. (Point to 6 5→.) What are the small numbers in this family again? (Touch.) *6 and 5.*
- What's the big number? (Signal.) *11.*
- Say the fact that starts with the first small number. (Signal.) *6 + 5 = 11.*
- Say the fact that starts with the other small number. (Signal.) *5 + 6 = 11.*
- Say the fact that goes backward along the arrow. (Signal.) *11 − 5 = 6.*
- Say the other minus fact. Get ready. (Signal.) *11 − 6 = 5.*
  (Repeat step b until firm.)

c. (Point to 6 6→.) What are the small numbers in this family? (Signal.) *6 and 6.*
- What's the big number? (Signal.) *12.*
  There's only one plus fact and one minus fact.
- Say the plus fact. (Signal.) *6 + 6 = 12.*
- Say the minus fact. Get ready. (Signal.) *12 − 6 = 6.*
  (Repeat step c until firm.)

d. (Display:)   [64:1B]

$$6 \xrightarrow{\quad 5 \quad} \_\_ \qquad 6 \xrightarrow{\quad 4 \quad} \_\_$$
$$4 \xrightarrow{\qquad} 6 \qquad \xrightarrow{\quad 5 \quad} 11$$
$$\xrightarrow{\quad 6 \quad} 12 \qquad 6 \xrightarrow{\quad 6 \quad} \_\_$$
$$6 \xrightarrow{\qquad} 9 \qquad 6 \xrightarrow{\qquad} 8$$

You're going to say the problem for the missing number in each family.
- (Point to 6 5→__.) Say the problem for the missing number. Get ready. (Touch.) *6 plus 5.*
- What's 6 plus 5? (Signal.) *11.*

e. (Point to 4→6.) Say the problem for the missing number. Get ready. (Touch.) *6 minus 4.*
- What's 6 minus 4? (Signal.) *2.*

f. (Repeat the following tasks for remaining families:)

(Point to __.)	Say the problem for the missing number.	What's __?	
→6,12	12 − 6	12 − 6	6
6→9	9 − 6	9 − 6	3
6 4→	6 + 4	6 + 4	10
→5,11	11 − 5	11 − 5	6
6 6→	6 + 6	6 + 6	12
6→8	8 − 6	8 − 6	2

(Repeat for families that were not firm.)

EXERCISE 2: SHAPES

INTRODUCTION OF TRIANGLES

a. (Display:) [64:2A]

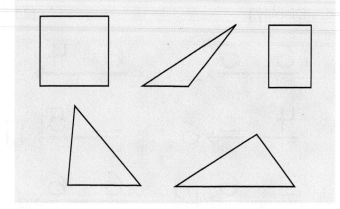

- Everybody, rectangles have how many sides? (Signal.) *4.*
- What do we call rectangles that have sides that are the same size? (Signal.) *Squares.*

b. (Point to △.) Is this a rectangle? (Touch.) *No.*
- How do you know? (Call on a child.) *It doesn't have 4 sides.*
- This shape is a triangle. Say **triangle.** (Touch.) *Triangle.*
- What shape is this? (Touch.) *(A) triangle.*

c. Listen: All triangles have 3 sides.
- Say the sentence. (Signal.) *All triangles have 3 sides.*
(Repeat until firm.)
- How many sides do all triangles have? (Signal.) *3.*

d. (Point to ▢.) How many sides does this shape have? (Touch.) *4.*
- So is this a triangle? (Signal.) *No.*

e. (Point to ⟋.) How many sides does this shape have? (Touch.) *3.*
- So is this a triangle? (Signal.) *Yes.*

f. (Repeat the following tasks for the remaining shapes:)
- How many sides does this shape have?
- So is this a triangle?

g. Look at all the shapes. Raise your hand when you know how many of these shapes are rectangles. ✔
- How many of these shapes are rectangles? (Signal.) *2.*

h. Raise your hand when you know how many of these shapes are squares. ✔
- How many of these shapes are squares? (Signal.) *1.*

i. Raise your hand when you know how many of these shapes are triangles. ✔
- How many of these shapes are triangles? (Signal.) *3.*

j. I'll touch each shape. Tell me triangle, rectangle, or square.
- (Point to ⟋.) What is this? (Touch.) *Triangle.*
- (Point to ▢.) What is this? (Touch.) *Rectangle.*
- (Point to ▢.) What is this? (Touch.) *Square.*
(To correct if students say rectangle:)
  - This is a rectangle. (Point to top and left sides.) Are these sides the same size? (Signal.) *Yes.*
  - So is this rectangle a square? (Signal.) *Yes.*
- (Point to ▢.) Tell me triangle, rectangle, or square. (Touch.) *Square.*
- (Point to △.) What is this? (Touch.) *Triangle.*
- (Point to △.) What is this? (Touch.) *Triangle.*
(Repeat step j until firm.)

EXERCISE 3: MIXED COUNTING

a. Your turn: Count by 25s to 100. Get ready. (Tap.) *25, 50, 75, 100.*
- Start with 100 and count by 25s to 200. Get 100 going. *One huuundred.* Count. (Tap.) *125, 150, 175, 200.*
(Repeat step a until firm.)

b. Start with 156 and count by ones to 166. Get 156 going. *One hundred fifty-siiix.* Count. (Tap.) *157, 158, 159, 160, 161, 162, 163, 164, 165, 166.*
(Repeat step b until firm.)

c. Start with 70 and count backward by ones to 60. Get 70 going. *Seventyyy.* Count backward. (Tap.) *69, 68, 67, 66, 65, 64, 63, 62, 61, 60.*
(Repeat step c until firm.)

d. Start with 70 and count by tens to 120. Get 70 going. *Seventyyy.* Count. (Tap.) *80, 90, 100, 110, 120.*
(Repeat step d until firm.)

e. Start with 100 and count backward by tens. Get 100 going. *One huuundred.* Count backward. (Tap.) *90, 80, 70, 60, 50, 40, 30, 20, 10.*
(Repeat step e until firm.)

f. Start at 70 and count by fives to 100. Get 70 going. *Seventyyy.* Count. (Tap.) *75, 80, 85, 90, 95, 100.*
(Repeat step f until firm.)

EXERCISE 4: FACTS

SUBTRACTION

a. (Distribute unopened workbooks to students.)
- Open your workbook to Lesson 64 and find part 1.
(Observe children and give feedback.)
(Teacher reference:)

a. $8-3$	d. $4-3$	h. $10-2$
b. $8-2$	e. $9-2$	i. $9-3$
c. $10-4$	f. $9-6$	j. $5-2$
	g. $10-6$	

These are minus problems for families you know. You're going to read each problem and say the answer. Then you'll go back and work them.
- Touch and read problem A. Get ready. (Signal.) *8 minus 3.*
- What's the answer? (Signal.) *5.*

b. (Repeat the following tasks for the remaining problems:)
- Touch and read problem __.
- What's the answer?

c. Complete the equations for all of the problems in part 1.
(Observe children and give feedback.)
(Answer key:)

a. $8-3=5$	d. $4-3=1$	h. $10-2=8$
b. $8-2=6$	e. $9-2=7$	i. $9-3=6$
c. $10-4=6$	f. $9-6=3$	j. $5-2=3$
	g. $10-6=4$	

d. Check your work. You'll touch and read each fact.
- Fact A. (Signal.) $8 - 3 = 5.$
- (Repeat for:) B, $8 - 2 = 6$; C, $10 - 4 = 6$; D, $4 - 3 = 1$; E, $9 - 2 = 7$; F, $9 - 6 = 3$; G, $10 - 6 = 4$; H, $10 - 2 = 8$; I, $9 - 3 = 6$; J, $5 - 2 = 3.$

EXERCISE 5: COINS

a. Find part 2 on worksheet 64. ✔
(Teacher reference:)   R Part

These rows have quarters in them.
- How much is a quarter worth? (Signal.) *25 cents.*
- Your turn to count by 25s to 100. Get ready. (Tap.) *25, 50, 75, 100.*

b. Row A has quarters and pennies. You'll touch and count the cents for the row.
- What do you count by for each quarter? (Signal.) *25.*
- Fingers ready. ✔
- Touch and count for the quarters. (Tap 3.) *25, 50, seventy-fiiive.* Count for the pennies. (Tap 3.) *76, 77, 78.*
(Repeat until firm.)
- How much is row A worth? (Signal.) *78 cents.*

c. Row B has quarters and nickels. You count by 25 for each quarter. What do you count by for each nickel? (Signal.) *5.*
- Fingers ready. ✔
- Touch and count for the quarters. Get ready. (Tap 3.) *25, 50, seventy-fiiive.* Count for the nickels. (Tap 3.) *80, 85, 90.*
(Repeat until firm.)
- How much is row B worth? (Signal.) *90 cents.*

d. Row C has quarters, dimes, and nickels. You count by 25 for each quarter and by five for each nickel. What do you count by for each dime? (Signal.) *10.*
- Fingers ready. ✔
- Touch and count for the quarters. Get ready. (Tap 2.) *25, fiftyyy.* Count for the dimes. (Tap 2.) *60, seventyyy.* Count for the nickels. (Tap 2.) *75, 80.*
(Repeat until firm.)
- How much is row C worth? (Signal.) *80 cents.*
Later, you'll count for these rows to yourself and write an equals and the number of cents for each row.

...n worksheet 64. ✔
...erence:)

$$9 \qquad + 3 =$$

b.

$$4 \quad + \quad 7 \quad =$$

c.

- Touch ruler A. ✔
  It shows centimeters.
- Touch the number of centimeters for the shaded part. ✔
- How many centimeters? (Signal.) *5.*
- Touch the number of centimeters for the unshaded part. ✔
- How many centimeters? (Signal.) *4.*

b. You're going to figure out how many centimeters there are in both parts. You'll touch the end of the shaded part and get the number going. Then you'll touch and count for the centimeters in the unshaded part.
- Touch the end of the shaded part. ✔
- What number will you get going? (Signal.) *5.*
- Get it going. *Fiiive.* Count for the lines. (Tap.)
  6, 7, 8, 9.
  (Repeat until firm.)

c. How many centimeters are there in both parts? (Signal.) *9.*
- Write 9. ✔
- Everybody, touch and read the fact for ruler A. Get ready. (Signal.) *5 + 4 = 9.*
  (Repeat until firm.)

d. Touch ruler B. ✔
- Touch the number of centimeters for the shaded part. ✔
- How many centimeters? (Signal.) *9.*
- Touch the number of centimeters for the unshaded part. ✔
- How many centimeters? (Signal.) *3.*

e. You're going to figure out how many centimeters there are in both parts.
- Touch the end of the shaded part. ✔
- What number will you get going? (Signal.) *9.*
- Get it going. *Niiine.* Count the lines. (Tap.)
  10, 11, 12.
  (Repeat until firm.)

f. How many centimeters are there in both parts? (Signal.) *12.*
- Write 12. ✔

- Everybody, touch and read the fact for ruler B. Get ready. (Signal.) *9 + 3 = 12.*
  (Repeat until firm.)

g. Touch ruler C. ✔
- Touch the number of centimeters for the shaded part. ✔
- How many centimeters? (Signal.) *4.*
- Touch the number of centimeters for the unshaded part. ✔
- How many centimeters? (Signal.) *7.*

h. You're going to figure out how many centimeters there are in both parts.
- Touch the end of the shaded part. ✔
- What number will you get going? (Signal.) *4.*
- Get it going. *Fouuur.* Count the lines. (Tap.)
  5, 6, 7, 8, 9, 10, 11.
  (Repeat until firm.)

i. How many centimeters are there in both parts? (Signal.) *11.*
- Write 11. ✔
- Everybody, touch and read the fact for ruler C. Get ready. (Signal.) *4 + 7 = 11.*
  (Repeat until firm.)

EXERCISE 7: 3 ADDENDS IN COLUMNS   [REMEDY]

a. Turn to the other side of worksheet 64 and find part 4. ✔
   (Teacher reference:)   [R | Part F]

$$
\begin{array}{llll}
\text{a.} & \begin{array}{r} 34 \\ 21 \\ +\ 23 \\ \hline \end{array} &
\text{b.} & \begin{array}{r} 12 \\ 11 \\ +\ 32 \\ \hline \end{array} &
\text{c.} & \begin{array}{r} 50 \\ 34 \\ +\ 12 \\ \hline \end{array}
\end{array}
$$

These problems plus 3 numbers.
- Problem A. Touch the symbols as I read problem A. 34 (children touch 34) plus (children touch +) 21 (children touch 21) plus 23 (children touch 23).

b. I'll say the problem in the ones column: 4 plus 1 plus 3.
- Your turn: Touch and say the problem in the ones column. (Signal.) *4 plus 1 plus 3.*
- Say the first problem you'll work. (Signal.) *4 plus 1.*
- What's the answer? (Signal.) *5.*
- Say the next problem. (Signal.) *5 plus 3.*
- What's the answer? (Signal.) *8.*
  (Repeat step b until firm.)

c. Write the answer in the ones column. Then work the problem in the tens column. Put your pencil down when you've completed the equation for A.
(Observe children and give feedback.)
• The problem for the tens column is 3 plus 2 plus 2. Everybody, what's the answer for the tens column? (Signal.) *7.*
• Touch and read equation A. Get ready. (Signal.) *34 + 21 + 23 = 78.*
Yes, 34 plus 21 plus 23 equals 78.

d. Work problem B. Put your pencil down when you've completed the equation.
(Observe children and give feedback.)
• Everybody, what does 12 plus 11 plus 32 equal? (Signal.) *55.*

e. Work problem C. Put your pencil down when you've completed the equation.
(Observe children and give feedback.)
• Everybody, what does 50 plus 34 plus 12 equal? (Signal.) *96.*

EXERCISE 8: COUNTING DOLLARS REMEDY

a. Find part 5 on worksheet 64. ✔
(Teacher reference:) R Part E

You're going to touch and count for the dollars in each row of bills. Row A has tens, fives, and a one.
• Fingers ready. ✔
• Touch and count for the tens. (Tap 2.) *10, twentyyy.* Count for the fives. (Tap 2.) *25, thirtyyy.* Count for the one. *(Signal.) 31.*
(Repeat until firm.)
• How much is row A worth? (Signal.)
31 dollars.

b. Row B has a 20, fives, and tens.
You'll touch and count for the bills in row B.
• Touch the first bill and get it going. *Twentyyy.* Count for the fives. (Tap 2.) *25, thirtyyy.* Count for the tens. (Tap 2.) *40, 50.*
(Repeat until firm.)
• How much is row B worth? (Signal.)
50 dollars.

c. Row C has a 50, tens, and fives.
You'll touch and count for the bills in row C.
• Touch the first bill and get it going. *Fiftyyy.* Count for the tens. (Tap 2.) *60, seventyyy.* Count for the fives. (Tap 3.) *75, 80, 85.*
(Repeat until firm.)
• How much is row C worth? (Signal.)
85 dollars.

d. Now you'll count for these rows to yourself and write an equals and the number of dollars for each row.
(Observe children and give feedback.)
• Figure out how many dollars row A is worth. Write equals and the number of dollars at the end of row A.
(Observe children and give feedback.)
• Everybody, what did you write at the end of row A? (Signal.) *Equals 31.*

e. Figure out how many dollars rows B and C are worth. Write the equals and the number of dollars at the end of each row.
(Observe children and give feedback.)

f. Check your work. Everybody, what did you write at the end of row B? (Signal.) *Equals 50.*
• Everybody, what did you write at the end of row C? (Signal.) *Equals 85.*

EXERCISE 9: WORD PROBLEMS (COLUMNS)

a. Find part 6. ✔
(Teacher reference:)

a. ▢ b. ▢ c. ▢

You're going to write the symbols for word problems in columns and work them. The equals bars are already shown.
• Touch where you'll write the symbols for problem A. ✔
Listen to problem A: There were 54 pancakes on the table. Then the girls ate 22 of those pancakes. How many pancakes were still on the table?
• Listen again and write the symbols for both parts: There were 54 pancakes on the table. Then the girls ate 22 of those pancakes. ✔
• Everybody, touch and read problem A. Get ready. (Signal.) *54 minus 22.*

b. Work problem A. Put your pencil down when you know how many pancakes were still on the table.
(Observe children and give feedback.)

- Read the problem and the answer you wrote for A. Get ready. (Signal.) *54 – 22 = 32.*
- How many pancakes were still on the table? (Signal.) *32.*

c. Touch where you'll write the symbols for problem B. ✔
Listen to problem B: Amos painted 26 houses. His brother painted 113 houses. How many houses did they paint in all?

- Listen again and write the symbols for both parts: Amos painted 26 houses. His brother painted 113 houses. ✔
- Everybody, touch and read problem B. Get ready. (Signal.) *26 plus 113.*

d. Work problem B. Put your pencil down when you know how many houses they painted in all.
(Observe children and give feedback.)

- Read the problem and the answer you wrote for B. Get ready. (Signal.) *26 plus 113 = 139.*
- How many houses did they paint in all? (Signal.) *139.*

e. Touch where you'll write the symbols for problem C. ✔
Listen to problem C: There were 428 birds on a wire. Then 120 of those birds flew away. How many birds were still on the wire?

- Listen again and write the symbols for both parts: There were 428 birds on a wire. Then 120 of those birds flew away. ✔
- Everybody, touch and read problem C. Get ready. (Signal.) *428 minus 120.*

f. Work problem C. Put your pencil down when you know how many birds were still on the wire.
(Observe children and give feedback.)
(Answer key:)

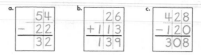

- Read the problem and the answer you wrote for C. Get ready. (Signal.) *428 – 120 = 308.*
- How many birds were still on the wire? (Signal.) *308.*

EXERCISE 10: INDEPENDENT WORK

a. Find part 7 on worksheet 64. ✔
(Teacher reference:)

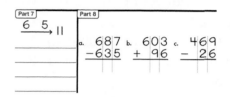

In the spaces below the family, you'll write the four facts for it. Then you'll work column problems in part 8.

b. Complete worksheet 64. Remember to write the equals and the number of cents to show what each row of coins in part 2 is worth.
(Observe children and mark incorrect responses on children's worksheets as you give feedback.)

Connecting Math Concepts

Lesson 65

EXERCISE 1: COINS EQUAL TO A DOLLAR

a. Listen: A dollar is worth 100 cents. How many cents is a dollar worth? (Signal.) *100.*
- Listen: A dollar is worth 100 cents. So how many cents is 2 dollars worth? (Signal.) *200.*
- How many cents is 5 dollars worth? (Signal.) *500.*
- How many cents is 9 dollars worth? (Signal.) *900.*
- How many cents is 1 dollar worth? (Signal.) *100.*
(Repeat step a until firm.)

b. (Display:)   [65:1A]

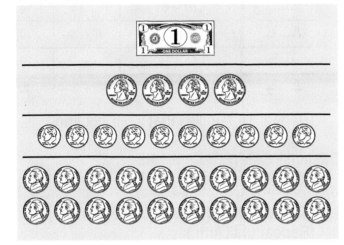

- (Point to 🂠.) This shows 100 cents in dollars.
- (Point to 🪙🪙🪙🪙.) This shows 100 cents in quarters.
- (Point to 🪙🪙🪙🪙🪙🪙🪙🪙🪙🪙.) This shows 100 cents in dimes.
- (Point to 🪙🪙🪙🪙🪙🪙🪙🪙🪙🪙🪙🪙🪙🪙🪙🪙🪙🪙🪙🪙.) This shows 100 cents in nickels.
- (Point to 🂠.) How many cents is a dollar? (Signal.) *100.*
So this row shows 100 cents.

c. (Point to 🪙.) How many cents is a quarter? (Signal.) *25.*
- Count by 25s for the quarters. Get ready. (Touch.) *25, 50, 75, 100.*
So we have 100 cents in quarters.

d. (Point to 🪙.) How many cents is a dime? (Signal.) *10.*
- Count by tens for the dimes. Get ready. (Touch.) *10, 20, 30, 40, 50, 60, 70, 80, 90, 100.* So we have 100 cents in dimes.
e. (Point to 🪙.) How many cents is a nickel? (Signal.) *5.*
- Count by fives for the nickels. Get ready. (Touch.) *5, 10, 15, 20, 25, 30, 35, 40, 45, 50, 55, 60, 65, 70, 75, 80, 85, 90, 95, 100.* So we have 100 cents in nickels.

EXERCISE 2: NUMBER FAMILIES   `REMEDY`

a. (Display:)   [65:2A]

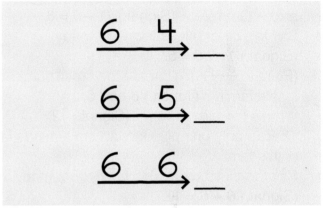

The big number is missing in these families. All these families have a small number of 6.
- (Point to 6 → 4.) What are the small numbers in this family? (Touch.) *6 and 4.*
- What's the big number? (Signal.) *10.*
b. (Point to 6 → 5.) What are the small numbers in this family? (Touch.) *6 and 5.*
- What's the big number? (Signal.) *11.*
c. (Point to 6 → 6.) What are the small numbers in this family? (Touch.) *6 and 6.*
- What's the big number? (Signal.) *12.*

d. (Point to $6 \xrightarrow{} 4$.) What are the small numbers in this family again? (Touch.) *6 and 4.*
- What's the big number? (Signal.) *10.*
- Say the fact that starts with the first small number. Get ready. (Signal.) *6 + 4 = 10.*
- Say the fact that starts with the other small number. Get ready. (Signal.) *4 + 6 = 10.*
- Say the fact that goes backward along the arrow. Get ready. (Signal.) *10 − 4 = 6.*
- Say the other minus fact. Get ready. (Signal.) *10 − 6 = 4.*

e. (Point to $6 \xrightarrow{} 5$.) What are the small numbers in this family? (Touch.) *6 and 5.*
- What's the big number? (Signal.) *11.*
- Say the fact that starts with the first small number. Get ready. (Signal.) *6 + 5 = 11.*
- Say the fact that starts with the other small number. Get ready. (Signal.) *5 + 6 = 11.*
- Say the fact that goes backward along the arrow. Get ready. (Signal.) *11 − 5 = 6.*
- Say the other minus fact. Get ready. (Signal.) *11 − 6 = 5.*

f. (Point to $6 \xrightarrow{} 6$.) What are the small numbers in this family? (Touch.) *6 and 6.*
- What's the big number? (Signal.) *12.* There's only one plus fact and one minus fact in this family.
- Say the plus fact for this family. Get ready. (Signal.) *6 + 6 = 12.*
- Say the minus fact for this family. Get ready. (Signal.) *12 − 6 = 6.*
(Repeat steps d through f until firm.)

g. (Display:) [65:2B]

6 ⟹ 11   6 ⟹ 10

= 6 ⟹ 16   6 3 ⟶ __

6 ⟹ 12   = 2 ⟹ 11

6 2 ⟶ __   6 5 ⟶ __

You're going to say the problem for the missing number in each family.

- (Point to $6 \xrightarrow{} 11$.) Say the problem for the missing number. Get ready. (Touch.) *11 minus 6.*
- What's 11 minus 6? (Signal.) *5.*

h. (Point to $\xrightarrow{6} 16$.) Say the problem for the missing number. Get ready. (Touch.) *16 minus 6.*
- What's 16 minus 6? (Signal.) *10.*

i. (Repeat the following tasks for remaining families:)

(Point to __.)	Say the problem for the missing number.	What's __?	
$6 \Rightarrow 12$	12 − 6	12 − 6	6
$6 \; 2 \Rightarrow$	6 + 2	6 + 2	8
$6 \Rightarrow 10$	10 − 6	10 − 6	4
$6 \; 3 \Rightarrow$	6 + 3	6 + 3	9
$= 2 \Rightarrow 11$	11 − 2	11 − 2	9
$6 \; 5 \Rightarrow$	6 + 5	6 + 5	11

(Repeat for families that were not firm.)

EXERCISE 3: MIXED COUNTING

a. Your turn: Count by 25s to 100. Get ready. (Tap.) *25, 50, 75, 100.*
(Repeat until firm.)
- Start with 100 and count by 25s to 200. Get 100 going. *One huuundred.* Count. (Tap.) *125, 150, 175, 200.*
(Repeat until firm.)
- Start with 200 and count by 25s to 300. Get 200 going. *Two huuundred.* Count. (Tap.) *225, 250, 275, 300.*
(Repeat until firm.)

b. Start with 34 and plus tens to 104. Get 34 going. *Thirty-fouuur.* Plus tens. (Tap.) *44, 54, 64, 74, 84, 94, 104.*

c. Start with 35 and count by fives to 100. Get 35 going. *Thirty-fiiive.* Count. (Tap.) *40, 45, 50, 55, 60, 65, 70, 75, 80, 85, 90, 95, 100.*
(Repeat step c until firm.)

d. You have 35. When you plus tens, what's the next number? (Signal.) *45.*
- You have 35. When you plus ones, what's the next number? (Signal.) *36.*
- You have 35. When you plus fives, what's the next number? (Signal.) *40.*

e. You have 70. When you minus ones, what's the next number? (Signal.) *69.*
• You have 70. When you plus ones, what's the next number? (Signal.) *71.*
• You have 70. When you plus tens, what's the next number? (Signal.) *80.*
• You have 70. When you plus fives, what's the next number? (Signal.) *75.*

EXERCISE 4: SHAPES

a. How many sides do all rectangles have? Get ready. (Signal.) *4.*
• How many sides do all triangles have? Get ready. (Signal.) *3.*
  Yes, all triangles have 3 sides.
  (Repeat step a until firm.)
b. (Display:) [65:4A]

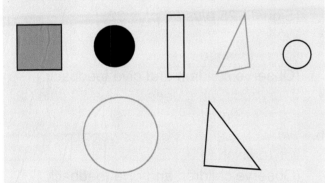

• Raise your hand when you know how many rectangles are in the picture. ✔
• How many rectangles are there? (Signal.) *2.*
c. Raise your hand when you know how many triangles are in the picture. ✔
• How many triangles are there? (Signal.) *2.*
d. Raise your hand when you know how many squares there are in the picture. ✔
• How many squares are there? (Signal.) *1.*
e. I'll touch some shapes. Tell me triangle, rectangle, or square.
• (Point to ■.) What is this? (Touch.) *Square.*
• (Point to ▯.) What is this? (Touch.) *Rectangle.*
• (Point to △.) What is this? (Touch.) *Triangle.*
• (Point to ◺.) What is this? (Touch.) *Triangle.*
  (Repeat step e until firm.)

EXERCISE 5: COINS   REMEDY

a. (Display:) [65:5A]

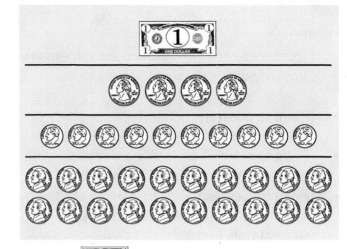

• (Point to 🉐.) This shows 100 cents in dollars.
b. (Repeat the following statement for quarters, dimes, and nickels:)
• (Point to __.) This shows 100 cents in __.
c. (Point to 🉐.) This shows 100 cents in dollars.
• How many dollars do you need to have 100 cents? (Signal.) *1.*
d. (Point to quarters.) This shows 100 cents in quarters.
• Raise your hand when you know how many quarters are worth 100 cents. ✔
• How many quarters do you need to have 100 cents? (Signal.) *4.*
e. (Point to dimes.) This shows 100 cents in dimes. I'll touch each dime. Count to yourself and figure out how many dimes you need to have 100 cents.
  (Touch dimes as children count to themselves.)
• (Point to dimes.) How many dimes do you need to have 100 cents? (Signal.) *10.*
f. (Point to quarters.) How many quarters do you need to have 100 cents? (Signal.) *4.*
• How many dimes do you need to have 100 cents? (Signal.) *10.*

g. (Point to nickels.) This shows 100 cents in nickels.
I'll touch each nickel. Count to yourself and figure out how many nickels you need to have 100 cents.
(Touch nickels as children count to themselves.)
- How many nickels do you need to have 100 cents? (Signal.) *20.*
- (Point to quarters.) How many quarters do you need to have 100 cents? (Signal.) *4.*
- (Point to dimes.) How many dimes do you need to have 100 cents? (Signal.) *10.*
- (Point to nickels.) How many nickels do you need to have 100 cents? (Signal.) *20.*
- (Repeat until firm.)

EXERCISE 6: FACTS
SUBTRACTION

a. (Distribute unopened workbooks to students.)
- Open your workbook to Lesson 65 and find part 1.
(Observe children and give feedback.)
(Teacher reference:)

a. $7-2$ d. $9-3$ h. $17-7$
b. $6-5$ e. $5-3$ i. $9-6$
c. $10-6$ f. $8-5$ j. $6-4$
          g. $10-2$

These are minus problems for families you know. You're going to read each problem and say the answer. Then you'll go back and work them.
- Touch and read problem A. Get ready. (Signal.) *7 minus 2.*
- What's the answer? (Signal.) *5.*
b. (Repeat the following tasks for problems B through J:)
- Touch and read problem __.
- What's the answer?
c. Complete the equations for all of the problems in part 1.
(Observe children and give feedback.)
(Answer key:)

a. $7-2=5$ d. $9-3=6$ h. $17-7=10$
b. $6-5=1$ e. $5-3=2$ i. $9-6=3$
c. $10-6=4$ f. $8-5=3$ j. $6-4=2$
            g. $10-2=8$

d. Check your work. You'll touch and read each fact.
- Fact A. (Signal.) $7-2=5.$
- (Repeat for:) B, $6-5=1$; C, $10-6=4$; D, $9-3=6$; E, $5-3=2$; F, $8-5=3$; G, $10-2=8$; H, $17-7=10$; I, $9-6=3$; J, $6-4=2.$

EXERCISE 7: COLUMN ADDITION   [REMEDY]

a. Find part 2 on worksheet 65. ✔
(Teacher reference:)   [R][Part G]

a. $\begin{array}{r} 25 \\ 30 \\ +\ 32 \\ \hline \end{array}$ b. $\begin{array}{r} 18 \\ 201 \\ +\ 40 \\ \hline \end{array}$ c. $\begin{array}{r} 356 \\ +620 \\ \hline \end{array}$ d. $\begin{array}{r} 61 \\ +320 \\ \hline \end{array}$

Some of these problems plus two numbers, some plus three numbers.
- Touch and read problem A. Get ready. (Signal.) *25 plus 30 plus 32.*
- Work the problem. Put your pencil down when you're finished.
(Observe children and give feedback.)
- Everybody, what does 25 plus 30 plus 32 equal? (Signal.) *87.*
b. Work problem B. Put your pencil down when you're finished.
(Observe children and give feedback.)
- Everybody, what does 18 plus 201 plus 40 equal? (Signal.) *259.*
c. Work problem C. Put your pencil down when you're finished.
(Observe children and give feedback.)
- Everybody, what does 356 plus 620 equal? (Signal.) *976.*
d. Work problem D. Put your pencil down when you're finished.
(Observe children and give feedback.)
- Everybody, what does 61 plus 320 equal? (Signal.) *381.*

EXERCISE 8: RULER SKILLS

a. Find part 3 on worksheet 65. ✔
   (Teacher reference:)

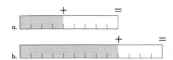

These rulers are supposed to show inches, but they are not the real size. The number of inches is not shown for each part.

- Touch ruler A. ✔
- Count the inches for the shaded part and write the number.
  (Observe children and give feedback.)
- Everybody, how many inches for the shaded part? (Signal.) *4.*
- Count the inches for the unshaded part and write the number.
  (Observe children and give feedback.)
- Everybody, how many inches for the unshaded part? (Signal.) *5.*
- Touch and read the problem for ruler A. Get ready. (Signal.) *4 plus 5 (equals).*

b. You're going to figure out how many inches there are for both parts.
- Touch the end of the shaded part. ✔
- What number will you get going? (Signal.) *4.*
- You'll get that number going. Then you'll touch and count for the inches in the unshaded part.
- Get it going. *Fouuur.* Touch and count the lines. (Tap 5.) *5, 6, 7, 8, 9.*
  (Repeat until firm.)
- How many inches are there in both parts? (Signal.) *9.*
- Write 9. ✔
- Everybody, touch and read the fact for ruler A. (Signal.) *4 + 5 = 9.*
- How long is ruler A supposed to be? (Signal.) *9 inches.*

c. Touch ruler B. ✔
- Count the inches for the shaded part and write the number.
  (Observe children and give feedback.)
- Everybody, how many inches for the shaded part? (Signal.) *9.*
- Count the inches for the unshaded part and write the number.
  (Observe children and give feedback.)

- Everybody, how many inches for the unshaded part? (Signal.) *4.*
- Touch and read the problem for ruler B. Get ready. (Signal.) *9 plus 4 (equals).*

d. You're going to figure out how many inches there are in both parts.
- Touch the end of the shaded part. ✔
- What number will you get going? (Signal.) *9.*
- Get it going. *Niiine.* Touch and count the lines. (Tap 4.) *10, 11, 12, 13.*
  (Repeat until firm.)
- How many inches are there in both parts? (Signal.) *13.*
- Write 13. ✔
- Everybody, touch and read the fact for ruler B. (Signal.) *9 + 4 = 13.*
- How many inches long is ruler B supposed to be? (Signal.) *13 inches.*

EXERCISE 9: WORD PROBLEMS (COLUMNS)

a. Find part 4 on worksheet 65. ✔
   (Teacher reference:)

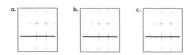

You're going to write the symbols for word problems in columns and work them. The equals bars are already shown, but only part of the dotted lines for the columns and rows are shown.

- Touch where you'll write the symbols for problem A. ✔
  Listen to problem A: There were 36 players on the Tigers. There were 101 players on the Spartans. How many players were there on both teams?
- Listen again and write the symbols for both parts: There were 36 players on the Tigers. There were 101 players on the Spartans. ✔
- Everybody, touch and read problem A. Get ready. (Signal.) *36 plus 101.*

b. Work problem A. Put your pencil down when you know how many players were on both teams.
  (Observe children and give feedback.)
- Read the problem and the answer you wrote for A. Get ready. (Signal.) *36 + 101 = 137.*
- How many players were on both teams? (Signal.) *137.*

c. Touch where you'll write the symbols for problem B. ✔
   Listen to problem B: 276 fish were in a lake. 214 of those fish swam out of the lake. How many fish were still in the lake?
   - Listen again and write the symbols for both parts: 276 fish were in a lake. 214 of those fish swam out of the lake. ✔
   - Everybody, touch and read problem B. Get ready. (Signal.) *276 minus 214.*

d. Work problem B. Put your pencil down when you know how many fish were still in the lake.
   (Observe children and give feedback.)
   - Read the problem and the answer you wrote for B. Get ready. (Signal.) *276 – 214 = 62.*
   - How many fish were still in the lake? (Signal.) *62.*

e. Touch where you'll write the symbols for problem C. ✔
   Listen to problem C: Dana had 59 apples. People ate 27 of those apples. How many apples did Dana end up with?
   - Listen again and write the symbols for both parts: Dana had 59 apples. People ate 27 of those apples. ✔
   - Everybody, touch and read problem C. Get ready. (Signal.) *59 minus 27.*

f. Work problem C. Put your pencil down when you know how many apples Dana ended up with.
   (Observe children and give feedback.)
   (Answer key:)

   - Read the problem and the answer you wrote for C. Get ready. (Signal.) *59 – 27 = 32.*
   - How many apples did Dana end up with? (Signal.) *32.*

EXERCISE 10: INDEPENDENT WORK

a. Find part 5 on worksheet 65. ✔
   (Teacher reference:)

   a. $700 + 0 + 9$ b. $\underline{\quad} + \underline{\quad} + \underline{\quad} = 518$

   You'll complete the place-value addition equations for this part.

b. Turn to the other side of worksheet 65 and find part 6. ✔
   (Teacher reference:)

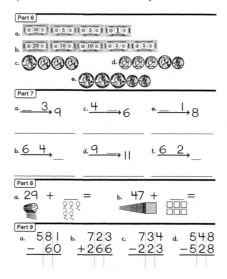

   There are rows of bills and rows of coins in part 6. You'll count for each row. You'll write the equals and the number of dollars or cents to show how much each row is worth.
   For part 7, you'll write the missing number in each family and write the fact for finding the missing number below.
   For part 8, you'll complete the equations for the balloons and for the boxes.
   You'll work column problems in part 9.

c. Your Turn: Complete worksheet 65. Remember to work the place-value equations in part 5.
   (Observe children and mark incorrect responses on children's worksheets as you give feedback.)

Lessons 66-70 Planning Page

	Lesson 66	**Lesson 67**	**Lesson 68**	**Lesson 69**	**Lesson 70**
Student Learning Objectives	**Exercises** 1. Say two addition and two subtraction facts for number families; find missing numbers in number families 2. Count forward or backward from a given number 3. Say how many of each coin equals a dollar 4. **Identify 2-dimensional shapes (circle)** 5. Solve subtraction facts 6. Solve 2-digit/3-digit addition and subtraction problems 7. **Use a centimeter ruler and count a part in a whole and count on for the whole to solve equations** 8. Solve addition and subtraction facts 9. Write the symbols for word problems in columns and solve 10. Complete work independently	**Exercises** 1. Say two addition and two subtraction facts for number families; find missing numbers in number families 2. Say how many of each coin equals a dollar 3. Identify tens digits and ones digits 4. Count forward or backward from a given number 5. **Identify 2-dimensional shapes (hexagon)** 6. Solve addition and subtraction facts 7. Use a centimeter ruler and count on for the whole to solve equations 8. Write the symbols for word problems in columns and solve 9. Solve addition facts 10. Complete work independently	**Exercises** 1. Say two addition and two subtraction facts for number families; find missing numbers in number families 2. Identify tens digits and ones digits 3. Count forward or backward from a given number 4. Identify 2-dimensional shapes 5. Solve addition and subtraction facts 6. Say how many of each coin equals a dollar 7. Solve problems with three addends in columns 8. Use a centimeter ruler and count on for the whole to solve equations 9. Write the symbols for word problems in columns and solve 10. Complete work independently	**Exercises** 1. Say addition and subtraction facts 2. Identify 2-dimensional shapes 3. Identify tens digits and ones digits 4. **Identify 3-dimensional objects (cube)** 5. Solve addition facts 6. Count and write the value of mixed groups of coins 7. Use a centimeter ruler and count on for the whole to solve equations 8. Solve subtraction facts 9. Solve problems with three addends in columns 10. Complete work independently	**Exercises** 1. **Learn about the dollar sign ($) and the decimal point** 2. Say addition and subtraction facts 3. Identify 3-dimensional objects 4. **Use a ruler to count backward from a whole to solve equations** 5. Solve subtraction facts 6. Use a centimeter ruler and count on for the whole to solve equations 7. Solve problems with three addends in columns 8. Solve addition facts 9. Write the symbols for word problems in columns and solve 10. Complete work independently
Common Core State Standards for Mathematics					
1.OA 3	✔	✔	✔		✔
1.OA 5	✔	✔	✔	✔	✔
1.OA 6	✔	✔	✔	✔	✔
1.OA 8	✔	✔	✔	✔	✔
1.NBT 1	✔	✔	✔	✔	✔
1.NBT 2	✔	✔	✔	✔	
1.NBT 4	✔	✔	✔	✔	✔
1.G 1	✔	✔	✔	✔	✔
1.G 2				✔	✔
Teacher Materials	Presentation Book 2, Board Displays CD or chalkboard				
Student Materials	Workbook 2, Pencil				
Additional Practice	Student Practice Software: Block 3: Activity 1 (1.NBT.2 and 1.NBT.4), Activity 2 (1.NBT.4 and 1.OA.6), Activity 3 (1.NBT.1), Activity 4 (1.NBT.1), Activity 5 (1.NBT.2), Activity 6 (1.MD.2)				
Mastery Test					Student Assessment Book (Present Mastery Test 7 following Lesson 70.)

Lesson 66

EXERCISE 1: NUMBER FAMILIES

a. (Display:)   [66:1A]

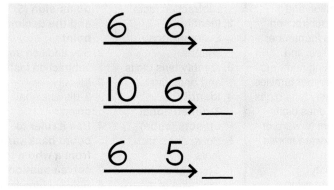

The big number is missing in these families. All these families have a small number of 6.
- (Point to 6 →6.) What are the small numbers in this family? (Touch.) *6 and 6.*
- What's the big number? (Signal.) *12.*

b. (Point to 10 →6.) What are the small numbers in this family? (Touch.) *10 and 6.*
- What's the big number? (Signal.) *16.*

c. (Point to 6 →5.) What are the small numbers in this family? (Touch.) *6 and 5.*
- What's the big number? (Signal.) *11.*

d. (Point to 6 →6.) What are the small numbers in this family again? (Touch.) *6 and 6.*
- What's the big number? (Signal.) *12.*
  There's only one plus fact and one minus fact in this family.
- Say the plus fact for this family. Get ready. (Signal.) *6 + 6 = 12.*
- Say the minus fact for this family. Get ready. (Signal.) *12 − 6 = 6.*

e. (Point to 10 →6.) What are the small numbers in this family? (Touch.) *10 and 6.*
- What's the big number? (Signal.) *16.*
- Say the fact that starts with the first small number. Get ready. (Signal.) *10 + 6 = 16.*
- Say the fact that starts with the other small number. Get ready. (Signal.) *6 + 10 = 16.*
- Say the fact that goes backward along the arrow. Get ready. (Signal.) *16 − 6 = 10.*
- Say the other minus fact. Get ready. (Signal.) *16 − 10 = 6.*

f. (Point to 6 →5.) What are the small numbers in this family? (Touch.) *6 and 5.*
- What's the big number? (Signal.) *11.*
- Say the fact that starts with the first small number. Get ready. (Signal.) *6 + 5 = 11.*

- Say the fact that starts with the other small number. Get ready. (Signal.) *5 + 6 = 11.*
- Say the fact that goes backward along the arrow. Get ready. (Signal.) *11 − 5 = 6.*
- Say the other minus fact. Get ready. (Signal.) *11 − 6 = 5.*
  (Repeat steps d through f until firm.)

g. (Display:)   [66:1B]

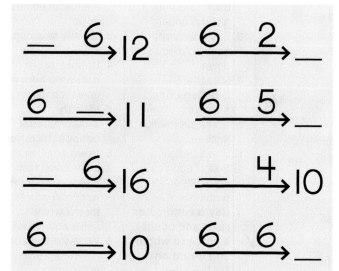

You're going to say the problem for the missing number in each family.

- (Point to =6→12.) Say the problem for the missing number. (Touch.) *12 minus 6.*
- What's 12 minus 6? (Signal.) *6.*

h. (Point to 6→11.) Say the problem for the missing number. (Touch.) *11 minus 6.*
- What's 11 minus 6? (Signal.) *5.*

i. (Repeat the following tasks for remaining families:)

(Point to ___.)	Say the problem for the missing number.	What's ___?	
=6→16	16 − 6	16 − 6	10
6 →10	10 − 6	10 − 6	4
6 2→	6 + 2	6 + 2	8
6 5→	6 + 5	6 + 5	11
=4→10	10 − 4	10 − 4	6
6 6→	6 + 6	6 + 6	12

(Repeat for families that were not firm.)

Connecting Math Concepts

EXERCISE 2: MIXED COUNTING

a. Start with 200 and count by 25s to 300. Get 200 going. *Two huuundred.* Count. **(Tap.)** *225, 250, 275, 300.*
  • Start with 47 and plus tens to 107. Get 47 going. *Forty-sevennn.* Plus tens. **(Tap.)** *57, 67, 77, 87, 97, 107.*
  • Start with 60 and count by fives to 100. Get 60 going. *Sixtyyy.* Count. **(Tap.)** *65, 70, 75, 80, 85, 90, 95, 100.*

b. Start with 100 and count backward by tens.
  • What number do you start with? **(Signal.)** *100.*
  • Get 100 going. *One huuundred.* Count backward. **(Tap.)** *90, 80, 70, 60, 50, 40, 30, 20, 10.*
  • Start with 70 and count backward by tens. Get 70 going. *Seventyyy.* Count backward. **(Tap.)** *60, 50, 40, 30, 20, 10.*
  (Repeat until firm.)

c. You have 75. When you count backward by ones, what's the next number? **(Signal.)** *74.*
  • You have 75. When you plus ones, what's the next number? **(Signal.)** *76.*
  • You have 75. When you plus fives, what's the next number? **(Signal.)** *80.*
  • You have 75. When you plus tens, what's the next number? **(Signal.)** *85.*
  • You have 75. When you plus 25s, what's the next number? **(Signal.)** *100.*

EXERCISE 3: COINS EQUAL TO A DOLLAR

a. Listen: A dollar is worth 100 cents. How many cents is a dollar worth? **(Signal.)** *100.*
  • Listen: A dollar is worth 100 cents, so how many cents is 2 dollars worth? **(Signal.)** *200.*
  • How many cents is 4 dollars worth? **(Signal.)** *400.*
  • How many cents is 6 dollars worth? **(Signal.)** *600.*
  • How many cents is 1 dollar worth? **(Signal.)** *100.*

b. (Display:)   [66:3A]

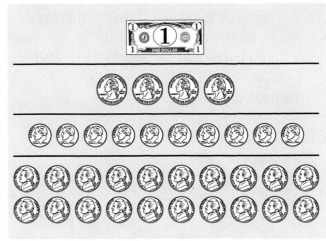

  • (Point to display.) Each group is worth 100 cents.
  • (Point to ⬤.) You need 4 quarters to have 100 cents. How many quarters do you need to have 100 cents? **(Signal.)** *4.*
  • (Point to ⬤.) You need 10 dimes to have 100 cents. How many dimes do you need to have 100 cents? **(Signal.)** *10.*
  • (Point to ⬤.) You need 20 nickels to have 100 cents. How many nickels do you need to have 100 cents? **(Signal.)** *20.*

c. Once more: How many quarters do you need to have 100 cents? **(Signal.)** *4.*
  • How many dimes do you need to have 100 cents? **(Signal.)** *10.*
  • How many nickels do you need to have 100 cents? **(Signal.)** *20.*

d. (Point to ⬤.) Listen: A dollar is worth 100 cents. So this group is worth a dollar.
  • How many dollars is this group worth? **(Touch.)** *1.*
  • How many cents is this group worth? **(Touch.)** *100.*
  • Again, how many dollars is this group worth? **(Touch.)** *1.*

e. (Repeat the following tasks for dimes and nickels:)
  • How many dollars is this group worth?
  • How many cents is this group worth?

f. (Point to ⬤.) How many quarters are in this group? **(Touch.)** *4.*
  • So how many quarters do you need to have 1 dollar? **(Signal.)** *4.*

g. (Repeat the following tasks for dimes and nickels:)
  • (Point to __.) How many __ are in this group?
  • So how many __ do you need to have 1 dollar?

h. (Point to .) Once more, how many quarters do you need to have 1 dollar? (Touch.) *4.*
- Yes, 4 quarters are worth 1 dollar. Say the statement. (Signal.) *4 quarters are worth 1 dollar.*
  (Repeat until firm.)

EXERCISE 4: SHAPES
INTRODUCTION OF CIRCLE

a. (Display:) [66:4A]

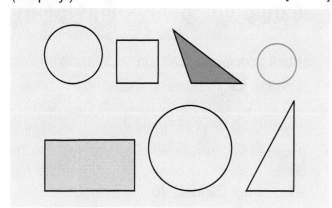

These shapes are rectangles, triangles, and circles.
- (Point to ◯.) This shape is a **circle.** What shape? (Touch.) *(A) circle.*
- Does a circle have 4 sides? (Signal.) *No.*
- What shape has 4 sides? (Signal.) *(A) rectangle.*
- Does a circle has 3 sides? (Signal.) *No.*
- What shape has 3 sides? (Signal.) *(A) triangle.*
b. Tell me rectangle, triangle, square, or circle.
- (Point to ☐.) What is this? (Touch.) *Square.*
- (Repeat for remaining shapes.)
  (Repeat step b until firm.)
c. Look at the shapes.
- Raise your hand when you know how many rectangles are in the picture. ✔
- How many rectangles are there? (Signal.) *2.*
- How many of the rectangles are squares? (Signal.) *1.*
d. Raise your hand when you know how many triangles are in the picture. ✔
- How many triangles? (Signal.) *2.*
e. Raise your hand when you know how many circles are in the picture. ✔
- How many circles? (Signal.) *3.*

EXERCISE 5: FACTS
SUBTRACTION

a. (Distribute unopened workbooks to students.)
- Open your workbook to Lesson 66 and find part 1.
  (Observe children and give feedback.)
  (Teacher reference:)

a. $12-10$ d. $6-2$ h. $8-7$
b. $8-3$ e. $14-10$ i. $8-5$
c. $9-6$ f. $8-2$ j. $11-9$
      g. $10-6$

These are minus problems for families you know. You're going to read each problem and say the answer. Then you'll go back and work them.
- Touch and read problem A. Get ready. (Signal.) *12 minus 10.*
- What's the answer? (Signal.) *2.*
b. (Repeat the following tasks for problems B through J:)
- Touch and read problem __.
- What's the answer?
c. Complete the equations for all of the problems in part 1.
  (Observe children and give feedback.)
  (Answer key:)

a. $12-10=2$ d. $6-2=4$ h. $8-7=1$
b. $8-3=5$ e. $14-10=4$ i. $8-5=3$
c. $9-6=3$ f. $8-2=6$ j. $11-9=2$
      g. $10-6=4$

d. Check your work. You'll touch and read each fact.
- Fact A. (Signal.) *12 – 10 = 2.*
- (Repeat for:) B, *8 – 3 = 5;* C, *9 – 6 = 3;* D, *6 – 2 = 4;* E, *14 – 10 = 4;* F, *8 – 2 = 6;* G, *10 – 6 = 4;* H, *8 – 7 = 1;* I, *8 – 5 = 3;* J, *11 – 9 = 2.*

EXERCISE 6: COLUMN PROBLEMS
ADDITION AND SUBTRACTION

a. Find part 2 on worksheet 66. ✔
(Teacher reference:)

a.	b.	c.	d.
33	374	25 441	289
+ 56	−314	+ 31	− 36

These are column problems. Some of them plus and some of them minus.

• Touch and read problem A. Get ready.
(Signal.) *33 plus 56.*
(Repeat until firm.)

• Work problem A. Put your pencil down when you're finished.
(Observe children and give feedback.)

• Check problem A. What does 33 plus 56 equal? (Signal.) *89.*

b. Touch and read problem B. Get ready.
(Signal.) *374 minus 314.*
(Repeat until firm.)

• Work problem B. Put your pencil down when you're finished.
(Observe children and give feedback.)

• Check problem B. What does 374 minus 314 equal? (Signal.) *60.*

c. Touch and read problem C. Get ready.
(Signal.) *25 plus 441 plus 31.*
(Repeat until firm.)

• Work problem C. Put your pencil down when you're finished.
(Observe children and give feedback.)

• Check problem C. What does 25 plus 441 plus 31 equal? (Signal.) *497.*

d. Touch and read problem D. Get ready.
(Signal.) *289 minus 36.*
(Repeat until firm.)

• Work problem D. Put your pencil down when you're finished.
(Observe children and give feedback.)

• Check problem D. What does 289 minus 36 equal? (Signal.) *253.*

EXERCISE 7: RULER SKILLS

a. Find part 3 on worksheet 66. ✔
(Teacher reference:)

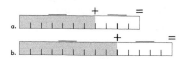

The numbers are not shown for each part.

• Touch ruler A. ✔

• Count the centimeters for the shaded part and write the number.
(Observe children and give feedback.)

• Everybody, how many centimeters for the shaded part? (Signal.) *7.*

• Count the centimeters for the unshaded part and write the number.
(Observe children and give feedback.)

• Everybody, how many centimeters for the unshaded part? (Signal.) *4.*

• Touch and read the problem for ruler A. Get ready. (Signal.) *7 plus 4 equals.*

b. You're going to figure out how many centimeters there are for both parts.

• Touch the end of the shaded part. ✔ You'll get that number going. Then you'll touch and count for the centimeters in the unshaded part.

• What number will you get going? (Signal.) *7.*

• Get it going. *Sevennn.* Touch and count the lines. (Tap.) *8, 9, 10, 11.*
(Repeat until firm.)

• How many centimeters are there for both parts? (Signal.) *11.*

• Write 11. ✔

• Everybody, touch and read the fact for ruler A. (Signal.) *7 + 4 = 11.*
(Repeat until firm.)

c. Touch ruler B. ✔

• Count the centimeters for the shaded part and write the number.
(Observe children and give feedback.)

• Everybody, how many centimeters for the shaded part? (Signal.) *9.*

• Count the centimeters for the unshaded part and write the number.
(Observe children and give feedback.)

• Everybody, how many centimeters for the unshaded part? (Signal.) *5.*

• Touch and read the problem for ruler B. Get ready. (Signal.) *9 plus 5 equals.*

d. You're going to figure out how many centimeters there are for both parts.
• Touch the end of the shaded part. ✔
• What number will you get going? (Signal.) *9.*
• Get it going. *Niiine.* Touch and count the lines. (Tap.) *10, 11, 12, 13, 14.*
  (Repeat until firm.)
• How many centimeters are there for both parts? (Signal.) *14.*
• Write 14. ✔
• Everybody, touch and read the fact for ruler B. (Signal.) *9 + 5 = 14.*
  (Repeat until firm.)

EXERCISE 8: FACTS
PLUS/MINUS MIX

a. Find part 4 on worksheet 66. ✔
   (Teacher reference:)

a. 3 + 5	d. 10 − 4	h. 8 − 3
b. 5 + 10	e. 2 + 6	i. 8 − 6
c. 6 + 3	f. 6 + 4	j. 2 + 8
	g. 14 − 10	

These problems are from families you know. You're going to tell me if the big number or a small number is missing. Then you'll tell me the answer.
• Touch and read problem A. Get ready. (Signal.) *3 plus 5.*
• Is the answer the big number or a small number? (Signal.) *The big number.*
• What's 3 plus 5? (Signal.) *8.*
b. Touch and read problem B. Get ready. (Signal.) *5 plus 10.*
• Is the answer the big number or a small number? (Signal.) *The big number.*
• What's 5 plus 10? (Signal.) *15.*
c. (Repeat the following tasks for problems C through J:)

Touch and read problem __.	Is the answer the big number or a small number?	What's __?	
C	6 + 3	*The big number.*	6 + 3	9
D	10 − 4	*A small number.*	10 − 4	6
E	2 + 6	*The big number.*	2 + 6	8
F	6 + 4	*The big number.*	6 + 4	10
G	14 − 10	*A small number.*	14 − 10	4
H	8 − 3	*A small number.*	8 − 3	5
I	8 − 6	*A small number.*	8 − 6	2
J	2 + 8	*The big number.*	2 + 8	10

(Repeat problems that were not firm.)

d. Complete the equations in part 4. Put your pencil down when you're finished.
   (Observe children and give feedback.)
   (Answer key:)

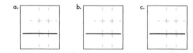

e. Check your work. You'll touch and read each fact.
• Fact A. (Signal.) *3 + 5 = 8.*
• (Repeat for:) B, *5 + 10 = 15;* C, *6 + 3 = 9;* D, *10 − 4 = 6;* E, *2 + 6 = 8;* F, *6 + 4 = 10;* G, *14 − 10 = 4;* H, *8 − 3 = 5;* I, *8 − 6 = 2;* J, *2 + 8 = 10.*

EXERCISE 9: WORD PROBLEMS (COLUMNS)

a. Find part 5 on worksheet 66. ✔
   (Teacher reference:)

You're going to write the symbols for word problems in columns and work them. The equals bars are already shown, but only part of the dotted lines and rows are shown.
• Touch where you'll write the symbols for problem A. ✔
  Listen to problem A: There were 147 apples in a barrel. A group of girls ate 45 of those apples. How many apples were still in the barrel?
• Listen again and write the symbols for both parts: There were 147 apples in a barrel. A group of girls ate 45 of those apples. ✔
• Everybody, touch and read problem A. Get ready. (Signal.) *147 minus 45.*
b. Work problem A. Put your pencil down when you know how many apples were still in the barrel.
   (Observe children and give feedback.)
• Read the problem and the answer you wrote for A. Get ready. (Signal.) *147 − 45 = 102.*
• How many apples were still in the barrel? (Signal.) *102.*

c. Touch where you'll write the symbols for problem B. ✔

Listen to problem B: There were 64 books in a case. 132 more books were put in the case. How many books ended up in the case?

- Listen again and write the symbols for both parts: There were 64 books in a case. 132 more books were put in the case. ✔
- Everybody, touch and read problem B. Get ready. (Signal.) *64 plus 132.*

d. Work problem B. Put your pencil down when you know how many books ended up in the case.

(Observe children and give feedback.)

- Read the problem and the answer you wrote for B. Get ready. (Signal.) *64 + 132 = 196.*
- How many books ended up in the case? (Signal.) *196.*

e. Touch where you'll write the symbols for problem C. ✔

Listen to problem C: Jim had 278 dollars. Jim spent 215 of those dollars. How many dollars did Jim end up with?

- Listen again and write the symbols for both parts: Jim had 278 dollars. Jim spent 215 of those dollars. ✔
- Everybody, touch and read problem C. Get ready. (Signal.) *278 minus 215.*

f. Work problem C. Put your pencil down when you know how many dollars Jim ended up with.

(Observe children and give feedback.)
(Answer key:)

- Read the problem and the answer you wrote for C. Get ready. (Signal.) *278 – 215 = 63.*
- How many dollars did Jim end up with? (Signal.) *63.*

EXERCISE 10: INDEPENDENT WORK

a. Turn to the other side of worksheet 66 and find part 6. ✔
(Teacher reference:)

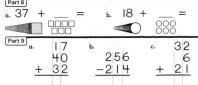

For each group in part 6, you'll count for the bills or coins. Write an equals and the dollars or cents each group is worth.

For each family in part 7, you'll write the missing number and the fact for finding it below.

For part 8, you'll complete the equations for the objects and work the problem.

You'll work column problems in part 9.

b. Complete worksheet 66.
(Observe children and mark incorrect responses on children's worksheets as you give feedback.)

Lesson 67

EXERCISE 1: NUMBER FAMILIES
REMEDY

a. (Display:) [67:1A]

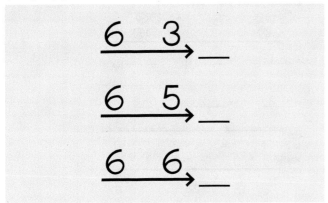

The big number is missing in these families. All these families have a small number of 6.

- (Point to $\overset{6\quad 3}{\longrightarrow}$.) What are the small numbers in this family? (Touch.) *6 and 3.*
- What's the big number? (Signal.) *9.*

b. (Point to $\overset{6\quad 5}{\longrightarrow}$.) What are the small numbers in this family? (Touch.) *6 and 5.*
- What's the big number? (Signal.) *11.*

c. (Point to $\overset{6\quad 6}{\longrightarrow}$.) What are the small numbers in this family? (Touch.) *6 and 6.*
- What's the big number? (Signal.) *12.*

d. (Point to $\overset{6\quad 3}{\longrightarrow}$.) What are the small numbers in this family again? (Touch.) *6 and 3.*
- What's the big number? (Signal.) *9.*
- Say the fact that starts with the first small number. Get ready. (Signal.) *6 + 3 = 9.*
- Say the fact that starts with the other small number. Get ready. (Signal.) *3 + 6 = 9.*
- Say the fact that goes backward along the arrow. Get ready. (Signal.) *9 − 3 = 6.*
- Say the other minus fact. Get ready. (Signal.) *9 − 6 = 3.*

e. (Point to $\overset{6\quad 5}{\longrightarrow}$.) What are the small numbers in this family? (Touch.) *6 and 5.*
- What's the big number? (Signal.) *11.*
- Say the fact that starts with the first small number. Get ready. (Signal.) *6 + 5 = 11.*
- Say the fact that starts with the other small number. Get ready. (Signal.) *5 + 6 = 11.*
- Say the fact that goes backward along the arrow. Get ready. (Signal.) *11 − 5 = 6.*
- Say the other minus fact. Get ready. (Signal.) *11 − 6 = 5.*

f. (Point to $\overset{6\quad 6}{\longrightarrow}$.) What are the small numbers in this family? (Touch.) *6 and 6.*

- What's the big number? (Signal.) *12.* There's only one plus fact and one minus fact in this family.
- Say the plus fact for this family. Get ready. (Signal.) *6 + 6 = 12.*
- Say the minus fact for this family. Get ready. (Signal.) *12 − 6 = 6.*
(Repeat steps d through f until firm.)

g. (Display:) [67:1B]

$$6 \Longrightarrow 10 \qquad \overset{2}{=\!\!\Longrightarrow} 11$$
$$\overset{10\quad 6}{\longrightarrow}\_ \qquad \overset{2}{=\!\!\Longrightarrow} 6$$
$$6 \Longrightarrow 11 \qquad 6 \Longrightarrow 12$$
$$6 \Longrightarrow 9 \qquad \overset{6\quad 6}{\longrightarrow}\_$$

You're going to say the problem for the missing number in each family.

- (Point to $6 \Longrightarrow 10$.) Say the problem for the missing number. Get ready. (Touch.) *10 minus 6.*
- What's 10 minus 6? (Signal.) *4.*

h. (Point to $\overset{10\quad 6}{\longrightarrow}\_$.) Say the problem for the missing number. (Touch.) *10 plus 6.*
- What's 10 plus 6? (Signal.) *16.*

i. (Repeat the following tasks for remaining families:)

(Point to __.)	Say the problem for the missing number.	What's __?	
$6 \Longrightarrow 11$	11 − 6	11 − 6	5
$6 \Longrightarrow 9$	9 − 6	9 − 6	3
$\overset{2}{=\!\!\Longrightarrow} 11$	11 − 2	11 − 2	9
$\overset{2}{=\!\!\Longrightarrow} 6$	6 − 2	6 − 2	4
$6 \Longrightarrow 12$	12 − 6	12 − 6	6
$\overset{6\quad 6}{\longrightarrow}\_$	6 + 6	6 + 6	12

(Repeat for families that were not firm.)

EXERCISE 2: COINS EQUAL TO A DOLLAR REMEDY

a. (Display:) [67:2A]

- (Point to coins.) Each group of coins is worth 100 cents.
- How much is a dollar worth? (Signal.) *100 cents.*
  So each group is worth 1 dollar.
- (Point to quarters.) 4 quarters equal 1 dollar. How many quarters equal 1 dollar? (Touch.) *4.*
- Say the sentence about 4 quarters. (Signal.) *4 quarters equal 1 dollar.*
b. (Point to dimes.) 10 dimes equal 1 dollar. How many dimes equal 1 dollar? (Signal.) *10.*
- Say the sentence about 10 dimes. (Signal.) *10 dimes equal 1 dollar.*
c. (Point to nickels.) 20 nickels equal 1 dollar. How many nickels equal 1 dollar? (Signal.) *20.*
- Say the sentence about 20 nickels. (Signal.) *20 nickels equal 1 dollar.*
d. Once more.
- (Point to quarters.) How many quarters equal 1 dollar? (Signal.) *4.*
- Say the sentence about 4 quarters. (Signal.) *4 quarters equal 1 dollar.*
e. (Point to dimes.) How many dimes equal 1 dollar? (Signal.) *10.*
- Say the sentence about 10 dimes. (Signal.) *10 dimes equal 1 dollar.*
f. (Point to nickels.) 20 nickels equal 1 dollar. How many nickels equal 1 dollar? (Signal.) *20.*
- Say the sentence about 20 nickels. (Signal.) *20 nickels equal 1 dollar.*
  (Repeat steps d though f until firm.)

EXERCISE 3: DIGITS

a. I'll say numbers that have two digits. You'll tell me the ones digit. Then you'll tell me the tens digit.
- Listen: 56. What number? (Signal.) *56.*
- The digits are 5 and 6. What are the digits of 56? (Signal.) *5 (and) 6.*
- What's the ones digit of 56? (Signal.) *6.*
- What's the tens digit? (Signal.) *5.*
  (Repeat until firm.)
b. New number: 18. What number? (Signal.) *18.*
  The digits are 1 and 8.
- What's the ones digit of 18? (Signal.) *8.*
- What's the tens digit? (Signal.) *1.*
  (Repeat step b until firm.)
c. (Repeat the following tasks for 35, 14, 60, 12, 97:)

New number: __. What number?	What are the digits of __?		What's the ones digit?	What's the tens digit?
35	35	3 (and) 5	5	3
14	14	1 (and) 4	4	1
60	60	6 (and) 0	0	6
12	12	1 (and) 2	2	1
97	97	9 (and) 7	7	9

(Repeat numbers that were not firm.)

EXERCISE 4: MIXED COUNTING

a. Start with 300 and count by 25s to 400. Get 300 going. *Three huuundred.* Count. (Tap.) *325, 350, 375, 400.*
  (Repeat until firm.)
- Start with 100 and count by 25s to 200. Get 100 going. *One huuundred.* Count. (Tap.) *125, 150, 175, 200.*
- Start with 45 and count by fives to 100. Get 45 going. *Forty-fiiive.* Count. (Tap.) *50, 55, 60, 65, 70, 75, 80, 85, 90, 95, 100.*
b. Start with 100 and count backward by tens.
- What number do you start with? (Signal.) *100.*
- Get 100 going. *One huuundred.* Count backward. (Tap.) *90, 80, 70, 60, 50, 40, 30, 20, 10.*
c. Start with 100 and count by hundreds to 1000. Get ready. (Tap.) *100, 200, 300, 400, 500, 600, 700, 800, 900, 1000.*
  (Repeat until firm.)

d. You have 63. When you count backward by ones, what's the next number? (Signal.) *62.*
• You have 63. When you plus ones, what's the next number? (Signal.) *64.*
• You have 63. When you plus tens, what's the next number? (Signal.) *73.*
e. You have 50. When you count backward by tens, what's the next number? (Signal.) *40.*
• You have 50. When you count backward by ones, what's the next number? (Signal.) *49.*
• You have 50. When you plus tens, what's the next number? (Signal.) *60.*
• You have 50. When you plus 25s, what's the next number? (Signal.) *75.*
• You have 50. When you plus fives, what's the next number? (Signal.) *55.*

EXERCISE 5: SHAPES
INTRODUCTION OF HEXAGON

a. (Display:)   [67:5A]

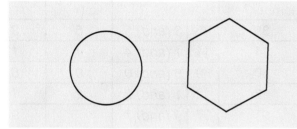

• (Point to ◯.) What shape is this? (Touch.) *(A) circle.*
b. (Point to ⬡.) Is this shape a triangle? (Touch.) *No.*
• How many sides does a triangle have? (Signal.) *3.*
• Does this shape have more than 3 sides or less than 3 sides? (Touch.) *More than 3 sides.*
c. A rectangle has more than 3 sides. Is this shape a rectangle? (Touch.) *No.*
• How many sides does a rectangle have? (Signal.) *4.*
• (Point to ⬡.) Does this shape have more than 4 sides or less than 4 sides? (Touch.) *More than 4 sides.*
d. I'll touch. You'll count the sides of this shape. Get ready. (Touch.) *1, 2, 3, 4, 5, 6.*
• How many sides? (Touch.) *6.*
e. The name of this shape is hard to say. **Hexagon**. Listen again: **hexagon**.
• What's this shape? (Signal.) *Hexagon.*
• Yes, hexagons have 6 sides. Say the sentence. (Signal.) *Hexagons have 6 sides.*

f. Say the sentence about rectangles. (Signal.) *Rectangles have 4 sides.*
• Say the sentence about triangles. (Signal.) *Triangles have 3 sides.*
• Say the sentence about hexagons. (Signal.) *Hexagons have 6 sides.*
(Repeat step f until firm.)
g. (Display:)   [67:5B]

These are street signs. You'll tell me the shape of each sign.
• (Point to yield.) What shape? (Touch.) *(A) triangle.*
• (Point to speed.) What shape? (Touch.) *(A) rectangle.*
• (Point to railroad crossing.) What shape? (Touch.) *(A) circle.*
(Repeat until firm.)

EXERCISE 6: FACTS
PLUS/MINUS MIX

a. (Distribute unopened workbooks to students.)
• Open your workbook to Lesson 67 and find part 1. ✔
(Observe children and give feedback.)
(Teacher reference:)

	d. $3+10$	h. $3+6$
a. $7-6$	e. $9-6$	i. $8-6$
b. $3+5$	f. $9+2$	j. $8-3$
c. $10-2$	g. $10-6$	

These problems are from families you know. You're going to tell me if the big number or a small number is missing. Then you'll tell me the answer.
• Touch and read problem A. Get ready. (Signal.) *7 minus 6.*
• Is the answer the big number or a small number? (Signal.) *A small number.*
• What's 7 minus 6? (Signal.) *1.*
b. Touch and read problem B. Get ready. (Signal.) *3 plus 5.*
• Is the answer the big number or a small number? (Signal.) *The big number.*
• What's 3 plus 5? (Signal.) *8.*

c. (Repeat the following tasks for problems C through J:)

Touch and read problem __.		Is the answer the big number or a small number?	What's __?	
C	10 − 2	A small number.	10 − 2	8
D	3 + 10	The big number.	3 + 10	13
E	9 − 6	A small number.	9 − 6	3
F	9 + 2	The big number.	9 + 2	11
G	10 − 6	A small number.	10 − 6	4
H	3 + 6	The big number.	3 + 6	9
I	8 − 6	A small number.	8 − 6	2
J	8 − 3	A small number.	8 − 3	5

d. Complete the equations in part 1. Put your pencil down when you're finished.
(Observe children and give feedback.)
(Answer key:)

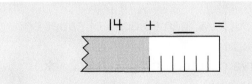

a. 7 − 6 = 1
b. 3 + 5 = 8
c. 10 − 2 = 8
d. 3 + 10 = 13
e. 9 − 6 = 3
f. 9 + 2 = 11
g. 10 − 6 = 4
h. 3 + 6 = 9
i. 8 − 6 = 2
j. 8 − 3 = 5

e. Check your work. You'll touch and read each fact.
• Fact A. (Signal.) *7 − 6 = 1.*
• (Repeat for:) B, *3 + 5 = 8;* C, *10 − 2 = 8;* D, *3 + 10 = 13;* E, *9 − 6 = 3;* F, *9 + 2 = 11;* G, *10 − 6 = 4;* H, *3 + 6 = 9;* I, *8 − 6 = 2;* J, *8 − 3 = 5.*

EXERCISE 7: RULER SKILLS

a. (Display:) W   [67:7A]

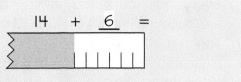

This is a new kind of problem. You can't see most of the shaded part. So you can't count all the lines for the shaded part. The number above the ruler tells how many centimeters the shaded part is.
• (Point to **14**.) How many centimeters is the shaded part? (Touch.) *14.*
• (Touch end of shaded part.) Count the centimeters for the unshaded part. Get ready. (Touch.) *1, 2, 3, 4, 5, 6.*

• How many centimeters is the unshaded part? (Signal.) *6.*
(Add to show:)   [67:7B]

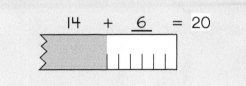

b. Now we count the centimeters for both groups. So I touch the end of the shaded part.
• (Touch the end of the shaded part.) What number do we get going? (Signal.) *14.*
• Get it going. *Fourteeen.* (Touch lines.) *15, 16, 17, 18, 19, 20.*
(Repeat until firm.)
• How many centimeters are there in both groups? (Touch.) *20.*
Yes, 14 plus 6 equals 20.
(Add to show:)   [67:7C]

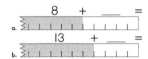

• Read the fact. Get ready. (Touch.) *14 + 6 = 20.*
c. Find part 2 on worksheet 67. ✔
(Teacher reference:)

a. [ruler diagram] 8 + ___ =
b. [ruler diagram] 13 + ___ =

These rulers show centimeters. You can't see most of the shaded part. So you can't count all the lines for the shaded part. The number above the ruler tells how many centimeters the shaded part is.
• Touch ruler A. ✔
• How many centimeters is the shaded part? (Signal.) *8.*
• Write the number for the unshaded part.
(Observe children and give feedback.)
• How many centimeters is the unshaded part? (Signal.) *5.*
d. Touch the end of the shaded part. ✔
• What number do you get going? (Signal.) *8.*
• Get it going. *Eieieight.* Touch and count. (Tap 5.) *9, 10, 11, 12, 13.*
(Repeat until firm.)

- How many centimeters long is ruler A? (Signal.) *13.*
- Write 13. ✔
- Everybody, touch and read the fact for ruler A. Get ready. (Signal.) *8 + 5 = 13.*

e. Touch ruler B. ✔
- How many centimeters is the shaded part? (Signal.) *13.*
- Write the number for the unshaded part. (Observe children and give feedback.)
- How many centimeters is the unshaded part? (Signal.) *4.*

f. Touch the end of the shaded part. ✔
- What number do you get going? (Signal.) *13.*
- Get it going. *Thirteeen.* Touch and count. (Tap 4.) *14, 15, 16, 17.* (Repeat until firm.)
- How many centimeters long is ruler B? (Signal.) *17.*
- Write 17. ✔
- Everybody, touch and read the fact for ruler B. Get ready. (Signal.) *13 + 4 = 17.*

EXERCISE 8: WORD PROBLEMS (COLUMNS)

a. Find part 3 on worksheet 67. ✔ (Teacher reference:)

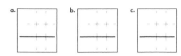

You're going to write the symbols for word problems in columns and work them. The equals bars are already shown, but only part of the dotted lines for the columns and the rows are shown.
- Touch where you'll write the symbols for problem A. ✔
  Listen to problem A: There were 126 people on a plane. 50 more people got on the plane. How many people ended up on the plane?
- Listen again and write the symbols for both parts: There were 126 people on a plane. 50 more people got on the plane. ✔
- Everybody, touch and read problem A. Get ready. (Signal.) *126 plus 50.*

b. Work problem A. Put your pencil down when you know how many people ended up on the plane. (Observe children and give feedback.)
- Read the problem and the answer you wrote for A. Get ready. (Signal.) *126 + 50 = 176.*
- So how many people ended up on the plane? (Signal.) *176.*

c. Touch where you'll write the symbols for problem B. ✔
  Listen to problem B: Amy had 859 dollars in her bank account. Amy took 56 of those dollars out of her account. How many dollars did Amy end up with in her bank account?
- Listen again and write the symbols for both parts: Amy had 859 dollars in her bank account. Amy took 56 of those dollars out of her account. ✔
- Everybody, touch and read problem B. Get ready. (Signal.) *859 minus 56.*

d. Work problem B. Put your pencil down when you know how many dollars Amy ended up with in her bank account. (Observe children and give feedback.)
- Read the problem and the answer you wrote for B. Get ready. (Signal.) *859 – 56 = 803.*
- So how many dollars did Amy end up with in her bank account? (Signal.) *803.*

e. Touch where you'll write the symbols for problem C. ✔
  Listen to problem C: There were 13 pieces of paper in a box. 476 more pieces of paper were put in that box. How many pieces of paper ended up in that box?
- Listen again and write the symbols for both parts: There were 13 pieces of paper in a box. 476 more pieces of paper were put in that box. ✔
- Everybody, touch and read problem C. Get ready. (Signal.) *13 plus 476.*

f. Work problem C. Put your pencil down when you know how many pieces of paper ended up in that box. (Observe children and give feedback.) (Answer key:)

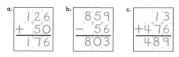

- Read the problem and the answer you wrote for C. Get ready. (Signal.) *13 + 476 = 489.*
- So how many pieces of paper ended up in that box? (Signal.) *489.*

EXERCISE 9: FACTS
ADDITION

a. Find part 4 on worksheet 67. ✔
   (Teacher reference:)

   a. 6 + 5   e. 6 + 3
   b. 4 + 6   f. 5 + 6
   c. 9 + 10   g. 2 + 6
   d. 6 + 6   h. 6 + 4

These are plus problems for families you know. You're going to read the problems. Then you'll go back and work them.
- Touch and read problem A. Get ready. (Signal.) *6 plus 5.*
- What's 6 plus 5? (Signal.) *11.*

b. Touch and read problem B. Get ready. (Signal.) *4 plus 6.*
- What's 4 plus 6? (Signal.) *10.*

c. (Repeat the following tasks for problems C through H:)
- Touch and read problem __.
- What's __?

d. Complete the equations for all of the problems in part 4.
   (Observe children and give feedback.)
   (Answer key:)

   a. 6 + 5 = 11   e. 6 + 3 = 9
   b. 4 + 6 = 10   f. 5 + 6 = 11
   c. 9 + 10 = 19   g. 2 + 6 = 8
   d. 6 + 6 = 12   h. 6 + 4 = 10

e. Check your work. You'll touch and read each fact.
- Fact A. (Signal.) *6 + 5 = 11.*
- (Repeat for:) B, *4 + 6 = 10;* C, *9 + 10 = 19;* D, *6 + 6 = 12;* E, *6 + 3 = 9;* F, *5 + 6 = 11;* G, *2 + 6 = 8;* H, *6 + 4 = 10.*

EXERCISE 10: INDEPENDENT WORK

a. Find part 5 on worksheet 67. ✔
   (Teacher reference:)

   $\dfrac{6\ \ 5}{} \longrightarrow 11$

You'll write the four facts for the family on the lines below it.

b. Turn to the other side of worksheet 67 and find part 6. ✔
   (Teacher reference:)

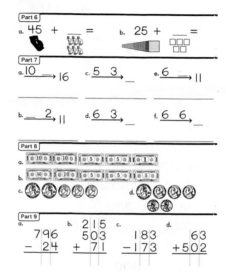

You'll complete the equations for the objects in part 6.
In part 7, you'll write the facts for the missing number in each family and complete the family.
In part 8, you'll write an equals and the number of dollars or the number of cents each group is worth.
You'll work the column problems in part 9.

c. Your turn: Complete worksheet 67. Remember to work part 5 on the other side of worksheet 67. (Observe children and mark incorrect responses on children's worksheets as you give feedback.)

Lesson

EXERCISE 1: NUMBER FAMILIES

a. (Display:)   [68:1A]

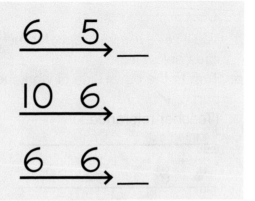

The big number is missing in these families.
All these families have a small number of 6.

- (Point to $\xrightarrow{6\ 5}$.) What are the small numbers in this family? (Touch.) *6 and 5.*
- What's the big number? (Signal.) *11.*

b. (Point to $\xrightarrow{10\ 6}$.) What are the small numbers in this family? (Touch.) *10 and 6.*
- What's the big number? (Signal.) *16.*

c. (Point to $\xrightarrow{6\ 6}$.) What are the small numbers in this family? (Touch.) *6 and 6.*
- What's the big number? (Signal.) *12.*

d. (Point to $\xrightarrow{6\ 5}$.) What are the small numbers in this family again? (Touch.) *6 and 5.*
- What's the big number? (Signal.) *11.*
- Say the fact that starts with the first small number. Get ready. (Signal.) *6 + 5 = 11.*
- Say the fact that starts with the other small number. Get ready. (Signal.) *5 + 6 = 11.*
- Say the fact that goes backward along the arrow. Get ready. (Signal.) *11 – 5 = 6.*
- Say the other minus fact. Get ready. (Signal.) *11 – 6 = 5.*

e. (Point to $\xrightarrow{10\ 6}$.) What are the small numbers in this family? (Touch.) *10 and 6.*
- What's the big number? (Signal.) *16.*
- Say the fact that starts with the first small number. Get ready. (Signal.) *10 + 6 = 16.*
- Say the fact that starts with the other small number. Get ready. (Signal.) *6 + 10 = 16.*
- Say the fact that goes backward along the arrow. Get ready. (Signal.) *16 – 6 = 10.*
- Say the other minus fact. Get ready. (Signal.) *16 – 10 = 6.*

f. (Point to $\xrightarrow{6\ 6}$.) What are the small numbers in this family? (Touch.) *6 and 6.*
- What's the big number? (Signal.) *12.*
There's only one plus fact and one minus fact in this family.
- Say the plus fact for this family. Get ready. (Signal.) *6 + 6 = 12.*
- Say the minus fact for this family. Get ready. (Signal.) *12 – 6 = 6.*
(Repeat steps d through f until firm.)

g. (Display:)   [68:1B]

You're going to say the problem for the missing number in each family.
- (Point to $\xrightarrow{10\ 4}$_.) Say the problem for the missing number. Get ready. (Touch.) *10 plus 4.*
- What's 10 plus 4? (Signal.) *14.*

h. (Point to $\xrightarrow{6}11$.) Say the problem for the missing number. (Touch.) *11 minus 6.*
- What's 11 minus 6? (Signal.) *5.*

i. (Repeat the following tasks for remaining families:)

(Point to __.)	Say the problem for the missing number.	What's __?	
$\xrightarrow{4}10$	10 – 4	10 – 4	6
$\xrightarrow{2}9$	9 – 2	9 – 2	7
$\xrightarrow{6\ 5}$_	6 + 5	6 + 5	11
$\xrightarrow{3}9$	9 – 3	9 – 3	6
$\xrightarrow{6\ 6}$_	6 + 6	6 + 6	12
$\xrightarrow{2}10$	10 – 2	10 – 2	8

(Repeat for families that were not firm.)

EXERCISE 2: DIGITS

a. I'll say numbers that have two digits. You'll tell me the number. Then you'll tell me the ones digit and the tens digit.

• Listen: 89. What number? (Signal.) *89.*
What are the digits of 89? (Signal.) *8 and 9.*
• What's the ones digit of 89? (Signal.) *9.*
• What's the tens digit of 89? (Signal.) *8.*
(Repeat until firm.)

b. New number: 13. What number? (Signal.) *13.*
• What's the ones digit? (Signal.) *3.*
• What's the tens digit? (Signal.) *1.*

c. (Repeat the following tasks for 37, 21, 12, 15, 13:)

New number: ___. What number?	What are the digits of ___?	What's the ones digit?	What's the tens digit?
37	37	3 *(and)* 7	7	3
21	21	2 *(and)* 1	1	2
12	12	1 *(and)* 2	2	1
15	15	1 *(and)* 5	5	1
13	13	1 *(and)* 3	3	1

(Repeat numbers that were not firm.)

EXERCISE 3: MIXED COUNTING

a. Start with 39 and plus tens to 109. Get 39 going. *Thirty-niiine.* Plus tens. (Tap.) *49, 59, 69, 79, 89, 99, 109.*
• Start with 11 and plus tens to 101. Get 11 going. *Elevennn.* Plus tens. (Tap.) *21, 31, 41, 51, 61, 71, 81, 91, 101.*
(Repeat step a until firm.)

b. Start with 58 and count backward by ones to 48. Get 58 going. *Fifty-eieieight.* Count backward. (Tap.) *57, 56, 55, 54, 53, 52, 51, 50, 49, 48.*
• Start with 100 and count backward by tens. Get 100 going. *One huuundred.* Count backward. (Tap.) *90, 80, 70, 60, 50, 40, 30, 20, 10.*
(Repeat step b until firm.)

c. Start with 55 and count by fives to 100. Get 55 going. *Fifty-fiiive.* Count. (Tap.) *60, 65, 70, 75, 80, 85, 90, 95, 100.*

d. You have 55. When you plus ones, what's the next number? (Signal.) *56.*
• You have 55. When you count backward by ones, what's the next number? (Signal.) *54.*
• You have 55. When you plus fives, what's the next number? (Signal.) *60.*
• You have 55. When you plus tens, what's the next number? (Signal.) *65.*

e. You have 60. When you plus ones, what's the next number? (Signal.) *61.*
• You have 60. When you count backward by ones, what's the next number? (Signal.) *59.*
• You have 60. When you plus fives, what's the next number? (Signal.) *65.*
• You have 60. When you plus tens, what's the next number? (Signal.) *70.*

EXERCISE 4: SHAPES

a. You learned a name for a shape that has 6 sides.
• Is a shape with 6 sides a rectangle? (Signal.) *No.*
• How many sides does a rectangle have? (Signal.) *4.*
• Is a shape with 6 sides a triangle? (Signal.) *No.*
• How many sides does a triangle have? (Signal.) *3.*
• Is a shape with 6 sides a hexagon? (Signal.) *Yes.*
• Yes. Hexagons have 6 sides. Say the sentence. (Signal.) *Hexagons have 6 sides.*

b. (Display:)   [68:4A]

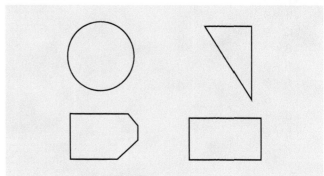

• (Point to ○.) What shape is this? (Touch.) *(A) circle.*

c. (Point to ◺.) What shape is this? (Touch.) *(A) triangle.*
• How many sides does a triangle have? (Signal.) *3.*

d. (Point to ▱.) What shape is this? (Touch.) *(A) hexagon.*
• How many sides does a hexagon have? (Signal.) *6.*
e. (Point to ▭.) What shape is this? (Touch.) *(A) rectangle.*
• How many sides does a rectangle have? (Signal.) *4.*
f. (Display:) [68:4B]

You'll tell me the shapes of these street signs.
• (Point to speed.) What shape? (Touch.) *(A) rectangle.*
• (Point to railroad crossing.) What shape? (Touch.) *(A) circle.*
• (Point to yield.) What shape? (Touch.) *(A) triangle.*
• (Repeat until firm.)
g. Everybody, a hexagon has how many sides? (Signal.) *6.*

EXERCISE 5: FACTS
PLUS/MINUS MIX

a. (Distribute unopened workbooks to students.)
• Open your workbook to Lesson 68 and find part 1.
(Observe children and give feedback.)
(Teacher reference:)

a. 11 − 5
b. 4 + 6
c. 12 − 6
d. 8 + 2
e. 9 − 6
f. 6 + 6
g. 8 − 6
h. 6 − 4
i. 5 + 6
j. 8 − 2

These problems are from families you know. You're going to tell me if the big number or a small number is missing. Then you'll tell me the answer.
• Touch and read problem A. Get ready. (Signal.) *11 minus 5.*

• Is the answer the big number or a small number? (Signal.) *A small number.*
• What's 11 minus 5? (Signal.) *6.*
b. Touch and read problem B. Get ready. (Signal.) *4 plus 6.*
• Is the answer the big number or a small number? (Signal.) *The big number.*
• What's 4 plus 6? (Signal.) *10.*
c. (Repeat the following tasks for problems C through J:)

Touch and read problem __.		Is the answer the big number or a small number?	What's __?	
C	12 − 6	A small number.	12 − 6	6
D	8 + 2	The big number.	8 + 2	10
E	9 − 6	A small number.	9 − 6	3
F	6 + 6	The big number.	6 + 6	12
G	8 − 6	A small number.	8 − 6	2
H	6 − 4	A small number.	6 − 4	2
I	5 + 6	The big number.	5 + 6	11
J	8 − 2	A small number.	8 − 2	6

(Repeat problems that were not firm.)
d. Complete the equations in part 1. Put your pencil down when you're finished.
(Observe children and give feedback.)
(Answer key:)

a. 11 − 5 = 6
b. 4 + 6 = 10
c. 12 − 6 = 6
d. 8 + 2 = 10
e. 9 − 6 = 3
f. 6 + 6 = 12
g. 8 − 6 = 2
h. 6 − 4 = 2
i. 5 + 6 = 11
j. 8 − 2 = 6

e. Check your work. You'll touch and read each fact.
• Fact A. (Signal.) *11 − 5 = 6.*
• (Repeat for:) B, *4 + 6 = 10;* C, *12 − 6 = 6;* D, *8 + 2 = 10;* E, *9 − 6 = 3;* F, *6 + 6 = 12;* G, *8 − 6 = 2;* H, *6 − 4 = 2;* I, *5 + 6 = 11;* J, *8 − 2 = 6.*

EXERCISE 6: COINS EQUAL TO A DOLLAR

a. (Display:) [68:6A]

- (Point to ⊙.) What coin is this? (Touch.)
  (A) quarter.
- What is a quarter worth? (Signal.) *25 cents.*
  You learned how many quarters equal a dollar.
- How many quarters equal a dollar? (Signal.) *4.*
- Say the sentence about 4 quarters. (Signal.)
  4 quarters equal 1 dollar.

b. (Point to ⊙.) What coin is this? (Touch.)
  (A) dime.
- What is a dime worth? (Signal.) *10 cents.*
  You learned how many dimes equal a dollar.
- How many dimes equal a dollar? (Signal.) *10.*
- Say the sentence about 10 dimes. (Signal.)
  10 dimes equal 1 dollar.

c. (Point to ⊙.) What coin is this? (Touch.)
  (A) nickel.
- What is a nickel worth? (Signal.) *5 cents.*
  You learned how many nickels equal a dollar.
- How many nickels equal a dollar? (Signal.) *20.*
- Say the sentence about 20 nickels. (Signal.)
  20 nickels equal 1 dollar.
  (Repeat steps a through c that were not firm.)

d. Find part 2 on worksheet 68. ✔
  (Teacher reference:)

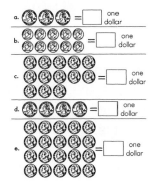

- Touch the coins in group A. ✔
- Touch the words after the box for group A. ✔
- The words are one dollar. What are they?
  (Signal.) *One dollar.*

e. If a group of coins is worth a dollar, circle the words, one dollar.
   What words will you circle if a group is worth one dollar? (Signal.) *One dollar.*

f. If a group is not worth a dollar, write the number of cents in the box.
- Where will you write the number of cents if a group is **not** worth one dollar? (Signal.)
  In the box.
  (Repeat steps e and f until firm.)

g. Count the cents for group A. Raise your hand when you know how much its worth. Do not circle or write anything.
   (Observe children and give feedback.)
- Everybody, how much is group A worth?
  (Signal.) *75 cents.*
- Will you circle the words one dollar?
  (Signal.) *No.*
- Will you write a number? (Signal.) *Yes.*
- What number? (Signal.) *75.*
- Write 75. ✔

h. Count the cents for group B. Circle the words one dollar if group B is worth one dollar. If it's not, write the number of cents in the box.
   (Observe children and give feedback.)
- Everybody, how much is group B worth?
  (Signal.) *One dollar.*

i. Count the cents for group C. Circle one dollar or write the cents.
   (Observe children and give feedback.)
- Everybody, how much is group C worth?
  (Signal.) *65 cents.*

j. Work the problems for groups D and E.
   (Observe children and give feedback.)

k. Check your work.
- Touch group D. ✔
- How much is group D worth? (Signal.)
  One dollar.
- Touch group E. ✔
- How much is group E worth? (Signal.)
  One dollar.

EXERCISE 7: 3 ADDENDS IN COLUMNS

a. Find part 3 on worksheet 68. ✔
(Teacher reference:)

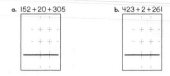

You're going to write column problems that plus 3 numbers. The equals bars and part of the dotted lines for the columns are shown.

b. Touch the symbols for problem A as I read them. (Students touch each symbol.) 152 plus 20 plus 305.
You're going to write a column problem for A.

c. Put your pencil where you'll write the first digit of 152.
(Observe children and give feedback.)
• Put your pencil where you'll write the plus sign. ✔
• Put your pencil in the space where you'll write the first digit of 20.
(Observe children and give feedback.)
• You only need to write one plus sign when you write the problem in a column. Put your pencil in the space where you'll write the first digit of 305.
(Observe children and give feedback.)
(Repeat step c until firm.)

d. Write 152 plus 20 plus 305 in a column. Put your pencil down when you've written the sign and all of the digits in the right places.
(Observe children and give feedback.)
(Teacher reference:)

• Touch and read the column problem you wrote for A. Get ready. (Signal.) *152 plus 20 plus 305.*
You'll work that problem later as part of your independent work.

e. Touch the symbols for B as I read them. (Students touch each symbol.) 423 plus 2 plus 261.
You're going to write problem B in a column.

f. Put your pencil where you'll write the first digit of 423.
(Observe children and give feedback.)
• Put your pencil where you'll write the plus sign. ✔
• Put your pencil in the space where you'll write 2. ✔
• You only need to write one plus sign when you write the problem in a column. Put your pencil where you'll write the first digit of 261. ✔
(Observe children and give feedback.)
(Repeat step f until firm.)

g. Write 423 plus 2 plus 261 in a column. Put your pencil down when you've written the sign and all of the digits in the right places.
(Observe children and give feedback.)
(Teacher reference:)

• Touch and read the column problem you wrote for B. Get ready. (Signal.) *423 plus 2 plus 261.* You'll work that problem later as part of your independent work.

EXERCISE 8: RULER SKILLS   [REMEDY]

a. Turn to the other side of worksheet 68 and find part 4. ✔
(Teacher reference:)   [R] [Part B]

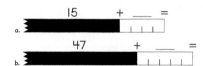

You can't see most of the shaded part. So you can't count all the lines for the shaded part. The number above the ruler tells how many centimeters the shaded part is.
• Touch ruler A. ✔
• How many centimeters is the shaded part? (Signal.) *15.*
• Count the centimeters for the unshaded part and write the number.
(Observe children and give feedback.)
• How many centimeters is the unshaded part? (Signal.) *4.*

b. Now you'll count for both parts.
- Touch the end of the shaded part. ✔
- What number do you get going? (Signal.) *15.*
- Get it going. *Fifteeen.* Touch and count. (Tap.)
  16, 17, 18, 19.
  (Repeat until firm.)
- How many centimeters long is ruler A?
  (Signal.) *19.*
  Yes, ruler A is 19 centimeters long.
- Write 19. ✔
- Everybody, touch and read the fact for ruler A.
  Get ready. (Signal.) *15 + 4 = 19.*

c. Touch ruler B. ✔.
- How many centimeters is the shaded part?
  (Signal.) *47.*
- Count the centimeters for the unshaded part
  and write the number.
  (Observe children and give feedback.)
- How many centimeters is the unshaded part?
  (Signal.) *5.*

d. Now you'll count for both parts.
- Touch the end of the shaded part. ✔
- What number do you get going? (Signal.) *47.*
- Get it going. *Forty-sevennn.* Touch and count.
  (Tap.) *48, 49, 50, 51, 52.*
  (Repeat until firm.)
- How many centimeters long is ruler B?
  (Signal.) *52.*
  Yes, ruler B is 52 centimeters long.
- Write 52. ✔
- Everybody, touch and read the fact for ruler B.
  Get ready. (Signal.) *47 + 5 = 52.*

EXERCISE 9: WORD PROBLEMS (COLUMNS)

a. Find part 5 on worksheet 68. ✔
  (Teacher reference:)

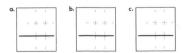

You're going to write the symbols for word
problems in columns and work them. The
equals bars are already shown, but only part
of the dotted lines for the columns and the
rows are shown.

b. Touch where you'll write the symbols for
  problem A. ✔
  Listen to problem A: There were 179 bricks
  in a pile. Then the workers took 150 bricks
  from that pile. How many bricks were still in
  the pile?
- Listen again and write the symbols for both
  parts: There were 179 bricks in a pile. Then
  the workers took 150 bricks from that pile.
- Everybody, touch and read problem A. Get
  ready. (Signal.) *179 minus 150.*
  (Repeat until firm.)

c. Work problem A. Put your pencil down when
  you know how many bricks were still in the pile.
  (Observe children and give feedback.)
- Read the problem and the answer you wrote
  for A. Get ready. (Signal.) *179 – 150 = 29.*
- How many bricks were still in the pile?
  (Signal.) *29.*

d. Listen to problem B: Fran rode her bike
  56 kilometers last week. Fran rode her
  bike 123 kilometers this week. How many
  kilometers did Fran ride altogether?
- Listen again and write the symbols for the both
  parts: Fran rode her bike 56 kilometers last
  week. Fran rode her bike 123 kilometers
  this week.
- Everybody, touch and read problem B. Get
  ready. (Signal.) *56 plus 123.*

e. Work problem B. Put your pencil down when
  you know how many kilometers Fran rode
  altogether.
  (Observe children and give feedback.)
- Read the problem and the answer you wrote
  for B. Get ready. (Signal.) *56 + 123 = 179.*
- How many kilometers did Fran ride
  altogether? (Signal.) *179.*

f. Touch where you'll write the symbols for
  problem C. ✔
  Listen to problem C: A company built
  496 doors. Then the company sold 76 of those
  doors. How many doors does the company
  still have?
- Listen again and write the symbols for both
  parts: A company built 496 doors. Then the
  company sold 76 of those doors.
- Everybody, touch and read problem C. Get
  ready. (Signal.) *496 minus 76.*

g. Work problem C. Put your pencil down when you know how many doors the company still has.
(Observe children and give feedback.)
(Answer key:)

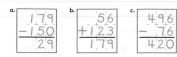

- Read the problem and the answer you wrote for C. Get ready. (Signal.) *496 – 76 = 420.*
- How many doors does the company still have? (Signal.) *420.*

EXERCISE 10: INDEPENDENT WORK

a. Find part 6 on worksheet 68. ✔
(Teacher reference:)

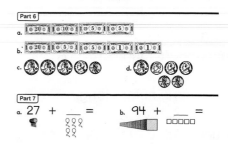

You'll write an equals and the number of dollars or cents each group is worth.
You'll complete the equations for the objects in part 7.

b. Complete worksheet 68. Remember to go back to part 3 and work both problems.
(Observe children and mark incorrect responses on children's worksheets as you give feedback.)

Lesson

EXERCISE 1: FACTS
PLUS/MINUS MIX

a. (Display:)   [69:1A]

6 − 2	10 − 2	8 − 1
10 − 4	6 + 2	3 + 6
6 + 10	11 − 6	12 − 6

These problems are from families you know.
- (Point to **6 − 2.**) Read this problem. Get ready. (Touch.) *6 minus 2.*
- Is the answer the big number or a small number? (Signal.) *A small number.*
- What's 6 minus 2? (Signal.) *4.*
- Say the fact. (Touch.) *6 − 2 = 4.*

b. (Point to **10 − 4.**) Read this problem. Get ready. (Touch.) *10 minus 4.*
- Is the answer the big number or a small number? (Signal.) *A small number.*
- What's 10 minus 4? (Signal.) *6.*
- Say the fact. (Touch.) *10 − 4 = 6.*

c. (Repeat the following tasks for remaining problems:)

(Point to __.) Read this problem.	Is the big number or a small number missing?	What's __?		Say the fact.
6 + 10	The big number.	6 + 10	16	6 + 10 = 16
10 − 2	A small number.	10 − 2	8	10 − 2 = 8
6 + 2	The big number.	6 + 2	8	6 + 2 = 8
11 − 6	A small number.	11 − 6	5	11 − 6 = 5
8 − 1	A small number.	8 − 1	7	8 − 1 = 7
3 + 6	The big number.	3 + 6	9	3 + 6 = 9
12 − 6	A small number.	12 − 6	6	12 − 6 = 6

(Repeat problems that were not firm.)

EXERCISE 2: SHAPES

a. I'll think of a shape. I'll tell you the number of sides the shape has. You'll tell me the name of the shape.
- Listen: The shape that I'm thinking of has 6 sides. What shape is this? (Signal.) *(A) hexagon.*
- New shape: This shape has 3 sides. What shape is this? (Signal.) *(A) triangle.*
- New shape: This shape has 4 sides. What shape is this? (Signal.) *(A) rectangle.*
- New shape: This shape is perfectly round and doesn't have any sides. What shape? (Signal.) *(A) circle.*

(Repeat tasks that were not firm.)

EXERCISE 3: DIGITS

a. I'll say numbers that have two digits. You'll tell me the number. Then you'll tell me the ones digit and the tens digit.
- Listen: 18. What number? (Signal.) *18.*
- What's the ones digit? (Signal.) *8.*
- What's the tens digit? (Signal.) *1.*
(Repeat until firm.)

b. New number: 12. What number? (Signal.) *12.*
- What's the ones digit? (Signal.) *2.*
- What's the tens digit? (Signal.) *1.*
(Repeat step b until firm.)

c. (Repeat the following tasks for 41, 75, 10, 80, 13:)
- New number:__. What number?
- What's the ones digit?
- What's the tens digit?
(Repeat numbers that were not firm.)

EXERCISE 4: 3-D OBJECTS
INTRODUCTION OF CUBE

a. (Display:) [69:4A]

- (Point to 🧊.) This is an ice cube. Say **cube**. (Touch.) *Cube.*
- Say **ice cube.** (Signal.) *Ice cube.* (Repeat until firm.)
  An ice cube is a cube that is made of ice.

b. (Point to F.) This is a cube, but it is not made of ice.
- Listen: The shape of each face of a cube is a square.
- What is the shape of each face of a cube? (Signal.) *(A) square.*

c. (Point to each letter.) There is a letter on each face we can see. The letters are T (touch), L (touch), and F (touch).
- Look at the letter on the top face. ✔
- What letter is on the top face? (Signal.) *T.*
- The way we see the top face, it doesn't look like a square, but what shape is the top face? (Signal.) *(A) square.*

d. (Point to **L.**) There is a letter on this face. What letter is on this face? (Touch.) *L.*
- The way we see this face, it doesn't look like a square, but what shape is it? (Touch.) *(A) square.*

e. (Point to **F.**) There is a letter on this face. What letter is on this face? (Touch.) *F.*
- This is the front face. What face? (Touch.) *The front face.*
- What letter is on the front face? (Touch.) *F.*
- What shape is the front face? (Touch.) *(A) square.*

f. Listen: A cube has 6 faces. Say that sentence. (Signal.) *A cube has 6 faces.*
- Each face is a square. Say that sentence. (Signal.) *Each face is a square.*

g. (Point to ▯.) This is not a cube.
- (Point to ▮.) Is this face a square? (Touch.) *No.*
- So is this object a cube? (Touch.) *No.*

h. (Point to ▭.) Is the front face a square? (Touch.) *Yes.*
- (Point to ▱.) Does this face look like it's a square? (Touch.) *Yes.*
- (Point to ▱.) Does the top face look like it's a square? (Touch.) *Yes.*
- All the faces are square, so is this object a cube? (Touch.) *Yes.*

i. Raise your hand if you remember how many faces a cube has. ✔
- Everybody, how many faces does a cube have? (Signal.) *6.*

EXERCISE 5: FACTS
ADDITION REMEDY

a. (Distribute unopened workbooks to students.)
- Open your workbook to Lesson 69 and find part 1.
  (Observe children and give feedback.)
  (Teacher reference:) R Test 8: Part G

  a. 6 + 4
  b. 2 + 9
  c. 3 + 6
  d. 5 + 2
  e. 6 + 6
  f. 6 + 5
  g. 8 + 2
  h. 3 + 5
  i. 2 + 6
  j. 9 + 1

  These are plus problems from families you know. You're going to read each problem and tell me the answer. Then you'll go back and work all of the problems.
- Touch and read problem A. Get ready. (Signal.) *6 plus 4.*
- What's 6 plus 4? (Signal.) *10.*

b. Touch and read problem B. (Signal.) *2 plus 9.*
- What's 2 plus 9? (Signal.) *11.*

c. (Repeat the following tasks for problems C through J:)
- Touch and read problem __.
- What's __?

d. Complete the equations for all of the problems in part 1.
(Observe children and give feedback.)
(Answer key:)

a. $6+4=10$
b. $2+9=11$
c. $3+6=9$
d. $5+2=7$
e. $6+6=12$
f. $6+5=11$
g. $8+2=10$
h. $3+5=8$
i. $2+6=8$
j. $9+1=10$

e. Check your work. You'll touch and read each fact.
- Fact A. (Signal.) $6 + 4 = 10$.
- (Repeat for:) B, $2 + 9 = 11$; C, $3 + 6 = 9$; D, $5 + 2 = 7$; E, $6 + 6 = 12$; F, $6 + 5 = 11$; G, $8 + 2 = 10$; H, $3 + 5 = 8$; I, $2 + 6 = 8$; J, $9 + 1 = 10$.

EXERCISE 6: COINS EQUAL TO A DOLLAR   [REMEDY]

a. Find part 2 on worksheet 69. ✔
(Teacher reference:)   R Test 8: Part I

- Touch the coins in group A. ✔
- Touch the words after the box for group A. ✔
- The words are one dollar. What are they? (Signal.) One dollar.

b. If a group of coins is worth one dollar, circle the words one dollar.
- What words will you circle if a group is worth one dollar? (Signal.) One dollar.

c. If a group of coins is not worth one dollar, write the number of cents in the box.
- Where will you write the number of cents if a group is **not** worth one dollar? (Signal.) In the box.

- You'll count for the cents in group A. First you'll touch and count for the nickels. Then you'll touch and count for the dimes.
- Fingers ready. ✔
- Touch and count for the nickels. (Tap 2.) 5, tennn. Count for the dimes. (Tap 9.) 20, 30, 40, 50, 60, 70, 80, 90, 100. (Repeat until firm.)
- Everybody, how many cents is group A worth? (Signal.) 100 cents.
- Will you circle the words one dollar? (Signal.) Yes.
- Circle the words one dollar. ✔

d. You'll count for the cents in group B. First you'll touch and count for the quarters. Then you'll touch and count for the nickels. Then you'll touch and count for the dimes.
- Fingers ready. ✔
- Touch and count for the quarters. (Tap 2.) 25, fiftyyy. Count for the nickels. (Tap 2.) 55, sixtyyy. Count for the dimes. (Tap 2.) 70, 80. (Repeat until firm.)
- Everybody, how many cents is group B worth? (Signal.) 80 cents.
- Will you circle the words one dollar? (Signal.) No.
- Write 80 in the box. ✔

e. You'll count for the cents in group C. First you'll touch and count for the quarters. Then you'll touch and count for the nickel. Then touch and count for the dimes.
- Fingers ready. ✔
- Touch and count for the quarters. (Tap 3.) 25, 50, seventy-fiiive. Count for the nickel. (Tap.) eightyyy. Count for the dimes. (Tap 2.) 90, 100. (Repeat until firm.)
- Everybody, how many cents is group C worth? (Signal.) 100 cents.
- Will you circle the words one dollar? (Signal.) Yes.
- Circle the words one dollar. ✔

f. Count the cents in group D. Circle the words one dollar if group D is worth one dollar. If it's not, write the cents in the box.
(Observe children and give feedback.)
- Everybody, how many cents is group D worth? (Signal.) 100 cents.
- Did you circle the words one dollar? (Signal.) Yes.

g. In group E there are quarters, a nickel, and a dime. Count the cents in group E. Circle the words one dollar if group E is worth one dollar. If it's not, write the cents in the box. (Observe children and give feedback.)
• Everybody, how much is group E worth? (Signal.) *90 cents.*
• Did you circle the words one dollar? (Signal.) *No.*
• Write 90 in the box. ✔

EXERCISE 7: RULER SKILLS

REMEDY

a. Find part 3 on worksheet 69. ✔ (Teacher reference:)

R Part C

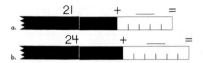

You can't see most of the shaded part. So you can't count all the lines for the shaded part. The number above the ruler tells how many centimeters the shaded part is.
• Touch ruler A. ✔
• How many centimeters is the shaded part? (Signal.) *21.*
• Count the centimeters for the unshaded part and write the number. (Observe children and give feedback.)
• How many centimeters is the unshaded part? (Signal.) *5.*
b. Now you'll count for both parts.
⌐• Touch the end of the shaded part. ✔
 • What number do you get going? (Signal.) *21.*
 • Get it going. *Twenty-wuuun.* Touch and count. (Tap.) *22, 23, 24, 25, 26.*
 └ (Repeat until firm.)
• How many centimeters long is ruler A? (Signal.) *26.*
   Yes, ruler A is 26 centimeters long.
• Write 26. ✔
• Everybody, touch and read the fact for ruler A. Get ready. (Signal.) *21+ 5 = 26.*
c. Touch ruler B. ✔
• How many centimeters is the shaded part? (Signal.) *24.*
• Count the centimeters for the unshaded part and write the number. (Observe children and give feedback.)
• How many centimeters is the unshaded part? (Signal.) *6.*

d. Now you'll count for both parts.
⌐• Touch the end of the shaded part. ✔
 • What number do you get going? (Signal.) *24.*
 • Get it going. *Twenty-fouuur.* Touch and count. (Tap.) *25, 26, 27, 28, 29, 30.*
 └ (Repeat until firm.)
• How many centimeters long is ruler B? (Signal.) *30.*
   Yes, ruler B is 30 centimeters long.
• Write 30. ✔
• Everybody, touch and read the fact for ruler B. Get ready. (Signal.) *24 + 6 = 30.*

EXERCISE 8: FACTS
SUBTRACTION

a. Find part 4 on worksheet 69. ✔ (Teacher reference:)

a. $8-2$ d. $10-4$ g. $8-3$
b. $11-6$ e. $9-8$ h. $12-6$
c. $19-9$ f. $9-6$ i. $10-2$

These are minus problems from families you know. You're going to read each problem and tell me the answer. Then you'll go back and work them.
• Touch and read problem A. Get ready. (Signal.) *8 minus 2.*
• What's the answer? (Signal.) *6.*
b. (Repeat the following tasks for problems B through I:)
• Touch and read problem __.
• What's the answer?
c. Complete the equations for all of the problems in part 4. (Observe children and give feedback.) (Answer key:)

a. $8-2=6$ d. $10-4=6$ g. $8-3=5$
b. $11-6=5$ e. $9-8=1$ h. $12-6=6$
c. $19-9=10$ f. $9-6=3$ i. $10-2=8$

d. Check your work. You'll touch and read each fact.
⌐• Fact A. (Signal.) *8 – 2 = 6.*
└• (Repeat for:) B, *11 – 6 = 5;* C, *19 – 9 = 10;* D, *10 – 4 = 6;* E, *9 – 8 = 1;* F, *9 – 6 = 3;* G, *8 – 3 = 5;* H, *12 – 6 = 6;* I, *10 – 2 = 8.*

EXERCISE 9: 3 ADDENDS IN COLUMNS

a. Turn to the other side of worksheet 69 and find
part 5. ✔
(Teacher reference:)

You're going to write column problems that
plus three numbers. The equals bars and part
of the dotted lines for the columns are shown.
- Touch the symbols for problem A as I read
them. (Students touch each symbol.) 50 plus
436 plus 13.
You're going to write a column problem for A.
b. Put your pencil in the space where you'll write
the first digit of 50.
(Observe children and give feedback.)
- Put your pencil where you'll write the plus sign. ✔
- Put your pencil in the space where you'll write
the first digit of 436.
(Observe children and give feedback.)
- You only need to write one plus sign when
you write the problem in a column. Put your
pencil in the space where you'll write the first
digit of 13.
(Observe children and give feedback.)
(Repeat step b until firm.)
c. Write 50 plus 436 plus 13 in a column. Put
your pencil down when you've written the sign
and all of the digits in the right places.
(Observe children and give feedback.)
(Teacher reference:)

- Touch and read the column problem you wrote
for A. Get ready. (Signal.) 50 plus 436 plus 13.
You'll work that problem later as part of your
independent work.
d. Touch the symbols for problem B as I read
them. (Students touch each symbol.) 206 plus
91 plus 702.

e. Write 206 plus 91 plus 702 in a column.
Remember, you only have to write one plus
sign. Put your pencil down when you've
written the sign and all of the digits in the
right places.
(Observe children and give feedback.)
(Teacher reference:)

- Touch and read the column problem you
wrote for B. Get ready. (Signal.) 206 plus 91
plus 702.
You'll work that problem later as part of your
independent work.

EXERCISE 10: INDEPENDENT WORK

a. Find part 6 on worksheet 69. ✔
(Teacher reference:)

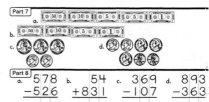

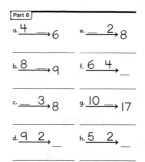

You'll write the fact for finding the missing
number in each family and complete the
family.
In part 7, you'll write an equals and the number
of dollars or the number of cents each group
is worth.
You'll work the column problems in part 8.
b. Complete worksheet 69. Remember to work
both problems in part 5.
(Observe children and mark incorrect responses
on children's worksheets as you give feedback.)

Lesson 70

EXERCISE 1: MONEY
WRITING DOLLAR AMOUNTS [REMEDY]

a. Listen: 5 dollars and 25 cents. Say that.
(Signal.) *5 dollars and 25 cents.*
• Listen: 6 dollars and 2 cents. Say that.
(Signal.) *6 dollars and 2 cents.*
• Listen: 3 dollars and 17 cents. Say that.
(Signal.) *3 dollars and 17 cents.*
• Listen: 4 dollars and 99 cents. Say that.
(Signal.) *4 dollars and 99 cents.*
• Listen: 12 dollars and 34 cents. Say that.
(Signal.) *12 dollars and 34 cents.*

b. Say, write 12 dollars. (Signal.) *Write 12 dollars.*
(Repeat until firm.)
(Display:) W   [70:1A]

$12.

Here's how we write 12 dollars.
• (Point to **$**.) This is the sign for dollars.
• (Point to **12**.) This is the number of dollars.
• (Point to • .) This is a dot after 12.

c. Say 12 dollars and 34 cents. (Signal.)
12 dollars and 34 cents.
• Tell me to write 12 dollars and 34 cents.
(Signal.) *Write 12 dollars and 34 cents.*
• First I write the dollar sign. What do I write
first? (Signal.) *The dollar sign.*
(Display:) W   [70:1B]

$

Then I write the number 12.
(Add to show:)   [70:1C]

$12

• After 12, I write the dot. What do I write after 12?
(Signal.) *The dot.*
Yes, the dot.
(Add to show:)   [70:1D]

$12.

• Now I write the number for 34 cents. What
number? (Signal.) *34.*
(Add to show:)   [70:1E]

$12.34

• (Point to **$12.34.**) Read this. (Touch.) *12 dollars
and 34 cents.*
(Change to show:)   [70:1F]

$12.51

• (Point to **$12.51.**) This isn't 12 dollars and
34 cents anymore. What is it? (Signal.)
12 dollars and 51 cents.

d. (Change to show:)   [70:1G]

$3.51

• (Point to **$3.51.**) This isn't 12 dollars and
51 cents anymore. What is it? (Signal.)
3 dollars and 51 cents.

e. (Change to show:)   [70:1H]

$3.17

• (Point to **$3.17.**) Read this. (Touch.) *3 dollars
and 17 cents.*
(Repeat steps c through e until firm.)

f. (Display:)   [70:1I]

$11.89 $1.17

$6.60 $20.99

Here are some dollar amounts.
• (Point to **$11.89.**) Read this. Get ready. (Touch.)
11 dollars and 89 cents.
• (Point to **$1.17.**) Read this. Get ready. (Touch.)
1 dollar and 17 cents.
• (Point to **$6.60.**) Read this. Get ready. (Touch.)
6 dollars and 60 cents.
• (Point to **$20.99.**) Read this. Get ready.
(Touch.) *20 dollars and 99 cents.*
(Repeat amounts that were not firm.)

EXERCISE 2: FACTS
PLUS/MINUS MIX

a. (Display:) [70:2A]

4 + 10	12 – 10	6 – 4
11 – 9	12 – 6	3 + 6
11 – 6	4 + 6	10 – 6

These problems are from number families you know.

- (Point to **4 + 10**.) Read this problem. Get ready. (Touch.) *4 plus 10.*
- Is the answer the big number or a small number? (Signal.) *The big number.*
- What's 4 plus 10? (Signal.) *14.*
- Say the fact. (Touch.) *4 + 10 =14.*

b. (Point to **11 – 9**.) Read this problem. Get ready. (Touch.) *11 minus 9.*
- Is the answer the big number or a small number? (Signal.) *A small number.*
- What's 11 minus 9? (Signal.) *2.*
- Say the fact. (Touch.) *11– 9 = 2.*

c. (Repeat the following tasks for remaining problems:)

(Point to __.) Read this problem.	Is the answer the big number or a small number?	What's __?		Say the fact.
11 – 6	A small number.	11 – 6	5	11 – 6 = 5
12 – 10	A small number.	12 – 10	2	12 – 10 = 2
12 – 6	A small number.	12 – 6	6	12 – 6 = 6
4 + 6	The big number.	4 + 6	10	4 + 6 = 10
6 – 4	A small number.	6 – 4	2	6 – 4 = 2
3 + 6	The big number.	3 + 6	9	3 + 6 = 9
10 – 6	A small number.	10 – 6	4	10 – 6 = 4

(Repeat problems that were not firm.)

EXERCISE 3: 3-D OBJECTS

a. (Display:) [70:3A]

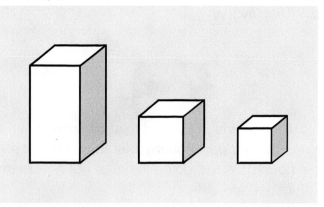

- (Point to objects.) Some objects in this picture are cubes.
- (Point to ▯.) Are all of the faces of this object shaped like a square? (Touch.) *No.*
- So is this a cube? (Touch.) *No.*

b. (Point to ▢.) Are all of the faces of this object shaped like a square? (Touch.) *Yes.*
- So is this a cube? (Touch.) *Yes.*

c. (Point to ▢.) Are all of the faces of this object shaped like a square? (Touch.) *Yes.*
- So is this a cube? (Touch.) *Yes.*

d. Raise your hand when you know how many objects in this picture are cubes. ✔
- How many objects are cubes? (Signal.) *2.*

e. Think about the number of faces a cube has. ✔
- How many faces does a cube have? (Signal.) *6.*
(Stop showing display.)

f. Now I'm going to ask you about shapes.
- A triangle has how many sides? Get ready. (Signal.) *3.*
- A rectangle has how many sides? Get ready. (Signal.) *4.*
- A hexagon has how many sides? Get ready. (Signal.) *6.*
- Tell me the name for a rectangle with sides that are the same size. Get ready. (Signal.) *(A) square.*

EXERCISE 4: RULERS

COUNT BACKWARD

REMEDY

a. (Display:) W   [70:4A]

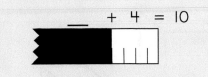

This ruler is supposed to show inches. It has a shaded part **(touch)** and an unshaded part **(touch)**. You're going to figure out how many inches are shaded.

- (Point to **4.**) How many inches is the unshaded part? (Touch.) *4.*
- (Point to **10.**) How many inches long is the ruler? (Touch.) *10.*

b. We can figure out how many inches are shaded by counting backward from the end of the ruler to the shaded part.
- (Point to the end of the ruler.) My turn: (touch the end of the ruler) *Tennn,* (touch lines) *9, 8, 7, 6.*

c. (Point to the end of the ruler.) Your turn to get ten going and count backward to the shaded part.
- (Touch the end of the ruler.) Get it going. *Tennn.* Count backward. (Touch lines.) *9, 8, 7, 6.* (Repeat until firm.)
- Everybody, how many inches is the shaded part? (Tap.) *6.*
(Add to show:)   [70:4B]

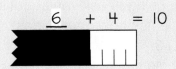

- (Point to **6.**) Read the equation. Get ready. (Touch.) *6 + 4 = 10.*
- (Point to **6.**) How many inches is the shaded part? (Touch.) *6.*
- (Point to **4.**) How many inches is the unshaded part? (Touch.) *4.*
- (Point to **10.**) How many inches is the whole ruler? (Touch.) *10.*

d. (Display:)   [70:4C]

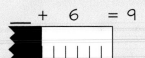

- (Point to ruler.) You're going to figure out how many inches are shaded for this ruler.
- (Point to **9.**) How many inches is this whole ruler? (Touch.) *9.*
- (Point to **6.**) How many inches is the unshaded part? (Touch.) *6.*

e. (Point to the end of the ruler.) I'll touch the end of the ruler. You'll get it going and count backward from the shaded part.
- (Touch the end of the ruler.) Get it going. *Niiine.* (Touch lines.) *8, 7, 6, 5, 4, 3.* (Repeat until firm.)
- Everybody, how many inches is the shaded part? (Tap.) *3.*
(Add to show:)   [70:4D]

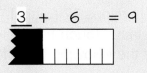

- (Point to **3.**) Read the equation. Get ready. (Touch.) *3 + 6 = 9.*
- How many inches are in the shaded part? (Touch.) *3.*
- How many inches are in the unshaded part? (Touch.) *6.*
- How many inches is the whole ruler. (Touch.) *9.*

f. (Display:)   [70:4E]

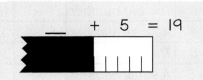

- (Point to ruler.) You're going to figure out how many inches are shaded for this ruler.
- (Point to **19.**) How many inches is this whole ruler? (Touch.) *19.*
- (Point to **5.**) How many inches is the unshaded part? (Touch.) *5.*
- (Point to the end of the ruler.) I'll touch the end of the ruler. You'll get it going and count backward from the shaded part.

- (Touch the end of the ruler.) Get it going. *Nineteeen.* (Touch lines.) *18, 17, 16, 15, 14.* (Repeat until firm.)
- Everybody, how many inches is the shaded part? (Tap.) *14.* (Add to show:) [70:4F]

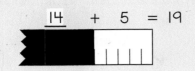

- (Point to **14**.) Read the equation. Get ready. (Touch.) *14 + 5 = 19.*
- How many inches are in the shaded part? (Touch.) *14.*
- How many inches are in the unshaded part? (Touch.) *5.*
- How many inches is the whole ruler? (Touch.) *19.*

EXERCISE 5: FACTS
SUBTRACTION

a. (Distribute unopened workbooks to students.)
- Open your workbook to Lesson 70 and find part 1.
  (Observe children and give feedback.)
  (Teacher reference:)

a. 9 − 3	d. 11 − 2	h. 8 − 6
b. 10 − 6	e. 12 − 6	i. 12 − 10
c. 7 − 5	f. 9 − 2	j. 6 − 1
	g. 11 − 5	

These are minus problems from families you know. You're going to read each problem and say the answer. Then you'll go back and work them.
- Touch and read problem A. Get ready. (Signal.) *9 minus 3.*
- What's the answer? (Signal.) *6.*
b. (Repeat the following tasks for problems B through J:)
- Touch and read problem __.
- What's the answer?
c. Complete the equations for all of the problems in part 1.
  (Observe children and give feedback.)

d. Check your work. You'll touch and read each fact.
- Fact A. (Signal.) *9 − 3 = 6.*
- (Repeat for:) B, *10 − 6 = 4*; C, *7 − 5 = 2*; D, *11 − 2 = 9*; E, *12 − 6 = 6*; F, *9 − 2 = 7*; G, *11 − 5 = 6*; H, *8 − 6 = 2*; I, *12 − 10 = 2*; J, *6 − 1 = 5.*

EXERCISE 6: RULER SKILLS

a. Find part 2 on worksheet 70. ✔
  (Teacher reference:)

a. 42 + __ =   b. 68 + __ =

The rulers in part 2 are supposed to show centimeters. They have a shaded part and an unshaded part. You can't see the whole shaded part, but the number above the shaded part tells how many centimeters long the shaded part is.
- Touch the number for the shaded part for ruler A. ✔
- How many centimeters is the shaded part for ruler A? (Signal.) *42 centimeters.*
- Count the centimeters for the unshaded part. Write the number and put your pencil down when you're finished.
  (Observe children and give feedback.)
- How many centimeters is the unshaded part? (Signal.) *4.*
b. Now you'll count for both parts of ruler A.
- Touch the end of the shaded part. ✔
- What number do you get going? (Signal.) *42.*
- Get it going. *Forty-twooo.* Touch and count. (Tap.) *43, 44, 45, 46.* (Repeat until firm.)
- How many centimeters long is the whole ruler? (Signal.) *46.*
- Write 46. ✔
- Touch and read the equation for ruler A. Get ready. (Signal.) *42 + 4 = 46.*
c. Touch the number for the shaded part for ruler B. ✔
- How many centimeters is the shaded part for ruler B? (Signal.) *68 centimeters.*
- Count the centimeters for the unshaded part. Write the number and put your pencil down when you're finished.
  (Observe children and give feedback.)
- How many centimeters is the unshaded part? (Signal.) *5.*

d. Now you'll count for both parts of ruler B.
- Touch the end of the shaded part. ✔
- What number do you get going? (Signal.) *68.*
- Get it going. *Sixty-eieieight.* Touch and count. (Tap.) *69, 70, 71, 72, 73.*
  (Repeat until firm.)
- How many centimeters long is the whole ruler? (Signal.) *73.*
- Write 73. ✔
- Touch and read the equation for ruler B. Get ready. (Signal.) *68 + 5 = 73.*

EXERCISE 7: 3 ADDENDS IN COLUMNS

a. Find part 3 on worksheet 70. ✔
   (Teacher reference:)

You're going to write column problems that plus three numbers. The problems you'll write are not written in rows. I'm going to tell you each of the problems you'll write.
- Touch where you'll write column problem A. ✔
  Listen to problem A: 23 plus 412 plus 51.
- Listen again: 23 plus 412 plus 51. Say problem A. (Signal.) *23 plus 412 plus 51.*
  (Repeat until firm.)
b. The first number is 23. Put your pencil where you'll write the beginning digit of 23. ✔
- Write 23. ✔
c. Put your pencil where you'll write the plus sign. ✔
- Write plus. ✔
d. The next number is 412. Put your pencil where you'll write the beginning digit of 412. ✔
- Write 412. ✔
e. The next number is 51. Put your pencil where you'll write the beginning digit of 51. ✔
- Write 51. ✔
f. Touch and read problem A. Get ready. (Signal.) *23 plus 412 plus 51.*
  Later, you'll work this problem as part of your independent work.

g. Touch where you'll write column problem B. ✔
   Listen to problem B: 16 plus 500 plus 33.
- Listen again: 16 plus 500 plus 33. Say problem B. (Signal.) *16 plus 500 plus 33.*
  (Repeat until firm.)
h. The first number is 16. Put your pencil where you'll write the beginning digit of 16. ✔
- Write 16. ✔
i. Put your pencil where you'll write the plus sign. ✔
- Write plus. ✔
j. The next number is 500. Put your pencil where you'll write the beginning digit of 500. ✔
- Write 500. ✔
k. The next number is 33. Put your pencil where you'll write the beginning digit of 33. ✔
- Write 33. ✔
l. Touch and read problem B. Get ready. (Signal.) *16 plus 500 plus 33.*
  Later, you'll work this problem as part of your independent work.

EXERCISE 8: FACTS
ADDITION

a. Find part 4 on worksheet 70. ✔
   (Teacher reference:)

a. $2 + 3$
b. $6 + 6$
c. $3 + 5$
d. $8 + 1$
e. $2 + 6$
f. $4 + 6$
g. $2 + 7$
h. $6 + 5$
i. $4 + 2$
j. $5 + 10$

These are plus problems from families you know. You're going to read each problem and tell me the answer. Then you'll go back and work all of the problems.
- Touch and read problem A. Get ready. (Signal.) *2 plus 3.*
- What's 2 plus 3? (Signal.) *5.*
b. Touch and read problem B. (Signal.) *6 plus 6.*
- What's 6 plus 6? (Signal.) *12.*
c. (Repeat the following tasks for problems C through J:)
- Touch and read problem __.
- What's __?

d. Complete the equations for all of the problems in part 4.
(Observe children and give feedback.)
e. Check your work. You'll touch and read each fact.
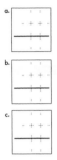
• Fact A. (Signal.) *2 + 3 = 5.*
• (Repeat for:) B, *6 + 6 = 12;* C, *3 + 5 = 8;* D, *8 + 1 = 9;* E, *2 + 6 = 8;* F, *4 + 6 = 10;* G, *2 + 7 = 9;* H, *6 + 5 = 11;* I, *4 + 2 = 6;* J, *5 + 10 = 15.*

EXERCISE 9: WORD PROBLEMS (COLUMNS)

a. Find part 5 on worksheet 70. ✔
(Teacher reference:)

You're going to write the symbols for word problems in columns and work them. The equals bars are already shown, but only part of the dotted lines for the columns and the rows are shown.
• Touch where you'll write the symbols for problem A. ✔
Listen to problem A: 17 ducks were on a lake. Then 352 more ducks came to the lake. How many ducks ended up on the lake?
• Listen again and write the symbols for both parts: 17 ducks were on a lake. Then 352 more ducks came to the lake. ✔
• Touch and read problem A. Get ready. (Signal.) *17 plus 352.*
b. Work problem A. Put your pencil down when you know how many ducks ended up on the lake.
(Observe children and give feedback.)
• Read the problem and the answer you wrote for A. Get ready. (Signal.) *17 + 352 = 369.*
• So how many ducks ended up on the lake? (Signal.) *369.*

c. Touch where you'll write the symbols for problem B. ✔
Listen to problem B: There were 853 people in a boat. Then 552 of those people jumped out of the boat. How many people were still in the boat?
• Listen again and write the symbols for both parts: There were 853 people in a boat. Then 552 of those people jumped out of the boat. ✔
• Touch and read problem B. Get ready. (Signal.) *853 minus 552.*
d. Work problem B. Put your pencil down when you know how many people were still in the boat.
(Observe children and give feedback.)
• Read the problem and the answer for B. Get ready. (Signal.) *853 – 552 = 301.*
• So how many were still in the boat? (Signal.) *301.*
e. Touch where you'll write the symbols for problem C. ✔
Listen to problem C: A bike store had 88 bikes. Then the store sold 81 of those bikes. How many bikes did the store still have?
• Listen again and write the symbols for both parts: A bike store had 88 bikes. Then the store sold 81 of those bikes. ✔
• Touch and read problem C. Get ready. (Signal.) *88 minus 81.*
f. Work problem C. Put your pencil down when you know how many bikes the store still had.
(Observe children and give feedback.)
(Answer key:)

• Read the problem and the answer you wrote for C. Get ready. (Signal.) *88 – 81 = 7.*
• So how many bikes did the store still have? (Signal.) *7.*

EXERCISE 10: INDEPENDENT WORK

a. Turn to the other side of worksheet 70 and find part 6. ✔

(Teacher reference:)

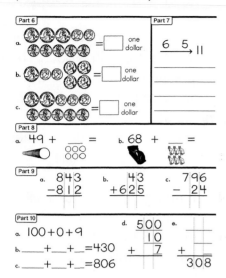

b. In part 7, you'll write four facts for the number family.
In part 8, you'll complete the equations for the objects.
In part 9, you'll work the column problems.
In part 10, you'll complete the place-value addition equations.

c. Complete worksheet 70. Remember to work the problems in part 3 on the other side of the worksheet.
(Observe children and mark incorrect responses on children's worksheets as you give feedback.)

You'll work these problems as part of your independent work. Remember, if the group is worth a dollar, circle the words one dollar. If the group is not worth a dollar, write the number of cents in the box.

Mastery Test 7

> *Note:* Mastery Tests are administered to all students in the group. Each student will need a pencil and a *Student Assessment Book.* Try to arrange students so they cannot look at other students' responses.

Teacher Presentation

a. Find Test 7 in your test booklet. ✔
- Touch part 1. ✔
(Teacher reference:)

a. 10 6 ,___
b. ___ 5 , 11
c. 6 6 , ___
d. ___ 2 , 6
e. 6 5 , ___
f. 6 ___ → 7
g. 6 2 , ___
h. ___ 6 , 12
i. 6 ___ → 8
j. 6 ___ → 11

6 is a number in the families for all of these problems. You'll write the missing number in each family.

b. Touch part 2. ✔
(Teacher reference:)

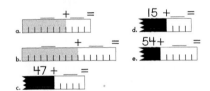

a. ___ + ___ =
b. ___ + ___ =
c. 47 + ___ =
d. 15 + ___ =
e. 54 + ___ =

You're going to complete the equation for each ruler in part 2. For rulers A and B, you'll write the number for the shaded parts and the unshaded parts. For rulers C, D, and E, you'll write the number for the unshaded parts. Then you'll count on and complete the equation for each ruler.

c. Touch part 3. ✔
(Teacher reference:)

a.
b.
c.
d.
e.

You'll count the dollars for each group of bills. Then you'll write an equals and the number to show how many dollars each group is worth.

d. Turn to the other side of Test 7 and touch part 4. ✔
(Teacher reference:)

a.
```
  34
  21
+ 23
```
b.
```
  50
  34
+ 12
```
c.
```
  32
   6
+ 21
```
d.
```
  18
  201
+ 40
```
e.
```
  274
   12
+ 102
```

The problems in part 4 plus three numbers. You'll read each problem to yourself and work it.

e. Touch part 5. ✔
(Teacher reference:)

a.
b.
c.
d.
e.

You'll count the cents for each group of coins. Then you'll write an equals and the number to show how many cents each group is worth.

f. Touch part 6. ✔
(Teacher reference:)

a.
```
  857
-  37
```
b.
```
  386
- 381
```
c.
```
  459
- 402
```
d.
```
  738
- 532
```
e.
```
   98
-  96
```

You'll work column problems in part 6.
- Your turn: Work all the problems on Test 7. ✔
(Direct students where to put their assessment books when they are finished.)

Scoring Notes

a. Collect test booklets. Use the Answer Key and Passing Criteria Table to score the tests.

Passing Criteria Table — Mastery Test 7			
Part	Score	Possible Score	Passing Score
1	2 for each problem	20	18
2	4 for each problem	20	16
3	3 for each problem	15	12
4	3 for each problem	15	12
5	3 for each problem	15	12
6	3 for each problem	15	9
	Total	100	

b. Complete the Mastery Test 7 Remedy Summary Sheet to determine whether group remedies are needed. Reproducible Remedy Summary Sheets are at the back of the Answer Key and at the back of the *Teacher's Guide.*

• If ¼ or more of the students did not pass a test part, present the remedy for that part before beginning Lesson 71. The Remedy Table follows and is also at the end of the Mastery Test 7 Answer Key. Remedies worksheets follow Mastery Test 7 in the *Student Assessment Book.*

Remedy Table — Mastery Test 7				
Part	Test Items	Remedy		Student Material Remedies Worksheet
		Lesson	Exercise	
1	New Number Families	64	1	—
		65	2	—
		67	1	—
2	Count On (Rulers)	60	3	—
		63	7	Part A
		68	8	Part B
		69	7	Part C
3	Bills	59	2	—
		60	4	—
		62	8	Part D
		64	8	Part E
4	3 Addends (Columns)	64	7	Part F
		65	7	Part G
5	Coins (Quarters)	61	6	Part H
		62	6	Part I
		64	5	Part J
6	Column Subtraction	62	2	—

Retest

Retest individual students on any part failed.

Lessons 71-75 Planning Page

	Lesson 71	Lesson 72	Lesson 73	Lesson 74	Lesson 75
Student Learning Objectives	**Exercises** 1. Say addition and subtraction facts 2. Identify 3-dimensional objects 3. **Write dollar amounts with the dollar sign ($) and the decimal point** 4. Count forward or backward from a given number 5. Solve subtraction facts 6. Use a ruler to count backward from the whole to solve equations 7. Solve problems with three addends in columns 8. Solve addition facts 9. **Use a ruler and count on for the whole to solve equations** 10. Complete work independently	**Exercises** 1. Find missing numbers in number families 2. **Identify 3-dimensional objects (cylinder)** 3. Count forward or backward from a given number 4. Solve subtraction facts 5. Write dollar amounts with the dollar sign ($) and the decimal point 6. Use a ruler to count backward from the whole to solve equations 7. Write the symbols for word problems in columns and solve 8. **Use a ruler or objects and count on for the whole to solve equations** 9. Solve addition facts 10. Complete work independently	**Exercises** 1. Find missing numbers in number families 2. Identify 3-dimensional objects 3. **Learn about carrying to the tens column** 4. Solve subtraction facts 5. Write dollar amounts with the dollar sign ($) and the decimal point 6. Use a ruler to count backward from the whole to solve equations 7. Solve problems with three addends 8. Solve addition facts 9. Write the symbols for word problems in columns and solve 10. Complete work independently	**Exercises** 1. Find missing numbers in number families 2. Learn about carrying to the tens column 3. Identify 3-dimensional objects 4. Solve addition facts 5. Count forward or backward from a given number 6. Use a ruler to count backward from the whole to solve equations 7. Write dollar amounts with the dollar sign ($) and the decimal point 8. Solve subtraction facts 9. Solve problems with three addends in columns 10. Complete work independently	**Exercises** 1. **Say subtraction facts** 2. **Decompose 2-dimensional figures** 3. Learn about carrying to the tens column 4. Count forward or backward from a given number 5. Find and write the missing number in a number family 6. Write dollar amounts with the dollar sign ($) and the decimal point 7. Use a ruler to count backward from the whole to solve equations 8. Solve addition and subtraction facts 9. Complete work independently

Common Core State Standards for Mathematics

	Lesson 71	Lesson 72	Lesson 73	Lesson 74	Lesson 75
1.OA 4	✔	✔	✔	✔	✔
1.OA 5	✔	✔	✔	✔	✔
1.OA 6	✔	✔	✔	✔	✔
1.OA 8	✔	✔	✔	✔	✔
1.NBT 1	✔	✔	✔	✔	✔
1.NBT 2	✔		✔	✔	✔
1.NBT 4	✔	✔	✔	✔	✔
1.G 1	✔	✔	✔	✔	✔
1.G 2					✔

Teacher Materials	Presentation Book 2, Board Displays CD or chalkboard
Student Materials	Workbook 2, Pencil
Additional Practice	• Student Practice Software: Block 3: Activity 1 (1.NBT.2 and 1.NBT.4), Activity 2 (1.NBT.4 and 1.OA.6), Activity 3 (1.NBT.1), Activity 4 (1.NBT.1), Activity 5 (1.NBT.2), Activity 6 (1.MD.2) • Math Fact Worksheets 33–34 (After Lesson 71), 35 (After Lesson 73), 36–40 (After Lesson 74)
Mastery Test	

Lesson 71

EXERCISE 1: FACTS
ADDITION/SUBTRACTION

a. (Display:) [71:1A]

2 + 7	12 − 6	9 − 3
6 − 2	11 − 6	6 + 6
4 + 6	7 − 5	10 − 6

These problems are from number families you know.

* (Point to **2 + 7.**) Read this problem. Get ready. (Touch.) *2 plus 7.*
* Is the answer the big number or a small number? (Signal.) *The big number.*
* What's 2 plus 7? (Signal.) *9.*
* Say the fact. (Touch.) *2 + 7 = 9.*

b. (Point to **6 − 2.**) Read this problem. Get ready. (Touch.) *6 minus 2.*
* Is the answer the big number or a small number? (Signal.) *A small number.*
* What's 6 minus 2? (Signal.) *4.*
* Say the fact. (Touch.) *6 − 2 = 4.*

c. (Repeat the following tasks for the remaining problems:)
* (Point to __.) Read the problem.
* Is the answer the big number or a small number?
* What's __?
* Say the fact.
(Repeat problems that were not firm.)

EXERCISE 2: 3-D OBJECTS

a. (Display:) [71:2A]

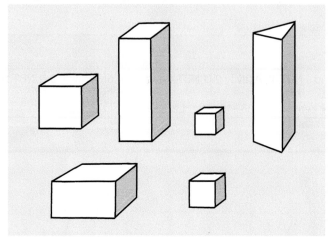

* (Point to objects.) Some objects in this picture are cubes.
* (Point to ☐.) Are all of the faces of this object shaped like a square? (Touch.) *Yes.*
* So is this a cube? (Touch.) *Yes.*

b. (Repeat the following tasks for the remaining objects:)
* (Point to __.) Are all of the faces of this object shaped like a square?
* So is this a cube?

c. Raise your hand when you know how many objects in this picture are cubes. ✔
* How many objects are cubes? (Signal.) *3.*

d. Think about the number of faces a cube has. ✔
* How many faces does a cube have? (Signal.) *6.*

e. Now I'm going to ask you about shapes.
* Tell me the shape that has **6** sides. Get ready. (Signal.) *(A) hexagon.*
* Tell me the shape that has **3** sides. Get ready. (Signal.) *(A) triangle.*
* Tell me the name for a rectangle with sides that are the same size. Get ready. (Signal.) *(A) square.*

f. Tell me the name of an object that has 6 faces. Get ready. (Signal.) *(A) cube.*
(Repeat steps e and f until firm.)

EXERCISE 3: MONEY
WRITING DOLLAR AMOUNTS

a. (Display:) [71:3A]

$3.18	$9.70
$5.19	$15.11
$11.40	$3.18

These are dollar amounts.

* (Point to **$3.18.**) This is 3 dollars and 18 cents. Read it. (Touch.) *3 dollars and 18 cents.*
* (Point to **$5.19.**) Read this amount. (Touch.) *5 dollars and 19 cents.*
* (Repeat for $11.40, $ 9.70, $15.11, $3.18.)
(Repeat amounts that were not firm.)

b. When you write dollar amounts, what do you write first? (Signal.) *The dollar sign.*
Yes, first you write the dollar sign.

- What do you write next? (Signal.) *The number for dollars.*
Yes, you write the number for dollars next.
- What do you write after the number for dollars? (Signal.) *(A) dot.*
Yes, a dot.
- What do you write after the dot? (Signal.) *The number for cents.*
Yes, you write the number for cents after the dot.
(Repeat step b until firm.)

c. I want to write 5 dollars and 24 cents.

- What do I want to write? (Signal.) *5 dollars and 24 cents.*
- What do I write first? (Signal.) *The dollar sign.*
(Display:) W [71:3B]

$

- What do I write after the dollar sign? (Signal.) *5.*
Yes, 5.
(Add to show:) [71:3C]

$5

- What do I write after the number of dollars? (Signal.) *(A) dot.*
Yes, a dot.
(Add to show:) [71:3D]

$5.

- I want to write 5 dollars and 24 cents. What do I write after the dot? (Signal.) *24.*
Yes, 24.
(Add to show:) [71:3E]

$5.24

- (Point to **$5.24.**) Read this amount. (Touch.) *5 dollars and 24 cents.*

Exercise 4: Mixed Counting

a. Start with 100 and count by fives to 150. Get 100 going. *One huuundred.* Count. (Tap.) *105, 110, 115, 120, 125, 130, 135, 140, 145, 150.*

- Start with 107 and plus tens to 157. Get 107 going. *One hundred sevennn.* Plus tens. (Tap.) *117, 127, 137, 147, 157.*
- Start with 80 and count backward by tens. Get 80 going. *Eightyyy.* Count backward. (Tap.) *70, 60, 50, 40, 30, 20, 10.*

b. Listen: You have 56. When you plus tens, what's the next number? (Signal.) *66.*

- Listen: You have 56. When you minus ones, what's the next number? (Signal.) *55.*
- Listen: You have 56. When you plus ones, what's the next number? (Signal.) *57.*

c. Listen: You have 75. When you plus fives, what's the next number? (Signal.) *80.*

- Listen: You have 75. When you plus 25s, what's the next number? (Signal.) *100.*
- Listen: You have 75. When you plus tens, what's the next number? (Signal.) *85.*

Exercise 5: Facts

Subtraction

REMEDY

a. (Distribute unopened workbooks to students.)

- Open your workbook to Lesson 71 and find part 1.
(Observe children and give feedback.)
(Teacher reference:) **R Part J**

a. 10 − 6	d. 12 − 6	h. 10 − 8
b. 11 − 9	e. 8 − 5	i. 9 − 3
c. 12 − 2	f. 11 − 5	j. 6 − 4
	g. 8 − 2	

These are minus problems from families you know. You're going to read each problem and say the answer. Then you'll go back and work them.

- Touch and read problem A. Get ready. (Signal.) *10 minus 6.*
- What's the answer? (Signal.) *4.*

b. (Repeat the following tasks for problems B through J:)

- Touch and read problem __.
- What's the answer?
(Repeat problems that were not firm.)

c. Complete the equations for all of the problems in part 1.
(Observe children and give feedback.)

d. Check your work. You'll touch and read each fact.

- Fact A. (Signal.) *10 − 6 = 4.*
- (Repeat for:) B, *11 − 9 = 2;* C, *12 − 2 = 10;* D, *12 − 6 = 6;* E, *8 − 5 = 3;* F, *11 − 5 = 6;* G, *8 − 2 = 6;* H, *10 − 8 = 2;* I, *9 − 3 = 6;* J, *6 − 4 = 2.*

EXERCISE 6: RULER
COUNT BACKWARD

a. (Display:) W [71:6A]

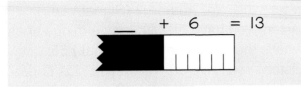

This ruler is supposed to show meters. It has a shaded part and an unshaded part. You're going to figure out how many meters are shaded.

- (Point to **6**.) How many meters is the unshaded part? (Touch.) *6.*
- (Point to **13**.) How many meters long is the ruler? (Touch.) *13.*

b. We can figure out how many meters are shaded by counting backward.

- (Point to the end of the ruler.) How many meters is this line from the beginning of the ruler. (Touch.) *13.*
- So what number do you get going for this line? (Touch.) *13.*
  I'll touch the lines. You get it going and count backward. Tell me to stop when I get to the end of the shaded part.
- (Touch end of ruler.) Get it going. *Thirteeen.* (Touch lines.) *12, 11, 10, 9, 8, 7, stop.*
- How many meters is the shaded part? (Signal.) *7.*
  (Repeat until firm.)
  (Add to show:) [71:6B]

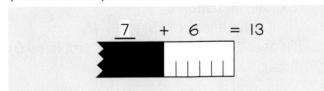

- (Point to **7**.) Read the equation for the ruler. (Signal.) *7 + 6 = 13.*

c. Find part 2 on worksheet 71. ✔
(Teacher reference:)

These rulers are supposed to show centimeters. You're going to figure out how many centimeters are shaded for each ruler.

- Touch ruler A. ✔

- Touch the number above the unshaded part. ✔
- How many centimeters is the unshaded part? (Signal.) *3.*
- Touch the number after the equals. ✔
- How many centimeters are both parts? (Signal.) *10.*

d. I'm going to get 10 going and count backward. You'll touch the correct lines. When you get to the end of the shaded part, tell me to stop.

- Touch the end of ruler A. ✔
- Tennn, (children touch 9th centimeter) 9, (children touch 8th) 8, (children touch 7th) 7. *Stop.*

e. Again. Touch the end of ruler A. ✔
- Tennn, 9, 8, 7. *Stop.*
  (Repeat step e until firm.)

f. How many centimeters is the shaded part? (Signal.) *7.*
Write 7. ✔

- Touch and read the equation for ruler A. Get ready. (Signal.) *7 + 3 = 10.*

g. Touch ruler B. ✔
- Touch the number above the unshaded part. ✔
- How many centimeters is the unshaded part? (Signal.) *6.*
- Touch the end of the ruler. ✔
- How many centimeters long is ruler B? (Signal.) *15.*

h. I'll get 15 going and count backward. You'll touch the correct lines and tell me when to stop.
- Fifteeen. (Children touch lines.) *14, 13, 12, 11, 10, 9. Stop.*
  (Repeat step h until firm.)

i. Your turn to touch and count backward.
- What number will you get going when you touch the end of the ruler? (Signal.) *15.*
- Touch the end of the ruler. ✔
- Get 15 going. *Fifteeen.* Count backward. (Tap.) *14, 13, 12, 11, 10, 9.*
  (Repeat until firm.)
- How many centimeters is the shaded part? (Signal.) *9.*
- Write 9. ✔
- Touch and read the equation. Get ready. (Signal.) *9 + 6 = 15.*

EXERCISE 7: 3 ADDENDS IN COLUMNS

a. Find part 3 on worksheet 71. ✔
   (Teacher reference:)

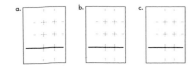

I'll tell you problems that plus three numbers.
You're going to write those problems in
columns.

- Touch where you'll write column problem A. ✔
  Listen to problem A: 62 plus 114 plus 3.
- Listen again: 62 plus 114 plus 3. Say problem
  A. (Signal.) *62 plus 114 plus 3.*
  (Repeat until firm.)

b. The first number is 62. Is the beginning digit of
   62 hundreds, tens, or ones? (Signal.) *Tens.*
- Write 62. ✔

c. Put your pencil where you'll write the plus
   sign. ✔
- Write plus. ✔

d. The next number is 114. Is the beginning
   digit of 114 hundreds, tens, or ones? (Signal.)
   Hundreds.
- Write 114. ✔

e. The next number is 3. Is the beginning digit of
   3 hundreds, tens, or ones? (Signal.) *Ones.*
- Write 3. ✔

f. Touch and read problem A. Get ready.
   (Signal.) *62 plus 114 plus 3.*
   Later, you'll work the problems in part 3
   independently.

g. Touch where you'll write column problem B. ✔
   Listen to problem B: 4 plus 20 plus 533.
- Listen again: 4 plus 20 plus 533. Say
  problem B. (Signal.) *4 plus 20 plus 533.*
  (Repeat step g until firm.)

h. The first number is 4. Is the beginning digit of
   4 hundreds, tens, or ones? (Signal.) *Ones.*
- The next symbol you write is plus. Put your
  pencil where you'll write the plus sign. ✔
- The next number is 20. Is the beginning digit
  of 20 hundreds, tens, or ones? (Signal.) *Tens.*
- The next number is 533. Is the beginning
  digit of 533 hundreds, tens, or ones? (Signal.)
  Hundreds.
  (Repeat step h until firm.)

i. Say the problem 4 plus 20 plus 533. Get
   ready. (Signal.) *4 plus 20 plus 533.*

j. Write 4 plus 20 plus 533 for problem B.
   (Observe children and give feedback.)
   (Teacher reference:)

k. Touch where you'll write column problem C.
   Listen to problem C: 224 plus 5 plus 70.
- Listen again: 224 plus 5 plus 70. Say
  problem C. (Signal.) *224 plus 5 plus 70.*
  (Repeat step k until firm.)

l. The first number is 224. Is the beginning digit
   of 224 hundreds, tens, or ones? (Signal.)
   Hundreds.
- The next symbol you write is plus. Put your
  pencil where you'll write the plus sign. ✔
- The next number is 5. Is the beginning digit of
  5 hundreds, tens, or ones? (Signal.) *Ones.*
- The next number is 70. Is the beginning digit
  of 70 hundreds, tens, or ones? (Signal.) *Tens.*
  (Repeat step l until firm.)

m. Say the problem 224 plus 5 plus 70 again. Get
   ready. (Signal.) *224 plus 5 plus 70.*

n. Write 224 plus 5 plus 70 for problem C.
   (Observe children and give feedback.)
   (Teacher reference:)

Later, you'll work the problems in part 3 as
part of your independent work.

EXERCISE 8: FACTS
ADDITION

a. Find part 4 on worksheet 71. ✔
   (Teacher reference:)

	d. 10 + 7	h. 2 + 5
a. 6 + 1	e. 6 + 6	i. 4 + 6
b. 5 + 6	f. 6 + 3	j. 2 + 9
c. 2 + 6	g. 8 + 2	

These are plus problems for families you
know. You're going to read each problem and
tell me the answer. Then you'll go back and
work all of the problems.

- Touch and read problem A. Get ready.
  (Signal.) *6 plus 1.*
- What's 6 plus 1? (Signal.) *7.*

b. Touch and read problem B. (Signal.) *5 plus 6.*
• What's 5 plus 6? (Signal.) *11.*
c. (Repeat the following tasks for problems C through J:)
• Touch and read problem __.
• What's __?
d. Complete the equations for all of the problems in part 4.
(Observe children and give feedback.)
e. Check your work. You'll touch and read each fact.
• Fact A. (Signal.) *6 + 1 = 7.*
• (Repeat for:) B, *5 + 6 = 11;* C, *2 + 6 = 8;* D, *10 + 7 = 17;* E, *6 + 6 = 12;* F, *6 + 3 = 9;* G, *8 + 2 = 10;* H, *2 + 5 = 7;* I, *4 + 6 = 10;* J, *2 + 9 = 11.*

EXERCISE 9: RULER SKILLS

a. Find part 5 on worksheet 71. ✔
(Teacher reference:)

The rulers in part 5 are supposed to show inches. They have a shaded part and an unshaded part. You can't see the whole shaded part, but the number above the shaded part tells how many inches long the shaded part is.
• Touch the number for the shaded part for ruler A. ✔.
• How many inches is the shaded part for ruler A? (Signal.) *74 inches.*
• Count the inches for the unshaded part. Write the number and put your pencil down when you're finished.
(Observe children and give feedback.)
• How many inches is the unshaded part? (Signal.) *6.*
b. Now you will count for both parts of ruler A.
• Touch the end of the shaded part. ✔
• What number do you get going? (Signal.) *74.*
• Get it going. *Seventy-fouuur.* Touch and count. (Tap.) *75, 76, 77, 78, 79, 80.*
• (Repeat until firm.)
• How many inches is the whole ruler? (Signal.) *80.*
• Write 80. ✔
• Touch and read the equation for ruler A. Get ready. (Signal.) *74 + 6 = 80.*

c. Touch the number for the shaded part for ruler B. ✔
• How many inches is the shaded part for ruler B? (Signal.) *18 inches.*
• Count the inches for the unshaded part. Write the number and put your pencil down when you're finished.
(Observe children and give feedback.)
• How many inches is the unshaded part? (Signal.) *4.*
d. Now you will count for both parts of ruler B.
• Touch the end of the shaded part. ✔
• What number do you get going? (Signal.) *18.*
• Get it going. *Eighteeen.* Touch and count. (Tap.) *19, 20, 21, 22.*
• (Repeat until firm.)
• How many inches is the whole ruler? (Signal.) *22.*
• Write 22. ✔
• Touch and read the equation for ruler B. Get ready. (Signal.) *18 + 4 = 22.*
e. Touch the number for the shaded part for ruler C. ✔
• How many inches is the shaded part for ruler C? (Signal.) *37 inches.*
• Count the inches for the unshaded part. Write the number and put your pencil down when you're finished.
(Observe children and give feedback.)
• How many inches is the unshaded part? (Signal.) *5.*
f. Now you will count for both parts of ruler C.
• Touch the end of the shaded part. ✔
• What number do you get going? (Signal.) *37.*
• Get it going. *Thirty-sevennn.* Touch and count. (Tap.) *38, 39, 40, 41, 42.*
• (Repeat until firm.)
• How many inches is the whole ruler? (Signal.) *42.*
• Write 42. ✔
• Touch and read the equation for ruler C. Get ready. (Signal.) *37 + 5 = 42.*

EXERCISE 10: INDEPENDENT WORK

a. Turn to the other side of worksheet 71 and find part 6. ✔
(Teacher reference:)

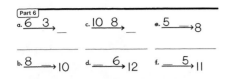

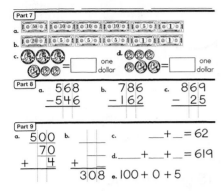

You'll write the fact for the missing number in each family and complete the family.

In part 7, you'll write an equals and the number of dollars to show how many each group of bills is worth. For each group of coins, you'll circle the words one dollar if the group is worth one dollar. If the group of coins doesn't equal one dollar, write the cents it equals in the box.

You'll work the column problems in part 8 and complete the place-value addition problems in part 9.

b. Your turn: Complete worksheet 71. Remember to work the problems in part 3 on the other side of your worksheet.

(Observe children and mark incorrect responses on children's worksheets as you give feedback.)

Lesson 72

EXERCISE 1: NUMBER FAMILIES
MISSING NUMBER IN FAMILY

a. (Display:) [72:1A]

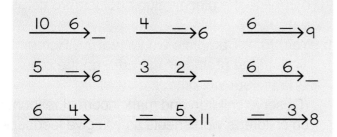

You're going to say the problem and the answer for the missing number in each family.

b. (Point to $\overset{10\quad 6}{\longrightarrow}$_.) Say the problem for the missing number. Get ready. (Touch.) *10 plus 6.*
• What's 10 plus 6? (Signal.) *16.*
c. (Point to $\overset{5}{\longrightarrow}$6.) Say the problem for the missing number. (Touch.) *6 minus 5.*
• What's 6 minus 5? (Signal.) *1.*
d. (Repeat the following tasks for remaining families:)

(Point to __.)	Say the problem for the missing number.	What's __?	
$\overset{6\quad 4}{\longrightarrow}$_	6 + 4	6 + 4	10
$\overset{4}{\longrightarrow}$6	6 − 4	6 − 4	2
$\overset{3\quad 2}{\longrightarrow}$_	3 + 2	3 + 2	5
$\overset{5}{\longrightarrow}$11	11 − 5	11 − 5	6
$\overset{6}{\longrightarrow}$9	9 − 6	9 − 6	3
$\overset{6\quad 6}{\longrightarrow}$_	6 + 6	6 + 6	12
$\overset{3}{\longrightarrow}$8	8 − 3	8 − 3	5

(Repeat for families that were not firm.)

EXERCISE 2: 3-D OBJECTS
CYLINDER

a. (Display:) [72:2A]

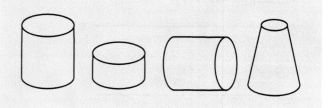

• (Point to [cans].) These objects are cans. What are they? (Touch.) *Cans.*
• Listen: The shape of these cans are cylinders. Say **cylinder**. (Signal.) *Cylinder.* A cylinder is an object that has a circle on each end, and the circles are the same size.
• (Point to [cone].) This object is not a cylinder because the circle on one end is not the same size as the circle on the other end.

b. (Display:) [72:2B]

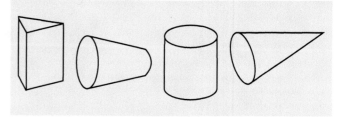

• (Point to [prism].) Does this object have a circle on each end? (Touch.) *No.*
• So is this a cylinder? (Signal.) *No.*
c. (Point to [cone].) Does this object have a circle on each end? (Touch.) *Yes.*
• Are the circles the same size? (Signal.) *No.*
• So is this a cylinder? (Signal.) *No.*
d. (Point to [cylinder].) Does this object have a circle on each end? (Touch.) *Yes.*
• Are the circles the same size? (Signal.) *Yes.*
• So is this a cylinder? (Signal.) *Yes.*
e. (Point to [cone/triangle].) Does this object have a circle on each end? (Touch.) *No.*
• So is this a cylinder? (Signal.) *No.* Remember, a cylinder has a circle at each end, and the circles are the same size.
• What object has a circle at each end and the circles are the same size? (Signal.) *(A) cylinder.*

EXERCISE 3: MIXED COUNTING

a. Start with 100 and count by 25s to 200. Get 100 going. *One huuundred.* Count. (Tap.) *125, 150, 175, 200.*

- Start with 100 and count by fives to 150. Get 100 going. *One huuundred.* Count. (Tap.) *105, 110, 115, 120, 125, 130, 135, 140, 145, 150.*

- Start with 100 and plus tens to 200. Get 100 going. *One huuundred.* Count. (Tap.) *110, 120, 130, 140, 150, 160, 170, 180, 190, 200.*

- Start with 100 and count backward by tens. Get 100 going. *One huuundred.* Count backward. (Tap.) *90, 80, 70, 60, 50, 40, 30, 20, 10.*

- Start with 100 and count backward by ones to 90. Get 100 going. *One huuundred.* Count backward. (Tap.) *99, 98, 97, 96, 95, 94, 93, 92, 91, 90.*

b. Listen: You have 55. When you minus ones, what's the next number? (Signal.) *54.*

- You have 55. When you plus tens, what's the next number? (Signal.) *65.*

- You have 55. When you plus ones, what's the next number? (Signal.) *56.*

- You have 55. When you plus fives, what's the next number? (Signal.) *60.*

EXERCISE 4: FACTS

SUBTRACTION

a. (Display:) [72:4A]

10 − 6	11 − 1	9 − 7
7 − 2	8 − 2	10 − 2
12 − 6	8 − 5	9 − 6

You're going to say the facts for all of these minus problems.

- (Point to **10 − 6.**) Read the problem. (Touch.) *10 minus 6.*

- What's 10 minus 6? (Signal.) *4.*

- Say the fact. (Signal.) *10 − 6 = 4.*

b. (Repeat the following tasks for the remaining problems:)

- (Point to __.) Read the problem.

- What's __?

- Say the fact.

(Repeat problems that were not firm.)

EXERCISE 5: MONEY

WRITING DOLLAR AMOUNTS   REMEDY

a. (Display:) [72:5A]

$2.10	$23.34
$1.89	$70.30
$14.15	

Remember how to read these dollar amounts. The number before the dot tells about dollars. Say **and** for the dot. The number after the dot tells about cents.

- (Point to **$2.10.**) Read this. (Touch.) *2 dollars and 10 cents.*

- (Repeat for remaining amounts.)
(Repeat amounts that were not firm.)

b. When you write dollar amounts, what do you write first? (Signal.) *The dollar sign.*
Yes, first you write the dollar sign.

- What do you write next? (Signal.) *The number for dollars.*
Yes, you write the number for dollars next.

- What do you write after the number for dollars? (Signal.) *(A) dot.*
Yes, a dot.

- What do you write after the dot? (Signal.) *The number for cents.*
Yes, the number for cents.

(Repeat step b until firm.)

c. I want to write 3 dollars and 17 cents.
- What do I want to write? (Signal.) *3 dollars and 17 cents.*
- What do I write first? (Signal.) *A dollar sign.*
(Display:) W [72:5B]

$$\$$$

- What do I write after the dollar sign?
(Signal.) *3.*
Yes, 3.
(Add to show:) [72:5C]

$$\$3$$

- What do I write after the number? (Signal.)
(A) dot.
Yes, a dot.
(Add to show:) [72:5D]

$$\$3.$$

- I want to write 3 dollars and 17 cents. What do I write after the dot? (Signal.) *17.*
Yes, 17.
(Add to show:) [72:5E]

$$\$3.17$$

- (Point to **$3.17.**) Read this. (Touch.) *3 dollars and 17 cents.*
d. (Distribute unopened workbooks to students.)
- Open your workbook to Lesson 72 and find part 1.
(Observe children and give feedback.)
(Teacher reference:) R Part A

ₐ. $_____ ᵦ. $_____ ᵨ. $_____

You're going to write dollar amounts. The dollar sign is already written for each amount.
- Touch space A. ✔
- You'll write 3 dollars and 17 cents in space A. What will you write? (Signal.) *3 dollars and 17 cents.*
- Touch the dollar sign for A. ✔
- What do you write after the dollar sign for 3 dollars and 17 cents? (Signal.) *3.*
- Touch where you'll write 3. ✔

e. What do you write next for 3 dollars and 17 cents. (Signal.) *(The) dot.*
- Touch where you'll write the dot for 3 dollars and 17 cents. ✔
- What do you write after the dot? (Signal.) *17.*
- Write 3 dollars and 17 cents.
(Observe children and give feedback.)
f. Check your work.
- Touch the dollar sign for A. ✔
- Touch the number 3. ✔
- Touch the dot. ✔
- Touch the number 17. ✔
- Did you do everything right? (Children respond.)
g. Touch the space for B. ✔
You'll write 9 dollars and 13 cents for space B.
- What will you write? (Signal.) *9 dollars and 13 cents.*
- Touch the dollar sign for B. ✔
- What do you write after the dollar sign for 9 dollars and 13 cents? (Signal.) *9.*
- Touch where you'll write 9. ✔
h. What do you write next for 9 dollars and 13 cents. (Signal.) *(The) dot.*
- What do you write after the dot? (Signal.) *13.*
- Write 9 dollars and 13 cents.
(Observe children and give feedback.)
i. Check your work.
- Touch the dollar sign for B. ✔
- Touch the number 9. ✔
- Touch the dot. ✔
- Touch the number 13. ✔
j. Touch the space for C. ✔
You'll write 10 dollars and 52 cents for space C.
- What will you write? (Signal.) *10 dollars and 52 cents.*
- What do you write after the dollar sign for 10 dollars and 52 cents? (Signal.) *10.*
k. What do you write next for 10 dollars and 52 cents. (Signal.) *(The) dot.*
- What do you write after the dot? (Signal.) *52.*
- Write 10 dollars and 52 cents.
(Observe children and give feedback.)
- Check your work.
- Touch the dollar sign for C. ✔
- Touch the number 10. ✔
- Touch the dot. ✔
- Touch the number 52. ✔

EXERCISE 6: RULER

COUNT BACKWARD

a. Find part 2 on worksheet 72. ✔
(Teacher reference:)

R Part D

These rulers are supposed to show
centimeters. We're going to figure out how
many centimeters are shaded for each ruler.

• Touch ruler A. ✔
• Touch the number above the unshaded part. ✔
• How many centimeters is the unshaded part?
(Signal.) 5.
• Touch the number after equals. ✔
• How many centimeters are both parts?
(Signal.) 17.
• Touch the end of the ruler. ✔
• I'll get 17 going and count backward. You'll
touch the correct lines. Tell me to stop when
you're touching the end of the shaded part.
(Children should be touching the end of the
ruler.) Seventeeen. (Children touch lines.)
16, 15, 14, 13, 12. Stop.
(Repeat until firm.)

b. Your turn to count backward.
• What number will you get going? (Signal.) 17.
• Touch the end of the ruler. ✔
• Get it going. Seventeeen. Touch and count.
(Tap.) 16, 15, 14, 13, 12.
(Repeat until firm.)

c. How many centimeters is the shaded part?
(Signal.) 12.
• Write 12. ✔
• Touch and read the equation. Get ready.
(Signal.) 12 + 5 = 17.

d. Touch ruler B. ✔
• Touch the number above the unshaded part. ✔
• How many centimeters is the unshaded part?
(Signal.) 8.
• Touch the number after the equals. ✔
• How many centimeters are both parts?
(Signal.) 16.
• You're going to get 16 going and count
backward. You'll touch the correct lines and
say stop when you get to the end of the
shaded part.

• Touch the end of ruler B. ✔
• Get it going. Sixteeen. Touch and count. (Tap.)
15, 14, 13, 12, 11, 10, 9, 8, stop.
(Repeat until firm.)

e. How many centimeters is the shaded part?
(Signal.) 8.
• Write 8. ✔
• Touch and read the equation. Get ready.
(Signal.) 8 + 8 = 16.

f. Touch ruler C. ✔
• Touch the number above the unshaded part. ✔
• How many centimeters is the unshaded part?
(Signal.) 5.
• Touch the number after the equals. ✔
• How many centimeters are both parts?
(Signal.) 14.
• You're going to get 14 going and count
backward. You'll touch the correct lines and
say stop when you get to the end of the
shaded part.

• Touch the end of ruler C. ✔
• Get it going. Fourteeen. Touch and count.
(Tap.) 13, 12, 11, 10, 9, stop.
(Repeat until firm.)

g. How many centimeters is the shaded part?
(Signal.) 9.
• Write 9. ✔
• Touch and read the equation. Get ready.
(Signal.) 9 + 5 = 14.

EXERCISE 7: WORD PROBLEMS (COLUMNS)

a. Find part 3 on worksheet 72. ✔
(Teacher reference:)

You're going to write the symbols for word
problems in columns and work them. Only
part of the column and row lines are shown
and the equals bars are dotted. You'll write the
numbers and the sign in the right places, and
you'll make the equals bar.

b. Touch where you'll write the symbols for problem A. ✔

Listen to problem A: Dennis had 319 quarters. Dennis lost 17 of those quarters. How many quarters did Dennis end up with?

- Listen again and write the symbols for both parts: Dennis had 319 quarters. Dennis lost 17 of those quarters.
- Everybody, touch and read problem A. Get ready. (Signal.) *319 minus 17.*

c. Work problem A. Put your pencil down when you know how many quarters Dennis ended up with.

(Observe children and give feedback.)

- Read the problem and the answer you wrote for A. Get ready. (Signal.) *319 − 17 = 302.*
- How many quarters did Dennis end up with? (Signal.) *302.*

d. Touch where you'll write the symbols for problem B. ✔

Listen to problem B: There were 43 people on a bus. There were 125 people on a train. How many people were there altogether?

- Listen again and write the symbols for both parts: There were 43 people on a bus. There were 125 people on a train.
- Everybody, touch and read problem B. Get ready. (Signal.) *43 plus 125.*

e. Work problem B. Put your pencil down when you know how many people there were altogether.

(Observe children and give feedback.)

- Read the problem and the answer you wrote for B. Get ready. (Signal.) *43 + 125 = 168.*
- How many people were there altogether? (Signal.) *168.*

f. Touch where you'll write the symbols for problem C. ✔

Listen to problem C: 564 fish were in a lake. A fisherman caught 62 of those fish. How many fish ended up in the lake?

- Listen again and write the symbols for both parts: 564 fish were in a lake. A fisherman caught 62 of those fish.
- Everybody, touch and read problem C. Get ready. (Signal.) *564 minus 62.*

g. Work problem C. Put your pencil down when you know how many fish ended up in the lake.
(Observe children and give feedback.)
(Answer key:)

- Read the problem and the answer you wrote for C. Get ready. (Signal.) *564 − 62 = 502.*
- How many fish ended up in the lake? (Signal.) *502.*

EXERCISE 8: COUNT ON
OBJECTS/RULERS

a. Find part 4 on worksheet 72. ✔
(Teacher reference:)

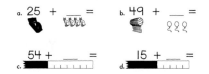

Some of these problems show objects. Some show rulers. The rulers are supposed to show inches.

- Touch problem A. ✔
- How many black cats are there? (Signal.) *25.*
- Count the number for the white cats and write it. ✔
- Now, you'll touch the number for black cats. Get it going to yourself. Then work the problem and complete the equation.
- Work problem A. Put your pencil down when you've completed the equation for cats.
(Observe children and give feedback.)
- Touch and read the equation for A. Get ready. (Signal.) *25 + 4 = 29.*
- How many cats are there altogether? (Signal.) *29.*

b. Touch problem B. ✔
- Write the number for the smaller balloons. Then work the problem and complete the equation.
(Observe children and give feedback.)
- Touch and read the equation for B. Get ready. (Signal.) *49 + 3 = 52.*
- How many balloons are there altogether? (Signal.) *52.*

c. Touch the ruler in problem C. ✔
• Write the number of inches for the unshaded part. Then work the problem and complete the equation.
  (Observe children and give feedback.)
• Touch and read the equation. (Signal.)
  54 + 7 = 61.
• How many inches long is ruler C? (Signal.) *61.*
d. Touch problem D. ✔
• Write the number of inches for the unshaded part. Then work the problem and complete the equation.
  (Observe children and give feedback.)
• Touch and read the equation for D. Get ready. (Signal.) *15 + 5 = 20.*
• How many inches long is ruler D? (Signal.) *20.*

EXERCISE 9: FACTS
ADDITION

a. Turn to the other side of worksheet 72 and find part 5. ✔
  (Teacher reference:)

a. 9+1	d. 2+5	h. 2+9
b. 5+6	e. 10+3	i. 4+6
c. 6+3	f. 6+5	j. 6+2
	g. 3+5	

These are plus problems for families you know. You're going to read each problem and tell me the answer. Then you'll go back and work all of the problems.
• Touch and read problem A. Get ready. (Signal.) *9 plus 1.*
• What's 9 plus 1? (Signal.) *10.*
b. Touch and read problem B. (Signal.) *5 plus 6.*
• What's 5 plus 6? (Signal.) *11.*
c. (Repeat the following tasks for problems C through J:)
• Touch and read problem __.
• What's __?
d. Complete the equations for all of the problems in part 5.
  (Observe children and give feedback.)
e. Check your work. You'll touch and read each fact.
• Fact A. (Signal.) *9 + 1 = 10.*
• (Repeat for:) B, *5 + 6 = 11;* C, *6 + 3 = 9;* D, *2 + 5 = 7;* E, *10 + 3 = 13;* F, *6 + 5 = 11;* G, *3 + 5 = 8;* H, *2 + 9 = 11;* I, *4 + 6 = 10;* J, *6 + 2 = 8.*

EXERCISE 10: INDEPENDENT WORK

a. Find part 6 on worksheet 72. ✔
  (Teacher reference:)

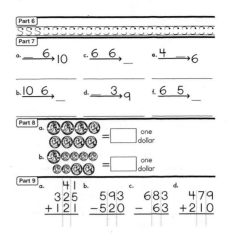

Part 6 has a row of capital Ss that are dotted. In a few lessons, you'll write the dollar sign for dollar amounts. You need to write a capital S when you make a dollar sign.
• Start at the big ball and follow the dotted line to make a capital S. ✔
• Make the next capital S. ✔
• The dotted line for the rest of the capital Ss is not completely shown. Complete the row of capital Ss. Make them look like the first Ss in the row.
  (Observe children and give feedback.)
b. Find part 7.
  In part 7, you'll write the fact for the missing number in each family and complete the family. In part 8, you'll circle the words one dollar if the group of coins is worth one dollar. If a group of coins doesn't equal one dollar, write the cents it equals in the box.
  You'll work the column problems in part 9.
c. Complete worksheet 72.
  (Observe children and mark incorrect responses on children's worksheets as you give feedback.)

Lesson

EXERCISE 1: NUMBER FAMILIES

MISSING NUMBER IN FAMILY

a. (Display:) [73:1A]

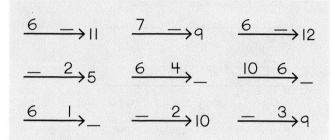

You're going to say the problem for the missing number in each family.

b. (Point to $6 \rightarrow$ 11.) Say the problem for the missing number. Get ready. (Touch.) *11 minus 6.*

• What's 11 minus 6? (Signal.) *5.*

c. (Point to $\rightarrow$ 5.) Say the problem for the missing number. (Touch.) *5 minus 2.*

• What's 5 minus 2? (Signal.) *3.*

d. (Repeat the following tasks for remaining families:)

(Point to __.)	Say the problem for the missing number.	What's __?	
$6 \quad \rightarrow$	6 + 1	6 + 1	7
$7 \rightarrow$ 9	9 – 7	9 – 7	2
$6 \quad 4 \rightarrow$	6 + 4	6 + 4	10
$\rightarrow$ 10	10 – 2	10 – 2	8
$6 \rightarrow$ 12	12 – 6	12 – 6	6
$10 \quad 6 \rightarrow$	10 + 6	10 + 6	16
$\rightarrow$ 9	9 – 3	9 – 3	6

(Repeat for families that were not firm.)

EXERCISE 2: 3-D OBJECTS

a. (Display:) [73:2A]

Last time you learned about cylinders. A cylinder has a circle on each end and the circles are the same size.

• (Point to ⬭.) Does this object have a circle on each end? (Touch.) *Yes.*

• Are the circles the same size? (Touch.) *Yes.*

• So is this a cylinder? (Touch.) *Yes.*

b. (Point to ⬭.) Does this object have a circle on each end? (Touch.) *Yes.*

• Are the circles the same size? (Touch.) *No.*

• So is this a cylinder? (Touch.) *No.*

c. Some of these objects are cylinders. Some are cubes.

• (Point to ⬭.) Is this a cylinder? (Touch.) *Yes.*

d. (Point to △.) Is this a cylinder? (Touch.) *No.*

• Is this a cube? (Touch.) *No.*

e. (Point to ▭.) Is this a cylinder? (Touch.) *No.*

• Is this a cube? (Touch.) *No.*

• Why not? (Call on a student. Idea: *The faces are not all squares.*)

f. (Point to ▭.) Is this a cylinder? (Touch.) *No.*

• Is this a cube? (Touch.) *Yes.*

g. (Point to ▯.) Is this a cylinder? (Touch.) *Yes.*

h. (Point to ⬭.) Is this a cylinder? (Touch.) *No.*

• Why not? (Call on a student. Idea: *The circles are not the same size.*)

i. (Point to ▭.) Is this a cylinder? (Touch.) *No.*

• What is it? (Touch.) *(A) cube.*

EXERCISE 3: COLUMN PROBLEMS
CARRYING PRESKILL

a. (Display:) W [73:3A]

$$\begin{array}{r} 5\,4 \\ +\,3\,9 \\ \hline \end{array}$$

Listen: The answer to the problems in the ones column is a two-digit number. I'll show you where to write the tens digit.

- (Point to **54**.) Read this problem. (Touch.) *54 plus 39.*
- (Point to the **4**.) Read the problem for the ones column. (Touch.) *4 plus 9.*
- The answer is 13. What's the answer? (Signal.) *13.*
- Is 13 a two-digit answer? (Signal.) *Yes.*
- What's the tens digit of 13? (Signal.) *1.*
- What's the ones digit of 13? (Signal.) *3.*
  I write the tens digit in the tens column, and the 3 in the ones column.
  Watch:
  (Add to show:) [73:3B]

$$\begin{array}{r} {\scriptstyle 1} \\ 5\,4 \\ +\,3\,9 \\ \hline 3 \end{array}$$

b. (Display:) W [73:3C]

$$\begin{array}{r} 2\,9 \\ +\,4\,1 \\ \hline \end{array}$$

- (Point to **29**.) Read this problem. (Touch.) *29 plus 41.*
- (Point to the **9**.) Read the problem for the ones column. (Touch.) *9 plus 1.*
- What's the answer? (Signal.) *10.*
- Is 10 a two-digit answer? (Signal.) *Yes.*
- What's the tens digit of 10? (Signal.) *1.*
  I write that digit in the tens column and the zero in the ones column.
  Watch:
  (Add to show:) [73:3D]

$$\begin{array}{r} {\scriptstyle 1} \\ 2\,9 \\ +\,4\,1 \\ \hline 0 \end{array}$$

c. (Display:) W [73:3E]

$$\begin{array}{r} 1\,2 \\ +\,6\,8 \\ \hline \end{array}$$

- (Point to **12**.) Read this problem. (Touch.) *12 plus 68.*
- (Point to the **2**.) Read the problem for the ones column. (Touch.) *2 plus 8.*
- What's the answer? (Signal.) *10.*
- Is 10 a two-digit answer? (Signal.) *Yes.*
- What's the tens digit of 10? (Signal.) *1.*
- Where do I write the tens digit? (Signal.) *In the tens column.*
- Where do I write zero? (Signal.) *In the ones column.*
  (Repeat until firm.)
  (Add to show:) [73:3F]

$$\begin{array}{r} {\scriptstyle 1} \\ 1\,2 \\ +\,6\,8 \\ \hline 0 \end{array}$$

d. (Display:) W [73:3G]

$$\begin{array}{r} 3\,6 \\ +\,1\,6 \\ \hline \end{array}$$

- (Point to **36**.) Read this problem. (Touch.) *36 plus 16.*
- (Point to the **6**.) Read the problem for the ones column. (Touch.) *6 plus 6.*
- What's the answer? (Signal.) *12.*
- Is 12 a two-digit answer? (Signal.) *Yes.*
- What's the tens digit of 12? (Signal.) *1.*
- Where do I write the tens digit? (Signal.) *In the tens column.*
- Where do I write 2? (Signal.) *In the ones column.*
  (Repeat until firm.)
  (Add to show:) [73:3H]

$$\begin{array}{r} {\scriptstyle 1} \\ 3\,6 \\ +\,1\,6 \\ \hline 2 \end{array}$$

Remember how to write two-digit answers for the ones column.

EXERCISE 4: FACTS
SUBTRACTION

REMEDY

a. (Display:)

[73:4A]

12 − 6	9 − 2	8 − 6
14 − 4	8 − 3	9 − 3
11 − 5	10 − 4	11 − 2

You're going to say the facts for all of these minus problems.

- (Point to **12 − 6.**) Read the problem. Get ready. (Touch.) *12 minus 6.*
- What's 12 minus 6? (Signal.) *6.*
- Say the fact. (Signal.) *12 − 6 = 6.*

b. (Repeat the following tasks for the remaining problems:)
- (Point to __.) Read the problem.
- What's __?
- Say the fact.
(Repeat problems that were not firm.)

EXERCISE 5: MONEY
WRITING DOLLAR AMOUNTS

a. (Display:)

[73:5A]

$3.18	$9.70
$5.98	$15.11
$11.40	

You'll read these dollar amounts.

- (Point to **$3.18.**) Read this. (Signal.) *3 dollars and 18 cents.*
- (Repeat for remaining amounts.)
(Repeat amounts that were not firm.)

b. (Distribute unopened workbooks to students.)
- Open your workbook to Lesson 73 and find part 1.
(Observe children and give feedback.)
(Teacher reference:)

a. $_____ b. $_____ c. $_____

You're going to write dollar amounts. The dollar sign is already written for each amount.

- Touch space A. ✔
- You'll write 6 dollars and 31 cents in space A. What will you write? (Signal.) *6 dollars and 31 cents.*
- Touch the dollar sign for A. ✔
- What do you write after the dollar sign for 6 dollars and 31 cents? (Signal.) *6.*
- What do you write next? (Signal.) *(The) dot.*
- What do you write after the dot? (Signal.) *31.*
- Write 6 dollars and 31 cents.
(Observe children and give feedback.)

c. Check your work.
- Touch the dollar sign for A. ✔
- Touch the number 6. ✔
- Touch the dot. ✔
- Touch the number 31. ✔
- Did you do everything right?

d. Touch the space for B. ✔
You'll write 8 dollars and 70 cents for space B.
- What will you write? (Signal.) *8 dollars and 70 cents.*
- What do you write after the dollar sign for 8 dollars and 70 cents? (Signal.) *8.*
- What do you write next for 8 dollars and 70 cents? (Signal.) *(The) dot.*
- What do you write after the dot? (Signal.) *70.*
- Write 8 dollars and 70 cents.
(Observe children and give feedback.)

e. Check your work.
- Touch the dollar sign for B. ✔
- What's the first thing you wrote? (Signal.) *8.*
- What's the next thing you wrote? (Signal.) *(A) dot.*
- What's the next thing you wrote? (Signal.) *70.*
- Read the amount for B. Get ready. (Signal.) *8 dollars and 70 cents.*

f. Touch the space for C. ✔
You'll write 13 dollars and 59 cents for space C.
- What will you write? (Signal.) *13 dollars and 59 cents.*
- What do you write after the dollar sign for 13 dollars and 59 cents? (Signal.) *13.*
- What do you write next for 13 dollars and 59 cents. (Signal.) *(The) dot.*
- What do you write after the dot? (Signal.) *59.*
- Write 13 dollars and 59 cents.
(Observe children and give feedback.)

g. Check your work.
- Touch the dollar sign for C. ✔
- What's the first thing you wrote? (Signal.) *13.*
- What's the next thing you wrote? (Signal.)
  (A) dot.
- What's the next thing you wrote? (Signal.) *59.*
- Read the amount for C. Get ready. (Signal.)
  13 dollars and 59 cents.

EXERCISE 6: RULER
COUNT BACKWARD

a. Find part 2 on worksheet 73. ✔
  (Teacher reference:)

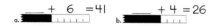

We're going to figure out how many
centimeters are shaded for each ruler.
- Touch ruler A. ✔
- Touch the number above the unshaded part. ✔
- How many centimeters is the unshaded part?
  (Signal.) *6.*
- Touch the number after equals. ✔
- How many centimeters is ruler A? (Signal.) *41.*
- Touch the end of the ruler. ✔
  I'll get 41 going and count backward. You'll
  touch the correct lines. Tell me to stop when
  you're touching the end of the shaded part.
  (Children should be touching the end of the
  ruler.) *Forty-wuuun.* (Children touch lines.)
  40, 39, 38, 37, 36, 35. Stop.
  (Repeat until firm.)
b. Your turn to count backward.
- What number will you get going? (Signal.) *41.*
- Touch the end of the ruler. ✔
- Get it going. *Forty-wuuun.* Touch and count.
  (Tap.) *40, 39, 38, 37, 36, 35, stop.*
  (Repeat step b until firm.)
c. How many centimeters is the shaded part?
  (Signal.) *35.*
- Write 35. ✔
- Touch and read the equation. Get ready.
  (Signal.) *35 + 6 = 41.*
d. Touch ruler B. ✔
- Touch the number above the unshaded part. ✔
- How many centimeters is the unshaded part?
  (Signal.) *4.*
- Touch the number after the equals. ✔
- How many centimeters are both parts?
  (Signal.) *26.*

- You're going to get 26 going and count
  backward. You'll touch the correct lines and
  say stop when you get to the end of the
  shaded part.
- Touch the end of ruler B. ✔
- Get it going. *Twenty-siiix.* Touch and count.
  (Tap.) *25, 24, 23, 22, stop.*
  (Repeat until firm.)
e. How many centimeters is the shaded part?
  (Signal.) *22.*
- Write 22. ✔
- Touch and read the equation. Get ready.
  (Signal.) *22 + 4 = 26.*

EXERCISE 7: 3 ADDENDS IN COLUMNS

a. Find part 3 on worksheet 73. ✔
  (Teacher reference:)

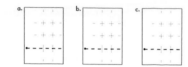

I'll tell you problems that plus three numbers.
You're going to write those problems in
columns. Only part of the column and row
lines are shown and the equals bar is dotted.
You'll write the digits and the signs in the right
place and make the equals bar.
- Touch where you'll write column problem A. ✔
  Listen to problem A: 12 plus 630 plus 7.
- Listen again: 12 plus 630 plus 7. Say
  problem A. (Signal.) *12 plus 630 plus 7.*
  (Repeat until firm.)
b. Put your pencil where you'll write the first
  digit of 12. ✔
- Put your pencil where you'll write the
  plus sign. ✔
- Put your pencil where you'll write the first
  digit of 630. ✔
- Put your pencil where you'll write 7. ✔
- Put your pencil where you'll start the
  equals bar.
c. Say the problem 12 plus 630 plus 7 for
  problem A. Get ready. (Signal.) *12 plus
  630 plus 7.*
- Write 12 plus 630 plus 7 for problem A.
  Remember to make the equals bar.
  (Observe children and give feedback.)
  (Teacher reference:)

d. Touch where you'll write column problem B. ✔
Listen to problem B: 532 plus 64 plus 3.

- Listen again: 532 plus 64 plus 3. Say problem B. (Signal.) *532 plus 64 plus 3.* (Repeat until firm.)
- Write 532 plus 64 plus 3 for problem B. Remember to make the equals bar. (Observe children and give feedback.) (Teacher reference:)

e. Touch where you'll write column problem C. ✔
Listen to problem C: 401 plus 2 plus 95.

- Listen again: 401 plus 2 plus 95. Say problem C. (Signal.) *401 plus 2 plus 95.* (Repeat until firm.)
- Write 401 plus 2 plus 95 for problem C. Remember to make the equals bar. (Observe children and give feedback.) (Teacher reference:)

Later, you'll work the problems in part 3 as part of your independent work.

EXERCISE 8: FACTS
ADDITION
[REMEDY]

a. Find part 4 on worksheet 73. ✔
(Teacher reference:)

[R Part H]

	d. 2 + 7	h. 6 + 10
a. 7 + 1	e. 6 + 6	i. 5 + 6
b. 4 + 6	f. 4 + 2	j. 6 + 2
c. 5 + 3	g. 3 + 6	

These are plus problems for families you know. You're going to read each problem and tell me the answer. Then you'll go back and work all of the problems.

- Touch and read problem A. Get ready. (Signal.) *7 plus 1.*
- What's 7 plus 1? (Signal.) *8.*

b. Touch and read problem B. (Signal.) *4 plus 6.*
- What's 4 plus 6? (Signal.) *10.*

c. (Repeat the following tasks for problems C through J:)
- Touch and read problem __.
- What's __?

d. Complete the equations for all of the problems in part 4.
(Observe children and give feedback.)

e. Check your work. You'll touch and read each fact.

- Fact A. (Signal.) *7 + 1 = 8.*
- (Repeat for:) B, *4 + 6 = 10;* C, *5 + 3 = 8;* D, *2 + 7 = 9;* E, *6 + 6 = 12;* F, *4 + 2 = 6;* G, *3 + 6 = 9;* H, *6 + 10 = 16;* I, *5 + 6 = 11;* J, *6 + 2 = 8.*

EXERCISE 9: WORD PROBLEMS (COLUMNS)

a. Find part 5 on worksheet 73. ✔
(Teacher reference:)

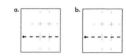

You're going to write the symbols for word problems in columns and work them. Only part of the column and row lines are shown and the equals bars are dotted. You'll write the numbers and the sign in the right places, and you'll make the equals bar.

b. Touch where you'll write the symbols for problem A. ✔
Listen to problem A: There were 580 apples on a tree. Jan and Jerry picked 430 of those apples. How many apples were still on the tree?

- Listen again and write the symbols for both parts: There were 580 apples on a tree. Jan and Jerry picked 430 of those apples.
- Everybody, touch and read problem A. Get ready. (Signal.) *580 minus 430.*

c. Work problem A. Put your pencil down when you know how many apples were still on the tree.
(Observe children and give feedback.)

- Read the problem and the answer you wrote for A. Get ready. (Signal.) *580 – 430 = 150.*
- How many apples were still on the tree? (Signal.) *150.*

d. Touch where you'll write the symbols for problem B. ✔

Listen to problem B: Mr. Briggs painted 113 pictures last year. He painted 46 this year. How many pictures did he paint altogether?

• Listen again and write the symbols for both parts: Mr. Briggs painted 113 pictures last year. He painted 46 this year.

• Everybody, touch and read problem B. Get ready. (Signal.) *113 plus 46.*

e. Work problem B. Put your pencil down when you know how many pictures Mr. Briggs painted altogether.

(Observe children and give feedback.)
(Answer key:)

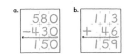

a.
$$\begin{array}{r} 5\,8\,0 \\ -4\,3\,0 \\ \hline 1\,5\,0 \end{array}$$

b.
$$\begin{array}{r} 1\,1\,3 \\ +\;4\,6 \\ \hline 1\,5\,9 \end{array}$$

• Read the problem and the answer you wrote for B. Get ready. (Signal.) *113 + 46 = 159.*

• How many pictures did Mr. Briggs paint altogether? (Signal.) *159.*

EXERCISE 10: INDEPENDENT WORK

a. Find part 6 on worksheet 73. ✔
(Teacher reference:)

$$\underline{6 \quad 6}\,{}_{\rightarrow}12$$
_____
_____

There are only two facts for this number family. You'll write the facts in the spaces below.

b. Turn to the other side of worksheet 73 and find part 7. ✔
(Teacher reference:)

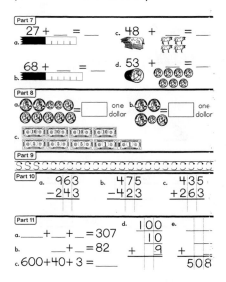

You'll complete the equations for the rulers and the objects.

In part 8, you'll circle the words one dollar or write the cents for each group of coins. You'll write an equals and the number of dollars to show what the group of bills is worth.

In part 9, you'll write the row of Ss.

You'll work the problems in parts 10 and 11.

c. Complete worksheet 73. Remember to work the problems in part 3 and write the facts for part 6 on the other side of it.

(Observe children and mark incorrect responses on children's worksheets as you give feedback.)

Lesson 74

EXERCISE 1: NUMBER FAMILIES
MISSING NUMBER IN FAMILY

a. (Display:) [74:1A]

$$\xrightarrow[\quad]{6 \quad 4} \_ \qquad \xrightarrow[\quad]{6 \quad 6} \_ \qquad \xrightarrow[\quad]{\_\_ \quad 2} 10$$

$$\xrightarrow[\quad]{\_\_ \quad 2} 9 \qquad \xrightarrow[\quad]{9 \quad \_\_} 11 \qquad \xrightarrow[\quad]{6 \quad 3} \_$$

$$\xrightarrow[\quad]{4 \quad \_\_} 6 \qquad \xrightarrow[\quad]{\_\_ \quad 5} 11 \qquad \xrightarrow[\quad]{10 \quad \_\_} 18$$

You're going to say the problem for the missing number in each family.

b. (Point to $\xrightarrow[\quad]{6 \quad 4}\_$.) Say the problem for the missing number. Get ready. (Touch.) *6 plus 4.*
* What's 6 plus 4? (Signal.) *10.*
c. (Point to $\xrightarrow[\quad]{\_\_ \quad 2}9$.) Say the problem for the missing number. (Touch.) *9 minus 2.*
* What's 9 minus 2? (Signal.) *7.*
d. (Repeat the following tasks for remaining families:)

(Point to __.)	Say the problem for the missing number. Get ready.	What's __?	
$\xrightarrow[\quad]{4 \quad \_}6$	6 – 4	6 – 4	2
$\xrightarrow[\quad]{6 \quad 6}\_$	6 + 6	6 + 6	12
$\xrightarrow[\quad]{9 \quad}11$	11 – 9	11 – 9	2
$\xrightarrow[\quad]{\_\quad 5}11$	11 – 5	11 – 5	6
$\xrightarrow[\quad]{\_\quad 2}10$	10 – 2	10 – 2	8
$\xrightarrow[\quad]{6 \quad 3}\_$	6 + 3	6 + 3	9
$\xrightarrow[\quad]{10 \quad}18$	18 – 10	18 – 10	8

(Repeat for families that were not firm.)

EXERCISE 2: COLUMN ADDITION
CARRYING   [REMEDY]

a. (Display:) [W] [74:2A]

$$\begin{array}{r} 2\,1 \\ +5\,9 \\ \hline \end{array} \qquad \begin{array}{r} 2\,6 \\ +5\,6 \\ \hline \end{array} \qquad \begin{array}{r} 1\,2 \\ +1\,8 \\ \hline \end{array}$$

You saw problems like these last time. The answer to the ones column for these problems is a two-digit number. You saw how to write the tens digit in the tens column and the ones digit in the ones column.

* (Point to **21**.) Read the problem. (Touch.) *21 plus 59.*
* (Point to $\begin{array}{r}2\overset{\downarrow}{1}\\+5\,9\end{array}$.) Read the problem for the ones column. (Touch.) *1 plus 9.*
* What's the answer? (Signal.) *10.*
* Is 10 a two-digit answer? (Signal.) *Yes.*
* What's the tens digit of 10? (Signal.) *1.*
* What's the ones digit of 10? (Signal.) *Zero.*
  (Repeat until firm.)
* Where do I write the tens digit? (Signal.) *In the tens column.*
  Yes, I write the tens digit in the tens column, and I write zero in the ones column.
  (Add to show:) [74:2B]

$$\begin{array}{r} {\scriptstyle 1} \\ 2\,1 \\ +5\,9 \\ \hline 0 \end{array}$$

b. (Point to **26**.) Read the problem. (Touch.) *26 plus 56.*
* (Point to $\begin{array}{r}2\overset{\downarrow}{6}\\+5\,6\end{array}$.) Read the problem for the ones column. (Touch.) *6 plus 6.*
* What's the answer? (Signal.) *12.*
* Is 12 a two-digit answer? (Signal.) *Yes.*
* What's the tens digit of 12? (Signal.) *1.*
* What's the ones digit of 12? (Signal.) *2.*
* Where do I write the tens digit? (Signal.) *In the tens column.*

- Where do I write the ones digit? (Signal.) *In the ones column.*
  (Add to show:) [74:2C]

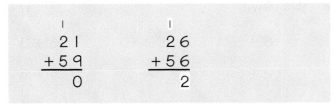

```
    2 1           2 6
  + 5 9         + 5 6
  ───────       ───────
      0             2
```

c. (Point to **12**.) Read the problem. (Touch.) *12 plus 18.*
- (Point to ¹²₊₁₈.) Read the problem for the ones column. (Touch.) *2 plus 8.*
- What's the answer? (Signal.) *10.*
- Is 10 a two-digit answer? (Signal.) *Yes.*
- What's the tens digit of 10? (Signal.) *1.*
- What's the ones digit of 10? (Signal.) *Zero.*
- Where do I write the tens digit? (Signal.) *In the tens column.*
- Where do I write the ones digit? (Signal.) *In the ones column.*
  (Add to show:) [74:2D]

```
    2 1           2 6           1 2
  + 5 9         + 5 6         + 1 8
  ───────       ───────       ───────
      0             2             0
```

d. We're going to work the rest of each problem.
- (Point to **21 + 59**.) Do we have an answer in the ones column for this problem? (Touch.) *Yes.*
- (Point to below 5.) Do we have an answer in the tens column? (Touch.) *No.*
e. I'll read the new problem in the tens column. (Touch.) 1 plus 2 plus 5.
- (Point to ²₊₅₉₀.) Your turn: Read the new problem in the tens column. (Touch.) *1 plus 2 plus 5.*
f. Listen: 1 plus 2 is 3, and 3 plus 5 is 8. What's the answer for the tens column? (Signal.) *8.*
  (Add to show:) [74:2E]

```
    2 1           2 6           1 2
  + 5 9         + 5 6         + 1 8
  ───────       ───────       ───────
    8 0             2             0
```

- We have the answer to the whole problem. What's 21 plus 59? (Signal.) *80.*

g. (Point to **26 + 56**.) Do we have an answer in the ones? (Touch.) *Yes.*
- (Point to below 5.) Do we have an answer in the tens? (Signal.) *No.*
- (Point to ²⁶₊₅₆₂.) Read the new problem for the tens column. (Touch.) *1 plus 2 plus 5.*
- What's 1 plus 2? (Signal.) *3.*
- What's 3 plus 5? (Signal.) *8.*
- What's the answer for the tens column? (Signal.) *8.*
  (Add to show:) [74:2F]

```
    2 1           2 6           1 2
  + 5 9         + 5 6         + 1 8
  ───────       ───────       ───────
    8 0           8 2             0
```

Look at the answer to the whole problem.
- What's 26 plus 56? (Signal.) *82.*
h. (Point to **12 + 18**.) Do we have an answer in the ones? (Touch.) *Yes.*
- (Point to below 1.) Do we have an answer in the tens? (Signal.) *No.*
- (Point to ¹²₊₁₈₀.) Read the new problem for the tens column. (Touch.) *1 plus 1 plus 1.*
- What's 1 plus 1? (Signal.) *2.*
- What's 2 plus 1? (Signal.) *3.*
- What's the answer for the tens column? (Signal.) *3.*
  (Add to show:) [74:2G]

```
    2 1           2 6           1 2
  + 5 9         + 5 6         + 1 8
  ───────       ───────       ───────
    8 0           8 2           3 0
```

Look at the answer to the whole problem.
- What's 12 plus 18? (Signal.) *30.*

EXERCISE 3: 3-D OBJECTS

CUBE, CYLINDER

a. (Display:) [74:3A]

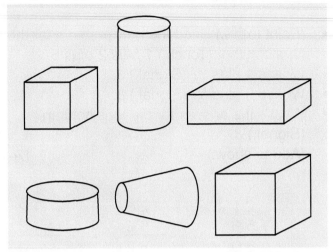

Some of these objects are cubes. Some of them are cylinders and some of them are not cubes or cylinders.

- (Point to ⬡.) Is this a cube? (Touch.) *Yes.*
- Is this a cylinder? (Touch.) *No.*
- What is this? (Touch.) *(A) cube.*

b. (Point to ⬡.) Is this a cube? (Touch.) *No.*
- Is this a cylinder? (Touch.) *Yes.*
- What is this? (Touch.) *(A) cylinder.*

c. (Point to ⬡.) Is this a cube? (Touch.) *No.*
- Is this a cylinder? (Touch.) *No.*

d. (Point to ⬡.) Is this a cube? (Touch.) *No.*
- Is this a cylinder? (Touch.) *Yes.*
- What is this? (Touch.) *(A) cylinder.*

e. (Point to ⬡.) Is this a cube? (Touch.) *No.*
- Is this a cylinder? (Touch.) *No.*

f. (Point to ⬡.) Is this a cube? (Touch.) *Yes.*
- Is this a cylinder? (Touch.) *No.*
- What is this? (Touch.) *(A) cube.*
  (Repeat objects that were not firm.)

EXERCISE 4: FACTS

ADDITION   `REMEDY`

a. (Display:) [74:4A]

6 + 5	4 + 6	6 + 3
2 + 6	6 + 1	5 + 6
6 + 10	6 + 6	6 + 4

All these plus problems are from families that have a small number of 6.

- (Point to **6 + 5.**) Read the problem. Get ready. (Touch.) *6 plus 5.*
- What's 6 plus 5? (Signal.) *11.*

b. (Repeat the following tasks for the remaining problems:)
- (Point to __.) Read the problem.
- What's __?
  (Repeat problems that were not firm.)

EXERCISE 5: MIXED COUNTING

a. Start with 100 and count by 25s to 200. Get 100 going. *One huuundred.* Count. (Tap.) *125, 150, 175, 200.*
- Start with 100 and count by tens to 200. Get 100 going. *One huuundred.* Count. (Tap.) *110, 120, 130, 140, 150, 160, 170, 180, 190, 200.*
- Start with 200 and count backward by tens to 100. Get 200 going. *Two huuundred.* Count backward. (Tap.) *190, 180, 170, 160, 150, 140, 130, 120, 110, 100.*
- Start with 100 and count by fives to 150. Get 100 going. *One huuundred.* Count. (Tap.) *105, 110, 115, 120, 125, 130, 135, 140, 145, 150.*

b. Listen: You have 105. When you plus tens, what's the next number? (Signal.) *115.*
- You have 105. When you minus ones, what's the next number? (Signal.) *104.*
- You have 105. When you plus fives, what's the next number? (Signal.) *110.*

EXERCISE 6: RULER

COUNT BACKWARD   REMEDY

a. (Distribute unopened workbooks to students.)
• Open your workbook to Lesson 74 and find part 1.
  (Observe children and give feedback.)
  (Teacher reference:)   R Part E

The rulers in part 1 are supposed to show centimeters. You'll figure out how many centimeters long the shaded part is. For each problem, you'll get the number going for the end of the ruler and count backward to the end of the shaded part.
• Touch ruler A. ✔
• Count and write the number above the unshaded part.
  (Observe children and give feedback.)
• How many centimeters is the unshaded part? (Signal.) 5.
• Touch the end of the ruler. ✔
  I'll get 37 going and count backward. You'll touch the correct lines. Tell me to stop when you're touching the end of the shaded part.
• (Children should be touching the end of the ruler.) Thirty-sevennn. (Children touch lines.) 36, 35, 34, 33, 32. *Stop.*
  (Repeat until firm.)
b. Your turn to count backward.
• What number will you get going? (Signal.) *37.*
• Touch the end of the ruler. ✔
• Get it going. *Thirty-sevennn.* (Tap.) *36, 35, 34, 33, 32, stop.*
  (Repeat step b until firm.)
c. How many centimeters is the shaded part? (Signal.) *32.*
• Write 32. ✔
• Touch and read the equation. Get ready. (Signal.) *32 + 5 = 37.*
d. Touch ruler B. ✔
• Count and write the number for the unshaded part.
  (Observe children and give feedback.)
• How many centimeters is the unshaded part? (Signal.) *4.*
• Touch the end of the ruler. ✔
  I'll get 21 going and count backward. You'll touch the correct lines. Tell me to stop when you're touching the end of the shaded part.

• (Children should be touching the end of the ruler.) Twenty-wuuun. (Children touch lines.) 20, 19, 18, 17. *Stop.*
  (Repeat until firm.)
e. Your turn to count backward.
• What number will you get going? (Signal.) *21.*
• Touch the end of the ruler. ✔
• Get it going. *Twenty-wuuun.* (Tap.) *20, 19, 18, 17, stop.*
  (Repeat step e until firm.)
f. How many centimeters is the shaded part? (Signal.) *17.*
• Write 17. ✔
• Touch and read the equation. Get ready. (Signal.) *17 + 4 = 21.*

EXERCISE 7: MONEY

WRITING DOLLAR AMOUNTS

a. (Display:)   [74:7A]

| $7.39 | $15.19 |
| $12.71 | $20.34 |

These numbers are dollar amounts. Remember how to read them.
• (Point to **$7.39**.) Read this. (Touch.) *7 dollars and 39 cents.*
• (Repeat for remaining amounts.)
  (Repeat amounts that were not firm.)
b. Find part 2 on worksheet 74. ✔
  (Teacher reference:)

  a. $_____   b. $_____   c. $_____

You're going to write dollar amounts. The dollar sign is already written for each amount.
• Touch space A. ✔
• You'll write 11 dollars and 91 cents in space A. What will you write? (Signal.) *11 dollars and 91 cents.*
• Write 11 dollars and 91 cents.
  (Observe children and give feedback.)
• Check your work. Touch the dollar sign for A. ✔
• What's the first thing you wrote? (Signal.) *11.*
• What's the next thing you wrote? (Signal.) *(A) dot.*
• What's the next thing you wrote? (Signal.) *91.*

c. Touch space B. ✔
- You'll write 8 dollars and 14 cents in space B. What will you write? (Signal.) *8 dollars and 14 cents.*
- Write 8 dollars and 14 cents.
  (Observe children and give feedback.)
- Check your work. Touch the dollar sign for B. ✔
- What's the first thing you wrote? (Signal.) *8.*
- What's the next thing you wrote? (Signal.) *(A) dot.*
- What's the next thing you wrote? (Signal.) *14.*

d. Touch space C. ✔
- You'll write 13 dollars and 18 cents in space C. What will you write? (Signal.) *13 dollars and 18 cents.*
- Write 13 dollars and 18 cents.
  (Observe children and give feedback.)
- Check your work. Touch the dollar sign for C. ✔
- What's the first thing you wrote? (Signal.) *13.*
- What's the next thing you wrote? (Signal.) *(A) dot.*
- What's the next thing you wrote? (Signal.) *18.*

EXERCISE 8: FACTS
SUBTRACTION

a. Find part 3 on worksheet 74. ✔
(Teacher reference:)

a. $5-2$	e. $8-2$	i. $9-6$
b. $20-10$	f. $12-6$	j. $6-4$
c. $10-4$	g. $10-8$	k. $5-4$
d. $4-2$	h. $7-5$	l. $11-6$

These are minus problems for families you know. You're going to read each problem and say the answer. Then you'll go back and work them.
- Touch and read problem A. Get ready. (Signal.) *5 minus 2.*
- What's the answer? (Signal.) *3.*

b. (Repeat the following tasks for problems B through L:)
- Touch and read problem __.
- What's the answer?

c. Complete the equations for all of the problems in part 3.
  (Observe children and give feedback.)

d. Check your work. You'll touch and read each fact.
- Fact A. (Signal.) *5 – 2 = 3.*
- (Repeat for:) B, *20 – 10 = 10;* C, *10 – 4 = 6;* D, *4 – 2 = 2;* E, *8 – 2 = 6;* F, *12 – 6 = 6;* G, *10 – 8 = 2;* H, *7 – 5 = 2;* I, *9 – 6 = 3;* J, *6 – 4 = 2;* K, *5 – 4 = 1;* L, *11 – 6 = 5.*

EXERCISE 9: COLUMN PROBLEMS
ADDITION AND SUBTRACTION

a. Find part 4 on worksheet 74. ✔
(Teacher reference:)

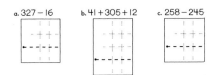

You'll read each problem written in a row and then write it in a column. For all of these problems, the equals bar is dotted. You'll make the equals bar for each problem you write.
- Touch and read the row problem for A. Get ready. (Signal.) *327 minus 16.*
- Write 327 minus 16 equals in a column. Don't work it, just write it. Remember to make the equals bar.
  (Observe children and give feedback.)

b. Touch and read the row problem for B. Get ready. (Signal.) *41 plus 305 plus 12.*
- Write the problem in a column for B.
  (Observe children and give feedback.)

c. Touch and read the row problem for C. Get ready. (Signal.) *258 minus 245.*
- Write the problem in a column for C.
  (Observe children and give feedback.)
  (Teacher reference:)

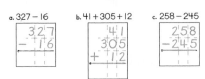

Later, you'll work the problems as part of your independent work.

a. Find part 5 on worksheet 74. ✔
(Teacher reference:)

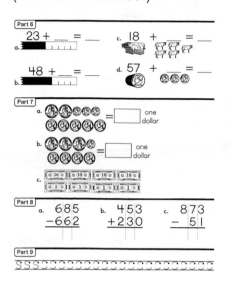

a. ____6__,12   b. 6 ___,8   c. 6 4 ,__
_____   _____   _____

You'll write the fact for the missing number in each family and complete the family.

b. Turn to the other side of worksheet 74 and find part 6. ✔
(Teacher reference:)

You'll complete the equations for the rulers and the objects.

In part 7, you'll circle the words one dollar or write the cents for each group of coins. You'll write an equals and the number of dollars to show what the group of bills is worth.

In part 8, you'll work the column problems. In part 9, you'll write a row of Ss.

c. Complete worksheet 74. Remember to work the problems in part 4 and write the families in part 5 on the other side of it.

(Observe children and mark incorrect responses on children's worksheets as you give feedback.)

Lesson

EXERCISE 1: FACTS

SUBTRACTION

REMEDY

a. (Display:)   [75:1A]

16 – 10	10 – 6	16 – 6
9 – 6	12 – 6	11 – 5
10 – 4	8 – 6	7 – 1

You're going to say the facts for all of these minus problems.

- (Point to **16 – 10.**) Read the problem. Get ready. (Touch.) *16 minus 10.*
- What's 16 minus 10? (Signal.) *6.*
- Say the fact. (Signal.) *16 – 10 = 6.*

b. (Repeat the following tasks for the remaining problems:)
- (Point to __.) Read the problem.
- What's __?
- Say the fact.
  (Repeat problems that were not firm.)

EXERCISE 2: 2-DIMENSIONAL FIGURES

DECOMPOSITION

a. (Display:) W   [75:2A]

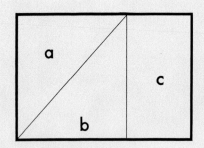

This is a big shape with parts. The thick line shows the big shape.

- (Trace outside of figure.) What is the big shape? (Signal.) *(A) rectangle.*
b. Look at part A. (Trace triangle.)
- What shape is part A? (Touch.) *(A) triangle.*
- (Point to **B.**) What shape is part B? (Touch.) *(A) triangle.*
- (Point to **C.**) What shape is part C? (Touch.) *(A) rectangle.*
- (Outline whole figure.) And what shape is the whole figure? (Touch.) *(A) rectangle.*
  (Repeat step b until firm.)

---

c. (Display:) W   [75:2B]

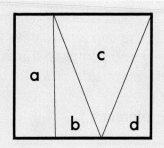

The thick line shows the whole figure.
- What shape is the whole figure? (Signal.) *(A) rectangle.* ✔
- Look at part A. ✔
- (Point to **A.**) What shape is part A? (Touch.) *(A) rectangle.*
- (Point to **B.**) What shape is part B? (Touch.) *(A) triangle.*
- (Point to **C.**) What shape is part C? (Touch.) *(A) triangle.*
- (Point to **D.**) What shape is part D? (Touch.) *(A) triangle.*

d. (Add to show:)   [75:2C]

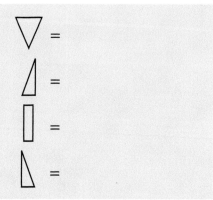

Here are shapes for the parts of the whole figure. We're going to figure out the letter for each part.

- (Point to ▽= and ▯.) Are these shapes the same? (Touch.) *No.*
- (Point to ▽= and ◺.) Are these shapes the same? (Touch.) *No.*
- (Point to ▽= and ▽ᶜ.) Are these shapes the same? (Touch.) *Yes.*
- This is part C. What part? (Touch.) *C.*
- So I write C after the equals.
  (Add to show:)   [75:2D]

$$\bigtriangledown = c$$

e. (Point to △=.) Here's another part of the figure.
- Raise your hand when you know the letter of this part. ✔
- Everybody, what letter? (Signal.) *D.*
  (Add to show:) [75:2E]

$$\triangle = d$$

f. (Point to ▯=.) Here's another part of the figure.
- Raise your hand when you know the letter of this part. ✔
- Everybody, what letter? (Signal.) *A.*
  (Add to show:) [75:2F]

$$\textstyle\square = a$$

g. (Point to ◺=.)
- Raise your hand when you know the letter of this part. ✔
- Everybody, what letter? (Signal.) *B.*
  (Add to show:) [75:2G]

$$\triangledown = c$$
$$\triangle = d$$
$$\square = a$$
$$\triangle = b$$

EXERCISE 3: COLUMN ADDITION
CARRYING
REMEDY

a. You learned that if the answer to the problem in the ones column has two digits, you write the ones digit in the ones column and the tens digit in the tens column.
- Where do you write the ones digit? (Signal.)
  In the ones column.
- Where do you write the tens digit? (Signal.)
  In the tens column.
b. Listen: Let's say the answer in the ones column is 14.
- What's the tens digit of 14? (Signal.) *1.*
- What's the ones digit of 14? (Signal.) *4.*
- Where do you write the tens digit? (Signal.)
  In the tens column.
- Where do you write the ones digit? (Signal.)
  In the ones column.

c. Listen: Let's say the answer in the ones column is 12.
- What's the tens digit of 12? (Signal.) *1.*
- What's the ones digit of 12? (Signal.) *2.*
- Where do you write the tens digit? (Signal.)
  In the tens column.
- Where do you write the ones digit? (Signal.)
  In the ones column.
d. Listen: Let's say the answer in the ones column is 18.
- What's the tens digit of 18? (Signal.) *1.*
- What's the ones digit of 18? (Signal.) *8.*
- Where do you write the tens digit? (Signal.)
  In the tens column.
- Where do you write the ones digit? (Signal.)
  In the ones column.
e. (Display:) W [75:3A]

$$\begin{array}{r} 2\,5 \\ +\,4\,9 \\ \hline \end{array}$$

- (Point to **25**.) Read the problem. (Signal.)
  25 plus 49.
- Say the problem for the ones. (Signal.)
  5 plus 9.
  The answer is 14.
- What's the answer? (Signal.) *14.*
- What's the tens digit of 14? (Signal.) *1.*
- What's the ones digit of 14? (Signal.) *4.*
  I write the tens digit in the tens column and the ones digit on the ones column.
f. (Add to show:) [75:3B]

$$\begin{array}{r} {}^{1} \\ 2\,5 \\ +\,4\,9 \\ \hline 4 \end{array}$$

- Read the new problem in the tens column.
  (Touch.) *1 plus 2 plus 4.*
  The answer is 7.
  (Add to show:) [75:3C]

$$\begin{array}{r} {}^{1} \\ 2\,5 \\ +\,4\,9 \\ \hline 7\,4 \end{array}$$

- Look at the answer. What's 25 plus 49?
  (Signal.) *74.*
- Read the problem we started with and the answer. (Touch.) *25 + 49 = 74.*

g. (Display:) [75:3D]

$$\begin{array}{r} 78 \\ +14 \\ \hline \end{array}$$

- (Point to **78**.) Read the problem. (Signal.)
  78 plus 14.
- Say the problem for the ones. (Signal.)
  8 plus 4.
  The answer is 12.
- What's the tens digit of 12? (Signal.) *1.*
- What's the ones digit of 12? (Signal.) *2.*
  I write the tens digit in the tens column and the
  ones digit on the ones column.

h. (Add to show:) [75:3E]

$$\begin{array}{r} 1 \\ 78 \\ +14 \\ \hline 2 \end{array}$$

- Read the new problem in the tens column.
  (Touch.) *1 plus 7 plus 1.*
  The answer is 9.
  (Add to show:) [75:3F]

$$\begin{array}{r} 1 \\ 78 \\ +14 \\ \hline 92 \end{array}$$

- Look at the answer. What's 78 plus 14?
  (Signal.) *92.*
- Read the problem we started with and the
  answer. (Touch.) *78 + 14 = 92.*

EXERCISE 4: MIXED COUNTING

a. Listen: Count by 25s to 200. Get ready. (Tap.)
   25, 50, 75, 100, 125, 150, 175, 200.
- Start with 56 and plus tens to 106. Get 56
  going. *Fifty-siiix.* Plus tens. (Tap.) *66, 76, 86,*
  96, 106.
- Start with 35 and count by fives to 80. Get 35
  going. *Thirty-fiiive.* Count. (Tap.) *40, 45, 50, 55,*
  60, 65, 70, 75, 80.

b. Start with 100 and count backward by tens.
   Get 100 going. *One huuundred.* Count
   backward. (Tap.) *90, 80, 70, 60, 50, 40, 30,*
   20, 10.
- Start with 158 and count backward by
  ones to 150. Get 158 going. *One hundred*
  fifty-eieieight. Count backward. (Tap.)
  157, 156, 155, 154, 153, 152, 151, 150.
- Start with 71 and plus tens to 101. Get 71
  going. *Seventy-wuuun.* Plus tens. (Tap.) *81,*
  91, 101.
c. Listen: You have 25. When you plus ones,
   what's the next number? (Signal.) *26.*
- You have 25. When you plus fives, what's the
  next number? (Signal.) *30.*
- You have 25. When you plus 25s, what's the
  next number? (Signal.) *50.*
- You have 25. When you plus tens, what's the
  next number? (Signal.) *35.*
- You have 25. When you minus ones, what's
  the next number? (Signal.) *24.*

EXERCISE 5: NUMBER FAMILIES
MISSING NUMBER IN FAMILY

a. (Distribute unopened workbooks to students.)
- Open your workbook to Lesson 75 and find
  part 1.
  (Observe children and give feedback.)
  (Teacher reference:)

You're going to say the problem for the
missing number in each family.
b. Family A. Say the problem for the missing
   number. Get ready. (Signal.) *6 minus 2.*
- What's 6 minus 2? (Signal.) *4.*
c. Family B. Say the problem for the missing
   number. Get ready. (Signal.) *11 minus 5.*
- What's 11 minus 5? (Signal.) *6.*

d. (Repeat the following tasks for families C through K:)

Family __.	Say the problem for the missing number.	What's __?	
C	8 – 5	8 – 5	3
D	12 – 6	12 – 6	6
E	3 + 1	3 + 1	4
F	14 – 10	14 – 10	4
G	5 – 3	5 – 3	2
H	9 – 6	9 – 6	3
I	9 + 2	9 + 2	11
J	10 – 6	10 – 6	4
K	6 – 4	6 – 4	2

(Repeat for families that were not firm.)

e. Write the missing number in each family. Put your pencil down when you've completed the families in part 1.
(Observe children and give feedback.)

f. Check your work. You'll tell me the number you wrote in each family.
• Family A. (Signal.) *4.*
• (Repeat for:) B, *6*; C, *3*; D, *6*; E, *4*; F, *4*; G, *2*; H, *3*; I, *11*; J, *4*; K, *2*.

EXERCISE 6: MONEY
WRITING DOLLAR AMOUNTS   REMEDY

a. You're going to write dollar amounts. The first thing you write is the dollar sign. I'll show you how to write dollar signs.
You start with a capital S.
(Display:) W   [75:6A]

S

Then you make a line that goes straight down through the S. Watch.
(Add to show:)   [75:6B]

$

b. Find part 2 on worksheet 75. ✔
(Teacher reference:)   R Part B

a. _____ b. _____ c. _____

You're going to write dollars and cents.
• Touch space A. ✔
• You'll write 4 dollars and 19 cents for space A. What will you write? (Signal.) *4 dollars and 19 cents.*

• Write the dollar sign for A. Write a capital S, then make a straight line down through it.
(Observe children and give feedback.)
• Now write the rest of 4 dollars and 19 cents. Remember the dot after the number of dollars.
(Observe children and give feedback.)
Check your work.
• Touch the dollar sign you wrote. ✔
• Touch the number of dollars. ✔
• What number? (Signal.) *4.*
• Touch the dot. ✔
• Touch the number of cents. ✔
• How many cents? (Signal.) *19.*
Did you do everything right?

c. Touch the space for B. ✔
• You'll write 9 dollars and 73 cents for space B. What will you write? (Signal.) *9 dollars and 73 cents.*
• Write the dollar sign. Remember, a capital S with a line straight down through it. ✔
• Now write the rest of 9 dollars and 73 cents. ✔
(Observe children and give feedback.)
• Touch and read the dollar amount you wrote for B. (Signal.) *9 dollars and 73 cents.*
• What's the first thing you wrote? (Signal.) *A dollar sign.*
• What's the next thing you wrote? (Signal.) *9.*
• What's the next thing you wrote? (Signal.) *(A) dot.*
• What's the next thing you wrote? (Signal.) *73.*

d. Touch the space for C. ✔
• You'll write 10 dollars and 12 cents for space C. What will you write? (Signal.) *10 dollars and 12 cents.*
• Write the dollar sign. Remember, capital S with a line straight down through it. ✔
• Now write the rest of 10 dollars and 12 cents. ✔
(Observe children and give feedback.)
• Touch and read the dollar amount you wrote for C. (Signal.) *10 dollars and 12 cents.*
• What's the first thing you wrote? (Signal.) *A dollar sign.*
• What's the next thing you wrote? (Signal.) *10.*
• What's the next thing you wrote? (Signal.) *(A) dot.*
• What's the next thing you wrote? (Signal.) *12.*
(Answer key:)

a. $4.19   b. $9.73   c. $10.12

EXERCISE 7: RULER

COUNT BACKWARD

REMEDY

a. Find part 3 on worksheet 75. ✔
(Teacher reference:)

R Part F

The rulers in part 3 are supposed to show centimeters. You'll figure out how many centimeters long the shaded part is. For each problem, you'll get the number going for the end of the ruler and count backward to the end of the shaded part.

• Touch ruler A. ✔
• Count and write the number above the unshaded part.
  (Observe children and give feedback.)
• How many centimeters is the unshaded part? (Signal.) *3.*
• Touch the end of the ruler. ✔
  I'll get 62 going and count backward. You'll touch the correct lines. Tell me to stop when you're touching the end of the shaded part.
• (Children should be touching end of the ruler.) Sixty-twooo. (Children touch lines.) *61, 60, 59. Stop.*
  (Repeat until firm.)

b. Your turn to count backward.
• What number will you get going? (Signal.) *62.*
• Touch the end of the ruler. ✔
• Get it going. *Sixty-twooo. Count.* (Tap 3.) *61, 60, 59, stop.*
  (Repeat step b until firm.)

c. How many centimeters is the shaded part? (Signal.) *59.*
• Write 59. ✔
• Touch and read the equation. Get ready. (Signal.) *59 + 3 = 62.*

d. Touch ruler B. ✔
• Count and write the number for the unshaded part.
  (Observe children and give feedback.)
• How many centimeters is the unshaded part? (Signal.) *5.*
• Touch the end of the ruler. ✔
  I'll get 19 going and count backward. You'll touch the correct lines. Tell me to stop when you're touching the end of the shaded part.

• (Children should be touching end of the ruler.) Nineteeen. (Children touch lines.) *18, 17, 16, 15, 14. Stop.*
  (Repeat until firm.)

e. Your turn to count backward.
• What number will you get going? (Signal.) *19.*
• Touch the end of the ruler. ✔
• Get it going. *Nineteeen.* (Tap 5.) *18, 17, 16, 15, 14, stop.*
  (Repeat step e until firm.)

f. How many centimeters is the shaded part? (Signal.) *14.*
• Write 14. ✔
• Touch and read the equation. Get ready. (Signal.) *14 + 5 = 19.*

EXERCISE 8: FACTS

PLUS/MINUS MIX

a. Find part 4 on worksheet 75. ✔
(Teacher reference:)

a. $6-2$	d. $5-3$	g. $10-4$
b. $6+3$	e. $6+10$	h. $6+2$
c. $8-3$	f. $6+6$	i. $6+5$

These problems are from number families you know.

• Touch and read problem A. (Signal.) *6 minus 2.*
• What's 6 minus 2? (Signal.) *4.*

b. (Repeat the following tasks for problems B through I:)
• Touch and read problem __.
• What's __?
  (Repeat problems that were not firm.)

c. Complete the equations for all the problems in part 4. Put your pencil down when you're finished.
  (Observe children and give feedback.)

d. Check your work. You'll touch and read each fact.
• Fact A. (Signal.) *6 – 2 = 4.*
• (Repeat for:) B, *6 + 3 = 9;* C, *8 – 3 = 5;* D, *5 – 3 = 2;* E, *6 + 10 = 16;* F, *6 + 6 = 12;* G, *10 – 4 = 6;* H, *6 + 2 = 8;* I, *6 + 5 = 11.*

EXERCISE 9: INDEPENDENT WORK

a. Find part 5 on worksheet 75. ✔
(Teacher reference:)

a. 235+51+102 b. 895−864 c. 335+3+60 d. 178−73

You're going to write each problem in a column. For all of these problems, the equals bars are dotted and only part of the column and row lines are shown. Some of the problems are minus problems. So make sure you write the right sign.

• Write the row problem for A in a column. Don't work it, just write the column problem. Remember to make the equals bar.
(Observe children and give feedback.)

• You'll write the column problems for the other row problems in part 5 and work them as part of your independent work. Remember to make the equals bar for each problem.

b. Turn to the other side of worksheet 75 and find part 6. ✔
(Teacher reference:)

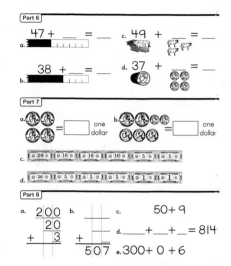

You'll complete the equations for the rulers and the objects.

In part 7, you'll circle the words one dollar or write the cents for each group of coins. You'll write an equals and the number of dollars to show what the group of bills is worth.

In part 8, you'll complete the place-value addition problems.

c. Complete worksheet 75. Remember to write the rest of the problems in part 5 in columns. Then work all the column problems.
(Observe children and mark incorrect responses on children's worksheets as you give feedback.)

Lessons 76–80 Planning Page

	Lesson 76	Lesson 77	Lesson 78	Lesson 79	Lesson 80
Student Learning Objectives	**Exercises** 1. Say addition and subtraction facts 2. Learn about carrying to the tens column 3. Identify 3-dimensional objects 4. Say subtraction facts 5. Decompose 2-dimensional figures 6. Write the symbols for word problems in columns and solve 7. Write dollar amounts with the dollar sign ($) and the decimal point 8. Solve subtraction facts 9. Complete work independently	**Exercises** 1. Identify 3-dimensional objects 2. Say subtraction facts 3. **Compare two lines for length** 4. Count forward or backward from a given number 5. Learn about carrying to the tens column 6. Decompose 2-dimensional figures 7. Solve addition and subtraction facts 8. **Solve problems with dollars and cents** 9. Complete work independently	**Exercises** 1. Say addition and subtraction facts 2. Compare two lines for length 3. Count forward or backward from a given number 4. **Solve addition problems with carrying** 5. Decompose 2-dimensional figures 6. Solve problems with dollars and cents 7. Find and write missing numbers in number families 8. Write the symbols for word problems in columns and solve 9. Complete work independently	**Exercises** 1. **Say two addition and two subtraction facts with a small number of 4** 2. Compare two lines for length 3. Count forward or backward from a given number 4. Solve addition problems with carrying 5. Solve addition and subtraction facts 6. Decompose 2-dimensional figures 7. Solve subtraction facts 8. **Solve money problems with a zero in the answer** 9. Complete work independently	**Exercises** 1. Say two addition and two subtraction facts with a small number of 4; find missing numbers in number families 2. **Learn how to write dollar amounts with a zero in the cents** 3. Count forward or backward from a given number 4. Solve addition and subtraction facts 5. **Solve addition problems with carrying and with 3 addends** 6. Compare two lines for length 7. Solve subtraction facts 8. **Solve money problems** 9. Complete work independently
Common Core State Standards for Mathematics					
1.OA 3		✔		✔	✔
1.OA 5	✔	✔	✔	✔	✔
1.OA 6	✔	✔	✔	✔	✔
1.OA 8	✔	✔	✔	✔	✔
1.NBT 1	✔	✔	✔	✔	✔
1.NBT 2	✔	✔	✔	✔	✔
1.NBT 4	✔	✔	✔	✔	✔
1.G 1	✔	✔			
1.G 2	✔	✔	✔	✔	✔
Teacher Materials	Presentation Book 2, Board Displays CD or chalkboard				
Student Materials	Workbook 2, Pencil				
Additional Practice	Student Practice Software: Block 3: Activity 1 (1.NBT.2 and 1.NBT.4), Activity 2 (1.NBT.4 and 1.OA.6), Activity 3 (1.NBT.1), Activity 4 (1.NBT.1), Activity 5 (1.NBT.2), Activity 6 (1.MD.2)				
Mastery Test					Student Assessment Book (Present Mastery Test 8 following Lesson 80.)

Lesson

EXERCISE 1: FACTS

PLUS/MINUS MIX

a. (Display:)   [76:1A]

3 + 6	3 + 2	6 + 1
9 − 1	4 + 6	4 + 2
7 − 5	11 − 2	9 − 2

These problems are from number families
you know.

- (Point to **3 + 6.**) Read this problem. Get ready.
  (Touch.) *3 plus 6.*
- Is the answer the big number or a small
  number? (Signal.) *The big number.*
- What's 3 plus 6? (Signal.) *9.*
- Say the fact. (Signal.) *3 + 6 = 9.*

b. (Point to **9 − 1.**) Read this problem. Get ready.
  (Touch.) *9 minus 1.*
- Is the answer the big number or a small
  number? (Signal.) *A small number.*
- What's 9 minus 1? (Signal.) *8.*
- Say the fact. (Signal.) *9 − 1 = 8.*

c. (Repeat the following tasks for the remaining
  problems:)
- (Point to __.) Read the problem.
- Is the answer the big number or a small
  number?
- What's __?
- Say the fact.
  (Repeat problems that were not firm.)

EXERCISE 2: COLUMN ADDITION

CARRYING

a. If the answer for the ones column has two
  digits, you write the tens digit in the tens
  column and the ones digit in the ones column.
- Where do you write the ones digit? (Signal.) *In
  the ones column.*
- Where do you write the tens digit? (Signal.) *In
  the tens column.*

b. (Display:) W   [76:2A]

$$\begin{array}{r} 72 \\ +18 \\ \hline \end{array}$$

- (Point to **72.**) Read the problem. (Touch.)
  72 plus 18.
- Say the problem for the ones. (Signal.)
  2 plus 8.
- What's the answer? (Signal.) *10.*
- What's the tens digit of 10? (Signal.) *1.*
- What's the ones digit of 10? (Signal.) *Zero.*
- Where do I write the tens digit? (Signal.) *In the
  tens column.*
- Where do I write the ones digit? (Signal.) *In
  the ones column.*
  (Add to show:)   [76:2B]

$$\begin{array}{r} {\scriptstyle 1} \\ 72 \\ +18 \\ \hline 0 \end{array}$$

c. Read the new problem in the tens column.
  (Touch.) *1 plus 7 plus 1.*
- What's 1 plus 7? (Signal.) *8.*
- What's 8 plus 1? (Signal.) *9.*
  That's the answer for the tens column.
  (Add to show:)   [76:2C]

$$\begin{array}{r} {\scriptstyle 1} \\ 72 \\ +18 \\ \hline 90 \end{array}$$

- Look at the answer. What's 72 plus 18?
  (Signal.) *90.*
- Read the problem and the answer. (Signal.)
  72 + 18 = 90.

d. (Display:) W   [76:2D]

$$\begin{array}{r} 27 \\ +\,58 \\ \hline \end{array}$$

- (Point to **27**.) Read the problem. (Touch.) *27 plus 58.*
- Say the problem for the ones. (Signal.) *7 plus 8.*
- 7 plus 8 equals 15. What's 7 plus 8? (Signal.) *15.*
  15 has a tens digit and a ones digit.
- What's the tens digit of 15? (Signal.) *1.*
- Where do I write the tens digit? (Signal.) *In the tens column.*
- What's the ones digit of 15? (Signal.) *5.*
- Where do I write the ones digit? (Signal.) *In the ones column.*
  (Add to show:)   [76:2E]

$$\begin{array}{r} {\scriptstyle 1} \\ 27 \\ +\,58 \\ \hline 5 \end{array}$$

e. Read the new problem in the tens column. (Signal.) *1 plus 2 plus 5.*
- What's 1 plus 2? (Signal.) *3.*
- What's 3 plus 5? (Signal.) *8.*
  (Add to show:)   [76:2F]

$$\begin{array}{r} {\scriptstyle 1} \\ 27 \\ +\,58 \\ \hline 85 \end{array}$$

- Look at the answer. What's 27 plus 58? (Signal.) *85.*
- Read the problem and the answer. (Signal.) *27 + 58 = 85.*

f. (Display:) W   [76:2G]

$$\begin{array}{r} 39 \\ +\,11 \\ \hline \end{array}$$

- (Point to **39**.) Read the problem. (Touch.) *39 plus 11.*
- Read the problem in the ones column. (Touch.) *9 plus 1.*
- What's the answer? (Signal.) *10.*
- What's the tens digit of 10? (Signal.) *1.*
- What's the ones digit of 10? (Signal.) *Zero.*
- Where do I write the tens digit? (Signal.) *In the tens column.*
- Where do I write the ones digit? (Signal.) *In the ones column.*
  (Add to show:)   [76:2H]

$$\begin{array}{r} {\scriptstyle 1} \\ 39 \\ +\,11 \\ \hline 0 \end{array}$$

g. Read the new problem in the tens column. (Touch.) *1 plus 3 plus 1.*
- What's 1 plus 3? (Signal.) *4.*
- What's 4 plus 1? (Signal.) *5.*
  (Add to show:)   [76:2I]

$$\begin{array}{r} {\scriptstyle 1} \\ 39 \\ +\,11 \\ \hline 50 \end{array}$$

- Look at the answer. What's 39 plus 11? (Signal.) *50.*
- Read the problem and the answer. (Signal.) *39 + 11 = 50.*

EXERCISE 3: 3-D OBJECTS
CUBE, CYLINDER

a. (Display:) [76:3A]

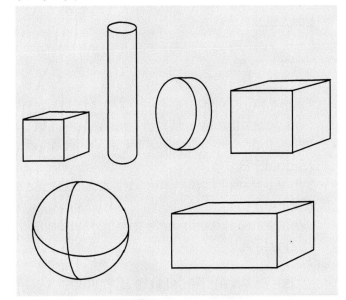

Tell me the names of these objects.

- (Point to ▱.) What's this object? (Touch.)
  (A) cube.

- (Point to ▯.) What's this object? (Touch.)
  (A) cylinder.

- (Point to ◑.) What's this object? (Touch.)
  (A) cylinder.

- (Point to ▱.) What's this object? (Touch.)
  (A) cube.
  (Repeat step a until firm.)

b. (Point to ⊕.) Is this a cylinder? (Touch.) *No.*
- Why not? (Call on a child. Idea: *It's not a cylinder because it doesn't have two ends that are the same size.*)

c. (Point to ▭.) Is this a cube? (Touch.) *No.*
- Why not? (Call on a child. Idea: *It's not a cube because it has faces that are not squares.*)

d. (Repeat steps b and c calling on different children.)

EXERCISE 4: FACTS
SUBTRACTION

a. (Display:) [76:4A]

12 − 1	11 − 6	8 − 6
10 − 2	9 − 6	12 − 6
9 − 3	11 − 5	10 − 6

You're going to say the facts for all of these minus problems.

- (Point to **12 − 1.**) Read the problem. Get ready. (Touch.) *12 minus 1.*
- What's 12 minus 1? (Signal.) *11.*
- Say the fact. (Signal.) *12 − 1 = 11.*

b. (Repeat the following tasks for the remaining problems:)

- (Point to __.) Read the problem.
- What's __?
- Say the fact.
  (Repeat problems that were not firm.)

EXERCISE 5: 2-DIMENSIONAL FIGURES
DECOMPOSITION

REMEDY

a. (Display:) W [76:5A]

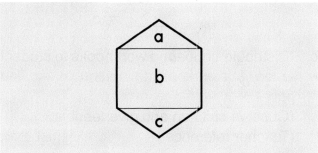

The thick line shows the whole figure. Raise your hand when you know if the whole figure is a triangle, a rectangle, or a hexagon. ✔

- What shape is the whole figure? (Touch.)
  (A) hexagon.
- (Point to **a.**) What shape is part A? (Touch.)
  (A) triangle.
- (Point to **b.**) What shape is part B? (Touch.)
  (A) rectangle.
- (Point to **c.**) What shape is part C? (Touch.)
  (A) triangle.

b. (Add to show:)   [76:5B]

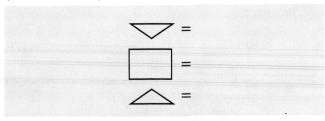

- (Point to ▽=.) Raise your hand when you know the letter for this part. ✔
- Everybody, what letter? (Signal.) *C.*
  (Add to show:)   [76:5C]

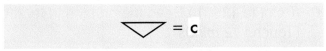

- (Point to ☐=.) Raise your hand when you know the letter for this part. ✔
- Everybody, what letter? (Signal.) *B.*
  (Add **b** to show.)   [76:5D]
- (Point to △=.) Raise your hand when you know the letter for this part. ✔
- Everybody, what letter? (Signal.) *A.*
  (Add to show:)   [76:5E]

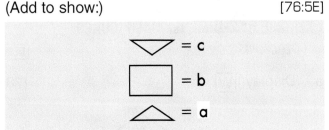

c. (Distribute unopened workbooks to students.)
- Open your workbook to Lesson 76 and find part 1.
  (Observe children and give feedback.)
  (Teacher reference:)   R Test 9: Part M

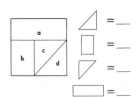

- Look at the whole figure. Is the shape for the whole figure a square or a hexagon? (Signal.) *A square.*
- Touch the first equals sign. ✔
  You're going to write the letter for that part.
- What shape is that part? (Signal.) *(A) triangle.*
- Find the letter for the correct triangle and write the letter.
  (Observe children and give feedback.)
- Everybody, what's the letter for the first equals? (Signal.) *D.*

d. Write letters for the rest of the parts.
   (Observe children and give feedback.)
   (Answer key:)

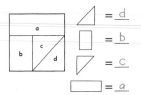

e. Check your work.
   You wrote the letter D for the first equals.
- What letter did you write for the next equals? (Signal.) *B.*
- What letter did you write for the next equals? (Signal.) *C.*
- What letter did you write for the last equals? (Signal.) *A.*

EXERCISE 6: WORD PROBLEMS (COLUMNS)

a. Find part 2 on worksheet 76. ✔
   (Teacher reference:)

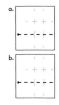

   You're going to write the symbols for word problems in columns and work them. Only part of the column and row lines are shown, and the equals bars are dotted. You'll write the numbers and the sign in the right places, and you'll make the equals bar.

b. Touch where you'll write the symbols for problem A. ✔
   Listen to problem A: There were 147 birds on the lake. Then 42 of those birds flew away. How many birds were still on the lake?
- Listen again and write the symbols for both parts: There were 147 birds on the lake. Then 42 of those birds flew away.
- Everybody, touch and read problem A. Get ready. (Signal.) *147 minus 42.*

c. Work problem A. Put your pencil down when you know how many birds were still on the lake.
(Observe children and give feedback.)
- Read the problem and the answer you wrote for A. Get ready. (Signal.) *147 – 42 = 105.*
- How many birds were still on the lake? (Signal.) *105.*
d. Touch where you'll write the symbols for problem B. ✔
Listen to problem B: Tim drove 67 miles on Saturday. He drove 32 more miles on Sunday. How many miles did Tim drive altogether?
- Listen again and write the symbols for both parts: Tim drove 67 miles on Saturday. He drove 32 more miles on Sunday.
- Everybody, touch and read problem B. Get ready. (Signal.) *67 + 32.*
e. Work problem B. Put your pencil down when you know how many miles he drove altogether.
(Observe children and give feedback.)
(Answer key:)

a.
```
 1 4 7
-  4 2
 1 0 5
```
b.
```
  6 7
+ 3 2
  9 9
```

- Read the problem and the answer you wrote for B. Get ready. (Signal.) *67 + 32 = 99.*
- How many miles did Tim drive altogether? (Signal.) *99.*

EXERCISE 7: MONEY
WRITING DOLLAR AMOUNTS   REMEDY

a. Find part 3 on worksheet 76. ✔
(Teacher reference:)   R|Part C

a. _____ b. _____ c. _____ d. _____

You're going to write dollars and cents.
- Touch space A. ✔

- You will write 5 dollars and 24 cents for space A. What will you write? (Signal.) *5 dollars and 24 cents.*
- Write the dollar sign for A. Write a capital S. Then make a straight line down through it.
(Observe children and give feedback.)
- Now write the rest of the 5 dollars and 24 cents. Remember the dot after the number of dollars.
(Observe children and give feedback.)
Check your work.
- Touch the dollar sign you wrote. ✔
- Touch the number of dollars. ✔
- What number? (Signal.) *5.*
- Touch the dot. ✔
- Touch the number of cents. ✔
- How many cents? (Signal.) *24.*
b. Touch the space for B. ✔
- You'll write 9 dollars and 53 cents for space B. What will you write? (Signal.) *9 dollars and 53 cents.*
- Write the dollar sign. Remember, a capital S with a line straight down through it. ✔
- Now write the rest of 9 dollars and 53 cents.
(Observe children and give feedback.)
- What's the first thing you wrote? (Signal.) *A dollar sign.*
- What's the next thing you wrote? (Signal.) *9.*
- What's the next thing you wrote? (Signal.) *(A) dot.*
- What's the last thing you wrote? (Signal.) *53.*
c. Touch the space for C. ✔
- You'll write 52 dollars and 10 cents for space C. What will you write? (Signal.) *52 dollars and 10 cents.*
- Write the dollar sign. Remember, capital S with a line straight down through it. ✔
- Now write the rest of 52 dollars and 10 cents.
(Observe children and give feedback.)
- What's the first thing you wrote? (Signal.) *A dollar sign.*
- What's the next thing you wrote? (Signal.) *52.*
- What's the next thing you wrote? (Signal.) *(A) dot.*
- What's the last thing you wrote? (Signal.) *10.*

d. Touch the space for D. ✔
- You'll write 3 dollars and 56 cents for space D. What will you write? (Signal.) *3 dollars and 56 cents.*
- Write the dollar sign. Remember, capital S with a line straight down through it.
- Now write the rest of 3 dollars and 56 cents. (Observe children and give feedback.)
- What's the first thing you wrote? (Signal.) *A dollar sign.*
- What's the next thing you wrote? (Signal.) *3.*
- What's the next thing you wrote? (Signal.) *(A) dot.*
- What's the last thing you wrote? (Signal.) *56.* (Answer key:)

  a. $5.24   b. $9.53   c. $52.10   d. $3.56

EXERCISE 8: FACTS
SUBTRACTION

a. Find part 4 on worksheet 76. ✔ (Teacher reference:)

a. $11-1$	e. $13-10$	i. $12-6$
b. $7-2$	f. $8-3$	j. $11-9$
c. $9-3$	g. $10-6$	k. $10-2$
d. $5-2$	h. $9-7$	l. $11-5$

These are minus problems from families you know. You're going to read each problem and tell me the answer. Then you'll go back and work them.
- Touch and read problem A. Get ready. (Signal.) *11 minus 1.*
- What's 11 minus 1? (Signal.) *10.*

b. (Repeat the following tasks for problems B through L:)
- Touch and read problem __.
- What's __?

c. Complete the equations for all of the problems in part 4. (Observe children and give feedback.)

d. Check your work.
[
- Touch and read fact A. (Signal.) *11 – 1 = 10.*
- (Repeat for:) B, *7 – 2 = 5;* C, *9 – 3 = 6;* D, *5 – 2 = 3;* E, *13 – 10 = 3;* F, *8 – 3 = 5;* G, *10 – 6 = 4;* H, *9 – 7 = 2;* I, *12 – 6 = 6;* J, *11 – 9 = 2;* K, *10 – 2 = 8;* L, *11 – 5 = 6.*

EXERCISE 9: INDEPENDENT WORK

a. Find part 5 on worksheet 76. ✔ (Teacher reference:)

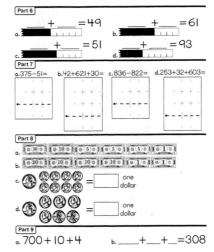

You'll write the fact for the missing number in each family and complete the family.

b. Turn to the other side of worksheet 76 and find part 6. ✔ (Teacher reference:)

You'll complete the equation for each ruler. You'll write the number for the unshaded part. Then you'll count backward from the number for the whole ruler to figure out the shaded part. In part 7, you'll write each problem in a column and work it. In part 8, you'll write an equals and the number of dollars to show what the group of bills is worth. You'll circle the words one dollar or write the cents for each group of coins. In part 9, you'll complete the place-value addition equations.

c. Complete worksheet 76. Remember to write the facts and complete the families in part 5 on the other side of your worksheet. (Observe children and mark incorrect responses on children's worksheets as you give feedback.)

Lesson 77

EXERCISE 1: 3-D OBJECTS

a. (Display:) [77:1A]

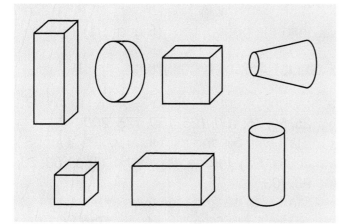

You're going to answer the questions about these objects.

- (Point to ▯.) Is this a cube? (Touch.) *No.*
- Why not? (Call on a child. Idea: *Not all the faces are squares.*)

b. (Point to ◯.) What's this object? (Touch.) *(A) cylinder.*

- (Point to ▱.) What's this object? (Touch.) *(A) cube.*

c. (Point to ◁.) Is this a cylinder? (Touch.) *No.*

- Why not? (Call on a child. Idea: *The circles aren't the same size.*)

d. (Point to ▱.) Is this a cube? (Touch.) *Yes.*

- (Point to ▭.) Is this a cube? (Touch.) *No.*
- Why not? (Call on a child. Idea: *Not all the faces are squares.*)

e. (Point to ▯.) Is this a cube? (Touch.) *No.*

- What is it? (Touch.) *(A) cylinder.*

EXERCISE 2: FACTS
SUBTRACTION

a. (Display:) [77:2A]

11 – 5	7 – 6	10 – 6
9 – 6	9 – 3	8 – 6
16 – 6	12 – 6	11 – 6

All these minus problems are from families that have a small number of 6.

- (Point to **11 – 5.**) 11 minus 5. What's the answer? (Signal.) *6.*
- Say the fact. (Signal.) *11 – 5 = 6.*

b. (Point to **9 – 6.**) 9 minus 6. What's the answer? (Signal.) *3.*

- Say the fact. (Signal.) *9 – 6 = 3.*

c. (Repeat the following tasks for the remaining problems:)

- (Point to __.) __. What's the answer?
- Say the fact.
(Repeat problems that were not firm.)

EXERCISE 3: COMPARISON
LENGTH

a. (Display:) [77:3A]

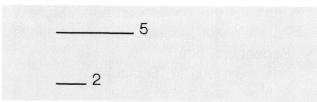

These lines are supposed to show inches, but they are not the right size. The numbers show how many inches long each line should be.

- (Point to 5 inch line.) This line is 5 inches long. How many inches long is this line? (Touch.) *5.*

b. (Point to **2.**) How many inches long is this line? (Touch.) *2.*

- (Point to 5 inch line.) Is this line longer than the other line? (Touch.) *Yes.*
- (Point to 2 inch line.) Is this line longer than the other line? (Touch.) *No.*

c. (Point to **5**.) You're going to figure out how much longer this line is than the other line. Here's how you do that. You write the number for the longer line. Then you minus the number for the shorter line.
- What's the number for the longer line? (Signal.) *5.*
- What number do you minus? (Signal.) *2.*
- Yes, 5 minus 2. Say the problem you work. (Signal.) *5 minus 2.*
- What's the answer? (Signal.) *3.*

d. Yes, the top line is 3 inches longer than the bottom line.
- Which line is longer? (Signal.) *The top line.*
- How much longer is it than the bottom line? (Signal.) *3 inches.*
(Repeat until firm.)

e. (Display:) [77:3B]

- Which line is longer, the top line or the bottom line? (Signal.) *The bottom line.*
- How many inches long is the top line? (Signal.) *3.*
- How many inches long is the bottom line? (Signal.) *9.*

f. We're going to figure out how much longer the bottom line is than the top line, so we start with the longer line and minus the shorter line.
- Which number do we start with? (Signal.) *9.*
- Say the minus problem. (Signal.) *9 minus 3.*
(Repeat until firm.)
- What's 9 minus 3? (Signal.) *6.*
- So how many inches longer is the bottom line than the top line? (Signal.) *6.*

g. (Display:) [77:3C]

- Which line is longer, the top line or the bottom line? (Signal.) *The top line.*
- How many inches long is the top line? (Signal.) *11.*

- How many inches long is the bottom line? (Signal.) *10.*
To figure out how much longer we start with the longer line and minus the shorter line.
- Which number do we start with? (Signal.) *11.*
- Say the minus problem. (Signal.) *11 minus 10.*
- What's 11 minus 10? (Signal.) *1.*
- So how many inches longer is the top line than the bottom line? (Signal.) *1.*

EXERCISE 4: MIXED COUNTING

a. Listen: Count by 25s to 200. Get ready. (Tap.) *25, 50, 75, 100, 125, 150, 175, 200.*
- Start with 53 and plus tens to 103. Get 53 going. *Fifty-threee.* Plus tens. (Tap.) *63, 73, 83, 93, 103.*
- Start with 45 and plus fives to 90. Get 45 going. *Forty-fiiive.* Plus fives. (Tap.) *50, 55, 60, 65, 70, 75, 80, 85, 90.*
- Start with 100 and count backward by tens. Get 100 going. *One huuundred.* Count backward. (Tap.) *90, 80, 70, 60, 50, 40, 30, 20, 10.*

b. Start with 239 and count backward by ones to 230. Get 239 going. *Two hundred thirty-niiine.* Count backward. (Tap.) *238, 237, 236, 235, 234, 233, 232, 231, 230.*

c. Start with 77 and plus tens to 107. Get 77 going. *Seventy-sevennn.* Plus tens. (Tap.) *87, 97, 107.*

d. Listen: You have 40. When you plus fives, what's the next number? (Signal.) *45.*
- You have 40. When you plus ones, what's the next number? (Signal.) *41.*
- You have 40. When you plus tens, what's the next number? (Signal.) *50.*
- You have 40. When you minus tens, what's the next number? (Signal.) *30.*
- You have 40. When you minus ones, what's the next number? (Signal.) *39.*

EXERCISE 5: COLUMN ADDITION
CARRYING

a. (Display:) W   [77:5A]

$$\begin{array}{r} 36 \\ +26 \\ \hline \end{array}$$

- (Point to **36**.) Read the problem. (Touch.)
  36 plus 26.
- Say the problem for the ones. (Signal.)
  6 plus 6.
- What's the answer? (Signal.) *12.*
- How many digits does 12 have? (Signal.) *2.*
- What's the tens digit? (Signal.) *1.*
- What's the ones digit? (Signal.) *2.*
- Where do I write the tens digit? (Signal.) *In the*
  tens column.
- Where do I write the ones digit? (Signal.) *In*
  the ones column.
  (Add to show:)   [77:5B]

$$\begin{array}{r} \overset{1}{3}6 \\ +26 \\ \hline 2 \end{array}$$

b. Read the new problem in the tens column.
   (Touch.) *1 plus 3 plus 2.*
- What's 1 plus 3? (Signal.) *4.*
- What's 4 plus 2? (Signal.) *6.*
  (Add to show:)   [77:5C]

$$\begin{array}{r} \overset{1}{3}6 \\ +26 \\ \hline 62 \end{array}$$

- Start with 36 and read the problem and the
  answer. Get ready. (Signal.) *36 + 26 = 62.*
c. (Distribute unopened workbooks to students.)
- Open your workbook to Lesson 77 and find
  part 1.
  (Observe children and give feedback.)
  (Teacher reference:)

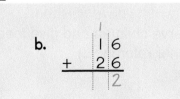

d. Touch and read problem A. Get ready.
   (Signal.) *56 plus 34.*
- Say the problem for the ones. (Signal.)
  6 plus 4.
- What's 6 plus 4? (Signal.) *10.*
- How many digits does 10 have? (Signal.) *2.*
- What's the tens digit of 10? (Signal.) *1.*
- What's the ones digit of 10? (Signal.) *Zero.*
- Write the tens digit at the top of the tens
  column. Then write the ones digit in the
  ones column.
  (Observe children and give feedback.)
e. (Display:) W   [77:5D]

$$\text{a.} \quad \begin{array}{r} \overset{1}{5}6 \\ +\;\;3\;4 \\ \hline 0 \end{array}$$

  Here's what you should have.
- Touch and read the new problem in the tens.
  (Signal.) *1 plus 5 plus 3.*
  You'll finish that problem later.
f. Touch and read problem B. (Signal.)
  16 plus 26.
- Say the problem for the ones. (Signal.)
  6 plus 6.
- What's the answer? (Signal.) *12.*
- How many digits does 12 have? (Signal.) *2.*
- What's the tens digit of 12? (Signal.) *1.*
- What's the ones digit of 12? (Signal.) *2.*
- Write the tens digit at the top of the tens
  column. Then write the ones digit in the
  ones column.
  (Observe children and give feedback.)
g. (Display:) W   [77:5E]

$$\text{b.} \quad \begin{array}{r} \overset{1}{1}6 \\ +\;\;2\;6 \\ \hline 2 \end{array}$$

  Here's what you should have.
- Touch and read the new problem in the tens.
  (Signal.) *1 plus 1 plus 2.*
  You'll finish that problem later.

h. Go back to problem A. ✔
- Touch and read the new problem in the tens column. (Signal.) *1 plus 5 plus 3.*
- What's 1 plus 5? (Signal.) *6.*
- What's 6 plus 3? (Signal.) *9.*
- Write 9 in the tens column. ✔
- Start with 56 and read problem A and the answer. (Signal.) *56 + 34 = 90.*
i. Touch problem B. ✔
- Touch and read the new problem in the tens column. (Signal.) *1 plus 1 plus 2.*
- Think about the answer. Everybody, what's 1 plus 1 plus 2. (Signal.) *4.*
- Write 4 in the answer. ✔
- Start with 16 and read problem B and the answer. (Signal.) *16 + 26 = 42.*

EXERCISE 6: 2-DIMENSIONAL FIGURES
DECOMPOSITION

a. Find part 2 on worksheet 77. ✔
(Teacher reference:)

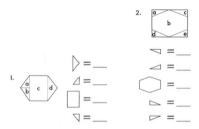

- Touch figure 1. ✔
- Figure 1 has parts. Raise your hand when you know the shape of the whole figure. ✔
- Everybody, what's the shape of figure 1? (Signal.) *(A) Hexagon.*
b. Look at the pictures of the parts. ✔
- Write the letter for each part. Put your pencil down when you're finished.
  (Observe children and give feedback.)
  (Teacher reference:)

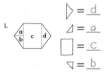

c. Check your work.
- Touch the first equals for figure 1. ✔
  What letter did you write? (Signal.) *D.*
- Touch the next equals. ✔
  What letter did you write? (Signal.) *A.*
- Touch the next equals. ✔
  What letter did you write? (Signal.) *C.*
- Touch the last equals. ✔
  What letter did you write? (Signal.) *B.*

d. Touch figure 2 and look at the whole figure. ✔
- What's the shape of the whole figure? (Signal.) *(A) rectangle.*
- Look at the pictures of the parts. ✔
- Write the letter for each part. Put your pencil down when you're finished.
  (Observe children and give feedback.)
  (Teacher reference:)

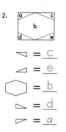

e. Check your work.
- Touch the first equals for figure 2. ✔
  What letter did you write? (Signal.) *C.*
- Touch the next equals. ✔
  What letter did you write? (Signal.) *E.*
- Touch the next equals. ✔
  What letter did you write? (Signal.) *B.*
- Touch the next equals. ✔
  What letter did you write? (Signal.) *D.*
- Touch the last equals. ✔
  What letter did you write? (Signal.) *A.*

EXERCISE 7: FACTS
PLUS/MINUS MIX

a. Find part 3 on worksheet 77. ✔
(Teacher reference:)

a. $5 - 1$	e. $9 - 6$	i. $10 - 4$
b. $6 + 6$	f. $10 - 6$	j. $14 - 4$
c. $3 + 5$	g. $6 - 4$	k. $12 - 6$
d. $11 - 6$	h. $5 + 6$	l. $6 - 2$

These problems are from number families you know.
- Touch and read problem A. Get ready. (Signal.) *5 minus 1.*
- Is the answer the big number or a small number? (Signal.) *A small number.*
- What's 5 minus 1? (Signal.) *4.*
b. (Repeat the following tasks for problems B through L:)
- Touch and read problem __.
- Is the answer the big number or a small number?
- What's __?
(Repeat problems that were not firm.)

c. Complete the equations for all the problems in part 3. Put your pencil down when you're finished.
(Observe children and give feedback.)
d. Check your work. You'll touch and read each fact.
- Fact A. Get ready. (Signal.) *5 – 1 = 4.*
- (Repeat for:) B, *6 + 6 = 12;* C, *3 + 5 = 8;* D, *11 – 6 = 5;* E, *9 – 6 = 3;* F, *10 – 6 = 4;* G, *6 – 4 = 2;* H, *5 + 6 = 11;* I, *10 – 4 = 6;* J, *14 – 4 = 10;* K, *12 – 6 = 6;* L, *6 – 2 = 4.*

EXERCISE 8: MONEY
WRITING PROBLEMS   REMEDY

a. (Display:) W   [77:8A]

$$\begin{array}{r} \$4.21 \\ +2.14 \\ \hline \end{array}$$

This problem adds dollars and cents amounts. My turn to read the problem: 4 dollars and 21 cents plus 2 dollars and 14 cents.
- (Point to **$4.21**.) Read the top amount. (Touch.) *4 dollars and 21 cents.*
- (Point to **2.14**.) Read the bottom amount. (Touch.) *2 dollars and 14 cents.*
- Everybody, read the problem. (Signal.) *4 dollars and 21 cents plus 2 dollars and 14 cents.*
- Before I work this problem, I write a dot in the answer.
(Add to show:)   [77:8B]

$$\begin{array}{r} \$4.21 \\ +2.14 \\ \hline . \end{array}$$

Now I just start with the cents and add.
- (Point to $\overset{\downarrow}{\underset{+2.14}{\$4.2\,1}}$.) What's 1 plus 4? (Signal.) *5.*
(Add to show:)   [77:8C]

$$\begin{array}{r} \$4.21 \\ +2.14 \\ \hline .\ 5 \end{array}$$

- (Point to $\overset{\downarrow}{\underset{.5.}{\$4.2\,1 \atop +2.14}}$.) What's 2 plus 1? (Signal.) *3.*
(Add to show:)   [77:8D]

$$\begin{array}{r} \$4.21 \\ +2.14 \\ \hline .35 \end{array}$$

- (Point to $\underset{.35.}{\$4.2\,1 \atop +2.14}$.) Now I do the dollars. What's 4 plus 2? (Signal.) *6.*
I write 6 and a dollar sign.
(Add to show:)   [77:8E]

$$\begin{array}{r} \$4.21 \\ +2.14 \\ \hline \$6.35 \end{array}$$

- So what's the answer to the whole problem? (Signal.) *6 dollars and 35 cents.*
b. Find part 4 on worksheet 77. ✔
(Teacher reference:)   R Part M

a. $\begin{array}{r} \$5.25 \\ +2.41 \\ \hline \end{array}$ b. $\begin{array}{r} \$4.25 \\ -3.15 \\ \hline \end{array}$

- Touch problem A. ✔
I'll read the problem. You touch the numbers. 5 dollars and 25 cents plus 2 dollars and 41 cents.
- Your turn: Touch and read problem A. Get ready. (Signal.) *5 dollars and 25 cents plus 2 dollars and 41 cents.*
(Repeat until firm.)
- What do you write in the answer before you plus? (Signal.) *The dot.*
- Write the dot. Then add the numbers for the cents.
(Observe children and give feedback.)
(Display:) W   [77:8F]

a. $\begin{array}{r} \$5.25 \\ +2.41 \\ \hline .66 \end{array}$

Here's what you should have: 66 cents.

- Now figure out the number of dollars and write the dollar sign and the answer.
(Observe children and give feedback.)
(Add to show:) [77:8G]

a. $5.25
   +2.41
   $7.66

Here's the answer for the problem.
- Everybody, read the answer. (Signal.)
7 dollars and 66 cents.
c. Touch and read problem B. Get ready.
(Signal.) *4 dollars and 25 cents minus 3 dollars and 15 cents.*
(Repeat until firm.)
- Work problem B. Remember to write the dot and the dollar sign in the answer.
(Observe children and give feedback.)
(Display:) W [77:8H]

b. $4.25
   −3.15
   $1.10

Here's what you should have.
- Everybody, read the answer. Get ready.
(Signal.) *1 dollar and 10 cents.*
Who got it right?

EXERCISE 9: INDEPENDENT WORK

a. Find part 5 on worksheet 77. ✔
(Teacher reference:)

6 4→ 10
___
___
___

You'll write four facts for the family on the lines below it.

b. Turn to the other side of worksheet 77 and find part 6. ✔
(Teacher reference:)

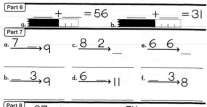

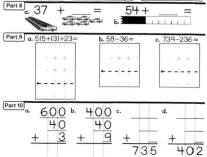

You'll complete the equation for each ruler. You'll write the number for the unshaded part. Then you'll count backward from the number for the whole ruler to figure out the shaded part.
In part 7, you'll write the fact for the missing number in each family and complete the family.
In part 8, you'll complete the equations for the objects and the ruler.
In part 9, you'll write each problem in a column and work it.
In part 10, you'll complete the place-value addition equations.

c. Complete worksheet 77. Remember to write the four facts for the family in part 5 on the other side of your worksheet.
(Observe children and mark incorrect responses on children's worksheets as you give feedback.)

Lesson 78

EXERCISE 1: FACTS
ADDITION/SUBTRACTION

a. (Display:)   [78:1A]

10 + 6	9 − 6	12 − 6
6 + 4	11 − 6	10 − 6
10 − 4	5 + 6	6 + 5

All these facts come from families that have a small number of 6.
- (Point to **10 + 6**.) Read the problem. (Touch.) *10 plus 6.*
- What's the answer? (Signal.) *16.*
- Say the fact. Get ready. (Signal.) *10 + 6 = 16.*

b. (Point to **6 + 4**.) Read the problem. (Touch.) *6 plus 4.*
- What's the answer? (Signal.) *10.*
- Say the fact. Get ready. (Signal.) *6 + 4 = 10.*

c. (Repeat the following tasks for remaining problems:)
- (Point to __.) Read the problem.
- What's the answer?
- Say the fact.
  (Repeat problems that were not firm.)

EXERCISE 2: COMPARISON
LENGTH

a. (Display:) W   [78:2A]

$$\frac{\quad 5 \quad}{\underline{\qquad 7 \qquad}}$$

The number above each line shows how many inches long it is supposed to be. You're going to figure out how much longer one of the lines is.
- (Point to **5**.) How many inches long is the top line? (Touch.) *5.*
- How many inches long is the bottom line? (Touch.) *7.*
- Which line is longer? (Signal.) *The bottom line.*
  To figure out how much longer, you start with the inches for the longer line and minus the inches for the other line.
- Which number do we start with? (Signal.) *7.*

- Start with 7 and say the minus problem. Get ready. (Signal.) *7 minus 5.*
  (Repeat until firm.)

b. (Add to show:)   [78:2B]

$$\begin{array}{r} 7 \\ -\ 5 \\ \hline \end{array}$$

- What's 7 minus 5? (Signal.) *2.*
  (Add to show:)   [78:2C]

$$\begin{array}{r} 7 \\ -\ 5 \\ \hline 2 \end{array}$$

- So how many inches longer is the bottom line than the top line? (Signal.) *2.*

c. (Display:) W   [78:2D]

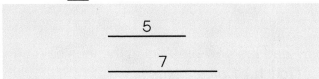

- (Point to **10**.) How many inches long is the top line? (Touch.) *10.*
- How many inches long is the bottom line? (Touch.) *2.*
  To figure out how much longer, you start with the longer line and minus the other line.
- Which number do we start with? (Signal.) *10.*
- Say the minus problem. Get ready. (Signal.) *10 minus 2.*
  (Repeat until firm.)

d. (Add to show:)   [78:2E]

$$\begin{array}{r} 1\ 0 \\ -\ 2 \\ \hline \end{array}$$

- What's 10 minus 2? (Signal.) *8.*
  (Add to show:)   [78:2F]

$$\begin{array}{r} 1\ 0 \\ -\ 2 \\ \hline 8 \end{array}$$

- So how many inches longer is the top line than the bottom line? (Signal.) *8.*

e. (Display:) W [78:2G]

$$\frac{3}{\qquad}$$
$$\frac{8}{\qquad}$$

- (Point to **3**.) How many inches long is the top line? (Touch.) *3*.
- How many inches long is the bottom line? (Touch.) *8*.
  To figure out how much longer, you start with the longer line and minus the other line.
- Which number do we start with? (Signal.) *8*.
- Say the minus problem. Get ready. (Signal.) *8 minus 3*.
  (Repeat until firm.)

f. (Add to show:) [78:2H]

$$\begin{array}{r} 8 \\ -\ 3 \\ \hline \end{array}$$

- What's 8 minus 3? (Signal.) *5*.
  (Add to show:) [78:2I]

$$\begin{array}{r} 8 \\ -\ 3 \\ \hline 5 \end{array}$$

- So how many inches longer is the bottom line than the top line? (Signal.) *5*.

EXERCISE 3: MIXED COUNTING

a. Start with 100 and count backward by tens. Get 100 going. *One huuundred.* Count. (Tap.) *90, 80, 70, 60, 50, 40, 30, 20, 10.*

b. Start with 68 and plus tens to 108. Get 68 going. *Sixty-eieieight.* Plus tens. (Tap.) *78, 88, 98, 108.*

c. Start with 45 and count by fives to 90. Get 45 going. *Forty-fiiive.* Count. (Tap.) *50, 55, 60, 65, 70, 75, 80, 85, 90.*

d. Count by 25s to 200. Get ready. (Signal.) *25, 50, 75, 100, 125, 150, 175, 200.*

e. Start with 199 and count backward to 190. Get 199 going. *One hundred ninety-niiine.* Count backward. (Tap.) *198, 197, 196, 195, 194, 193, 192, 191, 190.*

f. Start with 74 and plus tens to 104. Get 74 going. *Seventy-fouuur.* Count. (Tap.) *84, 94, 104.*

g. Count by hundreds to 1000. Get ready. (Tap.) *100, 200, 300, 400, 500, 600, 700, 800, 900, 1000.*

h. You have 95. When you plus ones, what's the next number? (Signal.) *96.*
- You have 95. When you plus tens, what's the next number? (Signal.) *105.*
- You have 95. When you plus fives, what's the next number? (Signal.) *100.*
- You have 95. When you minus ones, what's the next number? (Signal.) *94.*
- You have 95. When you minus fives, what's the next number? (Signal.) *90.*

EXERCISE 4: COLUMN ADDITION
CARRYING   REMEDY

a. (Display:) W [78:4A]

$$\begin{array}{r} 2\ 5 \\ +\ 5\ 6 \\ \hline \end{array}$$

- (Point to **25**.) Read the problem. (Signal.) *25 plus 56.*
- Say the problem for the ones. (Signal.) *5 plus 6.*
- What's the answer? (Signal.) *11.*
- How many digits does 11 have? (Signal.) *Two.*
- What's the tens digit of 11? (Signal.) *1.*
- What's the ones digit of 11? (Signal.) *1.*
- Where do I write the tens digit? (Signal.) *In the tens column.*
- Where do I write the ones digit? (Signal.) *In the ones column.*
  (Add to show:) [78:4B]

$$\begin{array}{r} {\scriptstyle 1} \\ 2\ 5 \\ +\ 5\ 6 \\ \hline 1 \end{array}$$

b. Read the new problem in the tens column. (Touch.) *1 plus 2 plus 5.*
- What's 1 plus 2? (Signal.) *3.*
- What's 3 plus 5? (Signal.) *8.*
  (Add to show:) [78:4C]

$$\begin{array}{r} {\scriptstyle 1} \\ 2\ 5 \\ +\ 5\ 6 \\ \hline 8\ 1 \end{array}$$

- Start with 25 and read the problem and the answer. Get ready. (Signal.) *25 + 56 = 81.*

c. (Distribute unopened workbooks to students.)
- Open your workbook to Lesson 78 and find part 1.
  (Observe children and give feedback.)
  (Teacher reference:)

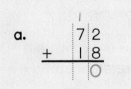

d. Touch and read problem A. Get ready. (Signal.) *72 plus 18.*
- Say the problem for the ones. (Signal.) *2 plus 8.*
- What's 2 plus 8? (Signal.) *10.*
- How many digits does 10 have? (Signal.) *Two.*
- What's the tens digit of 10? (Signal.) *1.*
- What's the ones digit of 10? (Signal.) *Zero.*
- Write the tens digit at the top of the tens column. Then write the ones digit in the ones column.
  (Observe children and give feedback.)
e. (Display:) W   [78:4D]

$$\begin{array}{r} \overset{1}{} \\ a. \quad 7\,2 \\ +\ 1\,8 \\ \hline 0 \end{array}$$

  Here's what you should have.
- Touch and read the new problem in the tens. (Signal.) *1 plus 7 plus 1.*
  You'll finish that problem later.
f. Touch and read problem B. (Signal.) *16 plus 46.*
- Say the problem for the ones. (Signal.) *6 plus 6.*
- What's 6 plus 6? (Signal.) *12.*
- How many digits does 12 have? (Signal.) *Two.*
- What's the tens digit of 12? (Signal.) *1.*
- What's the ones digit of 12? (Signal.) *2.*
- Write the tens digit at the top of the tens column. Then write the ones digit in the ones column.
  (Observe children and give feedback.)
g. (Display:) W   [78:4E]

$$\begin{array}{r} \overset{1}{} \\ b. \quad 1\,6 \\ +\ 4\,6 \\ \hline 2 \end{array}$$

  Here's what you should have.

- Touch and read the new problem in the tens. (Signal.) *1 plus 1 plus 4.*
  You'll finish that problem later.
h. Go back to problem A. ✔
- Touch and read the new problem in the tens column. (Signal.) *1 plus 7 plus 1.*
- What's 1 plus 7? (Signal.) *8.*
- What's 8 plus 1? (Signal.) *9.*
- Write 9 in the tens column. ✔
- Start with 72 and read the problem and the answer. (Signal.) *72 + 18 = 90.*
i. Touch problem B. ✔
- Touch and read the new problem in the tens column. (Signal.) *1 plus 1 plus 4.*
- What's 1 plus 1? (Signal.) *2.*
- What's 2 plus 4? (Signal.) *6.*
- Write 6 in the tens column. ✔
- Start with 16 and read the problem and the answer. (Signal.) *16 + 46 = 62.*

EXERCISE 5: 2-DIMENSIONAL FIGURES
DECOMPOSITION   REMEDY

a. (Display:) W   [78:5A]

> **Hexagon**
> **Rectangle**
> **Triangle**

  These are names of shapes.
- (Point to **Hexagon.**) This name is hexagon. What name? (Touch.) *Hexagon.*
- (Point to **Rectangle.**) This name is rectangle. What name? (Touch.) *Rectangle.*
- (Point to **Triangle.**) This name is triangle. What name? (Touch.) *Triangle.*
  Let's do those once more.
- (Point to **Hexagon.**) What name? (Touch.) *Hexagon.*
- (Point to **Rectangle.**) What name? (Touch.) *Rectangle.*
- (Point to **Triangle.**) What name? (Touch.) *Triangle.*
b. (Point to **Hexagon.**) The first letter in the word hexagon is H. What letter? (Signal.) *H.*
- (Point to **Rectangle.**) The first letter in the word rectangle is R. What letter? (Signal.) *R.*
- (Point to **Triangle.**) The first letter in the word triangle is T. What letter? (Signal.) *T.*

c. Find part 2 on worksheet 78. ✔
   (Teacher reference:) R Test 9: Part N

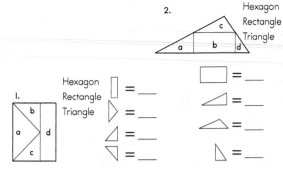

You can see figures with parts. Next to the
whole figures are the names.
Listen: Circle the name of the shape for
figure 1. ✔
• Everybody, what name did you circle for figure
  1? (Signal.) *(A) rectangle.*
  Yes, the whole figure is a rectangle.
• Look at the pictures of the parts. Write the
  letter for each part. Put your pencil down when
  you're finished.
  (Observe children and give feedback.)
  (Teacher reference:)

  ☐ = _d_
  ▷ = _a_
  ◸ = _c_
  ◺ = _b_

d. Check your work.
• Touch the first equals for figure 1. ✔
  What letter did you write? (Signal.) *D.*
• Touch the next equals. ✔
  What letter did you write? (Signal.) *A.*
• Touch the next equals. ✔
  What letter did you write? (Signal.) *C.*
• Touch the last equals. ✔
  What letter did you write? (Signal.) *B.*
e. Touch figure 2. ✔
• Circle the name of the whole figure. ✔
• Everybody, what name did you circle?
  (Signal.) *Triangle.*
  Yes, the whole figure is a triangle.

f. Look at the pictures of the parts. Write the
   letter for each part. Put your pencil down when
   you're finished.
   (Observe children and give feedback.)
   (Teacher reference:)

   ☐ = _b_
   ◸ = _a_
   △ = _c_
   ◺ = _d_

g. Check your work.
• Touch the first equals for figure 2. ✔
  What letter did you write? (Signal.) *B.*
• Touch the next equals. ✔
  What letter did you write? (Signal.) *A.*
• Touch the next equals. ✔
  What letter did you write? (Signal.) *C.*
• Touch the last equals. ✔
  What letter did you write? (Signal.) *D.*

EXERCISE 6: MONEY
DOLLAR PROBLEMS   REMEDY

a. Find part 3 on worksheet 78. ✔
   (Teacher reference:)   R Part N

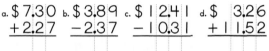

• Touch problem A. ✔
• Touch and read the first dollar amount.
  (Signal.) *7 dollars and 30 cents.*
• Read the next dollar amount. (Signal.)
  2 dollars and 27 cents.
• What do you write in the answer before you
  plus? (Signal.) *A dot.*
• Write the dot. Then work the problem.
  Remember to write a dollar sign in your
  answer.
  (Observe children and give feedback.)
  (Display:) W   [78:6A]

  $7.30
  +2.27
  $9.57

  Here's what you should have.
• Everybody, read the answer. (Signal.)
  9 dollars and 57 cents.

b. Touch and read problem B. Get ready.
(Signal.) *3 dollars and 89 cents minus 2 dollars and 37 cents.*
(Repeat until firm.)

• Work the problem. Remember to write the dot and the dollar sign in the answer.
(Observe children and give feedback.)
(Display:) W   [78:6B]

$$
\begin{array}{r}
\$3.89 \\
-2.37 \\
\hline
\$1.52
\end{array}
$$

Here's what you should have.

• Everybody, read the answer. Get ready.
(Signal.) *1 dollar and 52 cents.*

c. Touch and read problem C. Get ready.
(Signal.) *12 dollars and 41 cents minus 10 dollars and 31 cents.*
(Repeat until firm.)

• Work the problem. Remember to write the dot and the dollar sign in the answer.
(Observe children and give feedback.)
(Display:) W   [78:6C]

$$
\begin{array}{r}
\$12.41 \\
-10.31 \\
\hline
\$\ 2.10
\end{array}
$$

Here's what you should have.

• Everybody, read the answer. Get ready.
(Signal.) *2 dollars and 10 cents.*

d. Touch and read problem D. Get ready.
(Signal.) *3 dollars and 26 cents plus 11 dollars and 52 cents.*
(Repeat until firm.)

• Work the problem.
(Observe children and give feedback.)
(Display:) W   [78:6D]

$$
\begin{array}{r}
\$\ 3.26 \\
+11.52 \\
\hline
\$14.78
\end{array}
$$

Here's what you should have.

• Everybody, read the answer. Get ready.
(Signal.) *14 dollars and 78 cents.*

EXERCISE 7: NUMBER FAMILIES
MISSING NUMBER IN FAMILY

a. Find part 4 on worksheet 78. ✔
(Teacher reference:)

a. $\xrightarrow{\quad 5 \quad} 11$ e. $\xrightarrow{6 \quad} 11$ i. $\xrightarrow{6 \quad 5} \_$

b. $\xrightarrow{4 \quad 2} \_$ f. $\xrightarrow{6 \quad 6} \_$ j. $\xrightarrow{6 \quad 4} \_$

c. $\xrightarrow{6 \quad} 12$ g. $\xrightarrow{\quad 3 \quad} 9$ k. $\xrightarrow{6 \quad} 8$

d. $\xrightarrow{10 \quad 6} \_$ h. $\xrightarrow{\quad 4 \quad} 10$ l. $\xrightarrow{\quad 3 \quad} 8$

You're going to say the problem for the missing number in each family.

b. Family A. Say the problem for the missing number. Get ready. (Signal.) *11 minus 5.*

• What's 11 minus 5? (Signal.) *6.*

c. Family B. Say the problem for the missing number. Get ready. (Signal.) *4 plus 2.*

• What's 4 plus 2? (Signal.) *6.*

d. (Repeat the following tasks for families C through L:)

Family __.	Say the problem for the missing number.	What's __?	
C	12 − 6	12 − 6	6
D	10 + 6	10 + 6	16
E	11 − 6	11 − 6	5
F	6 + 6	6 + 6	12
G	9 − 3	9 − 3	6
H	10 − 4	10 − 4	6
I	6 + 5	6 + 5	11
J	6 + 4	6 + 4	10
K	8 − 6	8 − 6	2
L	8 − 3	8 − 3	5

(Repeat for families that were not firm.)

e. Write the missing number in each family. Put your pencil down when you've completed the families in part 3.
(Observe children and give feedback.)

f. Check your work. You'll read the number you wrote in each family.

• Family A. (Signal.) *6.*
• (Repeat for:) B, *6;* C, *6;* D, *16;* E, *5;* F, *12;* G, *6;* H, *6;* I, *11;* J, *10;* K, *2;* L, *5.*

EXERCISE 8: WORD PROBLEMS (COLUMNS)

a. Turn to the other side of worksheet 78 and find part 5. ✔
(Teacher reference:)

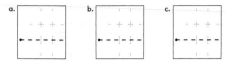

You're going to write symbols for word problems and work them.

b. Touch where you'll write symbols for A. ✔
• Listen to problem A: Tim had 64 cards. Hilary had 23 cards. How many cards did they have altogether?
• Listen again and write the symbols for the whole problem.
(Observe children and give feedback.)
• Everybody, touch and read the problem for A. Get ready. (Signal.) *64 + 23.*

c. Work problem A. Put your pencil down when you know how many cards Tim and Hilary had together.
(Observe children and give feedback.)
(Display:) W   [78:8A]

• Read the problem and the answer you wrote for A. Get ready. (Signal.) *64 + 23 = 87.*
• How many cards did they have altogether? (Signal.) *87.*

d. Listen to problem B: There were 183 people on a train. 123 of those people got off the train. How many people were still on the train?
• Listen again and write the symbols for the whole problem. There were 183 people on a train. 123 of those people got off the train.
(Observe children and give feedback.)
• Everybody, touch and read the problem for B. Get ready. (Signal.) *183 minus 123.*

e. Work problem B. Put your pencil down when you know how many people were still on the train.
(Observe children and give feedback.)
(Display:) W   [78:8B]

• Read the problem and the answer you wrote for B. Get ready. (Signal.) *183 – 123 = 60.*
• How many people were still on the train? (Signal.) *60.*

f. Listen to problem C. Mr. Johnson had 76 tools. He gave 55 of those tools to his sister. How many tools did Mr. Johnson still have?
• Listen again and write the symbols for the whole problem. Mr. Johnson had 76 tools. He gave 55 of those tools to his sister.
(Observe children and give feedback.)
• Everybody, touch and read the problem for C. Get ready. (Signal.) *76 minus 55.*

g. Work problem C. Put your pencil down when you know how many tools Mr. Johnson still had. (Display:) W   [78:8C]

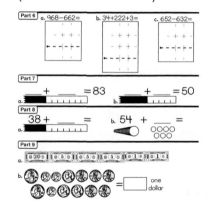

• Read the problem and the answer you wrote for C. Get ready. (Signal.) *76 – 55 = 21.*
• How many tools did Mr. Johnson still have? (Signal.) *21.*

EXERCISE 9: INDEPENDENT WORK

a. Find part 6 on worksheet 78. ✔
(Teacher reference:)

You'll write each problem in a column and work it.
In part 7, you'll complete the equation for each ruler. You'll write the number for the unshaded part. Then you'll count backward from the number for the whole ruler to figure out the shaded part.
In part 8, you'll complete the equations for the ruler and the objects.
In part 9, you'll write an equals and the number of dollars to show what the group of bills is worth. You'll circle the words one dollar or write the cents for the group of coins.

b. Complete worksheet 78.
(Observe children and mark incorrect responses on children's worksheets as you give feedback.)

Lesson  79

EXERCISE 1: NUMBER FAMILIES
SMALL NUMBER OF 4

a. (Display:)   [79:1A]

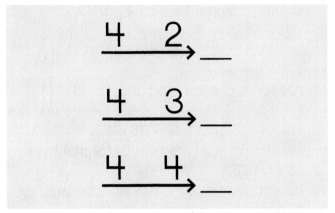

Here are two new families. The big number
is missing in these families. All these families
have a small number of 4.

- (Point to 4 —2→.) What are the small numbers
  in this family? (Touch.) *4 and 2.*
  What's the big number? (Signal.) *6.*
- (Point to 4 —3→.) What are the small numbers
  in this family? (Touch.) *4 and 3.*
  So what's the big number? (Signal.) *7.*
- (Point to 4 —4→.) What are the small numbers
  in this family? (Touch.) *4 and 4.*
  So what's the big number? (Signal.) *8.*
  (Repeat until firm.)

b. (Point to 4 —3→.) What are the small numbers
  in this family? (Touch.) *4 and 3.*
- What's the big number? (Signal.) *7.*
- Say the fact that starts with 4. (Signal.)
  4 + 3 = 7.
- Say the fact that starts with the other small
  number. (Signal.) *3 + 4 = 7.*
- Say the fact that goes backward down the
  arrow. (Signal.) *7 − 3 = 4.*
- Say the other minus fact. (Signal.) *7 − 4 = 3.*

c. (Point to 4 —4→.) What are the small numbers
  in this family? (Touch.) *4 and 4.*
- What's the big number? (Signal.) *8.*
  There's only one plus fact and one minus fact.
- Say the plus fact. (Signal.) *4 + 4 = 8.*
- Say the minus fact. (Signal.) *8 − 4 = 4.*
  Remember the facts for these families.

EXERCISE 2: COMPARISON
LENGTH   REMEDY

a. (Display:) W   [79:2A]

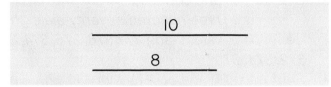

These lines are supposed to show inches.
- Which line is longer, the top line or the bottom
  line? (Signal.) *The top line.*
- How many inches long is the top line?
  (Signal.) *10.*
- How many inches long is the bottom line?
  (Signal.) *8.*

b. We're going to figure out how much longer
  the top line is than the bottom line, so we start
  with the inches for the longer line and minus
  the inches for the shorter line.
- Which number do we start with? (Signal.) *10.*
- Say the minus problem. (Signal.) *10 minus 8.*
- What's 10 minus 8? (Signal.) *2.*
- So how many inches longer is the top line
  than the bottom line? (Signal.) *2.*

c. (Display:) W   [79:2B]

These lines are supposed to show inches.
(top line: 1)
(bottom line: 9)

- Which line is longer, the top line or the bottom
  line? (Signal.) *The bottom line.*
- How many inches long is the top line?
  (Signal.) *1.*
- How many inches long is the bottom line?
  (Signal.) *9.*

d. We're going to figure out how much longer
  the bottom line is than the top line, so we start
  with the longer line and minus the shorter line.
- Which number do we start with? (Signal.) *9.*
- Say the minus problem. (Signal.) *9 minus 1.*
- What's 9 minus 1? (Signal.) *8.*
- So how many inches longer is the bottom line
  than the top line? (Signal.) *8.*

EXERCISE 3: MIXED COUNTING

a. Everybody, start with 66 and plus tens to 106. Get 66 going. *Sixty-siiix.* Count. (Tap.) *76, 86, 96, 106.*
- Count by hundreds to 1000. Get ready. (Signal.) *100, 200, 300, 400, 500, 600, 700, 800, 900, 1000.*
- Start at 378 and count backward to 370. Get 378 going. *Three hundred seventy-eieieight.* Count backward. (Tap.) *377, 376, 375, 374, 373, 372, 371, 370.*
- Start with 100 and count by tens to 170. Get 100 going. *One huuundred.* Count. (Tap.) *110, 120, 130, 140, 150, 160, 170.*
- Start with 100 and count backward by tens. Get 100 going. *One huuundred.* Count backward. (Tap.) *90, 80, 70, 60, 50, 40, 30, 20, 10.*
(Repeat tasks that were not firm.)

b. You have 75. When you plus tens, what's the next number? (Signal.) *85.*
- You have 75. When you plus 25s, what's the next number? (Signal.) *100.*
- You have 75. When you plus fives, what's the next number? (Signal.) *80.*
- You have 75. When you plus ones, what's the next number? (Signal.) *76.*
- You have 75. When you minus ones, what's the next number? (Signal.) *74.*
(Repeat tasks that were not firm.)

EXERCISE 4: COLUMN ADDITION
CARRYING

a. (Distribute unopened workbooks to children.)
- Open your workbook to Lesson 79 and find part 1.
(Observe children and give feedback.)
(Teacher reference:)

$$\begin{array}{r} \text{a.}\quad 44 \\ +\ 36 \\ \hline \end{array}$$

$$\begin{array}{r} \text{b.}\quad 16 \\ +\ 76 \\ \hline \end{array}$$

- Touch and read problem A. (Signal.) *44 plus 36.*

- Read the problem in the ones column. (Signal.) *4 plus 6.*
- What's the answer? (Signal.) *10.*
- What's the tens digit of 10? (Signal.) *1.*
- What's the ones digit of 10? (Signal.) *Zero.*
- Write the digits for 10 where they should go. (Observe children and give feedback.)
b. Touch and read the new problem for the tens column. (Signal.) *1 plus 4 plus 3.*
- What's 1 plus 4? (Signal.) *5.*
- What's 5 plus 3? (Signal.) *8.*
- Write 8 where it should go. (Observe children and give feedback.)
- Start with 44 and read the problem and the answer. (Signal.) *44 + 36 = 80.*
c. Touch and read problem B. (Signal.) *16 plus 76.*
- Read the problem in the ones column. (Signal.) *6 plus 6.*
- What's the answer? (Signal.) *12.*
- What's the tens digit of 12? (Signal.) *1.*
- What's the ones digit of 12? (Signal.) *2.*
- Write the digits for 12 where they should go. (Observe children and give feedback.)
d. Touch and read the problem for the tens column. (Signal.) *1 plus 1 plus 7.*
- What's 1 plus 1? (Signal.) *2.*
- What's 2 plus 7? (Signal.) *9.*
- Write 9 where it should go. (Observe children and give feedback.)
- Start with 16 and read the problem and the answer. (Signal.) *16 + 76 = 92.*

EXERCISE 5: FACTS
PLUS/MINUS MIX

a. Find part 2 on worksheet 79. ✔
(Teacher reference:)

	e. $5 + 3$
a. $6 - 4$	f. $11 - 9$
b. $9 - 2$	g. $5 + 6$
c. $6 + 4$	h. $12 - 2$
d. $6 + 6$	i. $8 - 6$

- Touch and read problem A. Get ready. (Signal.) *6 minus 4.*
- Is the answer the big number or a small number? (Signal.) *A small number.*
- What's 6 minus 4? (Signal.) *2.*

b. (Repeat the following tasks for problems B through I:)

Touch and read problem __.	Is the answer the big number or a small number?	What's __?	
B	A small number.	9 – 2	7
C	The big number.	6 + 4	10
D	The big number.	6 + 6	12
E	The big number.	5 + 3	8
F	A small number.	11 – 9	2
G	The big number.	5 + 6	11
H	A small number.	12 – 2	10
I	A small number.	8 – 6	2

(Repeat problems that were not firm.)

c. Write the answer to each fact. Put your pencil down when you've completed the equations for part 2.
(Observe children and give feedback.)

d. Check your work.
• Touch and read fact A. (Signal.) 6 – 4 = 2.
• (Repeat for:) B, 9 – 2 = 7; C, 6 + 4 = 10; D, 6 + 6 = 12; E, 5 + 3 = 8; F, 11 – 9 = 2; G, 5 + 6 = 11; H, 12 – 2 = 10; I, 8 – 6 = 2.

EXERCISE 6: 2-DIMENSIONAL FIGURES
DECOMPOSITION

a. Find part 3 on worksheet 79. ✔
(Teacher reference:)

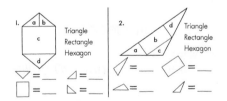

Next to figure 1 are the names triangle, rectangle, and hexagon.
• What's the first name? (Signal.) Triangle.
• What's the next name? (Signal.) Rectangle.
• What's the next name? (Signal.) Hexagon.
(Repeat until firm.)

b. Work figure 1. Circle the name of the whole figure and then write the letter for each part.
(Observe students and give feedback.)

c. Check your work for figure 1.
• Everybody, what name did you circle for the whole figure? (Signal.) Hexagon.

• What's the letter for the first equals? (Signal.) D.
• The next equals has a rectangle on one side. What letter did you write on the other side? (Signal.) C.
• The next equals has a triangle on one side. What letter did you write on the other side? (Signal.) A.
• What's the letter for the last equals? (Signal.) B.

d. There are three names next to figure 2.
• What's the first name? (Signal.) Triangle.
• What's the next name? (Signal.) Rectangle.
• What's the next name? (Signal.) Hexagon.
(Repeat step d until firm.)

e. Work figure 2. Circle the name of the whole figure and then write the letter for each part.
(Observe students and give feedback.)

f. Check your work for figure 2.
• Everybody, what name did you circle? (Signal.) Triangle.
• What's the letter for the first equals? (Signal.) D.
• The next equals has a triangle on one side. What letter did you write on the other side? (Signal.) A.
• What's the letter for the next equals? (Signal.) B.
• What's the letter for the last equals? (Signal.) C.

EXERCISE 7: FACTS
SUBTRACTION

a. Find part 4 on worksheet 79. ✔
(Teacher reference:)

a. 20 – 10   e. 8 – 2   i. 7 – 5
b. 4 – 2   f. 12 – 6   j. 10 – 8
c. 5 – 2   g. 9 – 6   k. 5 – 4
d. 10 – 4   h. 6 – 4   l. 11 – 6

These are minus problems from families you know. You're going to read each problem and tell me the answer. Then you'll go back and work them.
• Touch and read problem A. Get ready. (Signal.) 20 minus 10.
• What's 20 minus 10? (Signal.) 10.

b. (Repeat the following tasks for problems
   B through L:)
- Touch and read problem __.
- What's __?
  (Repeat problems that were not firm.)
c. Complete the equations for all of the problems
   in part 4.
   (Observe children and give feedback.)
d. Check your work.
- Touch and read fact A. (Signal.) *20 – 10 = 10.*
- (Repeat for:) B, *4 – 2 = 2;* C, *5 – 2 = 3;*
  D, *10 – 4 = 6;* E, *8 – 2 = 6;* F, *12 – 6 = 6;*
  G, *9 – 6 = 3;* H, *6 – 4 = 2;* I, *7 – 5 = 2;*
  J, *10 – 8 = 2;* K, *5 – 4 = 1;* L, *11 – 6 = 5.*

EXERCISE 8: MONEY
DOLLAR PROBLEMS

a. (Display:) W   [79:8A]

$$\begin{array}{r} \$14.37 \\ -\ 4.00 \\ \hline \end{array}$$

When you work this dollar problem you get a
zero in the answer.
- Read the problem. (Signal.) *14 dollars and
  37 cents minus 4 dollars.*
  I start by writing the dot in the answer.
  (Add to show:)   [79:8B]

$$\begin{array}{r} \$14.37 \\ -\ 4.00 \\ \hline .\ \ \ \end{array}$$

- (Point to **7.**) Say the problem for this cent
  column. (Signal.) *7 minus zero.*
- What's the answer? (Signal.) *7.*
  (Add to show:)   [79:8C]

$$\begin{array}{r} \$14.37 \\ -\ 4.00 \\ \hline .\ 7 \end{array}$$

- (Point to **3.**) Say the problem for this cent
  column. (Signal.) *3 minus zero.*
- What's the answer? (Signal.) *3.*
  (Add to show:)   [79:8D]

$$\begin{array}{r} \$14.37 \\ -\ 4.00 \\ \hline .37 \end{array}$$

- (Point to **4.**) Say the problem for this dollar
  column. (Signal.) *4 minus 4.*

- What's the answer? (Signal.) *Zero.*
  (Add to show:)   [79:8E]

$$\begin{array}{r} \$14.37 \\ +\ 4.00 \\ \hline 0.37 \end{array}$$

- (Point to **1.**) What do I write for this column?
  (Signal.) *1.*
  I write 1 and a dollar sign.
  (Add to show:)   [79:8F]

$$\begin{array}{r} \$14.37 \\ +\ 4.00 \\ \hline \$10.37 \end{array}$$

- Everybody, read the answer. (Signal.)
  10 dollars and 37 cents.
b. Turn to the other side of worksheet 79 and find
   part 5. ✔
   (Teacher reference:)

a. $\begin{array}{r} \$25.29 \\ -\ \ 5.16 \\ \hline \end{array}$ b. $\begin{array}{r} \$35.14 \\ +\ \ 1.51 \\ \hline \end{array}$ c. $\begin{array}{r} \$6.87 \\ -5.25 \\ \hline \end{array}$

- Touch and read problem A. (Signal.)
  *25 dollars and 29 cents minus 5 dollars and
  16 cents.*
- What do you write in the answer before you
  minus? (Signal.) *The dot.*
- What sign do you write after you work the
  problem? (Signal.) *(A) dollar sign.*
- Work problem A.
  (Observe children and give feedback.)
- Check your work. Everybody, touch and read
  the answer to problem A. (Signal.) *20 dollars
  and 13 cents.*
c. Touch and read problem B. (Signal.)
  *35 dollars and 14 cents plus 1 dollar and
  51 cents.*
- What do you write before you plus? (Signal.)
  The dot.
- Work problem B. Remember to write a
  dollar sign.
  (Observe children and give feedback.)
- Check your work. Everybody, touch and read
  the answer to problem B. (Signal.) *36 dollars
  and 65 cents.*

d. Touch and read problem C. (Signal.) *6 dollars and 87 cents minus 5 dollars and 25 cents.*
• What do you write before you minus? (Signal.) *The dot.*
• Work problem C. Remember to write a dollar sign.
(Observe children and give feedback.)
• Check your work. Everybody, touch and read the answer to problem C. (Signal.) *1 dollar and 62 cents.*
(Answer key:)

a. $\begin{array}{r} \$25.29 \\ -5.16 \\ \hline \end{array}$ b. $\begin{array}{r} \$35.14 \\ +1.51 \\ \hline \end{array}$ c. $\begin{array}{r} \$6.87 \\ -5.25 \\ \hline \end{array}$

You'll write each problem in a column and work it.
In part 7, you'll complete the place-value addition equations.
In part 8, you'll complete the equations for the rulers and the objects. For rulers A and B, you'll count backward to figure out the number for the shaded part.
b. Complete worksheet 79.
(Observe children and mark incorrect responses on children's worksheets as you give feedback.)

EXERCISE 9: INDEPENDENT WORK

a. Find part 6 on worksheet 79. ✔
(Teacher reference:)

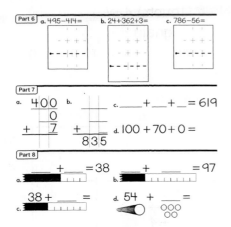

Part 6 a. 495−414=   b. 24+362+3=   c. 786−56=

Part 7
a. $\begin{array}{r} 400 \\ 0 \\ +7 \\ \hline \end{array}$ b. $\begin{array}{r} \\ + \\ \hline 835 \end{array}$ c. ___ + ___ + __ = 619   d. 100 + 70 + 0 =

Part 8
a. ___ + ___ = 38   b. ___ + ___ = 97
c. 38 + ___ =   d. 54 + ___ =

Lesson 80

EXERCISE 1: NUMBER FAMILIES

SMALL NUMBER OF 4   **REMEDY**

a. (Display:)   [80:1A]

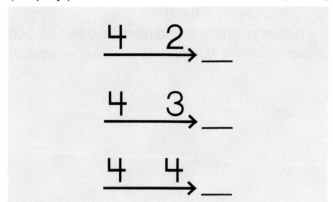

We worked with these number families
last time.

- (Point to $\overset{4\quad 2}{\longrightarrow}$.) What are the small numbers
  in this family? (Signal.) *4 and 2.*
  What's the big number? (Signal.) *6.*
- (Point to $\overset{4\quad 3}{\longrightarrow}$.) What are the small numbers
  in this family? (Signal.) *4 and 3.*
  What's the big number? (Signal.) *7.*
- (Point to $\overset{4\quad 4}{\longrightarrow}$.) What are the small numbers
  in this family? (Signal.) *4 and 4.*
  What's the big number? (Signal.) *8.*

b. (Point to $\overset{4\quad 3}{\longrightarrow}$.) The small numbers are 4
  and 3. What's the big number in this family?
  (Signal.) *7.*
- Say the fact that starts with 4. (Signal.)
  4 + 3 = 7.
- Say the other plus fact. (Signal.) *3 + 4 = 7.*
- Say the fact that goes backward down the
  arrow. (Signal.) *7 – 3 = 4.*
- Say the other minus fact. (Signal.) *7 – 4 = 3.*
  (Repeat step b until firm.)

c. (Point to $\overset{4\quad 4}{\longrightarrow}$.) The small numbers are 4
  and 4. What's the big number in this family?
  (Signal.) *8.*
  There's only one plus fact and one minus fact.
- Say the plus fact. (Signal.) *4 + 4 = 8.*
- Say the minus fact. (Signal.) *8 – 4 = 4.*
  (Repeat step c until firm.)

d. (Display:)   [80:1B]

$\overset{4}{\longrightarrow}7$	$\overset{\quad 4}{\Longrightarrow}8$	$\overset{\quad 3}{\Longrightarrow}7$
$\overset{4\quad 2}{\longrightarrow}\_$	$\overset{4}{\Longrightarrow}5$	$\overset{10\quad 4}{\longrightarrow}\_$
$\overset{\quad 4}{\Longrightarrow}14$	$\overset{4\quad 4}{\longrightarrow}\_$	$\overset{4\quad 3}{\longrightarrow}\_$

You're going to say the problem for the
missing number in each family.

e. (Point to $\overset{4}{\longrightarrow}7$.) Say the problem for the
  missing number. Get ready. (Touch.) *7 minus 4.*
- What's 7 minus 4? (Signal.) *3.*

f. (Point to $\overset{4\quad 2}{\longrightarrow}\_$.) Say the problem for the
  missing number. (Touch.) *4 plus 2.*
- What's 4 plus 2? (Signal.) *6.*

g. (Repeat the following tasks for remaining
  families:)

(Point to __.)	Say the problem for the missing number.	What's __?	
$\overset{\quad 4}{\Longrightarrow}14$	*14 – 4*	*14 – 4*	*10*
$\overset{\quad 4}{\Longrightarrow}8$	*8 – 4*	*8 – 4*	*4*
$\overset{4}{\longrightarrow}5$	*5 – 4*	*5 – 4*	*1*
$\overset{4\quad 4}{\longrightarrow}\_$	*4 + 4*	*4 + 4*	*8*
$\overset{\quad 3}{\Longrightarrow}7$	*7 – 3*	*7 – 3*	*4*
$\overset{10\quad 4}{\longrightarrow}\_$	*10 + 4*	*10 + 4*	*14*
$\overset{4\quad 3}{\longrightarrow}\_$	*4 + 3*	*4 + 3*	*7*

(Repeat for families that were not firm.)

EXERCISE 2: DOLLARS WITH 1-DIGIT CENTS [REMEDY]

a. I'm going to write dollars and cents amounts.
- Tell me to write 1 dollar and 26 cents. (Signal.) *Write 1 dollar and 26 cents.*
  (Display:) W̲   [80:2A]

$1.26

- Read this dollar amount. (Signal.) *1 dollar and 26 cents.*
b. Tell me to write 1 dollar and 16 cents. (Signal.) *Write 1 dollar and 16 cents.*
  (Change to show:)   [80:2B]

$1.16

- Read this dollar amount. (Signal.) *1 dollar and 16 cents.*
c. Tell me to write 1 dollar and 10 cents. (Signal.) *Write 1 dollar and 10 cents.*
  (Change to show:)   [80:2C]

$1.10

- Read this dollar amount. (Signal.) *1 dollar and 10 cents.*
d. Tell me to write 1 dollar and 9 cents. (Signal.) *Write 1 dollar and 9 cents.*
  I write two digits after the dot, zero and 9.
  (Change to show:)   [80:2D]

$1.09

- Read this dollar amount. (Signal.) *1 dollar and 9 cents.*
e. I can't do this.
  (Erase zero to show:)   [80:2E]

$1. 9

- I have to have two digits after the dot. How many digits have to be after the dot?
  (Signal.) *Two.*
  (Change to show:)   [80:2F]

$1.09

- Read this dollar amount. (Signal.) *1 dollar and 9 cents.*

f. (Change to show:)   [80:2G]

$1.08

- Read this dollar amount. (Signal.) *1 dollar and 8 cents.*
  (Change to show:)   [80:2H]

$1.80

- Read this dollar amount. (Signal.) *1 dollar and 80 cents.*
  (Change to show:)   [80:2I]

$1.70

- Read this dollar amount. (Signal.) *1 dollar and 70 cents.*
  (Change to show:)   [80:2J]

$1.07

- Read this dollar amount. (Signal.) *1 dollar and 7 cents.*
  (Change to show:)   [80:2K]

$1.03

- Read this dollar amount. (Signal.) *1 dollar and 3 cents.*
g. Remember, when you write cents numbers that are less than 10, you write zero before the number.
  If you want to write 5 cents, you write zero, 5 after the dot.
- If you want to write 8 cents, what do you write after the dot? (Signal.) *Zero, 8.*
- If you want to write 1 cent, what do you write after the dot? (Signal.) *Zero, 1.*
  (Repeat until firm.)

EXERCISE 3: MIXED COUNTING

a. Start with 125 and count by 25s to 200. Get 125 going. *One hundred twenty-fiiive.* Count. (Tap.) *150, 175, 200.*
- Start with 126 and plus tens to 196. Get 126 going. *One hundred twenty-siiix.* Plus tens. (Tap.) *136, 146, 156, 166, 176, 186, 196.*
b. Start with 196 and count backward to 190. Get 196 going. *One hundred ninety-siiix.* Count backward. (Tap.) *195, 194, 193, 192, 191, 190.*

c. Start at 115 and count by fives to 150. Get 115 going. *One hundred fifteeen.* Count. (Tap.) *120, 125, 130, 135, 140, 145, 150.*
- Start with 300 and plus tens to 400. Get 300 going. *Three huuundred.* Plus tens. (Tap.) *310, 320, 330, 340, 350, 360, 370, 380, 390, 400.*
d. You have 65. When you plus fives, what's the next number? (Signal.) *70.*
- You have 65. When you minus ones, what's the next number? (Signal.) *64.*
- You have 65. When you plus tens, what's the next number? (Signal.) *75.*
- You have 65. When you plus ones, what's the next number? (Signal.) *66.*

EXERCISE 4: FACTS
PLUS/MINUS MIX

a. (Distribute unopened workbooks to children.)
Open your workbook to Lesson 80 and find part 1.
(Observe children and give feedback.)
(Teacher reference:)

a. $10-4$ c. $6+6$ f. $8-3$
b. $5-3$ d. $6+3$ g. $6+10$
     e. $6-2$ h. $11-6$

- Touch and read problem A. Get ready. (Signal.) *10 minus 4.*
- Is the answer the big number or a small number? (Signal.) *A small number.*
- What's 10 minus 4? (Signal.) *6.*
b. (Repeat the following tasks for problems B through H:)

Touch and read problem __.	Is the answer the big number or a small number?	What's __?	
B	A small number.	5 − 3	2
C	The big number.	6 + 6	12
D	The big number.	6 + 3	9
E	A small number.	6 − 2	4
F	A small number.	8 − 3	5
G	The big number.	6 + 10	16
H	A small number.	11 − 6	5

(Repeat problems that were not firm.)
c. Write the answer to each fact. Put your pencil down when you've completed the equations for part 1.
(Observe children and give feedback.)

d. Check your work. You'll touch and read each fact.
- Fact A. (Signal.) *10 – 4 = 6.*
- (Repeat for:) B, *5 – 3 = 2;* C, *6 + 6 = 12;* D, *6 + 3 = 9;* E, *6 – 2 = 4;* F, *8 – 3 = 5;* G, *6 + 10 = 16;* H, *11 – 6 = 5.*

EXERCISE 5: COLUMN ADDITION
CARRYING
REMEDY

a. Find part 2 on worksheet 80. ✔
(Teacher reference:)
R Part L

a.
```
  34
  12
+ 36
```
b.
```
   8
  52
+ 37
```
c.
```
  76
+ 14
```

- Touch and read problem A. (Signal.) *34 plus 12 plus 36.*
- Read the problem in the ones column. (Signal.) *4 plus 2 plus 6.*
- What's 4 plus 2? (Signal.) *6.*
- What's 6 plus 6? (Signal.) *12.*
- Write the digits for 12 where they should go. (Observe children and give feedback.)
- Touch and read the problem for the tens column. (Signal.) *1 plus 3 plus 1 plus 3.*
- What's 1 plus 3? (Signal.) *4.*
- What's 4 plus 1? (Signal.) *5.*
- What's 5 plus 3? (Signal.) *8.*
- Write 8 where it should go. (Observe children and give feedback.)
- Start with 34 and read the problem and the answer. (Signal.) *34 + 12 + 36 = 82.*
b. Touch and read problem B. (Signal.) *8 plus 52 plus 37.*
- Read the problem in the ones column. (Signal.) *8 plus 2 plus 7.*
- What's 8 plus 2? (Signal.) *10.*
- What's 10 plus 7? (Signal.) *17.*
- Write the digits for 17 where they should go. (Observe children and give feedback.)
- Touch and read the problem for the tens column. (Signal.) *1 plus 5 plus 3.*
- What's 1 plus 5? (Signal.) *6.*
- What's 6 plus 3? (Signal.) *9.*
- Write 9 where it should go. (Observe children and give feedback.)
- Start with 8 and read the problem and the answer. (Signal.) *8 + 52 + 37 = 97.*

c. Touch and read problem C. (Signal.) *76 plus 14.*
- Read the problem in the ones column. (Signal.) *6 plus 4.*
- What's 6 plus 4? (Signal.) *10.*
- Write the digits for 10 where they should go. (Observe children and give feedback.)
- Touch and read the problem for the tens column. (Signal.) *1 plus 7 plus 1.*
- What's 1 plus 7? (Signal.) *8.*
- What's 8 plus 1? (Signal.) *9.*
- Write 9 where it should go. (Observe children and give feedback.)
- Start with 76 and read the problem and the answer. (Signal.) *76 + 14 = 90.*

EXERCISE 6: COMPARISON
LENGTH REMEDY

a. Find part 3 on worksheet 80. ✔
   (Teacher reference:) R Test 9: Part E

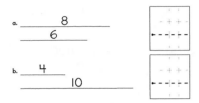

Each problem shows lines and how many centimeters long they are. You're going to write problems and figure out how much longer one line is than the other line.
- Touch the lines for problem A. ✔
- Which line is longer, the top line or the bottom line? (Signal.) *The top line.*
- How many centimeters long is the top line? (Signal.) *8.*
- How many centimeters long is the bottom line? (Signal.) *6.*
- Say the problem for figuring out how much longer the top line is. (Signal.) *8 minus 6.*
- Write a column problem and work it. Remember the equals bar. (Observe children and give feedback.) (Check equals bar.)
- Touch and read the problem and the answer for A. (Signal.) *8 − 6 = 2.*
- How many centimeters longer is the top line than the bottom line? (Signal.) *2.*

b. Touch the lines for problem B. ✔
- Which line is longer, the top line or the bottom line? (Signal.) *The bottom line.*
- How many centimeters long is the bottom line? (Signal.) *10.*
- Say the problem that starts with 10. (Signal.) *10 minus 4.*
- Write a column problem and the answer. (Observe children and give feedback.) (Check equals bar.)
- Touch and read the problem and the answer for B. (Signal.) *10 − 4 = 6.*
- How many centimeters longer is the bottom line than the top line? (Signal.) *6.*

EXERCISE 7: FACTS
SUBTRACTION REMEDY

a. Find part 4 on worksheet 80. ✔
   (Teacher reference:) R Part K

a. 3 − 2	e. 6 − 2	i. 8 − 5
b. 12 − 6	f. 8 − 6	j. 11 − 2
c. 14 − 10	g. 9 − 7	k. 11 − 5
d. 4 − 1	h. 10 − 6	l. 9 − 6

These are minus problems from families you know. You're going to read each problem and tell me the answer. Then you'll go back and work them.
- Touch and read problem A. Get ready. (Signal.) *3 minus 2.*
- What's 3 minus 2? (Signal.) *1.*

b. (Repeat the following tasks for problems B through L:)
- Touch and read problem __.
- What's __?
(Repeat problems that were not firm.)

c. Complete the equations for all of the problems in part 4. (Observe children and give feedback.)

d. Check your work.
- Touch and read fact A. (Signal.) *3 − 2 = 1.*
- (Repeat for:) B, *12 − 6 = 6;* C, *14 − 10 = 4;* D, *4 − 1 = 3;* E, *6 − 2 = 4;* F, *8 − 6 = 2;* G, *9 − 7 = 2;* H, *10 − 6 = 4;* I, *8 − 5 = 3;* J, *11 − 2 = 9;* K, *11 − 5 = 6;* L, *9 − 6 = 3.*

EXERCISE 8: MONEY

DOLLAR PROBLEMS

a. Turn to the other side of worksheet 80 and find part 5. ✔
(Teacher reference:)

a. $\begin{array}{r} \$\ 2\ 7.5\ 5 \\ -\quad 3.2\ 4 \end{array}$ b. $\begin{array}{r} \$\ 5.2\ 1 \\ +\ 3.1\ 8 \end{array}$

You'll touch and read each dollar problem.

- Problem A. Get ready. (Signal.) *27 dollars and 55 cents minus 3 dollars and 24 cents.*
- Problem B. Get ready. (Signal.) *5 dollars and 21 cents plus 3 dollars and 18 cents.*

b. Work the problems in part 5. Remember, write a dollar sign and a dot in each answer.
(Observe children and give feedback.)

c. (Display:)   [80:8A]

a. $\begin{array}{r} \$\ 2\ 7.5\ 5 \\ -\quad 3.2\ 4 \\ \hline \$\ 2\ 4.3\ 1 \end{array}$	b. $\begin{array}{r} \$\ 5.2\ 1 \\ +\ 3.1\ 8 \\ \hline \$\ 8.3\ 9 \end{array}$

Here's what you should have.

- Look at the answer for problem A. Everybody, what's 27 dollars and 55 cents minus 3 dollars and 24 cents? (Signal.) *24 dollars and 31 cents.*
- Look at the answer for problem B. Everybody, what's 5 dollars and 21 cents plus 3 dollars and 18 cents? (Signal.) *8 dollars and 39 cents.*

EXERCISE 9: INDEPENDENT WORK

a. Find part 6 on worksheet 80. ✔
(Teacher reference:)

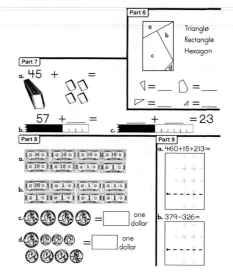

You'll do this part when you do your independent work.
Remember, circle the name for the whole figure. Then write the letter for each part.
In part 7, you'll complete the equations for the objects and the rulers. For ruler C, you'll count backward to figure out the number for the shaded part.
In part 8, you'll write an equals and the number of dollars to show what the group of bills is worth. You'll circle the words one dollar, or write the cents for the group of coins.
In part 9, you'll write each problem in a column and work it.

b. Complete worksheet 80.
(Observe children and mark incorrect responses on children's worksheets as you give feedback.)

Mastery Test ⑧

Note: Mastery Tests are administered to all students in the group. Each student will need a pencil and a *Student Assessment Book.* Try to arrange students so they cannot look at other students' responses.

Teacher Presentation

a. Find Test 8 in your test booklet. ✔
- Touch part 1. ✔
  (Teacher reference:)

  a. _____ b. _____ c. _____ d. _____ e. _____

  You're going to write symbols for dollars and cents.
- Touch the space for A. ✔
- You'll write 9 dollars and 13 cents in space A. What will you write in space A? (Signal.)
  9 dollars and 13 cents.
- Write the dollar sign. Remember, capital S with a straight line down through it. ✔
- Now write the rest of 9 dollars and 13 cents. (Observe children.)
b. Touch the space for B. ✔
- You'll write 10 dollars and 52 cents in space B. What will you write in space B? (Signal.)
  10 dollars and 52 cents.
- Write the dollar sign. Then write the rest of 10 dollars and 52 cents. (Observe children.)
c. Touch the space for C. ✔
- You'll write 3 dollars and 56 cents in space C. What will you write in space C? (Signal.)
  3 dollars and 56 cents.
- Write the dollar sign. Then write the rest of 3 dollars and 56 cents. (Observe children.)
d. Touch the space for D. ✔
- You'll write 3 dollars and 17 cents in space D. What will you write in space D? (Signal.)
  3 dollars and 17 cents.
- Write the dollar sign. Then write the rest of 3 dollars and 17 cents. (Observe children.)

e. Touch the space for E. ✔
- You'll write 4 dollars and 19 cents in space E. What will you write in space E? (Signal.)
  4 dollars and 19 cents.
- Write the dollar sign. Then write the rest of 4 dollars and 19 cents. (Observe children.)
f. Touch part 2 on test sheet 8. ✔
  (Teacher reference:)

  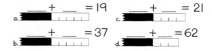

  You're going to complete the equation for each ruler.
- For each ruler, count and write the number for the unshaded part. (Observe children.)
g. Now you'll figure out how many centimeters the shaded part of each ruler is. You'll touch the end of each ruler and get the centimeters for both parts going. Then you'll count backward and touch the lines until you get to the end of the shaded part.
- Figure out the centimeters for the shaded part for the rulers in part 2. Put your pencil down when you've completed the equation for each ruler. (Observe children.)
h. Touch part 3 on test sheet 8. ✔
  (Teacher reference:)

	d. 7 + 2	h. 6 + 5
a. 3 + 6	e. 3 + 5	i. 2 + 10
b. 2 + 9	f. 6 + 4	j. 6 + 3
c. 6 + 6	g. 2 + 8	

  All of the plus problems in part 3 are from families you know. You'll complete equations in part 3.

i. Touch part 4 on test sheet 8. ✔
    (Teacher reference:)

a. 🪙🪙🪙🪙🪙🪙 = ☐ one dollar
b. 🪙🪙🪙🪙🪙🪙🪙🪙 = ☐ one dollar
c. 🪙🪙🪙🪙🪙🪙
   🪙🪙🪙🪙🪙🪙 = ☐ one dollar
d. 🪙🪙🪙🪙🪙🪙🪙 = ☐ one dollar
e. 🪙🪙🪙🪙🪙🪙🪙 = ☐ one dollar

• Touch the words after the box for group A. ✔
• The words are one dollar. What are they?
  (Signal.) *One dollar.*
• If a group of coins is worth one dollar, circle
  the words one dollar. What words will you
  circle if a group is worth one dollar?
  (Signal.) *One dollar.*
• If a group of coins is not worth one dollar, write
  the number of cents in the box. Where will you
  write the number of cents if a group of coins is
  not worth one dollar? (Signal.) *In the box.*

j. Count the cents for each group of coins. If the
    group is worth one dollar, circle the words one
    dollar. If the group isn't worth one dollar, write
    the number of cents in the box.
    (Observe children.)

k. Turn to the other side of test sheet 8 and touch
    part 5. ✔
    (Teacher reference:)

a. 11 − 5   d. 9 − 7   h. 10 − 4
b. 11 − 9   e. 8 − 5   i. 10 − 2
c. 9 − 3   f. 12 − 6   j. 11 − 6
               g. 12 − 10

All of the minus problems in part 5 are from
families you know.
You'll complete equations in part 5.

l. Touch part 6. ✔
    (Teacher reference:)

a.   b.   c.   d.
   72   16   76   8
  +18   +46   +14   52
                                      +37

You'll work column problems in part 6.

m. Touch part 7. ✔
    (Teacher reference:)

a. $4.25   b. $5.25   c. $ 3.26   d. $12.41
   −3.15   +2.41   +11.52   −10.31

In part 7, you'll work the money problems. For
each problem, the first thing you'll write in the
answer is a dot.

• What's the first thing you'll write in the answer
  for each problem? (Signal.) *(A) dot.*
  After you write the dot for each problem, you'll
  work it. Remember to write a dollar sign in
  your answer.
• Your turn: Work the rest of the problems on
  Test 8. ✔
  (Direct students where to put their
  assessment books when they are finished.)

Scoring Notes

a. Collect test booklets. Use the Answer Key and
    Passing Criteria Table to score the tests.

Passing Criteria Table — Mastery Test 8			
Part	Score	Possible Score	Passing Score
1	2 for each dollar amount	10	8
2	3 for each problem	12	9
3	2 for each problem	20	18
4	2 for each problem	10	8
5	2 for each problem	20	18
6	4 for each problem	16	12
7	3 for each problem	12	9
	Total	100	

b. Complete the Mastery Test 8 Remedy
    Summary Sheet to determine whether group
    remedies are needed. Reproducible Remedy
    Summary Sheets are at the back of the
    Answer Key and at the back of the *Teacher's
    Guide.*

• If ¼ or more of the students did not pass a test
  part, present the remedy for that part before
  beginning Lesson 81. The Remedy Table
  follows and is also at the end of the Mastery
  Test 8 Answer Key. Remedies worksheets
  follow Mastery Test 8 in the *Student
  Assessment Book.*

Remedy Table — Mastery Test 8				
Part	Test Items	Remedy		Student Material Remedies Worksheet
		Lesson	Exercise	
1	Writing Dollar Amounts	70	1	—
		72	5	Part A
		75	6	Part B
		76	7	Part C
2	Count Backward (Rulers)	70	4	—
		72	6	Part D
		74	6	Part E
		75	7	Part F
3	Facts (Addition)	69	5	Part G
		73	8	Part H
		74	4	—
4	Coins (Equal to a Dollar)	65	5	—
		67	2	—
		69	6	Part I
5	Facts (Subtraction)	71	5	Part J
		73	4	—
		75	1	—
		80	7	Part K
6	Column Addition (Carrying)	74	2	—
		75	3	—
		78	4	—
		80	5	Part L
7	Money Problems	77	8	Part M
		78	6	Part N

Retest

Retest individual students on any part failed.

Lesson 41 Side 1 Name _____

Part 1

a. $10 - 1 = 9$ c. $5 - 4 = 1$ e. $8 - 1 = 7$
b. $1 + 7 = 8$ d. $9 + 1 = 10$ f. $7 - 6 = 1$

Part 2

a. $\underline{800} + \underline{10} + \underline{3} = 813$

b. $\underline{500} + \underline{0} + \underline{0} = 500$

c. $\underline{400} + \underline{0} + \underline{6} = 406$

d. $\underline{100} + \underline{70} + \underline{0} = 170$

Part 3

a.
$$\begin{array}{r} 106 \\ - 5 \\ \hline 101 \end{array}$$

b.
$$\begin{array}{r} 60 \\ + 15 \\ \hline 75 \end{array}$$

Part 4

a. $38 + 10 = 48$

b. $26 + 10 = 36$

c. $15 + 10 = 25$

d. $13 + 10 = 23$

Part 5

a.			3
b.	1	6	5
c.		1	2
d.	4	7	0
e.		3	9

Part 6

a. $10 + 3 + 1 = 14$ b. $\underline{2} + \underline{6} + \underline{1} = 9$

Connecting Math Concepts   Lesson 41 59

Lesson 41 Side 2 Name _____

Side 2 work is independent.

Part 7

a. $6 \quad 2 \rightarrow 8$
$6 + 2 = 8$
$2 + 6 = 8$
$8 - 2 = 6$
$8 - 6 = 2$

b. $3 \quad 2 \rightarrow 5$
$3 + 2 = 5$
$2 + 3 = 5$
$5 - 2 = 3$
$5 - 3 = 2$

Part 8

a. $7 + 1 = 8$ c. $1 + 1 = 2$ e. $8 + 1 = 9$
$7 + 2 = 9$ $1 + 2 = 3$ $8 + 2 = 10$
$97 + 2 = 99$ $51 + 2 = 53$ $88 + 2 = 90$

b. $5 + 1 = 6$ d. $3 + 1 = 4$ f. $4 + 1 = 5$
$5 + 2 = 7$ $3 + 2 = 5$ $4 + 2 = 6$
$15 + 2 = 17$ $33 + 2 = 35$ $64 + 2 = 66$

Part 9

a. $400 + 50 + 0 = 450$ d. $200 + 0 + 9 = 209$
b. $100 + 10 + 3 = 113$ e. $700 + 10 + 6 = 716$
c. $70 + 2 = 72$ f. $600 + 0 + 0 = 600$

60   Lesson 41   Connecting Math Concepts

Lesson 42 Side 1 Name _____

Part 1

a. $11 - 1 = 10$ d. $5 - 4 = 1$ g. $11 - 10 = 1$
b. $10 - 9 = 1$ e. $10 + 1 = 11$ h. $4 - 1 = 3$
c. $1 + 3 = 4$ f. $10 - 1 = 9$

Part 2

a. $\underline{800} + \underline{20} + \underline{5} = 825$

b. $700 + 0 + 3 = \underline{703}$

c. $\underline{300} + \underline{50} + \underline{0} = 350$

d. $\underline{400} + \underline{0} + \underline{5} = 405$

Part 3

a.
$$\begin{array}{r} 39 \\ + 720 \\ \hline 759 \end{array}$$

b.
$$\begin{array}{r} 894 \\ - 84 \\ \hline 810 \end{array}$$

c.
$$\begin{array}{r} 402 \\ + 7 \\ \hline 409 \end{array}$$

d.
$$\begin{array}{r} 581 \\ + 208 \\ \hline 789 \end{array}$$

Part 4

a. $17 + 62$

$$\begin{array}{r} 17 \\ + 62 \\ \hline 79 \end{array}$$

b. $36 - 15$

$$\begin{array}{r} 36 \\ - 15 \\ \hline 21 \end{array}$$

INDEPENDENT WORK

Part 5

a. $3 + 2 + 1 = 6$

b. $7 + 1 + 2 = 10$

Connecting Math Concepts   Lesson 42 61

Lesson 42 Side 2 Name _____

Side 2 work is independent.

Part 6

a. $6 \quad 2 \rightarrow 8$
$6 + 2 = 8$
$2 + 6 = 8$
$8 - 2 = 6$
$8 - 6 = 2$

b. $9 \quad 2 \rightarrow 11$
$9 + 2 = 11$
$2 + 9 = 11$
$11 - 2 = 9$
$11 - 9 = 2$

Part 7

a. $13 + 10 = 23$ d. $51 + 10 = 61$ g. $28 + 10 = 38$
b. $47 + 10 = 57$ e. $14 + 10 = 24$ h. $75 + 10 = 85$
c. $62 + 10 = 72$ f. $86 + 10 = 96$ i. $11 + 10 = 21$

Part 8

a. $100 + 10 + 1 = 111$ e. $800 + 20 + 0 = 820$
b. $30 + 0 = 30$ f. $800 + 0 + 2 = 802$
c. $400 + 0 + 8 = 408$ g. $800 + 10 + 2 = 812$
d. $10 + 6 = 16$ h. $700 + 0 + 0 = 700$

Part 9

a. $\underline{30} + \underline{9} = 39$ c. $\underline{10} + \underline{5} = 15$

b. $\underline{10} + \underline{3} = 13$ d. $\underline{40} + \underline{0} = 40$

62   Lesson 42   Connecting Math Concepts

Lesson 43 Side 1   Name _____

Part 1

a. $1 + 9 + 5 = 15$ b. $1 + 5 + 1 = 7$

Part 2

a. $\underline{700} + \underline{40} + \underline{1} = 741$

b. $500 + 60 + 0 = \underline{560}$

c. $300 + 0 + 9 = \underline{309}$

d. $\underline{200} + \underline{10} + \underline{7} = 217$

Part 3

a. $\underset{5}{\overset{3\quad 2}{\longrightarrow}}$ d. $\underset{7}{\overset{6\quad 1}{\longrightarrow}}$ g. $\underset{11}{\overset{9\quad 2}{\longrightarrow}}$

b. $\underset{6}{\overset{4\quad 2}{\longrightarrow}}$ e. $\underset{\underline{6}}{\overset{5\quad 1}{\longrightarrow}}$ h. $\underset{9}{\overset{7\quad 2}{\longrightarrow}}$

c. $\underset{\underline{10}}{\overset{8\quad 2}{\longrightarrow}}$ f. $\underset{4}{\overset{3\quad 1}{\longrightarrow}}$

Part 4

a. $50 + 318$

```
   50
+ 318
  368
```
INDEPENDENT WORK

b. $479 - 61$

```
  479
-  61
  418
```

Lesson 43 Side 2   Name _____

Part 5

a. $9 - 7 = 2$ c. $6 - 2 = 4$ e. $2 + 7 = 9$

b. $5 + 2 = 7$ d. $7 - 5 = 2$ f. $8 - 2 = 6$

Part 6

a. $\underline{50 + 7 = 57}$ b. $\underline{6 - 5 = 1}$ c. $\underline{7 + 2 = 9}$

INDEPENDENT WORK

Part 7

a. $\underset{8}{\overset{7\quad 1}{\longrightarrow}}$ b. $\underset{9}{\overset{7\quad 2}{\longrightarrow}}$

$7 + 1 = 8$ $7 + 2 = 9$

$1 + 7 = 8$ $2 + 7 = 9$

$8 - 1 = 7$ $9 - 2 = 7$

$8 - 7 = 1$ $9 - 7 = 2$

Part 8

a.
```
   39
+ 720
  759
```

b.
```
  894
-  84
  810
```

c.
```
  402
+   7
  409
```

Part 9

a. $13 + 10 = 23$ c. $32 + 10 = 42$

b. $45 + 10 = 55$ d. $18 + 10 = 28$

Part 10

a. $1 + 6 = 7$ c. $9 - 1 = 8$

b. $5 - 4 = 1$ d. $3 - 2 = 1$

Lesson 44 Side 1   Name _____

Part 1

a. $1 + 10 + 1 = 12$ c. $1 + 9 + 2 = 12$

b. $2 + 6 + 1 = 9$ d. $5 + 2 + 2 = 9$

Part 2

a.
```
   62
+ 14
  76
```

b.
```
   45
- 21
  24
```

Part 3

a. $\underset{\underline{10}}{\overset{8\quad 2}{\longrightarrow}}$ c. $\underset{\underline{12}}{\overset{10\quad 2}{\longrightarrow}}$

$10 - 8 = 2$ $12 - 10 = 2$

b. $\underset{\underline{6}}{\overset{4\quad 2}{\longrightarrow}}$ d. $\underset{\underline{9}}{\overset{7\quad 2}{\longrightarrow}}$

$6 - 4 = 2$ $9 - 7 = 2$

Part 4

a. $500 + 20 + 9 = \underline{529}$

b. $\underline{200} + \underline{0} + \underline{4} = 204$

c. $\underline{60} + \underline{8} = 68$

d. $700 + 10 + 3 = \underline{713}$

e. $\underline{90} + \underline{5} = 95$

Part 5

a. $11 - 9 = 2$

b. $10 - 2 = 8$

c. $4 + 2 = 6$

d. $7 - 2 = 5$

e. $8 - 6 = 2$

f. $2 + 9 = 11$

Lesson 44 Side 2   Name _____

Side 2 work is independent.

Part 6

a. $\underset{8}{\overset{6\quad 2}{\longrightarrow}}$ b. $\underset{5}{\overset{3\quad 2}{\longrightarrow}}$

$6 + 2 = 8$ $3 + 2 = 5$

$2 + 6 = 8$ $2 + 3 = 5$

$8 - 2 = 6$ $5 - 2 = 3$

$8 - 6 = 2$ $5 - 3 = 2$

Part 7

a.
```
   62
+ 525
  587
```

b.
```
  478
-  52
  426
```

c.
```
  249
- 120
  129
```

Part 8

a. $12 + 10 = 22$ c. $41 + 10 = 51$ e. $21 + 10 = 31$

b. $66 + 10 = 76$ d. $18 + 10 = 28$ f. $34 + 10 = 44$

Part 9

a. $10 + 1 = 11$ c. $5 - 4 = 1$ e. $1 + 7 = 8$

b. $1 + 10 = 11$ d. $9 - 1 = 8$ f. $8 - 7 = 1$

Lesson 45 Side 1 Name _____

Part 1
a. $1 + 10 + 2 = 13$
b. $2 + 9 + 1 = 12$
c. $1 + 8 + 2 = 11$

Part 2
a.
```
  52
+ 17
  69
```
b.
```
  680
- 170
  510
```

Part 3
a. $9 \xrightarrow{\quad 2 \quad} 11$
$11 - 9 = 2$

b. $8 \xrightarrow{\quad 1 \quad} 9$
$9 - 1 = 8$

c. $9 \xrightarrow{\quad 2 \quad} 11$
$9 + 2 = 11$

d. $7 \xrightarrow{\quad 2 \quad} 9$
$9 - 2 = 7$

Part 4
a. $200 + 0 + 7 = 207$
b. $200 + 0 + 0 = 200$
c. $60 + 1 = 61$
d. $400 + 10 + 6 = 416$

Part 5
a. $7 - 2 = 5$
b. $6 - 1 = 5$
c. $8 - 6 = 2$
d. $9 - 8 = 1$
e. $10 - 2 = 8$
f. $4 - 1 = 3$
g. $7 - 5 = 2$
h. $6 - 2 = 4$

Lesson 45 Side 2 Name _____

Side 2 work is independent.

Part 6
a. $9 \xrightarrow{\quad 1 \quad} 10$
$10 - 9 = 1$

b. $7 \xrightarrow{\quad 1 \quad} 8$
$7 + 1 = 8$

c. $4 \xrightarrow{\quad 1 \quad} 5$
$5 - 1 = 4$

d. $8 \xrightarrow{\quad 1 \quad} 9$
$9 - 1 = 8$

e. $2 \xrightarrow{\quad 1 \quad} 3$
$2 + 1 = 3$

f. $7 \xrightarrow{\quad 1 \quad} 8$
$8 - 7 = 1$

Part 7
a.
```
  512
- 411
  101
```
b.
```
  368
+  21
  389
```
c.
```
  459
- 158
  301
```

Part 8
a. $36 + 10 = 46$
b. $11 + 10 = 21$
c. $59 + 10 = 69$
d. $77 + 10 = 87$
e. $15 + 10 = 25$
f. $23 + 10 = 33$

Part 9
a. $7 + 1 = 8$
$7 + 2 = 9$
$27 + 2 = 29$

b. $9 + 1 = 10$
$9 + 2 = 11$
$59 + 2 = 61$

c. $3 + 1 = 4$
$3 + 2 = 5$
$43 + 2 = 45$

Lesson 46 Side 1 Name _____

Part 1
a. $1 + 7 + 2 = 10$
b. $2 + 8 + 5 = 15$
c. $1 + 9 + 3 = 13$

Part 2
a.
```
  54
- 13
  41
```
b.
```
  162
+ 215
  377
```

Part 3
a. $3 \xrightarrow{\quad 2 \quad} 5$
b. $6 \xrightarrow{\quad 2 \quad} 8$
c. $10 \xrightarrow{\quad 1 \quad} 11$
d. $2 \xrightarrow{\quad 2 \quad} 4$
e. $1 \xrightarrow{\quad 1 \quad} 2$
f. $9 \xrightarrow{\quad 2 \quad} 11$
g. $8 \xrightarrow{\quad 1 \quad} 9$
h. $7 \xrightarrow{\quad 2 \quad} 9$

Part 4
a. $300 + 0 + 6 = 306$
b. $200 + 10 + 9 = 219$
c. $200 + 30 + 1 = 231$
d. $200 + 10 + 3 = 213$
e. $70 + 8 = 78$

Lesson 46 Side 2 Name _____

Part 5
a. $10 - 8 = 2$
b. $10 - 1 = 9$
c. $12 - 10 = 2$
d. $8 - 2 = 6$
e. $7 - 6 = 1$
f. $5 - 2 = 3$
g. $7 - 5 = 2$
h. $9 - 2 = 7$

INDEPENDENT WORK

Part 6
a. $6 \xrightarrow{\quad 2 \quad} 8$
$6 + 2 = 8$
$2 + 6 = 8$
$8 - 2 = 6$
$8 - 6 = 2$

b. $2 \xrightarrow{\quad 1 \quad} 3$
$2 + 1 = 3$
$1 + 2 = 3$
$3 - 1 = 2$
$3 - 2 = 1$

Part 7
a.
```
   36
+ 202
  238
```
b.
```
  875
- 255
  620
```
c.
```
  924
+  71
  995
```

Part 8
a. $67 + 10 = 77$
b. $43 + 10 = 53$
c. $56 + 10 = 66$
d. $12 + 10 = 22$
e. $88 + 10 = 98$
f. $34 + 10 = 44$

Copyright © The McGraw-Hill Companies, Inc.

Lesson 47 — Side 1 Name _____

Part 1

a. $9 - 8 = 1$ e. $8 - 7 = 1$ i. $2 + 7 = 9$

b. $6 - 1 = 5$ f. $3 - 2 = 1$ j. $6 - 4 = 2$

c. $5 - 3 = 2$ g. $1 + 10 = 11$

d. $1 + 4 = 5$ h. $7 - 1 = 6$

Part 2

a. $\underline{700} + \underline{30} + \underline{5} = 735$

b. $\underline{400} + \underline{10} + \underline{2} = 412$

c. $\underline{200} + \underline{0} + \underline{8} = 208$

d. $\underline{600} + \underline{90} + \underline{1} = 691$

Part 3

a. $\underline{10\ \ 4} \to \underline{14}$

b. $\underline{10\ \ 8} \to \underline{18}$

c. $\underline{10\ \ 6} \to \underline{16}$

d. $\underline{10\ \ 9} \to \underline{19}$

e. $\underline{10\ \ 3} \to \underline{13}$

f. $\underline{10\ \ 5} \to \underline{15}$

Part 4

a.
$$\begin{array}{r} 13 \\ + 42 \\ \hline 55 \end{array}$$

b.
$$\begin{array}{r} 276 \\ - 166 \\ \hline 110 \end{array}$$

c.
$$\begin{array}{r} 39 \\ - \ \ 8 \\ \hline 31 \end{array}$$

Connecting Math Concepts

Lesson 47 **71**

Lesson 47 — Side 2 Name _____

Part 5

a. $1 + 6 + 2 = 9$ c. $1 + 9 + 4 = 14$

b. $8 + 2 + 6 = 16$ d. $3 + 2 + 1 = 6$

INDEPENDENT WORK

Part 6

a. $\underline{3\ \ 1} \to 4$ $4 - 3 = 1$

d. $\underline{6\ \ 1} \to \underline{7}$ $6 + 1 = 7$

b. $\underline{2\ \ 2} \to \underline{4}$ $2 + 2 = 4$

e. $\underline{5\ \ 2} \to 7$ $7 - 2 = 5$

c. $\underline{8\ \ 1} \to 9$ $9 - 1 = 8$

f. $\underline{3\ \ 2} \to 5$ $5 - 3 = 2$

Part 7

a.
$$\begin{array}{r} 758 \\ - 107 \\ \hline 651 \end{array}$$

b.
$$\begin{array}{r} 463 \\ + 210 \\ \hline 673 \end{array}$$

c.
$$\begin{array}{r} 892 \\ - \ \ 82 \\ \hline 810 \end{array}$$

72 Lesson 47

Connecting Math Concepts

Lesson 48 — Side 1 Name _____

Part 1

a. $2 + 6 + 1 = 9$ c. $10 + 0 + 6 = 16$

b. $1 + 9 + 3 = 13$ d. $1 + 5 + 2 = 8$

Part 2

a. $9 - 1 = 8$ d. $3 - 2 = 1$ g. $7 - 5 = 2$

b. $7 - 6 = 1$ e. $10 - 2 = 8$ h. $6 - 1 = 5$

c. $9 - 7 = 2$ f. $12 - 2 = 10$

Part 3

a.  $= 34$

b. $= 23$

INDEPENDENT WORK

Part 4

a.
$$\begin{array}{r} 140 \\ + \ 29 \\ \hline 169 \end{array}$$

b.
$$\begin{array}{r} 7 \\ + 32 \\ \hline 39 \end{array}$$

c.
$$\begin{array}{r} 258 \\ - \ 17 \\ \hline 241 \end{array}$$

Connecting Math Concepts

Lesson 48 **73**

Lesson 48 — Side 2 Name _____

Part 5

a. $\underline{10\ \ 3} \to \underline{13}$ c. $\underline{10\ \ 5} \to \underline{15}$ e. $\underline{10\ \ 6} \to \underline{16}$

b. $\underline{10\ \ 7} \to \underline{17}$ d. $\underline{10\ \ 2} \to \underline{12}$ f. $\underline{10\ \ 1} \to \underline{11}$

Part 6

a. $7 - 2 = 5$ c. $8 - 6 = 2$ f. $4 - 1 = 3$

b. $6 - 1 = 5$ d. $9 - 8 = 1$ g. $7 - 5 = 2$

e. $10 - 2 = 8$ h. $6 - 2 = 4$

INDEPENDENT WORK

Part 7

a. $\underline{9\ \ 2} \to \underline{11}$ $9 + 2 = 11$

d. $\underline{3\ \ 2} \to 5$ $5 - 2 = 3$

b. $\underline{7\ \ 2} \to 9$ $9 - 7 = 2$

e. $\underline{5\ \ 2} \to \underline{7}$ $5 + 2 = 7$

c. $\underline{5\ \ 1} \to 6$ $6 - 1 = 5$

f. $\underline{7\ \ 1} \to 8$ $8 - 7 = 1$

Part 8

a. $800 + 0 + 3 = \underline{803}$ c. $\underline{300} + \underline{10} + \underline{7} = 317$

b. $\underline{10} + \underline{4} = 14$ d. $\underline{500} + \underline{40} + \underline{0} = 540$

74 Lesson 48

Connecting Math Concepts

Connecting Math Concepts

Answer Key **269**

Lesson 49 · Side 1   Name _____

Part 1

c. $5 - 3 = 2$ f. $10 - 9 = 1$

a. $2 + 9 = 11$ d. $8 + 1 = 9$ g. $1 + 7 = 8$

b. $7 + 2 = 9$ e. $7 - 2 = 5$ h. $6 - 2 = 4$

Part 2

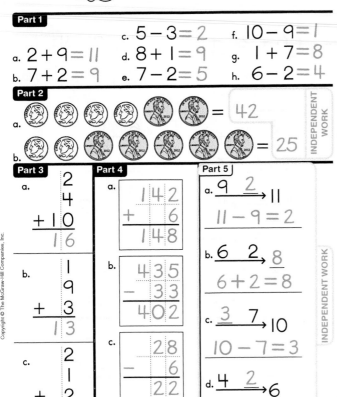

a. $= 42$

b. $= 25$

INDEPENDENT WORK

Part 3

a.
$$\begin{array}{r} 2 \\ 4 \\ +10 \\ \hline 16 \end{array}$$

b.
$$\begin{array}{r} 1 \\ 9 \\ +3 \\ \hline 13 \end{array}$$

c.
$$\begin{array}{r} 2 \\ 1 \\ +2 \\ \hline 5 \end{array}$$

Part 4

a.
$$\begin{array}{r} 142 \\ + 6 \\ \hline 148 \end{array}$$

b.
$$\begin{array}{r} 435 \\ - 33 \\ \hline 402 \end{array}$$

c.
$$\begin{array}{r} 28 \\ - 6 \\ \hline 22 \end{array}$$

Part 5

a. $9 \quad 2 \rightarrow 11$
$11 - 9 = 2$

b. $6 \quad 2 \rightarrow 8$
$6 + 2 = 8$

c. $3 \quad 7 \rightarrow 10$
$10 - 7 = 3$

d. $4 \quad 2 \rightarrow 6$
$6 - 4 = 2$

INDEPENDENT WORK

Connecting Math Concepts   Lesson 49   **75**

Lesson 49 · Side 2   Name _____

Side 2 work is independent.

Part 6

a.
$$\begin{array}{r} 61 \\ +427 \\ \hline 488 \end{array}$$

b.
$$\begin{array}{r} 398 \\ -227 \\ \hline 171 \end{array}$$

c.
$$\begin{array}{r} 516 \\ - 14 \\ \hline 502 \end{array}$$

d.
$$\begin{array}{r} 207 \\ +251 \\ \hline 458 \end{array}$$

Part 7

a. $800 + 10 + 5 = 815$ d. $70 + 8 = 78$

b. $600 + 20 + 5 = 625$ e. $400 + 50 + 0 = 450$

c. $700 + 0 + 3 = 703$ f. $60 + 0 = 60$

Part 8

a. $17 + 10 = 27$ d. $54 + 10 = 64$ g. $65 + 10 = 75$

b. $85 + 10 = 95$ e. $23 + 10 = 33$ h. $45 + 10 = 55$

c. $73 + 10 = 83$ f. $12 + 10 = 22$ i. $19 + 10 = 29$

Part 9

a. $4 - 4 = 0$ d. $2 - 2 = 0$ g. $9 - 0 = 9$

b. $5 - 0 = 5$ e. $7 - 1 = 6$ h. $9 - 9 = 0$

c. $6 - 1 = 5$ f. $4 - 2 = 2$ i. $5 - 1 = 4$

76   Lesson 49   Connecting Math Concepts

Lesson 50 · Side 1   Name _____

Part 1

a. $5 \quad 3 \rightarrow 8$ c. $5 \quad 3 \rightarrow 8$ e. $5 \quad 3 \rightarrow 8$

b. $5 \quad 1 \rightarrow 6$ d. $6 \quad 2 \rightarrow 8$ f. $5 \quad 2 \rightarrow 7$

Part 2

a.
$$\begin{array}{r} 146 \\ - 32 \\ \hline 114 \end{array}$$

b.
$$\begin{array}{r} 107 \\ + 31 \\ \hline 138 \end{array}$$

c.
$$\begin{array}{r} 98 \\ - 7 \\ \hline 91 \end{array}$$

Part 3

a. $10 - 2 = 8$ e. $1 + 7 = 8$

b. $4 - 2 = 2$ f. $9 - 7 = 2$

c. $2 + 6 = 8$ g. $6 - 4 = 2$

d. $5 - 2 = 3$ h. $2 + 8 = 10$

Part 4

a.
$$\begin{array}{r} 5 \\ 2 \\ +1 \\ \hline 8 \end{array}$$

c.
$$\begin{array}{r} 7 \\ 1 \\ +1 \\ \hline 9 \end{array}$$

b.
$$\begin{array}{r} 1 \\ 4 \\ +2 \\ \hline 7 \end{array}$$

d.
$$\begin{array}{r} 2 \\ 3 \\ +2 \\ \hline 7 \end{array}$$

Part 5

$7 \quad 2 \rightarrow 9$
$7 + 2 = 9$
$2 + 7 = 9$
$9 - 2 = 7$
$9 - 7 = 2$

INDEPENDENT WORK

Connecting Math Concepts   Lesson 50   **77**

Lesson 50 · Side 2   Name _____

Side 2 work is independent.

Part 6

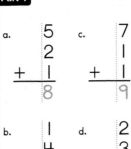

a. $= 34$ b. $= 33$

Part 7

a. $10 + 3 = 13$ d. $70 + 2 = 72$

b. $700 + 0 + 5 = 705$ e. $100 + 90 + 0 = 190$

c. $200 + 80 + 4 = 284$ f. $600 + 30 + 0 = 630$

Part 8

a.
$$\begin{array}{r} 238 \\ - 31 \\ \hline 207 \end{array}$$

b.
$$\begin{array}{r} 61 \\ +527 \\ \hline 588 \end{array}$$

c.
$$\begin{array}{r} 375 \\ + 20 \\ \hline 395 \end{array}$$

d.
$$\begin{array}{r} 794 \\ -524 \\ \hline 270 \end{array}$$

Part 9

a. $6 \quad 1 \rightarrow 7$
$7 - 6 = 1$

c. $1 \quad 1 \rightarrow 2$
$2 - 1 = 1$

e. $5 \quad 2 \rightarrow 7$
$5 + 2 = 7$

b. $8 \quad 2 \rightarrow 10$
$8 + 2 = 10$

d. $3 \quad 2 \rightarrow 5$
$5 - 3 = 2$

f. $6 \quad 2 \rightarrow 8$
$8 - 2 = 6$

78   Lesson 50   Connecting Math Concepts

Lesson 51 Side 1 Name _____

Part 1

a.
```
  2
  8
+ 4
─────
 14
```
c.
```
  3
  2
+ 1
─────
  6
```

b.
```
  1
  6
+ 1
─────
  8
```
d.
```
  1
  9
+ 3
─────
 13
```

Part 2

a. $8 - 2 = 6$ e. $10 - 8 = 2$
b. $6 + 1 = 7$ f. $2 + 6 = 8$
c. $11 - 2 = 9$ g. $4 - 2 = 2$
d. $4 - 3 = 1$ h. $2 + 8 = 10$

Part 3

a. $\underline{2} + \underline{4} = 6$

Part 4

a. $7 \xrightarrow{\underline{1}} 8$ c. $6 \xrightarrow{2} 8$ e. $5 \xrightarrow{\underline{1}} 6$

b. $\underline{5} \xrightarrow{3} 8$ d. $5 \xrightarrow{3} \underline{8}$ f. $5 \xrightarrow{\underline{3}} 8$

Part 5

a.
```
 148
-  41
─────
 107
```
b.
```
 224
+  52
─────
 276
```
c.
```
  15
+ 42
─────
  57
```

Part 6 INDEPENDENT WORK

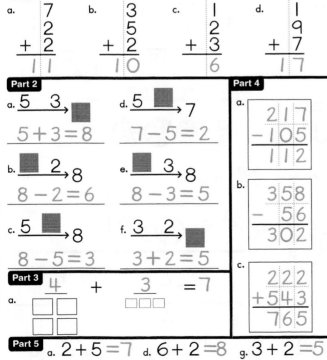

a. = 24

b. = 42

Connecting Math Concepts

Lesson 51 Side 2 Name _____

Side 2 work is independent.

Part 7

a. $\xrightarrow{9 \quad 2} 11$
$9 + 2 = 11$
$2 + 9 = 11$
$11 - 2 = 9$
$11 - 9 = 2$

b. $\xrightarrow{10 \quad 7} 17$
$10 + 7 = 17$
$7 + 10 = 17$
$17 - 7 = 10$
$17 - 10 = 7$

Part 8

a.
```
 673
-252
─────
 421
```
b.
```
  83
+412
─────
 495
```
c.
```
 285
-183
─────
 102
```

Part 9

a. $700 + 30 + 5 = 735$ d. $300 + 80 + 4 = 384$
b. $800 + 0 + 6 = 806$ e. $30 + 0 = 30$
c. $60 + 9 = 69$ f. $500 + 10 + 2 = 512$

Part 10

a.
```
  9
+ 2
───
 11
```
b.
```
  7
+ 2
───
  9
```
c.
```
  7
- 2
───
  5
```
d.
```
  2
+ 5
───
  7
```
e.
```
  8
- 6
───
  2
```
f.
```
  8
+ 2
───
 10
```
g.
```
 11
+ 2
───
  9
```

Connecting Math Concepts

Lesson 52 Side 1 Name _____

Part 1

a.
```
  7
  2
+ 2
───
 11
```
b.
```
  3
  5
+ 2
───
 10
```
c.
```
  1
  2
+ 3
───
  6
```
d.
```
  1
  9
+ 7
───
 17
```

Part 2

a. $5 \xrightarrow{3} \blacksquare$
$5 + 3 = 8$

b. $\blacksquare \xrightarrow{2} 8$
$8 - 2 = 6$

c. $5 \xrightarrow{\blacksquare} 8$
$8 - 5 = 3$

d. $5 \xrightarrow{\blacksquare} 7$
$7 - 5 = 2$

e. $\blacksquare \xrightarrow{3} 8$
$8 - 3 = 5$

f. $3 \xrightarrow{2} \blacksquare$
$3 + 2 = 5$

Part 3

a. $\underline{4} + \underline{3} = 7$

Part 4

a.
```
 217
-105
─────
 112
```
b.
```
 358
-  56
─────
 302
```
c.
```
 222
+543
─────
 765
```

Part 5

a. $2 + 5 = 7$ d. $6 + 2 = 8$ g. $3 + 2 = 5$
b. $2 + 7 = 9$ e. $8 + 2 = 10$ h. $5 + 3 = 8$
c. $3 + 5 = 8$ f. $3 + 10 = 13$ i. $2 + 10 = 12$

Connecting Math Concepts

Lesson 52 Side 2 Name _____

Side 2 work is independent.

Part 6

a. = 34

b. = 21

Part 7

a.
```
 564
-  63
─────
 501
```
b.
```
 815
+172
─────
 987
```
c.
```
  52
+634
─────
 686
```

Part 8

a. $900 + 0 + 6 = 906$ e. $90 + 3 = 93$
b. $500 + 10 + 7 = 517$ f. $200 + 0 + 8 = 208$
c. $60 + 2 = 62$ g. $500 + 40 + 1 = 541$
d. $100 + 80 + 0 = 180$ h. $600 + 40 + 0 = 640$

Part 9

a.
```
 19
- 9
───
 10
```
b.
```
  9
+ 2
───
 11
```
c.
```
  9
- 2
───
  7
```
d.
```
  7
- 6
───
  1
```
e.
```
  5
- 5
───
  0
```
f.
```
  8
+ 2
───
 10
```

g.
```
 11
- 9
───
  2
```
h.
```
 10
+ 5
───
 15
```
i.
```
  5
- 3
───
  2
```
j.
```
 10
- 2
───
  8
```
k.
```
  3
+ 5
───
  8
```
l.
```
  2
+ 4
───
  6
```

Connecting Math Concepts

Connecting Math Concepts

Lesson 53 · Side 1   Name _____

Part 1

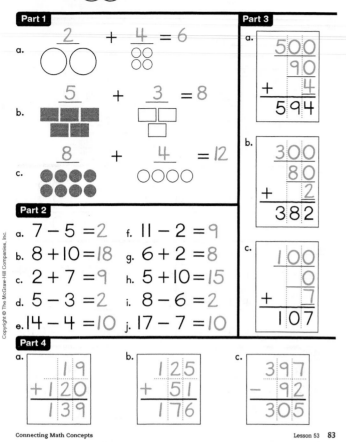

a. $2 + 4 = 6$

b. $5 + 3 = 8$

c. $8 + 4 = 12$

Part 2

a. $7 - 5 = 2$ f. $11 - 2 = 9$
b. $8 + 10 = 18$ g. $6 + 2 = 8$
c. $2 + 7 = 9$ h. $5 + 10 = 15$
d. $5 - 3 = 2$ i. $8 - 6 = 2$
e. $14 - 4 = 10$ j. $17 - 7 = 10$

Part 3

a.
$$\begin{array}{r} 500 \\ 90 \\ + 4 \\ \hline 594 \end{array}$$

b.
$$\begin{array}{r} 300 \\ 80 \\ + 2 \\ \hline 382 \end{array}$$

c.
$$\begin{array}{r} 100 \\ 0 \\ + 7 \\ \hline 107 \end{array}$$

Part 4

a.
$$\begin{array}{r} 19 \\ +120 \\ \hline 139 \end{array}$$

b.
$$\begin{array}{r} 125 \\ + 51 \\ \hline 176 \end{array}$$

c.
$$\begin{array}{r} 397 \\ - 92 \\ \hline 305 \end{array}$$

Connecting Math Concepts   Lesson 53   **83**

Lesson 53 · Side 2   Name _____

Part 5

a. $\begin{array}{r}1\\7\\+2\\\hline 10\end{array}$ b. $\begin{array}{r}8\\2\\+2\\\hline 12\end{array}$ c. $\begin{array}{r}1\\9\\+5\\\hline 15\end{array}$

Part 6

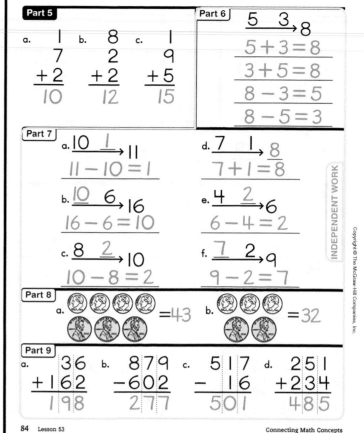

$5 \enspace 3 \rightarrow 8$

$5 + 3 = 8$
$3 + 5 = 8$
$8 - 3 = 5$
$8 - 5 = 3$

Part 7

a. $10 \enspace 1 \rightarrow 11$
$11 - 10 = 1$

b. $10 \enspace 6 \rightarrow 16$
$16 - 6 = 10$

c. $8 \enspace 2 \rightarrow 10$
$10 - 8 = 2$

d. $7 \enspace 1 \rightarrow 8$
$7 + 1 = 8$

e. $4 \enspace 2 \rightarrow 6$
$6 - 4 = 2$

f. $7 \enspace 2 \rightarrow 9$
$9 - 2 = 7$

Part 8

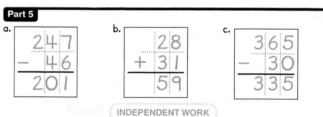

a. $= 43$ b. $= 32$

Part 9

a. $\begin{array}{r}36\\+162\\\hline 198\end{array}$ b. $\begin{array}{r}879\\-602\\\hline 277\end{array}$ c. $\begin{array}{r}517\\-16\\\hline 501\end{array}$ d. $\begin{array}{r}251\\+234\\\hline 485\end{array}$

INDEPENDENT WORK

84   Lesson 53   Connecting Math Concepts

Lesson 54 · Side 1   Name _____

Part 1

a. $\begin{array}{r}700\\60\\+0\\\hline 760\end{array}$ b. $\begin{array}{r}600\\10\\+9\\\hline 619\end{array}$ c. $\begin{array}{r}100\\80\\+1\\\hline 181\end{array}$ d. $\begin{array}{r}500\\0\\+3\\\hline 503\end{array}$

Part 2

a. $20 + 5 = 25$ e. $40 + 5 = 45$
b. $35 + 5 = 40$ f. $15 + 5 = 20$
c. $15 + 10 = 25$ g. $5 + 10 = 15$
d. $27 + 10 = 37$ h. $10 + 5 = 15$

Part 3

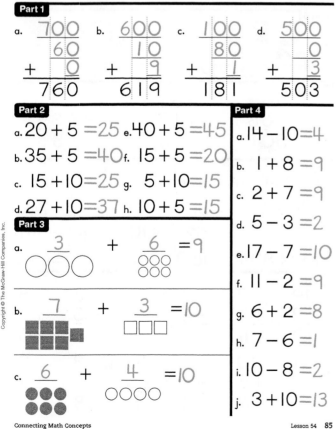

a. $3 + 6 = 9$

b. $7 + 3 = 10$

c. $6 + 4 = 10$

Part 4

a. $14 - 10 = 4$
b. $1 + 8 = 9$
c. $2 + 7 = 9$
d. $5 - 3 = 2$
e. $17 - 7 = 10$
f. $11 - 2 = 9$
g. $6 + 2 = 8$
h. $7 - 6 = 1$
i. $10 - 8 = 2$
j. $3 + 10 = 13$

Connecting Math Concepts   Lesson 54   **85**

Lesson 54 · Side 2   Name _____

Part 5

a. $\begin{array}{r}247\\-46\\\hline 201\end{array}$ b. $\begin{array}{r}28\\+31\\\hline 59\end{array}$ c. $\begin{array}{r}365\\-30\\\hline 335\end{array}$

INDEPENDENT WORK

Part 6

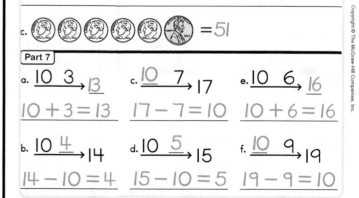

a. $= 36$

b. $= 12$

c. $= 51$

Part 7

a. $10 \enspace 3 \rightarrow 13$
$10 + 3 = 13$

b. $10 \enspace 4 \rightarrow 14$
$14 - 10 = 4$

c. $10 \enspace 7 \rightarrow 17$
$17 - 7 = 10$

d. $10 \enspace 5 \rightarrow 15$
$15 - 10 = 5$

e. $10 \enspace 6 \rightarrow 16$
$10 + 6 = 16$

f. $10 \enspace 9 \rightarrow 19$
$19 - 9 = 10$

86   Lesson 54   Connecting Math Concepts

Lesson 55 — Side 1   Name _____

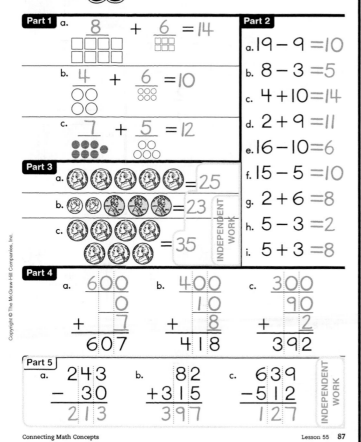

Part 1
a. $\underline{8} + \underline{6} = 14$
b. $\underline{4} + \underline{6} = 10$
c. $\underline{7} + \underline{5} = 12$

Part 2
a. $19 - 9 = 10$
b. $8 - 3 = 5$
c. $4 + 10 = 14$
d. $2 + 9 = 11$
e. $16 - 10 = 6$
f. $15 - 5 = 10$
g. $2 + 6 = 8$
h. $5 - 3 = 2$
i. $5 + 3 = 8$

INDEPENDENT WORK

Part 3
a. $= 25$
b. $= 23$
c. $= 35$

Part 4
a. $600 + 0 + 7 = 607$
b. $400 + 10 + 8 = 418$
c. $300 + 90 + 2 = 392$

Part 5
a. $243 - 30 = 213$
b. $82 + 315 = 397$
c. $639 - 512 = 127$

INDEPENDENT WORK

Connecting Math Concepts   Lesson 55 **87**

Lesson 55 — Side 2   Name _____

Side 2 work is independent.

Part 6
a. $\underline{10}\ \underline{9} \rightarrow 19$ $19 - 9 = 10$
c. $\underline{10}\ \underline{5} \rightarrow 15$ $10 + 5 = 15$
e. $\underline{10}\ \underline{4} \rightarrow 14$ $14 - 10 = 4$
b. $\underline{10}\ \underline{1} \rightarrow 11$ $11 - 10 = 1$
d. $\underline{10}\ \underline{7} \rightarrow 17$ $17 - 7 = 10$
f. $\underline{10}\ \underline{10} \rightarrow 20$ $20 - 10 = 10$

Part 7

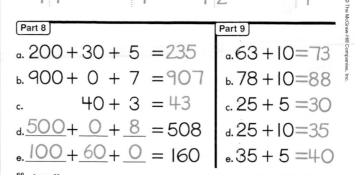

a. $8 + 1 + 2 = 11$
b. $3 + 2 + 2 = 7$
c. $1\ 9 + 2 = 12$
d. $6 + 2 + 1 = 9$

Part 8
a. $200 + 30 + 5 = 235$
b. $900 + 0 + 7 = 907$
c. $40 + 3 = 43$
d. $500 + 0 + 8 = 508$
e. $100 + 60 + 0 = 160$

Part 9
a. $63 + 10 = 73$
b. $78 + 10 = 88$
c. $25 + 5 = 30$
d. $25 + 10 = 35$
e. $35 + 5 = 40$

88 Lesson 55   Connecting Math Concepts

Lesson 56 — Side 1   Name _____

Part 1
a. $2 + 6 = 8$
d. $9 - 2 = 7$
g. $2 + 9 = 11$
b. $10 - 8 = 2$
e. $4 - 3 = 1$
h. $2 + 8 = 10$
c. $8 - 5 = 3$
f. $7 - 1 = 6$
i. $3 + 5 = 8$

Part 2
a. $= 28$
b. $= 33$
c. $= 42$
d. $= 19$

Part 3
a. $516 - 506 = 10$
b. $269 - 219 = 50$

Part 4
a. $10 + 6 = 16$
e. $2 + 3 = 5$
h. $19 - 10 = 9$
b. $2 + 9 = 11$
f. $5 - 3 = 2$
i. $2 + 5 = 7$
c. $20 - 10 = 10$
g. $2 + 7 = 9$
j. $8 - 6 = 2$
d. $17 - 7 = 10$

Connecting Math Concepts   Lesson 56 **89**

Lesson 56 — Side 2   Name _____

Part 5

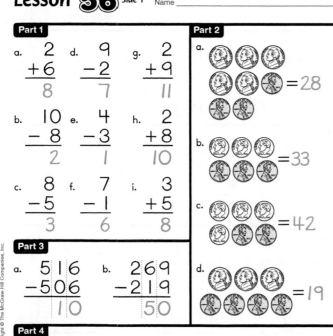

a. $\underline{6} + \underline{5} = 11$
b. $\underline{8} + \underline{4} = 12$
c. $\underline{4} + \underline{2} = 6$

Part 6
a. $27 + 42 = 69$
c. $66 + 202 = 268$
b. $376 - 126 = 250$

INDEPENDENT WORK

Part 7
a. $700 + 0 + 8 = 708$
b. $500 + 90 + 0 = 590$
c. $300 + 10 + 6 = 316$
d. $400 + 20 + 7 = 427$

Part 8
a. $\underline{10}\ \underline{3} \rightarrow 13$ $13 - 10 = 3$
c. $\underline{9}\ \underline{2} \rightarrow 11$ $9 + 2 = 11$
e. $\underline{4}\ \underline{2} \rightarrow 6$ $6 - 2 = 4$
b. $\underline{8}\ \underline{2} \rightarrow 10$ $10 - 2 = 8$
d. $\underline{7}\ \underline{2} \rightarrow 9$ $9 - 2 = 7$
f. $\underline{3}\ \underline{2} \rightarrow 5$ $5 - 3 = 2$

90 Lesson 56   Connecting Math Concepts

Connecting Math Concepts   Answer Key **273**

Lesson 57 Side 1 Name _____

Part 1

a. $6+4=10$ d. $3+6=9$ g. $6+3=9$

b. $5+3=8$ e. $2+5=7$ h. $5+2=7$

c. $5+10=15$ f. $1+6=7$ i. $4+6=10$

Part 2

 a.  $=40$ b. $=30$

c. $=55$ INDEPENDENT WORK

Part 3

a. $\begin{array}{r} 58 \\ -\ 57 \\ \hline 1 \end{array}$ b. $\begin{array}{r} 364 \\ -343 \\ \hline 21 \end{array}$ c. $\begin{array}{r} 496 \\ -416 \\ \hline 80 \end{array}$

Part 4

a. $18 + \underline{4} = 22$

   oooo

b. $37 + \underline{6} = 43$

Connecting Math Concepts   Lesson 57 **91**

Lesson 57 Side 2   Name _____

Part 5

a. $8-2=6$ f. $2+6=8$

b. $8+10=18$ g. $4-2=2$

c. $11-2=9$ h. $9+2=11$

d. $10-2=8$ i. $17-7=10$

e. $13-3=10$ j. $6-4=2$

Part 6

a. $\begin{array}{r} 765 \\ -562 \\ \hline 203 \end{array}$

b. $\begin{array}{r} 83 \\ +215 \\ \hline 298 \end{array}$

Part 7   INDEPENDENT WORK

a. $200+0+5=205$

b. $\underline{100}+\underline{80}+\underline{4}=184$

c. $\underline{10}+\underline{6}=16$

d. $\begin{array}{r} 700 \\ 20 \\ +\ \ 1 \\ \hline 721 \end{array}$

e. $\begin{array}{r} 40 \\ +\ 9 \\ \hline 49 \end{array}$

Part 8

a. $\begin{array}{r} 265 \\ +213 \\ \hline 478 \end{array}$ b. $\begin{array}{r} 857 \\ -255 \\ \hline 602 \end{array}$ c. $\begin{array}{r} 416 \\ +\ 83 \\ \hline 499 \end{array}$

92 Lesson 57   Connecting Math Concepts

Lesson 58 Side 1   Name _____

Part 1

a. $\begin{array}{r} 10 \\ -\ 6 \\ \hline 4 \end{array}$ b. $\begin{array}{r} 8 \\ -3 \\ \hline 5 \end{array}$ c. $\begin{array}{r} 7 \\ -2 \\ \hline 5 \end{array}$ d. $\begin{array}{r} 9 \\ -6 \\ \hline 3 \end{array}$ e. $\begin{array}{r} 15 \\ -10 \\ \hline 5 \end{array}$ f. $\begin{array}{r} 10 \\ -8 \\ \hline 2 \end{array}$ g. $\begin{array}{r} 8 \\ -1 \\ \hline 7 \end{array}$ h. $\begin{array}{r} 6 \\ -4 \\ \hline 2 \end{array}$

Part 2

a. $28 + \underline{6} = 34$ b. $16 + \underline{6} = 22$

c. $73 + \underline{7} = 80$

Part 3

a. $\begin{array}{r} 725 \\ -705 \\ \hline 20 \end{array}$ b. $\begin{array}{r} 53 \\ +821 \\ \hline 874 \end{array}$ c. $\begin{array}{r} 96 \\ -91 \\ \hline 5 \end{array}$ d. $\begin{array}{r} 364 \\ -\ 54 \\ \hline 310 \end{array}$

Part 4

a. $=48$

b. $=32$ INDEPENDENT WORK

Part 5

a. $10-8=2$ d. $7-2=5$ g. $10-6=4$

b. $8-5=3$ e. $9-3=6$ h. $3+5=8$

c. $10+2=12$ f. $11-9=2$

Connecting Math Concepts   Lesson 58 **93**

Lesson 58 Side 2   Name _____

Side 2 work is independent.

Part 6

a. $\underline{700}+\underline{10}+\underline{4}=714$

b. $800+0+5=805$

c. $\underline{200}+\underline{30}+\underline{0}=230$

d. $\begin{array}{r} 900 \\ 60 \\ +\ \ 1 \\ \hline 961 \end{array}$

e. $\begin{array}{r} 40 \\ +\ 8 \\ \hline 48 \end{array}$

Part 7

a. $\begin{array}{r} 7 \\ 2 \\ +10 \\ \hline 19 \end{array}$ b. $\begin{array}{r} 1 \\ 9 \\ +\ 6 \\ \hline 16 \end{array}$

c. $5+3+2=10$

d. $9+2+4=15$

e. $4+6+8=18$

Part 8

a. $7 \xrightarrow{\ 2\ } 9$ $9-7=2$

d. $8 \xrightarrow{\ 1\ } 9$ $9-8=1$

g. $4 \xrightarrow{\ 2\ } 6$ $6-4=2$

b. $9 \xrightarrow{\ 2\ } 11$ $9+2=11$

e. $5 \xrightarrow{\ 3\ } 8$ $8-3=5$

h. $6 \xrightarrow{\ 2\ } 8$ $6+2=8$

c. $1 \xrightarrow{\ 9\ } 10$ $10-9=1$

f. $5 \xrightarrow{\ 1\ } 6$ $5+1=6$

i. $9 \xrightarrow{\ 2\ } 11$ $11-9=2$

94 Lesson 58   Connecting Math Concepts

Lesson 59 Side 1   Name _____

Part 1

a. $5+3=8$ f. $3+6=9$
b. $8-2=6$ g. $2+4=6$
c. $11-9=2$ h. $9-3=6$
d. $6+4=10$ i. $9+10=19$
e. $13-3=10$ j. $10-6=4$

Part 2   INDEPENDENT WORK

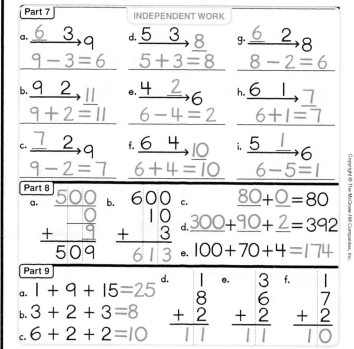

a. = 42
b. = 51

Part 3

a. $\begin{array}{r} 13 \\ -\ 3 \\ \hline 10 \end{array}$
b. $\begin{array}{r} 6 \\ -4 \\ \hline 2 \end{array}$
c. $\begin{array}{r} 9 \\ -8 \\ \hline 1 \end{array}$
d. $\begin{array}{r} 9 \\ -6 \\ \hline 3 \end{array}$
e. $\begin{array}{r} 8 \\ -3 \\ \hline 5 \end{array}$

f. $\begin{array}{r} 12 \\ -10 \\ \hline 2 \end{array}$
g. $\begin{array}{r} 5 \\ -3 \\ \hline 2 \end{array}$
h. $\begin{array}{r} 8 \\ -2 \\ \hline 6 \end{array}$
i. $\begin{array}{r} 15 \\ -\ 5 \\ \hline 10 \end{array}$
j. $\begin{array}{r} 10 \\ -\ 4 \\ \hline 6 \end{array}$

Part 4

a. $\begin{array}{r} 56 \\ +132 \\ \hline 188 \end{array}$

b. $\begin{array}{r} 187 \\ -127 \\ \hline 60 \end{array}$

Part 5

a. 34 + 6 = 40
b. 17 + 4 = 21
c. 58 + 5 = 63
d. 66 + 9 = 75

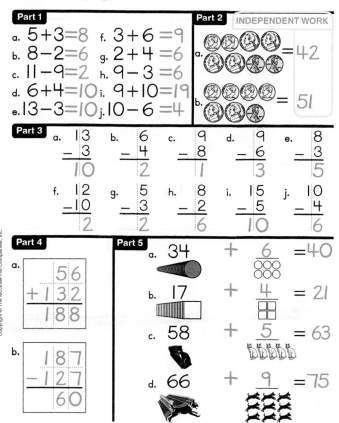

Connecting Math Concepts   Lesson 59 95

Lesson 59 Side 2   Name _____

Part 6

a. $\begin{array}{r} 856 \\ -650 \\ \hline 206 \end{array}$
b. $\begin{array}{r} 764 \\ -724 \\ \hline 40 \end{array}$
c. $\begin{array}{r} 207 \\ +\ 51 \\ \hline 258 \end{array}$
d. $\begin{array}{r} 69 \\ -57 \\ \hline 12 \end{array}$

Part 7   INDEPENDENT WORK

a. $6 \quad 3 \rightarrow 9$ $9-3=6$
d. $5 \quad 3 \rightarrow 8$ $5+3=8$
g. $6 \quad 2 \rightarrow 8$ $8-2=6$

b. $9 \quad 2 \rightarrow 11$ $9+2=11$
e. $4 \quad 2 \rightarrow 6$ $6-4=2$
h. $6 \quad 1 \rightarrow 7$ $6+1=7$

c. $7 \quad 2 \rightarrow 9$ $9-2=7$
f. $6 \quad 4 \rightarrow 10$ $6+4=10$
i. $5 \quad 1 \rightarrow 6$ $6-5=1$

Part 8

a. $\begin{array}{r} 500 \\ 0 \\ +\ 9 \\ \hline 509 \end{array}$
b. $\begin{array}{r} 600 \\ 10 \\ +\ 3 \\ \hline 613 \end{array}$
c. $80+0=80$
d. $300+90+2=392$
e. $100+70+4=174$

Part 9

a. $1+9+15=25$
b. $3+2+3=8$
c. $6+2+2=10$
d. $\begin{array}{r} 1 \\ 8 \\ +2 \\ \hline 11 \end{array}$
e. $\begin{array}{r} 3 \\ 6 \\ +2 \\ \hline 11 \end{array}$
f. $\begin{array}{r} 1 \\ 7 \\ +2 \\ \hline 10 \end{array}$

96 Lesson 59   Connecting Math Concepts

Lesson 60 Side 1   Name _____

Part 1

a. $6+4=10$ f. $5+10=15$
b. $7+2=9$ g. $2+9=11$
c. $3+6=9$ h. $4+6=10$
d. $10+8=18$ i. $5+2=7$
e. $2+3=5$ j. $3+5=8$

Part 2

a. $\begin{array}{r} 176 \\ -\ 62 \\ \hline 114 \end{array}$

b. $\begin{array}{r} 316 \\ +\ 43 \\ \hline 359 \end{array}$

Part 3

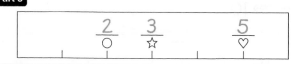

2   3   5
○   ☆   ♡

Part 4

a. $10-9=1$ d. $9-6=3$ g. $10-8=2$
b. $7-5=2$ e. $8-7=1$ h. $4-2=2$
c. $10-4=6$ f. $8-3=5$

Part 5

a. $\begin{array}{r} 26 \\ +433 \\ \hline 459 \end{array}$
b. $\begin{array}{r} 468 \\ -425 \\ \hline 43 \end{array}$
c. $\begin{array}{r} 879 \\ -357 \\ \hline 522 \end{array}$

Connecting Math Concepts   Lesson 60 97

Lesson 60 Side 2   Name _____

Side 2 work is independent.

Part 6

a. $6 \quad 4 \rightarrow 10$ $6+4=10$
d. $4 \quad 2 \rightarrow 6$ $6-4=2$
g. $10 \quad 7 \rightarrow 17$ $10+7=17$

b. $6 \quad 3 \rightarrow 9$ $9-3=6$
e. $9 \quad 2 \rightarrow 11$ $9+2=11$
h. $6 \quad 4 \rightarrow 10$ $10-6=4$

c. $8 \quad 1 \rightarrow 9$ $9-8=1$
f. $6 \quad 2 \rightarrow 8$ $8-2=6$
i. $5 \quad 3 \rightarrow 8$ $5+3=8$

Part 7

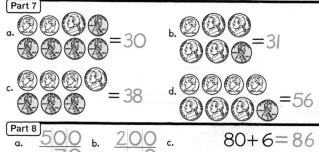

a. = 30
b. = 31
c. = 38
d. = 56

Part 8

a. $\begin{array}{r} 500 \\ 70 \\ +\ 6 \\ \hline 576 \end{array}$
b. $\begin{array}{r} 200 \\ 0 \\ +\ 9 \\ \hline 209 \end{array}$
c. $80+6=86$
d. $700+10+3=713$
e. $400+50+0=450$

98 Lesson 60   Connecting Math Concepts

Lesson 61 Side 1   Name _____

Part 1
a. $3+5=8$ e. $3+6=9$

b. $2+8=10$ f. $7+10=17$

c. $6+4=10$ g. $2+3=5$

d. $9+2=11$ h. $6+3=9$

Part 2
a. 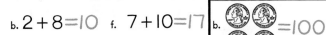 $=50$

b. $=100$

c. $=40$

Part 3
a. $5-2=3$ e. $8-3=5$

b. $10-4=6$ f. $10-2=8$

c. $9-7=2$ g. $9-6=3$

d. $19-9=10$ h. $8-7=1$

Part 5
a.
$$\begin{array}{r} 487 \\ -456 \\ \hline 31 \end{array}$$

b.
$$\begin{array}{r} 93 \\ +206 \\ \hline 299 \end{array}$$

c.
$$\begin{array}{r} 827 \\ -317 \\ \hline 510 \end{array}$$

Part 4
$$\frac{2}{\bigcirc} \qquad \frac{4}{\heartsuit} \qquad \frac{5}{\star}$$

Connecting Math Concepts

Lesson 61   1

Lesson 61 Side 2   Name _____

Side 2 work is independent.

Part 6
a. $\dfrac{5\quad 3}{}\!\!\rightarrow 8$ c. $\dfrac{10\quad 3}{}\!\!\rightarrow 13$ e. $\dfrac{6\quad 3}{}\!\!\rightarrow 9$

$5+3=8$ $13-3=10$ $6+3=9$

b. $\dfrac{7\quad 2}{}\!\!\rightarrow 9$ d. $\dfrac{6\quad 2}{}\!\!\rightarrow 8$ f. $\dfrac{4\quad 2}{}\!\!\rightarrow 6$

$9-7=2$ $8-2=6$ $6-4=2$

Part 7
a. $37 + \underline{7} = 44$ b. $75 + \underline{6} = 81$

Part 8
a.
$$\begin{array}{r} 500 \\ 0 \\ +7 \\ \hline 507 \end{array}$$

b.
$$\begin{array}{r} 90 \\ 6 \\ \hline 96 \end{array}$$

c. $800+40+0=840$

d. $200+10+6=216$

e. $400+50+1=451$

Part 9
a. $4+10+2=16$

b. $9+2+6=17$

c. $1+8+2=11$

d.
$$\begin{array}{r} 1 \\ 9 \\ +7 \\ \hline 17 \end{array}$$

e.
$$\begin{array}{r} 5 \\ 3 \\ +1 \\ \hline 9 \end{array}$$

2   Lesson 61

Connecting Math Concepts

Lesson 62 Side 1   Name _____

Part 1
a. $13-10=3$ d. $10-4=6$ h. $7-6=1$

b. $9-6=3$ e. $11-9=2$ i. $8-3=5$

c. $9-7=2$ f. $8-5=3$ j. $9-2=7$

g. $10-8=2$

Part 2
a. $=90$

b. $=80$

c. $=55$

INDEPENDENT WORK

Part 3
a.
$$\begin{array}{r} 178 \\ -27 \\ \hline 151 \end{array}$$

b.
$$\begin{array}{r} 13 \\ +12 \\ \hline 25 \end{array}$$

c.
$$\begin{array}{r} 37 \\ -5 \\ \hline 32 \end{array}$$

Part 4
a. $=50$

b. $=65$

c. $=32$

Connecting Math Concepts

Lesson 62   3

Lesson 62 Side 2   Name _____

Part 5
a. $2+8=10$ f. $2+4=6$

b. $3+6=9$ g. $6+4=10$

c. $2+9=11$ h. $3+5=8$

d. $4+6=10$ i. $6+3=9$

e. $7+10=17$ j. $2+2=4$

INDEPENDENT WORK

Part 6
$\dfrac{6\quad 3}{}\!\!\rightarrow 9$

$6+3=9$

$3+6=9$

$9-3=6$

$9-6=3$

Part 7
a. $\dfrac{9\quad 1}{}\!\!\rightarrow 10$ c. $\dfrac{6\quad 2}{}\!\!\rightarrow 8$ e. $\dfrac{3\quad 2}{}\!\!\rightarrow 5$

$10-9=1$ $8-6=2$ $3+2=5$

b. $\dfrac{6\quad 4}{}\!\!\rightarrow 10$ d. $\dfrac{4\quad 2}{}\!\!\rightarrow 6$ f. $\dfrac{6\quad 1}{}\!\!\rightarrow 7$

$6+4=10$ $6-2=4$ $7-6=1$

Part 8
a.
$$\begin{array}{r} 25 \\ +301 \\ \hline 326 \end{array}$$

b.
$$\begin{array}{r} 437 \\ -335 \\ \hline 102 \end{array}$$

c.
$$\begin{array}{r} 583 \\ -32 \\ \hline 551 \end{array}$$

d.
$$\begin{array}{r} 628 \\ -622 \\ \hline 6 \end{array}$$

Part 9
$36 + \underline{8} = 44$

4   Lesson 62

Connecting Math Concepts

Connecting Math Concepts

Lesson 63 · Side 1   Name _____

Part 1
d. $9 - 6 = 3$ h. $11 - 2 = 9$
a. $7 - 2 = 5$ e. $8 - 5 = 3$ i. $9 - 3 = 6$
b. $6 - 4 = 2$ f. $10 - 6 = 4$ j. $8 - 6 = 2$
c. $14 - 10 = 4$ g. $10 - 9 = 1$

Part 2

a. $= 70$
b. $= 65$
c. $= 60$

INDEPENDENT WORK

Part 3
d. $2 + 3 = 5$ h. $8 + 10 = 18$
a. $6 + 3 = 9$ e. $9 + 2 = 11$ i. $6 + 4 = 10$
b. $8 + 2 = 10$ f. $5 + 3 = 8$ j. $4 + 2 = 6$
c. $10 + 6 = 16$ g. $6 + 2 = 8$

Part 4
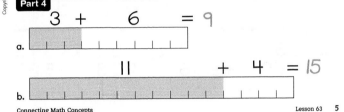
a. $3 + 6 = 9$
b. $11 + 4 = 15$

Connecting Math Concepts   Lesson 63   5

Lesson 63 · Side 2   Name _____

Part 5

a.
$$471 + 216 = 687$$

b.
$$179 - 53 = 126$$

c.
$$85 + 113 = 198$$

Part 6
a. $= 31$
b. $= 75$

INDEPENDENT WORK

Part 7
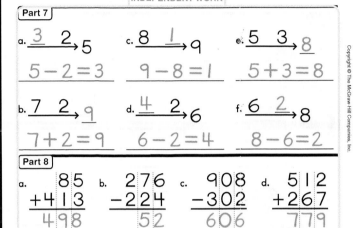

a. $\dfrac{3 \quad 2}{\ } \to 5$ $5 - 2 = 3$
c. $\dfrac{8 \quad 1}{\ } \to 9$ $9 - 8 = 1$
e. $\dfrac{5 \quad 3}{\ } \to 8$ $5 + 3 = 8$

b. $\dfrac{7 \quad 2}{\ } \to 9$ $7 + 2 = 9$
d. $\dfrac{4 \quad 2}{\ } \to 6$ $6 - 2 = 4$
f. $\dfrac{6 \quad 2}{\ } \to 8$ $8 - 6 = 2$

Part 8
a. $85 + 413 = 498$
b. $276 - 224 = 52$
c. $908 - 302 = 606$
d. $512 + 267 = 779$

6   Lesson 63   Connecting Math Concepts

Lesson 64 · Side 1   Name _____

Part 1
d. $4 - 3 = 1$ h. $10 - 2 = 8$
a. $8 - 3 = 5$ e. $9 - 2 = 7$ i. $9 - 3 = 6$
b. $8 - 2 = 6$ f. $9 - 6 = 3$ j. $5 - 2 = 3$
c. $10 - 4 = 6$ g. $10 - 6 = 4$

Part 2
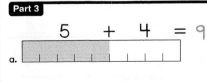
a. $= 78$
b. $= 90$
c. $= 80$

INDEPENDENT WORK

Part 3
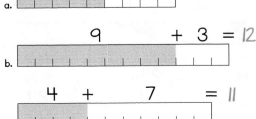
a. $5 + 4 = 9$
b. $9 + 3 = 12$
c. $4 + 7 = 11$

Connecting Math Concepts   Lesson 64   7

Lesson 64 · Side 2   Name _____

Part 4
a.
$$34 \\ 21 \\ + 23 \\ \overline{78}$$

b.
$$12 \\ 11 \\ + 32 \\ \overline{55}$$

c.
$$50 \\ 34 \\ + 12 \\ \overline{96}$$

Part 5
a. $= 31$
b. $= 50$
c. $= 85$

Part 6
a.
$$54 - 22 = 32$$

b.
$$26 + 113 = 139$$

c.
$$428 - 120 = 308$$

Part 7
$\dfrac{6 \quad 5}{\ } \to 11$
$6 + 5 = 11$
$5 + 6 = 11$
$11 - 5 = 6$
$11 - 6 = 5$

Part 8
INDEPENDENT WORK
a. $687 - 635 = 52$
b. $603 + 96 = 699$
c. $469 - 26 = 443$

8   Lesson 64   Connecting Math Concepts

Connecting Math Concepts   Answer Key   277

Lesson 65 Side 1 Name _____

Part 1

d. $9-3=6$ h. $17-7=10$

a. $7-2=5$ e. $5-3=2$ i. $9-6=3$

b. $6-5=1$ f. $8-5=3$ j. $6-4=2$

c. $10-6=4$ g. $10-2=8$

Part 2

a.
$$\begin{array}{r} 25 \\ 30 \\ +32 \\ \hline 87 \end{array}$$

b.
$$\begin{array}{r} 18 \\ 201 \\ +40 \\ \hline 259 \end{array}$$

c.
$$\begin{array}{r} 356 \\ +620 \\ \hline 976 \end{array}$$

d.
$$\begin{array}{r} 61 \\ +320 \\ \hline 381 \end{array}$$

Part 3

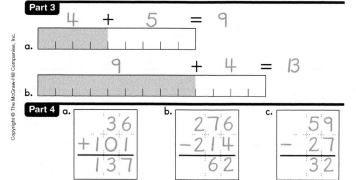

a. $4 + 5 = 9$

b. $9 + 4 = 13$

Part 4

a.
$$\begin{array}{r} 36 \\ +101 \\ \hline 137 \end{array}$$

b.
$$\begin{array}{r} 276 \\ -214 \\ \hline 62 \end{array}$$

c.
$$\begin{array}{r} 59 \\ -27 \\ \hline 32 \end{array}$$

Part 5 INDEPENDENT WORK

a. $700+0+9=709$ b. $500+10+8=518$

Connecting Math Concepts Lesson 65 9

Lesson 65 Side 2 Name _____

Side 2 work is independent.

Part 6

a. $=61$

b. $=50$

c. $=40$ d. $=36$

e. $=77$

Part 7

a. $\underset{\longrightarrow}{6 \quad 3} 9$ $9-3=6$

c. $\underset{\longrightarrow}{4 \quad 2} 6$ $6-4=2$

e. $\underset{\longrightarrow}{7 \quad 1} 8$ $8-1=7$

b. $\underset{\longrightarrow}{6 \quad 4} 10$ $6+4=10$

d. $\underset{\longrightarrow}{9 \quad 2} 11$ $11-9=2$

f. $\underset{\longrightarrow}{6 \quad 2} 8$ $6+2=8$

Part 8

a. $29 + 5 = 34$ b. $47 + 6 = 53$

Part 9

a.
$$\begin{array}{r} 581 \\ -60 \\ \hline 521 \end{array}$$

b.
$$\begin{array}{r} 723 \\ +266 \\ \hline 989 \end{array}$$

c.
$$\begin{array}{r} 734 \\ -223 \\ \hline 511 \end{array}$$

d.
$$\begin{array}{r} 548 \\ -528 \\ \hline 20 \end{array}$$

10 Lesson 65 Connecting Math Concepts

Lesson 66 Side 1 Name _____

Part 1

d. $6-2=4$ h. $8-7=1$

a. $12-10=2$ e. $14-10=4$ i. $8-5=3$

b. $8-3=5$ f. $8-2=6$ j. $11-9=2$

c. $9-6=3$ g. $10-6=4$

Part 2

a.
$$\begin{array}{r} 33 \\ +56 \\ \hline 89 \end{array}$$

b.
$$\begin{array}{r} 374 \\ -314 \\ \hline 60 \end{array}$$

c.
$$\begin{array}{r} 25 \\ 441 \\ +31 \\ \hline 497 \end{array}$$

d.
$$\begin{array}{r} 289 \\ -36 \\ \hline 253 \end{array}$$

Part 3

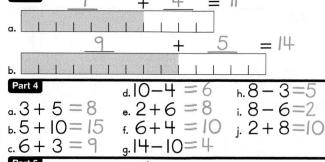

a. $7 + 4 = 11$

b. $9 + 5 = 14$

Part 4

d. $10-4=6$ h. $8-3=5$

a. $3+5=8$ e. $2+6=8$ i. $8-6=2$

b. $5+10=15$ f. $6+4=10$ j. $2+8=10$

c. $6+3=9$ g. $14-10=4$

Part 5

a.
$$\begin{array}{r} 147 \\ -45 \\ \hline 102 \end{array}$$

b.
$$\begin{array}{r} 64 \\ +132 \\ \hline 196 \end{array}$$

c.
$$\begin{array}{r} 278 \\ -215 \\ \hline 63 \end{array}$$

Connecting Math Concepts Lesson 66 11

Lesson 66 Side 2 Name _____

Side 2 work is independent.

Part 6

a. $=80$

b. $=32$

c. $=60$ d. $=36$

Part 7

a. $\underset{\longrightarrow}{4 \quad 2} 6$ $6-2=4$

c. $\underset{\longrightarrow}{9 \quad 2} 11$ $11-9=2$

e. $\underset{\longrightarrow}{6 \quad 4} 10$ $10-6=4$

b. $\underset{\longrightarrow}{6 \quad 2} 8$ $6+2=8$

d. $\underset{\longrightarrow}{3 \quad 2} 5$ $5-2=3$

f. $\underset{\longrightarrow}{5 \quad 3} 8$ $5+3=8$

Part 8

a. $37 + 7 = 44$ b. $18 + 6 = 24$

Part 9

a.
$$\begin{array}{r} 17 \\ 40 \\ +32 \\ \hline 89 \end{array}$$

b.
$$\begin{array}{r} 256 \\ -214 \\ \hline 42 \end{array}$$

c.
$$\begin{array}{r} 32 \\ 6 \\ +21 \\ \hline 59 \end{array}$$

12 Lesson 66 Connecting Math Concepts

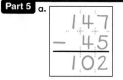
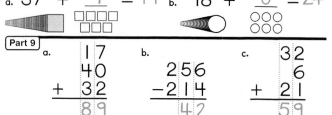
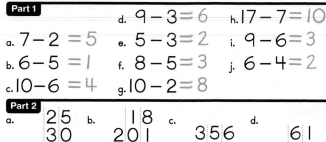
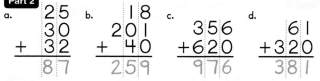

Lesson 67 Side 1   Name _____

Part 1

d. $3+10=13$ h. $3+6=9$

a. $7-6=1$ e. $9-6=3$ i. $8-6=2$

b. $3+5=8$ f. $9+2=11$ j. $8-3=5$

c. $10-2=8$ g. $10-6=4$

Part 2

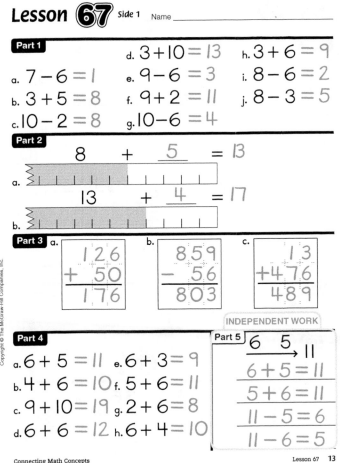

a. $8 + \underline{5} = 13$

b. $13 + \underline{4} = 17$

Part 3

a.
$$\begin{array}{r} 126 \\ +\ 50 \\ \hline 176 \end{array}$$

b.
$$\begin{array}{r} 859 \\ -\ 56 \\ \hline 803 \end{array}$$

c.
$$\begin{array}{r} 13 \\ +476 \\ \hline 489 \end{array}$$

Part 4

a. $6+5=11$ e. $6+3=9$

b. $4+6=10$ f. $5+6=11$

c. $9+10=19$ g. $2+6=8$

d. $6+6=12$ h. $6+4=10$

INDEPENDENT WORK

Part 5

$6 \quad 5 \longrightarrow 11$

$6+5=11$

$5+6=11$

$11-5=6$

$11-6=5$

Connecting Math Concepts   Lesson 67   **13**

Lesson 67 Side 2   Name _____

Side 2 work is independent.

Part 6

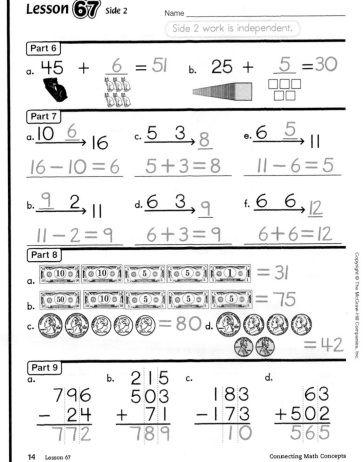

a. $45 + \underline{6} = 51$ b. $25 + \underline{5} = 30$

Part 7

a. $10 \quad 6 \longrightarrow 16$ c. $5 \quad 3 \longrightarrow 8$ e. $6 \quad 5 \longrightarrow 11$

$16-10=6$ $5+3=8$ $11-6=5$

b. $9 \quad 2 \longrightarrow 11$ d. $6 \quad 3 \longrightarrow 9$ f. $6 \quad 6 \longrightarrow 12$

$11-2=9$ $6+3=9$ $6+6=12$

Part 8

a. [10][10][5][5][1] $=31$

b. [50][10][5][5][5] $=75$

c. ⓠⓠⓠⓠⓠⓠ $=80$ d. ⓠⓠⓠⓠ ⓟⓟ $=42$

Part 9

a.
$$\begin{array}{r} 796 \\ -\ 24 \\ \hline 772 \end{array}$$

b.
$$\begin{array}{r} 215 \\ 503 \\ +\ 71 \\ \hline 789 \end{array}$$

c.
$$\begin{array}{r} 183 \\ -173 \\ \hline 10 \end{array}$$

d.
$$\begin{array}{r} 63 \\ +502 \\ \hline 565 \end{array}$$

14   Lesson 67   Connecting Math Concepts

Lesson 68 Side 1   Name _____

Part 1

a. $11-5=6$

b. $4+6=10$

c. $12-6=6$

d. $8+2=10$

e. $9-6=3$

f. $6+6=12$

g. $8-6=2$

h. $6-4=2$

i. $5+6=11$

j. $8-2=6$

Part 2

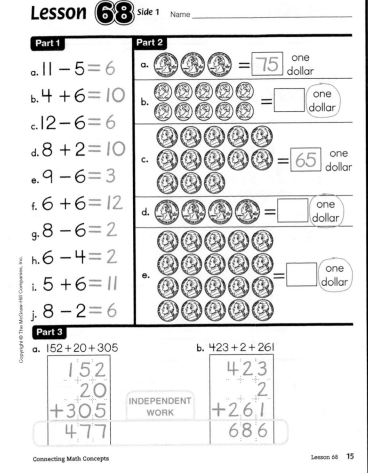

a. ⓠⓠⓠ $=\boxed{75}$ one dollar

b. ⓠⓠⓠⓠ ⓠⓠⓠⓠ $=\boxed{\ }$ one dollar

c. ⓝⓝⓝⓝⓝ ⓝⓝⓝⓝⓝ ⓝⓝⓝ $=\boxed{65}$ one dollar

d. ⓠⓠⓠⓠ $=\boxed{\ }$ one dollar

e. ⓝⓝⓝⓝ ⓝⓝⓝⓝ ⓝⓝⓝⓝ $=\boxed{\ }$ one dollar

Part 3

a. $152+20+305$

$$\begin{array}{r} 152 \\ 20 \\ +305 \\ \hline 477 \end{array}$$

b. $423+2+261$

$$\begin{array}{r} 423 \\ 2 \\ +261 \\ \hline 686 \end{array}$$

INDEPENDENT WORK

Connecting Math Concepts   Lesson 68   **15**

Lesson 68 Side 2   Name _____

Part 4

a. $15 + \underline{4} = 19$

b. $47 + \underline{5} = 52$

Part 5

a.
$$\begin{array}{r} 179 \\ -150 \\ \hline 29 \end{array}$$

b.
$$\begin{array}{r} 56 \\ +123 \\ \hline 179 \end{array}$$

c.
$$\begin{array}{r} 496 \\ -\ 76 \\ \hline 420 \end{array}$$

INDEPENDENT WORK

Part 6

a. [20][10][5][5] $=40$

b. [20][5][5][1][1] $=32$

c. ⓠⓠⓠ ⓟ $=81$ d. ⓠⓠⓠ ⓝ ⓟⓟ $=47$

Part 7

a. $27 + \underline{5} = 32$ b. $94 + \underline{5} = 99$

16   Lesson 68   Connecting Math Concepts

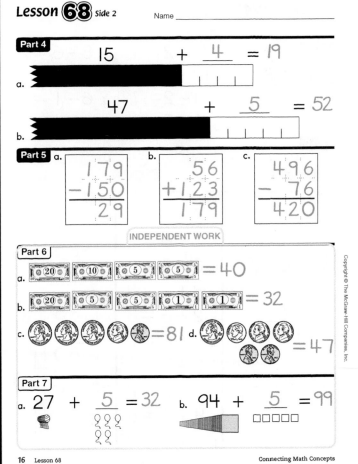

Lesson 69 Side 1   Name _____

Part 1

a. $6+4=10$
b. $2+9=11$
c. $3+6=9$
d. $5+2=7$
e. $6+6=12$
f. $6+5=11$
g. $8+2=10$
h. $3+5=8$
i. $2+6=8$
j. $9+1=10$

Part 2

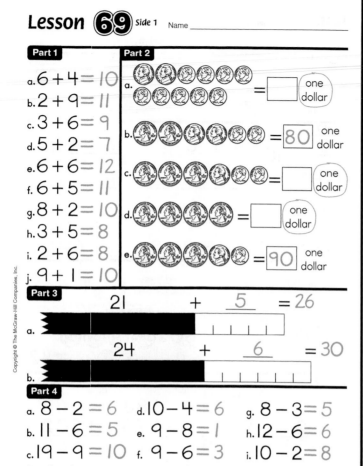

a. [ ] one dollar
b. 80 one dollar
c. [ ] one dollar
d. [ ] one dollar
e. 90 one dollar

Part 3

a. $21 + 5 = 26$
b. $24 + 6 = 30$

Part 4

a. $8-2=6$ d. $10-4=6$ g. $8-3=5$
b. $11-6=5$ e. $9-8=1$ h. $12-6=6$
c. $19-9=10$ f. $9-6=3$ i. $10-2=8$

Connecting Math Concepts   Lesson 69   17

Lesson 69 Side 2   Name _____

Part 5

a. $50+436+13$
```
  50
 436
+ 13
 499
```
b. $206+91+702$
```
 206
  91
+702
 999
```

Part 6

a. $4 \to 2 \to 6$ $6-4=2$
b. $8 \to 1 \to 9$ $9-8=1$
c. $5 \to 3 \to 8$ $8-3=5$
d. $9 \to 2 \to 11$ $9+2=11$
e. $6 \to 2 \to 8$ $8-2=6$
f. $6 \to 4 \to 10$ $6+4=10$
g. $10 \to 7 \to 17$ $17-10=7$
h. $5 \to 2 \to 7$ $5+2=7$

Part 7

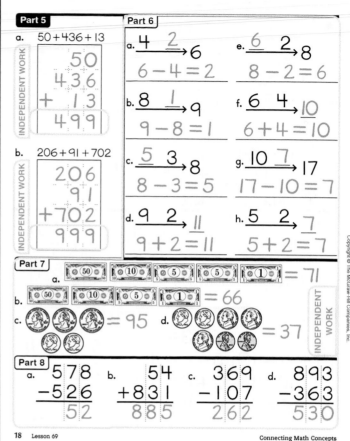

a. [ ] = 71
b. = 66
c. = 95   d. = 37

Part 8

a.
```
 578
-526
  52
```
b.
```
  54
+831
 885
```
c.
```
 369
-107
 262
```
d.
```
 893
-363
 530
```

18   Lesson 69   Connecting Math Concepts

Lesson 70 Side 1   Name _____

Part 1

a. $9-3=6$ d. $11-2=9$ h. $8-6=2$
b. $10-6=4$ e. $12-6=6$ i. $12-10=2$
c. $7-5=2$ f. $9-2=7$ j. $6-1=5$
             g. $11-5=6$

Part 2

a. $42 + 4 = 46$ b. $68 + 5 = 73$

Part 3

a.
```
  23
 412
+ 51
 486
```
b.
```
  16
 500
+ 33
 549
```

Part 4

a. $2+3=5$
b. $6+6=12$
c. $3+5=8$
d. $8+1=9$
e. $2+6=8$
f. $4+6=10$
g. $2+7=9$
h. $6+5=11$
i. $4+2=6$
j. $5+10=15$

Part 5

a.
```
  17
+352
 369
```
b.
```
 853
-552
 301
```
c.
```
  88
- 81
   7
```

Connecting Math Concepts   Lesson 70   19

Lesson 70 Side 2   Name _____

Side 2 work is independent.

Part 6

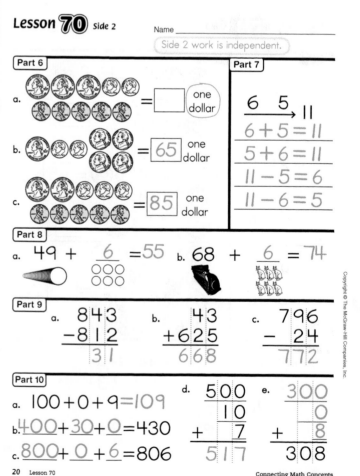

a. [ ] one dollar
b. 65 one dollar
c. 85 one dollar

Part 7

$6 \to 5 \to 11$
$6+5=11$
$5+6=11$
$11-5=6$
$11-6=5$

Part 8

a. $49 + 6 = 55$
b. $68 + 6 = 74$

Part 9

a.
```
 843
-812
  31
```
b.
```
  43
+625
 668
```
c.
```
 796
- 24
 772
```

Part 10

a. $100+0+9=109$
b. $400+30+0=430$
c. $800+0+6=806$
d.
```
 500
  10
+  7
 517
```
e.
```
 300
   0
+  8
 308
```

20   Lesson 70   Connecting Math Concepts

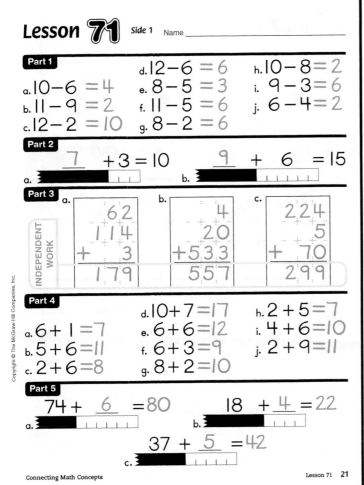

Lesson 71 Side 1 — Name _____

Part 1
a. $10-6=4$
b. $11-9=2$
c. $12-2=10$
d. $12-6=6$
e. $8-5=3$
f. $11-5=6$
g. $8-2=6$
h. $10-8=2$
i. $9-3=6$
j. $6-4=2$

Part 2
a. $\underline{7}+3=10$
b. $\underline{9}+6=15$

Part 3 (INDEPENDENT WORK)
a.
$$\begin{array}{r} 62 \\ 114 \\ +\;\;3 \\ \hline 179 \end{array}$$
b.
$$\begin{array}{r} 4 \\ 20 \\ +533 \\ \hline 557 \end{array}$$
c.
$$\begin{array}{r} 224 \\ 5 \\ +\;\;70 \\ \hline 299 \end{array}$$

Part 4
a. $6+1=7$
b. $5+6=11$
c. $2+6=8$
d. $10+7=17$
e. $6+6=12$
f. $6+3=9$
g. $8+2=10$
h. $2+5=7$
i. $4+6=10$
j. $2+9=11$

Part 5
a. $74+\underline{6}=80$
b. $18+\underline{4}=22$
c. $37+\underline{5}=42$

Connecting Math Concepts
Lesson 71 **21**

Lesson 71 Side 2 — Name _____
Side 2 work is independent.

Part 6
a. $6\quad 3 \rightarrow 9$; $6+3=9$
b. $8\quad 2 \rightarrow 10$; $10-8=2$
c. $10\quad 8 \rightarrow 18$; $10+8=18$
d. $6\quad 6 \rightarrow 12$; $12-6=6$
e. $5\quad 3 \rightarrow 8$; $8-5=3$
f. $6\quad 5 \rightarrow 11$; $11-5=6$

Part 7
a. (bills) $=86$
b. (bills) $=37$
c. (coins) $=\boxed{}$ one dollar
d. (coins) $=\boxed{50}$ one dollar

Part 8
a.
$$\begin{array}{r} 568 \\ -546 \\ \hline 22 \end{array}$$
b.
$$\begin{array}{r} 786 \\ -162 \\ \hline 624 \end{array}$$
c.
$$\begin{array}{r} 869 \\ -\;\;25 \\ \hline 844 \end{array}$$

Part 9
a.
$$\begin{array}{r} 500 \\ 70 \\ +\;\;4 \\ \hline 574 \end{array}$$
b.
$$\begin{array}{r} 300 \\ 0 \\ +\;\;8 \\ \hline 308 \end{array}$$
c. $60+2=62$
d. $600+10+9=619$
e. $100+0+5=105$

22 Lesson 71
Connecting Math Concepts

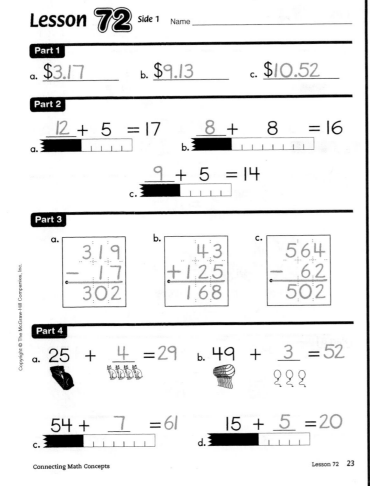

Lesson 72 Side 1 — Name _____

Part 1
a. $\$3.17$
b. $\$9.13$
c. $\$10.52$

Part 2
a. $\underline{12}+5=17$
b. $\underline{8}+8=16$
c. $\underline{9}+5=14$

Part 3
a.
$$\begin{array}{r} 319 \\ -\;17 \\ \hline 302 \end{array}$$
b.
$$\begin{array}{r} 43 \\ +125 \\ \hline 168 \end{array}$$
c.
$$\begin{array}{r} 564 \\ -\;\;62 \\ \hline 502 \end{array}$$

Part 4
a. $25+\underline{4}=29$
b. $49+\underline{3}=52$
c. $54+\underline{7}=61$
d. $15+\underline{5}=20$

Connecting Math Concepts
Lesson 72 **23**

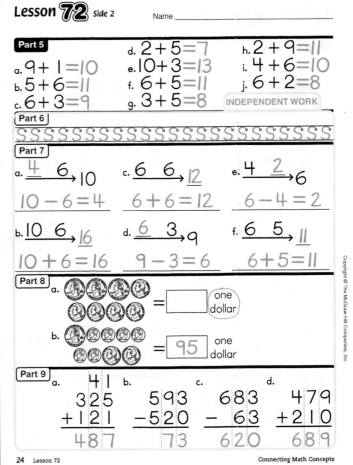

Lesson 72 Side 2 — Name _____

Part 5
a. $9+1=10$
b. $5+6=11$
c. $6+3=9$
d. $2+5=7$
e. $10+3=13$
f. $6+5=11$
g. $3+5=8$
h. $2+9=11$
i. $4+6=10$
j. $6+2=8$

INDEPENDENT WORK

Part 6
SSSSSSSSSSSSSSSSSSSSSSSS

Part 7
a. $4\quad 6 \rightarrow 10$; $10-6=4$
b. $10\quad 6 \rightarrow 16$; $10+6=16$
c. $6\quad 6 \rightarrow 12$; $6+6=12$
d. $6\quad 3 \rightarrow 9$; $9-3=6$
e. $4\quad 2 \rightarrow 6$; $6-4=2$
f. $6\quad 5 \rightarrow 11$; $6+5=11$

Part 8
a. (coins) $=\boxed{}$ one dollar
b. (coins) $=\boxed{95}$ one dollar

Part 9
a.
$$\begin{array}{r} 41 \\ 325 \\ +121 \\ \hline 487 \end{array}$$
b.
$$\begin{array}{r} 593 \\ -520 \\ \hline 73 \end{array}$$
c.
$$\begin{array}{r} 683 \\ -\;\;63 \\ \hline 620 \end{array}$$
d.
$$\begin{array}{r} 479 \\ +210 \\ \hline 689 \end{array}$$

24 Lesson 72
Connecting Math Concepts

Lesson 73 Side 1 — Name _____

Part 1
a. $6.31   b. $8.70   c. $13.59

Part 2
a. 35 + 6 = 41   b. 22 + 4 = 26

Part 3

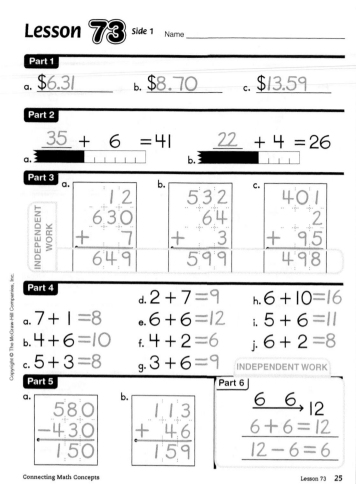

a.
```
  12
 630
+  7
───
 649
```
b.
```
 532
  64
+  3
───
 599
```
c.
```
 401
   2
+ 95
───
 498
```

Part 4
a. 7 + 1 = 8
b. 4 + 6 = 10
c. 5 + 3 = 8
d. 2 + 7 = 9
e. 6 + 6 = 12
f. 4 + 2 = 6
g. 3 + 6 = 9
h. 6 + 10 = 16
i. 5 + 6 = 11
j. 6 + 2 = 8

INDEPENDENT WORK

Part 5
a.
```
 580
-430
───
 150
```
b.
```
 113
+ 46
───
 159
```

Part 6
6   6 → 12
6 + 6 = 12
12 − 6 = 6

Connecting Math Concepts

Lesson 73   25

Lesson 73 Side 2 — Name _____

Side 2 work is independent.

Part 7

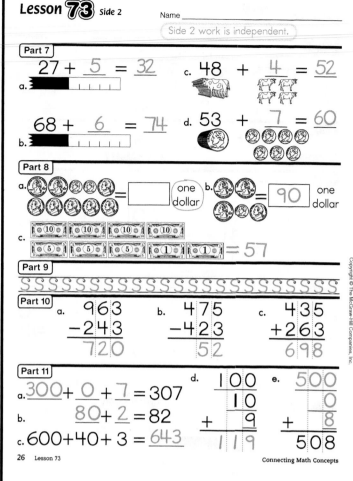

a. 27 + 5 = 32
b. 68 + 6 = 74
c. 48 + 4 = 52
d. 53 + 7 = 60

Part 8
a. = ☐ one dollar
b. = 90 one dollar
c. = 57

Part 9
SSSSSSSSSSSSSSSSSSSSSSSSSSS

Part 10
a.
```
 963
-243
───
 720
```
b.
```
 475
-423
───
  52
```
c.
```
 435
+263
───
 698
```

Part 11
a. 300 + 0 + 7 = 307
b. 80 + 2 = 82
c. 600 + 40 + 3 = 643
d.
```
 100
  10
+  9
───
 119
```
e.
```
 500
   0
+  8
───
 508
```

26   Lesson 73

Connecting Math Concepts

Lesson 74 Side 1 — Name _____

Part 1
a. 32 + 5 = 37   b. 17 + 4 = 21

Part 2
a. $11.91   b. $8.14   c. $13.18

Part 3
a. 5 − 2 = 3
b. 20 − 10 = 10
c. 10 − 4 = 6
d. 4 − 2 = 2
e. 8 − 2 = 6
f. 12 − 6 = 6
g. 10 − 8 = 2
h. 7 − 5 = 2
i. 9 − 6 = 3
j. 6 − 4 = 2
k. 5 − 4 = 1
l. 11 − 6 = 5

Part 4

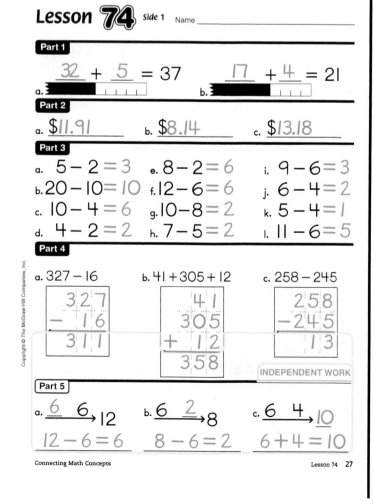

a. 327 − 16
```
 327
- 16
───
 311
```
b. 41 + 305 + 12
```
  41
 305
+ 12
───
 358
```
c. 258 − 245
```
 258
-245
───
  13
```

INDEPENDENT WORK

Part 5
a.
6   6 → 12
12 − 6 = 6
b.
6   2 → 8
8 − 6 = 2
c.
6   4 → 10
6 + 4 = 10

Connecting Math Concepts

Lesson 74   27

Lesson 74 Side 2 — Name _____

Side 2 work is independent.

Part 6

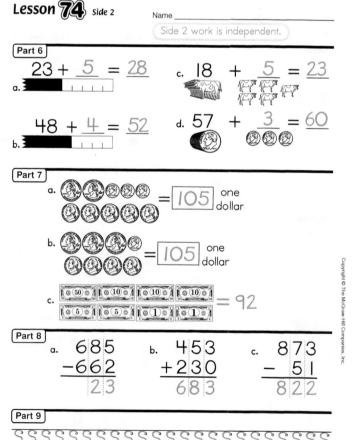

a. 23 + 5 = 28
b. 48 + 4 = 52
c. 18 + 5 = 23
d. 57 + 3 = 60

Part 7
a. = 105 one dollar
b. = 105 one dollar
c. = 92

Part 8
a.
```
 685
-662
───
  23
```
b.
```
 453
+230
───
 683
```
c.
```
 873
- 51
───
 822
```

Part 9
SSSSSSSSSSSSSSSSSSSSSSSSSSS

28   Lesson 74

Connecting Math Concepts

Connecting Math Concepts

Lesson 75 Side 1 Name_____

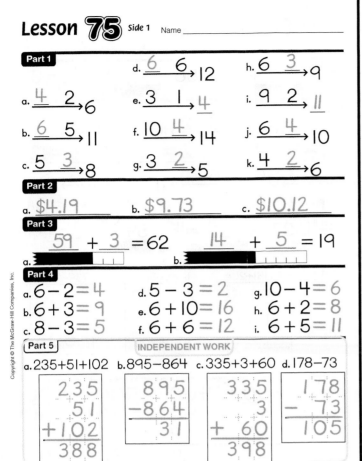

Part 1

d. 6 6 →12   h. 6 3 →9

a. 4 2 →6   e. 3 1 →4   i. 9 2 →11

b. 6 5 →11   f. 10 4 →14   j. 6 4 →10

c. 5 3 →8   g. 3 2 →5   k. 4 2 →6

Part 2

a. $4.19   b. $9.73   c. $10.12

Part 3

a. 59 + 3 = 62   b. 14 + 5 = 19

Part 4

a. 6 − 2 = 4   d. 5 − 3 = 2   g. 10 − 4 = 6

b. 6 + 3 = 9   e. 6 + 10 = 16   h. 6 + 2 = 8

c. 8 − 3 = 5   f. 6 + 6 = 12   i. 6 + 5 = 11

Part 5 INDEPENDENT WORK

a. 235+51+102   b. 895−864   c. 335+3+60   d. 178−73

```
  235        895       335       178
   51       -864         3      - 73
+ 102        ───      + 60       ───
  ───         31       ───       105
  388                  398
```

Connecting Math Concepts   Lesson 75   29

Lesson 75 Side 2 Name_____

Side 2 work is independent.

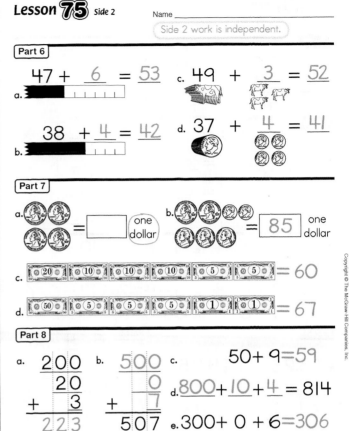

Part 6

a. 47 + 6 = 53   c. 49 + 3 = 52

b. 38 + 4 = 42   d. 37 + 4 = 41

Part 7

a. = [ ] one dollar   b. = 85 one dollar

c. = 60

d. = 67

Part 8

a.
```
  200
   20
 +  3
  ───
  223
```
b.
```
  500
    0
 +  7
  ───
  507
```
c. 50 + 9 = 59

d. 800 + 10 + 4 = 814

e. 300 + 0 + 6 = 306

30   Lesson 75   Connecting Math Concepts

Lesson 76 Side 1 Name_____

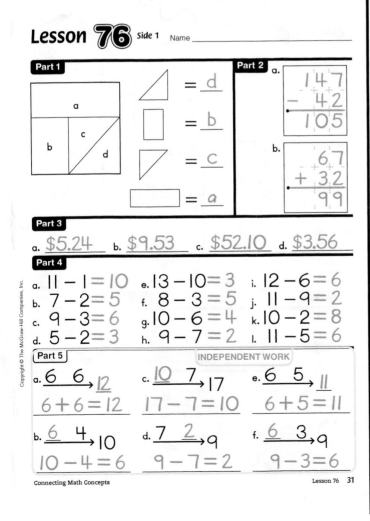

Part 1

(figure)

Part 2

a.
```
  147
 - 42
  ───
  105
```
b.
```
   67
 + 32
  ───
   99
```

Part 1 answers: = d, = b, = c, = a

Part 3

a. $5.24   b. $9.53   c. $52.10   d. $3.56

Part 4

a. 11 − 1 = 10   e. 13 − 10 = 3   i. 12 − 6 = 6

b. 7 − 2 = 5   f. 8 − 3 = 5   j. 11 − 9 = 2

c. 9 − 3 = 6   g. 10 − 6 = 4   k. 10 − 2 = 8

d. 5 − 2 = 3   h. 9 − 7 = 2   l. 11 − 5 = 6

Part 5 INDEPENDENT WORK

a. 6 6 →12   c. 10 7 →17   e. 6 5 →11

  6 + 6 = 12   17 − 7 = 10   6 + 5 = 11

b. 6 4 →10   d. 7 2 →9   f. 6 3 →9

  10 − 4 = 6   9 − 7 = 2   9 − 3 = 6

Connecting Math Concepts   Lesson 76   31

Lesson 76 Side 2 Name_____

Side 2 work is independent.

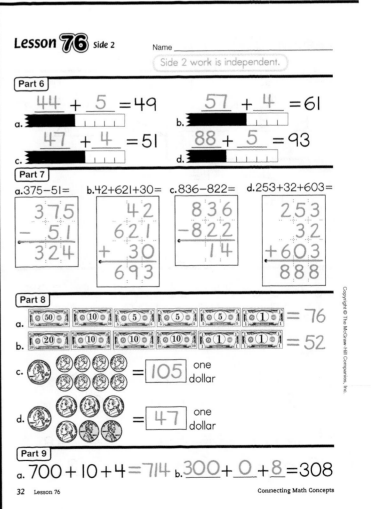

Part 6

a. 44 + 5 = 49   b. 57 + 4 = 61

c. 47 + 4 = 51   d. 88 + 5 = 93

Part 7

a. 375 − 51 =
```
  375
 - 51
  ───
  324
```
b. 42 + 621 + 30 =
```
   42
  621
 + 30
  ───
  693
```
c. 836 − 822 =
```
  836
 -822
  ───
   14
```
d. 253 + 32 + 603 =
```
  253
   32
 +603
  ───
  888
```

Part 8

a. = 76

b. = 52

c. = 105 one dollar

d. = 47 one dollar

Part 9

a. 700 + 10 + 4 = 714   b. 300 + 0 + 8 = 308

32   Lesson 76   Connecting Math Concepts

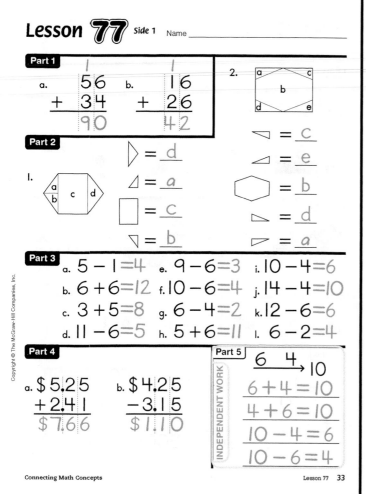

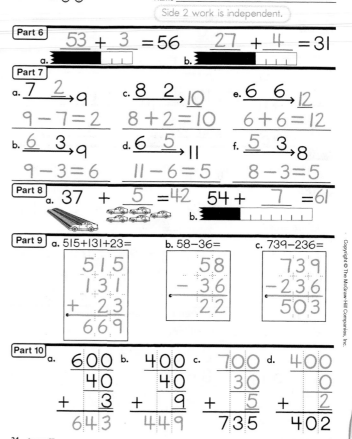

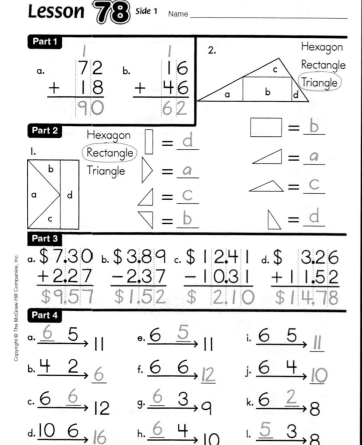

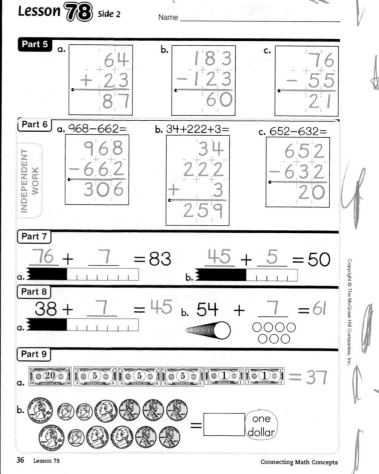

Part 1

a.
$$\begin{array}{r} 1 \\ 44 \\ +\ 36 \\ \hline 80 \end{array}$$

b.
$$\begin{array}{r} 1 \\ 16 \\ +\ 76 \\ \hline 92 \end{array}$$

Part 2

a. $6-4=2$
b. $9-2=7$
c. $6+4=10$
d. $6+6=12$
e. $5+3=8$
f. $11-9=2$
g. $5+6=11$
h. $12-2=10$
i. $8-6=2$

Part 3

1.
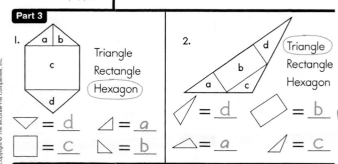

Triangle
Rectangle
(Hexagon)

$\bigtriangledown = \underline{d}$ $\triangle = \underline{a}$

$\square = \underline{c}$ $\triangle = \underline{b}$

2.
(Triangle)
Rectangle
Hexagon

$\square = \underline{d}$ $\square = \underline{b}$

$\triangle = \underline{a}$ $\triangle = \underline{c}$

Part 4

a. $20-10=10$
b. $4-2=2$
c. $5-2=3$
d. $10-4=6$
e. $8-2=6$
f. $12-6=6$
g. $9-6=3$
h. $6-4=2$
i. $7-5=2$
j. $10-8=2$
k. $5-4=1$
l. $11-6=5$

Connecting Math Concepts   Lesson 79   **37**

Part 5

a.
$$\begin{array}{r} \$25.29 \\ -\ 5.16 \\ \hline \$20.13 \end{array}$$

b.
$$\begin{array}{r} \$35.14 \\ +\ 1.51 \\ \hline \$36.65 \end{array}$$

c.
$$\begin{array}{r} \$6.87 \\ -5.25 \\ \hline \$1.62 \end{array}$$

INDEPENDENT WORK

Part 6

a. $495-414=$
$$\begin{array}{r} 495 \\ -414 \\ \hline 81 \end{array}$$

b. $24+362+3=$
$$\begin{array}{r} 24 \\ 362 \\ +\ 3 \\ \hline 389 \end{array}$$

c. $786-56=$
$$\begin{array}{r} 786 \\ -\ 56 \\ \hline 730 \end{array}$$

Part 7

a.
$$\begin{array}{r} 400 \\ 0 \\ +\ 7 \\ \hline 407 \end{array}$$

b.
$$\begin{array}{r} 800 \\ 30 \\ +\ 5 \\ \hline 835 \end{array}$$

c. $600+10+9=619$

d. $100+70+0=170$

Part 8

a. $\underline{34}+\underline{4}=38$ $\underline{89}+\underline{8}=97$

b.

c. $38+\underline{6}=44$ d. $54+\underline{5}=59$

38   Lesson 79   Connecting Math Concepts

Part 1

a. $10-4=6$
b. $5-3=2$
c. $6+6=12$
d. $6+3=9$
e. $6-2=4$
f. $8-3=5$
g. $6+10=16$
h. $11-6=5$

Part 2

a.
$$\begin{array}{r} 1 \\ 34 \\ 12 \\ +\ 36 \\ \hline 82 \end{array}$$

b.
$$\begin{array}{r} 1 \\ 8 \\ 52 \\ +\ 37 \\ \hline 97 \end{array}$$

c.
$$\begin{array}{r} 1 \\ 76 \\ +\ 14 \\ \hline 90 \end{array}$$

Part 3

a.   8
     6

$$\begin{array}{r} 8 \\ -\ 6 \\ \hline 2 \end{array}$$

b.   4
        10

$$\begin{array}{r} 10 \\ -\ 4 \\ \hline 6 \end{array}$$

Part 4

a. $3-2=1$
b. $12-6=6$
c. $14-10=4$
d. $4-1=3$
e. $6-2=4$
f. $8-6=2$
g. $9-7=2$
h. $10-6=4$
i. $8-5=3$
j. $11-2=9$
k. $11-5=6$
l. $9-6=3$

Connecting Math Concepts   Lesson 80   **39**

Part 5

a.
$$\begin{array}{r} \$27.55 \\ -\ 3.24 \\ \hline \$24.31 \end{array}$$

b.
$$\begin{array}{r} \$5.21 \\ +3.18 \\ \hline \$8.39 \end{array}$$

Part 6   INDEPENDENT WORK

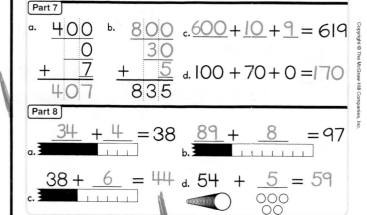

Triangle
(Rectangle)
Hexagon

$\square = \underline{b}$ $\square = \underline{c}$

$\triangle = \underline{a}$ $\triangle = \underline{d}$

Part 7

a. $45+\underline{4}=49$

b. $57+\underline{4}=61$

c. $17+\underline{6}=23$

Part 8

a. $=106$

b. $=39$

c. $\bigcirc\bigcirc\bigcirc\bigcirc = \boxed{}$ one dollar

d. $\bigcirc\bigcirc\bigcirc\bigcirc\bigcirc\bigcirc\bigcirc\bigcirc\bigcirc = \boxed{71}$ one dollar

Part 9

a. $460+15+213=$
$$\begin{array}{r} 460 \\ 15 \\ +213 \\ \hline 688 \end{array}$$

b. $379-326=$
$$\begin{array}{r} 379 \\ -326 \\ \hline 53 \end{array}$$

40   Lesson 80   Connecting Math Concepts

Mastery Test 5

Passing Criteria Table — Mastery Test 5			
Part	Score	Possible Score	Passing Score
1	2 for each number	10	10
2	3 for each problem	9	6
3	3 for each problem	15	12
4	2 for each problem	8	6
5	2 for each problem	8	6
6	3 for each equation	15	12
7	1 for each problem	9	8
8	2 for each equation	8	6
9	2 for each equation	18	14
	Total	100	

Mastery Test 5   Name _____

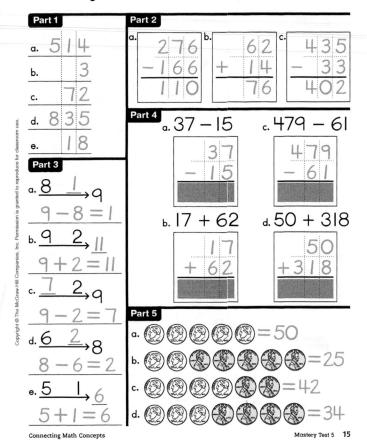

Part 1

a. 514
b. __3
c. _72
d. 835
e. _18

Part 3

a. 8 →1→ 9
9 − 8 = 1

b. 9 →2→ 11
9 + 2 = 11

c. 7 →2→ 9
9 − 2 = 7

d. 6 →2→ 8
8 − 6 = 2

e. 5 →1→ 6
5 + 1 = 6

Part 2

a. 276 − 166 = 110
b. 62 + 14 = 76
c. 435 − 33 = 402

Part 4

a. 37 − 15   37 − 15
c. 479 − 61   479 − 61
b. 17 + 62   17 + 62
d. 50 + 318   50 + 318

Part 5

a. = 50
b. = 25
c. = 42
d. = 34

Connecting Math Concepts   Mastery Test 5   **15**

Mastery Test 5   Name _____

Part 6

a. 200 + 0 + 9 = 209
b. 800 + 10 + 3 = 813
c. 500 + 0 + 0 = 500
d. 400 + 0 + 6 = 406
e. 600 + 20 + 0 = 620

Part 7

a. 10 →7→ 17
b. 10 →1→ 11
c. 10 →8→ 18
d. 10 →3→ 13
e. 10 →5→ 15
f. 10 →2→ 12
g. 10 →9→ 19
h. 10 →6→ 16
i. 10 →4→ 14

Part 8

a. 1 + 7 + 2 = 10
b. 9 + 1 + 6 = 16
c. 1 + 4 + 2 = 7
d. 1 + 9 + 5 = 15

Part 9

a. 5 + 2 = 7
b. 9 − 7 = 2
c. 6 − 2 = 4
d. 2 + 6 = 8
e. 7 − 5 = 2
f. 8 − 2 = 6
g. 9 + 2 = 11
h. 5 − 3 = 2
i. 2 + 7 = 9

16   Mastery Test 5   Connecting Math Concepts

Remedies   Name _____

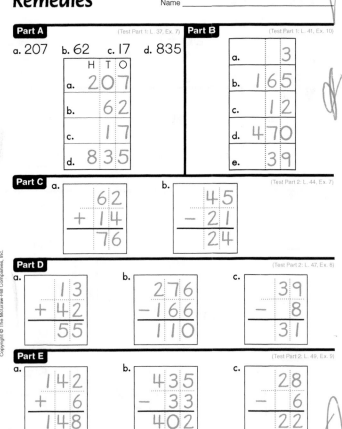

Part A   (Test Part 1: L. 37, Ex. 7)

a. 207   b. 62   c. 17   d. 835

	H	T	O
a.	2	0	7
b.		6	2
c.		1	7
d.	8	3	5

Part B   (Test Part 1: L. 41, Ex. 10)

a.	3
b.	165
c.	12
d.	470
e.	39

Part C   (Test Part 2: L. 44, Ex. 7)

a. 62 + 14 = 76
b. 45 − 21 = 24

Part D   (Test Part 2: L. 47, Ex. 8)

a. 13 + 42 = 55
b. 276 − 166 = 110
c. 39 − 8 = 31

Part E   (Test Part 2: L. 49, Ex. 9)

a. 142 + 6 = 148
b. 435 − 33 = 402
c. 28 − 6 = 22

Connecting Math Concepts   Mastery Test 5 *Remedies*   **17**

---

286   **Answer Key**   *Connecting Math Concepts*

Remedies CONTINUED   Name _____

Part F   (Test Part 3: L. 45, Ex. 8)

a. $\underline{9}\ \underline{2}\rightarrow 11$ e. $\underline{5}\ \underline{1}\rightarrow 6$

b. $\underline{8}\ \ 1\rightarrow 9$ f. $\underline{2}\ \ 2\rightarrow 4$

c. $\underline{9}\ \ 2\rightarrow 11$ g. $4\ \ 2\rightarrow \underline{6}$

d. $\underline{7}\ \ 2\rightarrow 9$ h. $\underline{8}\ \ \underline{2}\rightarrow 10$

Part G   (Test Part 4: L. 42, Ex. 9)

a. 17 + 62

```
   17
 + 62
   79
```

b. 36 − 15

```
   36
 − 15
   21
```

Part H   (Test Part 4: L. 43, Ex. 7)

a. 50 + 318

```
    50
 + 318
   368
```

b. 479 − 61

```
   479
 −  61
   418
```

Part I   (Test Part 5: L. 48, Ex. 7)

a. 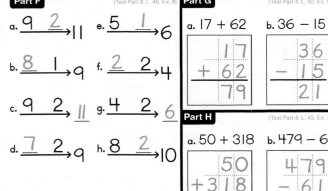 = 34

b. = 23

Part J   (Test Part 5: L. 49, Ex. 7)

a.  = 42

b.  = 25

Remedies CONTINUED   Name _____

Part K   (Test Part 6: L. 37, Ex. 6)

a. $\underline{700}+\underline{80}+\underline{2}=782$

b. $\underline{200}+\underline{30}+\underline{0}=230$

c. $\underline{100}+\underline{90}+\underline{3}=193$

d. $\underline{800}+\underline{50}+\underline{0}=850$

Part L   (Test Part 6: L. 39, Ex. 8)

a. $\underline{500}+\underline{0}+\underline{6}=506$

b. $\underline{600}+\underline{20}+\underline{0}=620$

c. $\underline{200}+\underline{0}+\underline{9}=209$

d. $\underline{300}+\underline{0}+\underline{0}=300$

Part M   (Test Part 6: L. 40, Ex. 7)

a. $\underline{100}+\underline{0}+\underline{2}=102$

b. $\underline{200}+\underline{40}+\underline{0}=240$

c. $\underline{700}+\underline{10}+\underline{1}=711$

Part N   (Test Part 6: L. 41, Ex. 7)

a. $\underline{800}+\underline{10}+\underline{3}=813$

b. $\underline{500}+\underline{0}+\underline{0}=500$

c. $\underline{400}+\underline{0}+\underline{6}=406$

d. $\underline{100}+\underline{70}+\underline{0}=170$

Part O   (Test Part 7: L. 47, Ex. 7)

a. $\underline{10}\ \underline{4}\rightarrow 14$ c. $\underline{10}\ \underline{6}\rightarrow 16$ e. $\underline{10}\ \underline{3}\rightarrow 13$

b. $\underline{10}\ \underline{8}\rightarrow 18$ d. $\underline{10}\ \underline{9}\rightarrow 19$ f. $\underline{10}\ \underline{5}\rightarrow 15$

Remedies CONTINUED   Name _____

Part P   (Test Part 7: L. 48, Ex. 9)

a. $\underline{10}\ \underline{3}\rightarrow 13$ c. $\underline{10}\ \underline{5}\rightarrow 15$ e. $\underline{10}\ \underline{6}\rightarrow 16$

b. $\underline{10}\ \underline{7}\rightarrow 17$ d. $\underline{10}\ \underline{2}\rightarrow 12$ f. $\underline{10}\ \underline{1}\rightarrow 11$

Part Q   (Test Part 8: L. 42, Ex. 10)

a. $\underline{3}+\underline{2}+1=6$

b. $\underline{7}+\underline{1}+2=10$

Part R   (Test Part 8: L. 43, Ex. 4)

a. $1+9+5=15$

b. $1+5+1=7$

Part S   (Test Part 9: L. 43, Ex. 8)

a. $9-7=2$ c. $6-2=4$ e. $2+7=9$

b. $5+2=7$ d. $7-5=2$ f. $8-2=6$

Part T   (Test Part 9: L. 47, Ex. 5)

a. $9-8=1$ e. $8-7=1$ i. $2+7=9$

b. $6-1=5$ f. $3-2=1$ j. $6-4=2$

c. $5-3=2$ g. $1+10=11$

d. $1+4=5$ h. $7-1=6$

Remedy Table — Mastery Test 5

Part	Test Items	Remedy Lesson	Remedy Exercise	Student Material Remedies Worksheet
1	Writing Numbers in Columns	35	2	—
		37	7	Part A
		41	10	Part B
2	Word Problems	44	7	Part C
		47	8	Part D
		49	9	Part E
3	Number Families (Writing Facts to Find Missing Numbers)	37	3	—
		42	4	—
		45	8	Part F
4	Writing Column Problems	42	9	Part G
		43	7	Part H
5	Coins	42	5	—
		44	4	—
		48	7	Part I
		49	7	Part J
6	Place-Value Equations	37	6	Part K
		39	8	Part L
		40	7	Part M
		41	7	Part N
7	Number Family (Small Number of 10)	47	3	—
		47	7	Part O
		48	1	—
		48	9	Part P
8	3 Addends	37	2	—
		42	10	Part Q
		43	4	Part R
9	Facts (Mix) (Small Number of 2)	43	1	—
		43	8	Part S
		46	1	—
		47	5	Part T

Mastery Test 6

Part	Score	Possible Score	Passing Score
1	3 for each problem	9	6
2	3 for each problem	15	12
3	2 for each fact	20	18
4	3 for each group of coins	12	9
5	2 for each fact	20	18
6	3 for each equation	12	9
7	3 for each problem	12	9
	Total	100	

Passing Criteria Table — Mastery Test 6

Mastery Test 6   Name _____

Part 1

a.
$$\begin{array}{r} 187 \\ -127 \\ \hline 60 \end{array}$$

b.
$$\begin{array}{r} 182 \\ +215 \\ \hline 397 \end{array}$$

c.
$$\begin{array}{r} 765 \\ -562 \\ \hline 203 \end{array}$$

Part 2

a. $58 + 5 = 63$

b. $17 + 4 = 21$

c. $28 + 6 = 34$

d. $73 + 7 = 80$

e. $37 + 6 = 43$

Part 3

a. $6 - 4 = 2$

b. $9 + 2 = 11$

c. $13 - 3 = 10$

d. $11 - 2 = 9$

e. $8 + 10 = 18$

f. $8 - 2 = 6$

g. $10 - 8 = 2$

h. $2 + 6 = 8$

i. $17 - 7 = 10$

j. $4 - 2 = 2$

Connecting Math Concepts   Mastery Test 6   21

Mastery Test 6   Name _____

Copyright © The McGraw-Hill Companies, Inc. Permission is granted to reproduce for classroom use.

Part 4

a. = 51
b. = 42
c. = 48
d. = 32

Part 5

d. $9-3=6$ h. $3+6=9$

a. $8-3=5$ e. $6+4=10$ i. $10-4=6$

b. $10-6=4$ f. $9-6=3$ j. $5+3=8$

c. $3+5=8$ g. $8-5=3$

Part 6

a.
$$300$$
$$80$$
$$+\ \ 2$$
$$382$$

b.
$$100$$
$$0$$
$$+\ \ 7$$
$$107$$

c.
$$600$$
$$10$$
$$+\ \ 9$$
$$619$$

d.
$$700$$
$$60$$
$$+\ \ 0$$
$$760$$

Part 7

a.
$$265$$
$$-210$$
$$55$$

b.
$$265$$
$$+210$$
$$475$$

c.
$$468$$
$$-426$$
$$42$$

d.
$$364$$
$$-\ 54$$
$$310$$

Remedies   Name _____

Copyright © The McGraw-Hill Companies, Inc.

Part A   (Test Part 1: L. 57, Ex. 10)

a.
$$765$$
$$-562$$
$$203$$

b.
$$83$$
$$+215$$
$$298$$

Part B   (Test Part 1: L. 59, Ex. 8)

a.
$$56$$
$$+132$$
$$188$$

b.
$$187$$
$$-127$$
$$60$$

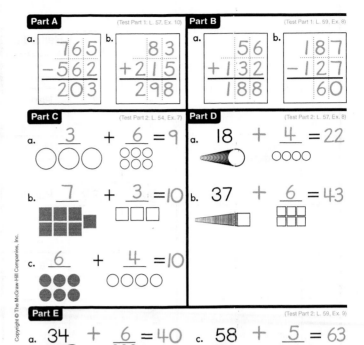

Part C   (Test Part 2: L. 54, Ex. 7)

a. $3 + 6 = 9$

b. $7 + 3 = 10$

c. $6 + 4 = 10$

Part D   (Test Part 2: L. 57, Ex. 8)

a. $18 + 4 = 22$

b. $37 + 6 = 43$

Part E   (Test Part 2: L. 59, Ex. 9)

a. $34 + 6 = 40$ c. $58 + 5 = 63$

b. $17 + 4 = 21$ d. $66 + 9 = 75$

Remedies Continued   Name _____

Copyright © The McGraw-Hill Companies, Inc.

Part F   (Test Part 3: L. 52, Ex. 8)

a. $2+5=7$ d. $6+2=8$ g. $3+2=5$

b. $2+7=9$ e. $8+2=10$ h. $5+3=8$

c. $3+5=8$ f. $3+10=13$ i. $2+10=12$

Part G   (Test Part 4: L. 55, Ex. 9)

a. = 25   b. = 23

c. = 35

Part H   (Test Part 4: L. 58, Ex. 8)

a. = 48

b. = 32

Part I   (Test Part 4: L. 59, Ex. 6)

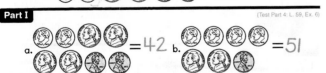

a. = 42   b. = 51

Part J   (Test Part 5: L. 60, Ex. 6)

d. $10+8=18$

a. $6+4=10$ e. $2+3=5$ h. $4+6=10$

b. $7+2=9$ f. $5+10=15$ i. $5+2=7$

c. $3+6=9$ g. $2+9=11$ j. $3+5=8$

Remedies Continued   Name _____

Part K   (Test Part 6: L. 53, Ex. 7)

a.
$$500$$
$$90$$
$$+\ \ 4$$
$$594$$

b.
$$300$$
$$80$$
$$+\ \ 2$$
$$382$$

c.
$$100$$
$$0$$
$$+\ \ 7$$
$$107$$

Part L   (Test Part 6: L. 54, Ex. 5)

a.
$$700$$
$$60$$
$$+\ \ 0$$
$$760$$

b.
$$600$$
$$10$$
$$+\ \ 9$$
$$619$$

c.
$$100$$
$$80$$
$$+\ \ 1$$
$$181$$

d.
$$500$$
$$0$$
$$+\ \ 3$$
$$503$$

Part M   (Test Part 7: L. 57, Ex. 7)

a.
$$58$$
$$-\ 57$$
$$1$$

b.
$$364$$
$$-343$$
$$21$$

c.
$$496$$
$$-416$$
$$80$$

Part N   (Test Part 7: L. 58, Ex. 7)

a.
$$725$$
$$-705$$
$$20$$

b.
$$53$$
$$+821$$
$$874$$

c.
$$96$$
$$-\ 91$$
$$5$$

d.
$$364$$
$$-\ 54$$
$$310$$

Remedy Table — Mastery Test 6

Part	Test Items	Remedy Lesson	Remedy Exercise	Student Material Remedies Worksheet
1	Word Problems	57	10	Part A
		59	8	Part B
2	Count-on	50	6	—
		54	7	Part C
		57	8	Part D
		59	9	Part E
3	Facts (Small Numbers of 2, 10)	49	1	—
		51	3	—
		52	8	Part F
		55	3	—
4	Coins	55	9	Part G
		58	8	Part H
		59	6	Part I
5	Facts (Small Numbers of 5, 6)	54	3	—
		58	2	—
		60	6	Part J
6	Place-Value Equations	53	7	Part K
		54	5	Part L
7	Column Problems	55	4	—
		55	6	—
		57	7	Part M
		58	7	Part N

Mastery Test 7

Passing Criteria Table — Mastery Test 7			
Part	Score	Possible Score	Passing Score
1	2 for each problem	20	18
2	4 for each problem	20	16
3	3 for each problem	15	12
4	3 for each problem	15	12
5	3 for each problem	15	12
6	3 for each problem	15	9
	Total	100	

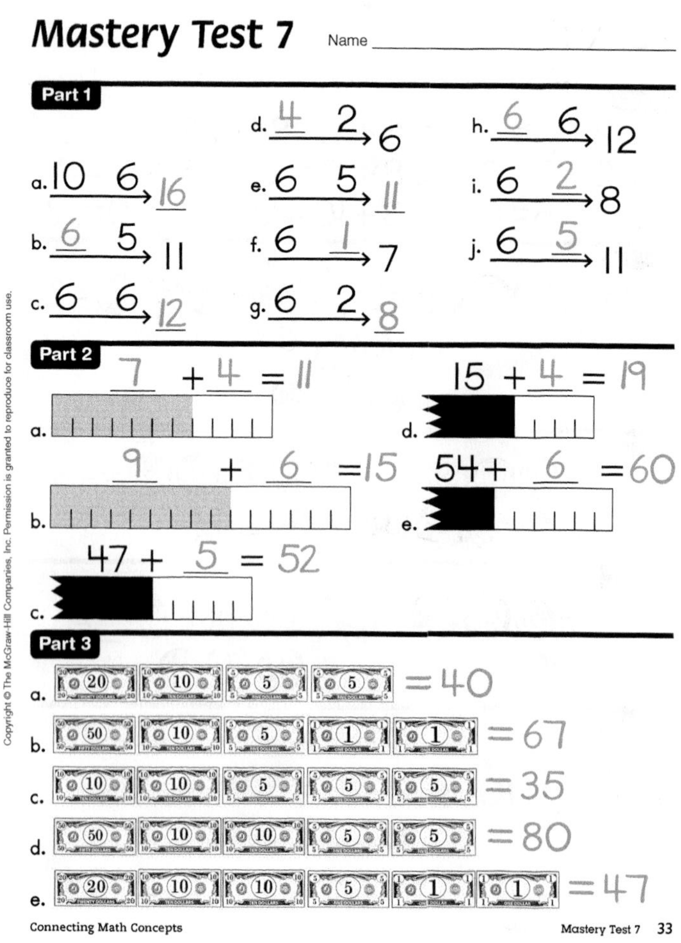

Mastery Test 7   Name _____

Part 1

a. $10 \quad 6 \rightarrow \underline{16}$

b. $\underline{6} \quad 5 \rightarrow 11$

c. $6 \quad 6 \rightarrow \underline{12}$

d. $\underline{4} \quad 2 \rightarrow 6$

e. $6 \quad 5 \rightarrow 11$

f. $6 \quad \underline{1} \rightarrow 7$

g. $6 \quad 2 \rightarrow \underline{8}$

h. $\underline{6} \quad 6 \rightarrow 12$

i. $6 \quad \underline{2} \rightarrow 8$

j. $6 \quad \underline{5} \rightarrow 11$

Part 2

a. $\underline{7} + \underline{4} = 11$

b. $\underline{9} + \underline{6} = 15$

c. $47 + \underline{5} = 52$

d. $15 + \underline{4} = 19$

e. $54 + \underline{6} = 60$

Part 3

a. $= 40$

b. $= 67$

c. $= 35$

d. $= 80$

e. $= 47$

Mastery Test 7   33

Mastery Test 7   Name _____

Part 4

a.
$$
\begin{array}{r}
3\,4 \\
2\,1 \\
+\ 2\,3 \\
\hline
7\,8
\end{array}
$$

b.
$$
\begin{array}{r}
5\,0 \\
3\,4 \\
+\ 1\,2 \\
\hline
9\,6
\end{array}
$$

c.
$$
\begin{array}{r}
3\,2 \\
6 \\
+\ 2\,1 \\
\hline
5\,9
\end{array}
$$

d.
$$
\begin{array}{r}
1\,8 \\
2\,0\,1 \\
+\ 4\,0 \\
\hline
2\,5\,9
\end{array}
$$

e.
$$
\begin{array}{r}
2\,7\,4 \\
1\,2 \\
+\,1\,0\,2 \\
\hline
3\,8\,8
\end{array}
$$

Part 5

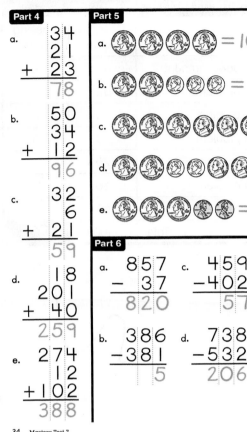

a. = 100
b. = 80
c. = 90
d. = 80
e. = 77

Part 6

a.
$$
\begin{array}{r}
8\,5\,7 \\
-\ 3\,7 \\
\hline
8\,2\,0
\end{array}
$$

c.
$$
\begin{array}{r}
4\,5\,9 \\
-4\,0\,2 \\
\hline
5\,7
\end{array}
$$

e.
$$
\begin{array}{r}
9\,8 \\
-\,9\,6 \\
\hline
2
\end{array}
$$

b.
$$
\begin{array}{r}
3\,8\,6 \\
-3\,8\,1 \\
\hline
5
\end{array}
$$

d.
$$
\begin{array}{r}
7\,3\,8 \\
-5\,3\,2 \\
\hline
2\,0\,6
\end{array}
$$

Remedies   Name _____

Part A   (Test Part 2: L. 63, Ex. 7)

$$3 \ + \ 6 \ = 9$$

a.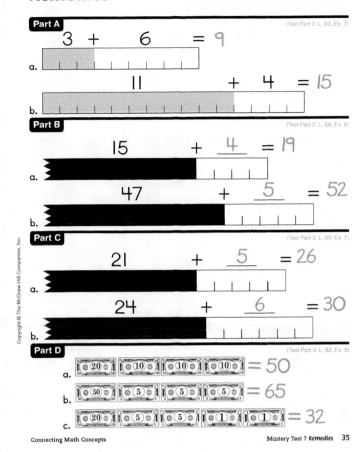

$$11 \ + \ 4 \ = 15$$

b.

Part B   (Test Part 2: L. 68, Ex. 8)

a. $$15 \ + \ 4 \ = 19$$

b. $$47 \ + \ 5 \ = 52$$

Part C   (Test Part 2: L. 69, Ex. 7)

a. $$21 \ + \ 5 \ = 26$$

b. $$24 \ + \ 6 \ = 30$$

Part D   (Test Part 3: L. 62, Ex. 8)

a. = 50
b. = 65
c. = 32

Remedies CONTINUED   Name _____

Part E   (Test Part 3: L. 64, Ex. 8)

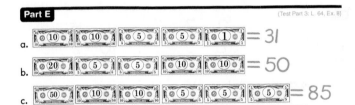

a. = 31
b. = 50
c. = 85

Part F   (Test Part 4: L. 64, Ex. 7)

a.
$$
\begin{array}{r}
3\,4 \\
2\,1 \\
+\ 2\,3 \\
\hline
7\,8
\end{array}
$$

b.
$$
\begin{array}{r}
1\,2 \\
1\,1 \\
+\ 3\,2 \\
\hline
5\,5
\end{array}
$$

c.
$$
\begin{array}{r}
5\,0 \\
3\,4 \\
+\ 1\,2 \\
\hline
9\,6
\end{array}
$$

Part G   (Test Part 4: L. 65, Ex. 7)

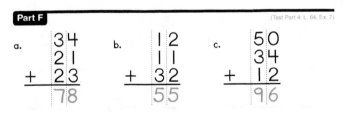

a.
$$
\begin{array}{r}
2\,5 \\
3\,0 \\
+\ 3\,2 \\
\hline
8\,7
\end{array}
$$

b.
$$
\begin{array}{r}
1\,8 \\
2\,0\,1 \\
+\ 4\,0 \\
\hline
2\,5\,9
\end{array}
$$

c.
$$
\begin{array}{r}
3\,5\,6 \\
+6\,2\,0 \\
\hline
9\,7\,6
\end{array}
$$

d.
$$
\begin{array}{r}
6\,1 \\
+3\,2\,0 \\
\hline
3\,8\,1
\end{array}
$$

Remedies CONTINUED   Name _____

Part H   (Test Part 5: L. 61, Ex. 6)

a. = 50
b. = 100
c. = 40

Part I   (Test Part 5: L. 62, Ex. 6)

a. = 90
b. = 80
c. = 55

Part J   (Test Part 5: L. 64, Ex. 5)

a. = 78
b. = 90
c. = 80

Remedy Table — Mastery Test 7

Part	Test Items	Remedy Lesson	Remedy Exercise	Student Material Remedies Worksheet
1	New Number Families	64	1	—
		65	2	—
		67	1	—
2	Count On (Rulers)	60	3	—
		63	7	Part A
		68	8	Part B
		69	7	Part C
3	Bills	59	2	—
		60	4	—
		62	8	Part D
		64	8	Part E
4	3 Addends (Columns)	64	7	Part F
		65	7	Part G
5	Coins (Quarters)	61	6	Part H
		62	6	Part I
		64	5	Part J
6	Column Subtraction	62	2	—

Mastery Test 8

Passing Criteria Table — Mastery Test 8			
Part	Score	Possible Score	Passing Score
1	2 for each dollar amount	10	8
2	3 for each problem	12	9
3	2 for each problem	20	18
4	2 for each problem	10	8
5	2 for each problem	20	18
6	4 for each problem	16	12
7	3 for each problem	12	9
	Total	100	

Mastery Test 8   Name _____

Part 1

a. $9.13 b. $10.52 c. $3.56 d. $3.17 e. $4.19

Part 2

a. $14 + 5 = 19$

c. $17 + 4 = 21$

b. $32 + 5 = 37$

d. $59 + 3 = 62$

Part 3

a. $3 + 6 = 9$

b. $2 + 9 = 11$

c. $6 + 6 = 12$

d. $7 + 2 = 9$

e. $3 + 5 = 8$

f. $6 + 4 = 10$

g. $2 + 8 = 10$

h. $6 + 5 = 11$

i. $2 + 10 = 12$

j. $6 + 3 = 9$

Part 4

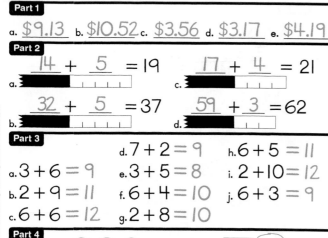

a. ____ one dollar

b. 85 one dollar

c. ____ one dollar

d. 90 one dollar

e. ____ one dollar

Mastery Test 8   Name _____

Part 5

a. $11 - 5 = 6$

b. $11 - 9 = 2$

c. $9 - 3 = 6$

d. $9 - 7 = 2$

e. $8 - 5 = 3$

f. $12 - 6 = 6$

g. $12 - 10 = 2$

h. $10 - 4 = 6$

i. $10 - 2 = 8$

j. $11 - 6 = 5$

Part 6

a. $\begin{array}{r} 1 \\ 72 \\ + 18 \\ \hline 90 \end{array}$

b. $\begin{array}{r} 1 \\ 16 \\ + 46 \\ \hline 62 \end{array}$

c. $\begin{array}{r} 1 \\ 76 \\ + 14 \\ \hline 90 \end{array}$

d. $\begin{array}{r} 1 \\ 8 \\ 52 \\ + 37 \\ \hline 97 \end{array}$

Part 7

a. $\begin{array}{r} \$4.25 \\ -3.15 \\ \hline \$1.10 \end{array}$

b. $\begin{array}{r} \$5.25 \\ +2.41 \\ \hline \$7.66 \end{array}$

c. $\begin{array}{r} \$3.26 \\ +11.52 \\ \hline \$14.78 \end{array}$

d. $\begin{array}{r} \$12.41 \\ -10.31 \\ \hline \$2.10 \end{array}$

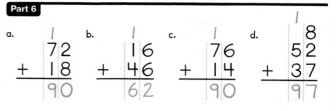

Remedies

Name _____

Part A
(Test Part 1: L. 72, Ex. 5)

a. **$3.17**   b. **$9.13**   c. **$10.52**

Part B
(Test Part 1: L. 75, Ex. 6)

a. **$4.19**   b. **$9.73**   c. **$10.12**

Part C
(Test Part 1: L. 76, Ex. 7)

a. **$5.24**   b. **$9.53**   c. **$52.10**   d. **$3.56**

Part D
(Test Part 2: L. 72, Ex. 6)

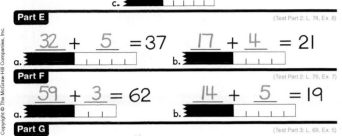

a. $12 + 5 = 17$

b. $8 + 8 = 16$

c. $9 + 5 = 14$

Part E
(Test Part 2: L. 74, Ex. 6)

a. $32 + 5 = 37$

b. $17 + 4 = 21$

Part F
(Test Part 2: L. 75, Ex. 7)

a. $59 + 3 = 62$

b. $14 + 5 = 19$

Part G
(Test Part 3: L. 69, Ex. 5)

a. $6 + 4 = 10$ d. $5 + 2 = 7$ h. $3 + 5 = 8$

b. $2 + 9 = 11$ e. $6 + 6 = 12$ i. $2 + 6 = 8$

c. $3 + 6 = 9$ f. $6 + 5 = 11$ j. $9 + 1 = 10$

g. $8 + 2 = 10$

Remedies CONTINUED

Name _____

Part H
(Test Part 3: L. 73, Ex. 8)

a. $7 + 1 = 8$ d. $2 + 7 = 9$ h. $6 + 10 = 16$

b. $4 + 6 = 10$ e. $6 + 6 = 12$ i. $5 + 6 = 11$

c. $5 + 3 = 8$ f. $4 + 2 = 6$ j. $6 + 2 = 8$

g. $3 + 6 = 9$

Part I
(Test Part 4: L. 69, Ex. 6)

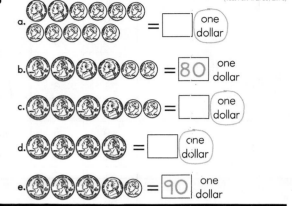

a. = ☐ one dollar

b. = 80 one dollar

c. = ☐ one dollar

d. = ☐ one dollar

e. = 90 one dollar

Part J
(Test Part 5: L. 71, Ex. 5)

a. $10 - 6 = 4$ d. $12 - 6 = 6$ h. $10 - 8 = 2$

b. $11 - 9 = 2$ e. $8 - 5 = 3$ i. $9 - 3 = 6$

c. $12 - 2 = 10$ f. $11 - 5 = 6$ j. $6 - 4 = 2$

g. $8 - 2 = 6$

Remedies CONTINUED

Name _____

Part K
(Test Part 5: L. 80, Ex. 7)

a. $3 - 2 = 1$ e. $6 - 2 = 4$ i. $8 - 5 = 3$

b. $12 - 6 = 6$ f. $8 - 6 = 2$ j. $11 - 2 = 9$

c. $14 - 10 = 4$ g. $9 - 7 = 2$ k. $11 - 5 = 6$

d. $4 - 1 = 3$ h. $10 - 6 = 4$ l. $9 - 6 = 3$

Part L
(Test Part 6: L. 80, Ex. 5)

a.
$$\begin{array}{r} 3\overset{1}{4} \\ 12 \\ + 36 \\ \hline 82 \end{array}$$

b.
$$\begin{array}{r} \overset{1}{8} \\ 52 \\ + 37 \\ \hline 97 \end{array}$$

c.
$$\begin{array}{r} \overset{1}{7}6 \\ + 14 \\ \hline 90 \end{array}$$

Part M
(Test Part 7: L. 77, Ex. 8)

a.
$$\begin{array}{r} \$5.25 \\ + 2.41 \\ \hline \$7.66 \end{array}$$

b.
$$\begin{array}{r} \$4.25 \\ - 3.15 \\ \hline \$1.10 \end{array}$$

Part N
(Test Part 7: L. 78, Ex. 6)

a.
$$\begin{array}{r} \$7.30 \\ + 2.27 \\ \hline \$9.57 \end{array}$$

b.
$$\begin{array}{r} \$3.89 \\ - 2.37 \\ \hline \$1.52 \end{array}$$

c.
$$\begin{array}{r} \$12.41 \\ - 10.31 \\ \hline \$2.10 \end{array}$$

d.
$$\begin{array}{r} \$\ \ 3.26 \\ + 11.52 \\ \hline \$14.78 \end{array}$$

Remedy Table — Mastery Test 8				
Part	Test Items	Remedy		Student Material Remedies Worksheet
		Lesson	Exercise	
1	Writing Dollar Amounts	70	1	—
		72	5	Part A
		75	6	Part B
		76	7	Part C
2	Count Backward (Rulers)	70	4	—
		72	6	Part D
		74	6	Part E
		75	7	Part F
3	Facts (Addition)	69	5	Part G
		73	8	Part H
		74	4	—
4	Coins (Equal to a Dollar)	65	5	—
		67	2	—
		69	6	Part I
5	Facts (Subtraction)	71	5	Part J
		73	4	—
		75	1	—
		80	7	Part K
6	Column Addition (Carrying)	74	2	—
		75	3	—
		78	4	—
		80	5	Part L
7	Money Problems	77	8	Part M
		78	6	Part N

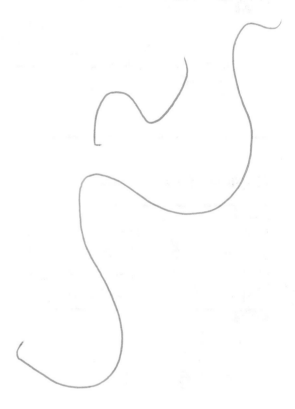

Remedy Summary—Group Summary of Test Performance

Note: Test remedies are specified in the Answer Key.

Name	Test 5									Total %	Test 6							Total %	Test 7						Total %	Test 8							Total %
	Check parts not passed										Check parts not passed								Check parts not passed							Check parts not passed							
	1	2	3	4	5	6	7	8	9		1	2	3	4	5	6	7		1	2	3	4	5	6		1	2	3	4	5	6	7	
1.																																	
2.																																	
3.																																	
4.																																	
5.																																	
6.																																	
7.																																	
8.																																	
9.																																	
10.																																	
11.																																	
12.																																	
13.																																	
14.																																	
15.																																	
16.																																	
17.																																	
18.																																	
19.																																	
20.																																	
21.																																	
22.																																	
23.																																	
24.																																	
25.																																	
26.																																	
27.																																	
28.																																	
29.																																	
30.																																	

Number of students
Not Passed = NP

Total number
of students = T

Remedy needed if
NP/T = 25% or more

Level B Correlation to Grade 1 Common Core State Standards for Mathematics

Operations and Algebraic Thinking (1.OA)

Represent and solve problems involving addition and subtraction.

1. Use addition and subtraction within 20 to solve word problems involving situations of adding to, taking from, putting together, taking apart, and comparing, with unknowns in all positions, e.g., by using objects, drawings, and equations with a symbol for the unknown number to represent the problem.

Lesson	43
Exercise	43.9

Operations and Algebraic Thinking (1.OA)

Represent and solve problems involving addition and subtraction.

2. Solve word problems that call for addition of three whole numbers whose sum is less than or equal to 20, e.g., by using objects, drawings, and equations with a symbol for the unknown number to represent the problem.

This standard is addressed in the following activities of the Student Practice Software:

- **Block 6:** Activity 4
- **Block 6:** Activity 6

Operations and Algebraic Thinking (1.OA)

Understand and apply properties of operations and the relationship between addition and subtraction.

3. Apply properties of operations as strategies to add and subtract. Examples: If 8 + 3 = 11 is known, then 3 + 8 = 11 is also known. (Commutative property of addition.) To add 2 + 6 + 4, the second two numbers can be added to make a ten, so 2 + 6 + 4 = 2 + 10 = 12. (Associative property of addition.)

Lesson	41	42	43	44	46	47	48	49	50	51
Exercise	IW41.12	IW42.11	IW43.10	IW44.11	IW46.10	47.3	48.4	49.4, IW49.10	50.3, 50.7, IW50.11	51.7, IW51.10

Lesson	52	53	54	55	56	57	58	61	62	63
Exercise	IW52.9	IW53.10	54.3	55.5	56.1	57.2	58.2	61.1	IW62.10	63.1

Lesson	64	65	66	67	68	70	77	79	80
Exercise	64.1, IW64.10	65.2	66.1	67.1, IW67.10	68.1	70.1	IW77.9	79.1	80.1

Operations and Algebraic Thinking (1.OA)

Understand and apply properties of operations and the relationship between addition and subtraction.

4. Understand subtraction as an unknown-addend problem. For example, subtract 10 – 8 by finding the number that makes 10 when added to 8.

Lesson	71	72	73	74	75
Exercise	71.6	72.6	73.6	74.6	75.7

Operations and Algebraic Thinking (1.OA)

Add and subtract within 20.

5. Relate counting to addition and subtraction (e.g., by counting on 2 to add 2).

Lesson	41	42	47	48	49	50	51	52	53	54
Exercise	41.2	42.2	47.2	48.2	49.3, 49.5	50.6	51.6	52.6	53.1, 53.5	54.2, 54.7

Lesson	55	56	57	58	59	62	63	64	65	66
Exercise	55.7	56.8	57.8	58.6	59.9	62.1	63.7	64.6	65.8	66.7

Lesson	67	68	69	70	71	72	73	74	75	76
Exercise	67.7	68.8	69.7	70.6	71.6, 71.9	72.6, 72.8	73.6, IW73.10	74.6, IW74.10	75.7, IW75.9	IW76.9

Lesson	77	78	79	80
Exercise	IW77.9	IW78.9	IW79.9	IW80.9

Operations and Algebraic Thinking (1.OA)

Add and subtract within 20.

6. Add and subtract within 20, demonstrating fluency for addition and subtraction within 10. Use strategies such as counting on; making ten (e.g., 8 + 6 = 8 + 2 + 4 = 10 + 4 = 14); decomposing a number leading to a ten (e.g., 13 – 4 = 13 – 3 – 1 = 10 – 1 = 9); using the relationship between addition and subtraction (e.g., knowing that 8 + 4 = 12, one knows 12 – 8 = 4); and creating equivalent but easier or known sums (e.g., adding 6 + 7 by creating the known equivalent 6 + 6 + 1 = 12 + 1 = 13).

Lesson	41	42	43	44	45	46	47	48	49	50
Exercise	41.1, 41.4, 41.6, 41.11, IW41.12	42.1, 42.4, 42.6, 42.10	43.1, 43.4, 43.6, 43.8, 43.9, IW43.10	44.2, 44.5, 44.6, 44.8, 44.10, IW44.11	45.2, 45.3, 45.5, 45.6, 45.8, 45.10, IW45.11	46.1, 46.4, 46.5, 46.7, 46.9	47.1, 47.3, 47.5, 47.7, 47.9, IW47.10	48.1, 48.3, 48.5, 48.6, 48.9, 48.10, IW48.11	49.1, 49.4, 49.5, 49.6, 49.8, IW49.10	50.1, 50.3, 50.5, 50.6, 50.7, 50.9, 50.10, IW50.11

Lesson	51	52	53	54	55	56	57	58	59	60
Exercise	51.1, 51.3, 51.4, 51.5, 51.6, 51.7, IW51.10	52.1, 52.3, 52.4, 52.5, 52.6, 52.8, IW52.9	53.2, 53.4, 53.5, 53.6, 53.9, IW53.10	54.1, 54.3, 54.6, 54.7, 54.8, IW54.10	55.1, 55.3, 55.5, 55.7, 55.8, IW55.11	56.1, 56.4, 56.7, 56.8, IW56.10	57.2, 57.5, 57.9	58.2, 58.5, 58.9, IW58.10	59.5, 59.7, IW59.11	60.1, 60.6, 60.9, IW60.11

Lesson	61	62	63	64	65	66	67	68	69	70
Exercise	61.1, 61.5, 61.7, IW61.10	62.1, 62.5, 62.9, IW62.10	63.1, 63.4, 63.6, 63.7, IW63.10	64.1, 64.4, 64.6, IW64.10	65.2, 65.6, 65.8, IW65.10	66.1, 66.5, 66.7, 66.8, IW66.10	67.1, 67.6, 67.7, 67.9, IW67.10	68.1, 68.5	69.1, 69.5, 69.8, IW69.10	70.2, 70.5, 70.8, IW70.10

Lesson	71	72	73	74	75	76	77	78	79	80
Exercise	71.1, 71.5, 71.6, 71.8, IW71.10	72.1, 72.4, 72.6, 72.9, IW72.10	73.1, 73.4, 73.8, IW73.10	74.1, 74.4, 74.8, IW74.10	75.1, 75.5, 75.7, 75.8	76.1, 76.4, 76.8, IW76.9	77.2, 77.3, 77.7, IW77.9	78.1, 78.2, 78.7	79.1, 79.2, 79.5, 79.7	80.1, 80.4, 80.6, 80.7

Operations and Algebraic Thinking (1.OA)

Work with addition and subtraction equations.

7. Understand the meaning of the equal sign, and determine if equations involving addition and subtraction are true or false. For example, which of the following equations are true and which are false? $6 = 6$, $7 = 8 - 1$, $5 + 2 = 2 + 5$, $4 + 1 = 5 + 2$.

Lesson	62	63
Exercise	62.1	63.7

Operations and Algebraic Thinking (1.OA)

Work with addition and subtraction equations.

8. Determine the unknown whole number in an addition or subtraction equation relating to three whole numbers. For example, determine the unknown number that makes the equation true in each of the equations $8 + ? = 11$, $5 = \_\_ - 3$, $6 + 6 = \_\_$.

Lesson	41	42	43	44	45	46	47	48	49	50
Exercise	41.4, 41.6, 41.8, 41.9, 41.11, IW41.12	42.3, 42.6, 42.7, 42.8, 42.9, 42.10, IW42.11	43.1, 43.4, 43.5, 43.7, 43.8, 43.9, IW43.10	44.2, 44.6, 44.8, 44.9, 44.10, IW44.11	45.2, 45.6, 45.7, 45.9, 45.10, IW45.11	46.1, 46.5, 46.6, 46.9, IW46.10	47.1, 47.5, 47.8, 47.9, IW47.10	48.3, 48.5, 48.6, 48.8, 48.10, IW48.11	49.1, 49.6, 49.8, 49.9, IW49.10	50.1, 50.6, 50.8, 50.9, 50.10, IW50.11

Lesson	51	52	53	54	55	56	57	58	59	60
Exercise	51.1, 51.4, 51.5, 51.6, 51.8, IW51.10	52.1, 52.3, 52.4, 52.6, 52.7, 52.8, IW52.9	53.2, 53.4, 53.5, 53.6, 53.8, 53.9, IW53.10	54.1, 54.6, 54.7, 54.8, 54.9	55.1, 55.3, 55.6, 55.7, 55.8, IW55.11	56.4, 56.6, 56.7, 56.8, 56.9, IW56.10	57.5, 57.8, 57.9, 57.10, IW57.11	58.5, 58.6, 58.7, 58.9, IW58.10	59.5, 59.7, 59.8, 59.9, 59.10, IW59.11	60.6, 60.7, 60.9, 60.10, IW60.11

Lesson	61	62	63	64	65	66	67	68	69	70
Exercise	61.1, 61.7, 61.9, IW61.10	62.2, 62.5, 62.7, 62.9, IW62.10	63.4, 63.6, 63.7, 63.8, IW63.10	64.4, 64.5, 64.7, 64.9, IW64.10	65.6, 65.7, 65.8, 65.9, IW65.10	66.5, 66.6, 66.7, 66.8, 66.9, IW66.10	67.6, 67.7, 67.8, 67.9, IW67.10	68.5, 68.7, 68.8, 68.9, IW68.10	69.5, 69.7, 69.8, 69.9, IW69.10	70.4, 70.5, 70.6, 70.7, 70.8, 70.9, IW70.10

Lesson	71	72	73	74	75	76	77	78	79	80
Exercise	71.5, 71.6, 71.7, 71.8, 71.9, IW71.10	72.4, 72.6, 72.7, 72.8, 72.9, IW72.10	73.3, 73.4, 73.6, 73.7, 73.8, 73.9, IW73.10	74.2, 74.4, 74.6, 74.8, 74.9, IW74.10	75.3, 75.7, 75.8, IW75.9	76.2, 76.6, 76.8, IW76.9	77.5, 77.7, 77.8, IW77.9	78.2, 78.4, 78.6, 78.8, IW78.9	79.4, 79.5, 79.7, 79.8, IW79.9	80.4, 80.5, 80.6, 80.7, 80.8, IW80.9

Number and Operations in Base Ten (1.NBT)

Extend the counting sequence.

1. Count to 120, starting at any number less than 120. In this range, read and write numerals and represent a number of objects with a written numeral.

Lesson	41	42	43	44	45	46	47	48	49	50
Exercise	41.1, 41.3, 41.4, 41.5, 41.6, 41.7, 41.8, 41.9, 41.10, 41.11, IW41.12	42.1, 42.3, 42.4, 42.5, 42.6, 42.7, 42.8, 42.9, 42.10, IW42.11	43.1, 43.2, 43.4, 43.5, 43.6, 43.7, 43.8, 43.9, IW43.10	44.2, 44.3, 44.4, 44.5, 44.6, 44.7, 44.8, 44.9, 44.10, IW44.11	45.1, 45.2, 45.4, 45.5, 45.6, 45.7, 45.8, 45.9, 45.10, IW45.11	46.1, 46.3, 46.4, 46.5, 46.6, 46.7, 46.8, 46.9, IW46.10	47.1, 47.3, 47.4, 47.5, 47.6, 47.7, 47.8, 47.9, IW47.10	48.1, 48.3, 48.4, 48.5, 48.6, 48.7, 48.8, 48.9, 48.10, IW48.11	49.1, 49.2, 49.4, 49.5, 49.6, 49.7, 49.8, 49.9, IW49.10	50.1, 50.2, 50.3, 50.4, 50.6, 50.7, 50.8, 50.9, 50.10, IW50.11

Lesson	51	52	53	54	55	56	57	58	59	60
Exercise	51.1, 51.2, 51.3, 51.4, 51.5, 51.6, 51.7, 51.8, 51.9, IW51.10	52.1, 52.2, 52.3, 52.4, 52.5, 52.6, 52.7, 52.8, IW52.9	53.2, 53.4, 53.5, 53.6, 53.7, 53.8, 53.9, IW53.10	54.1, 54.3, 54.4, 54.5, 54.6, 54.7, 54.8, 54.9, IW54.10	55.1, 55.3, 55.4, 55.5, 55.6, 55.7, 55.8, 55.9, 55.10, IW55.11	56.1, 56.3, 56.4, 56.5, 56.6, 56.7, 56.8, 56.9, IW56.10	57.1, 57.2, 57.4, 57.5, 57.6, 57.7, 57.8, 57.9, 57.10, IW57.11	58.1, 58.2, 58.5, 58.6, 58.7, 58.8, 58.9, IW58.10	59.1, 59.2, 59.3, 59.5, 59.6, 59.7, 59.8, 59.9, 59.10, IW59.11	60.1, 60.4, 60.6, 60.7, 60.8, 60.9, 60.10, IW60.11

Lesson	61	62	63	64	65	66	67	68	69	70
Exercise	61.1, 61.4, 61.5, 61.7, 61.8, 61.9, IW61.10	62.1, 62.2, 62.5, 62.7, 62.8, 62.9, IW62.10	63.1, 63.4, 63.6, 63.7, 63.8, IW63.10	64.1, 64.4, 64.5, 64.6, 64.7, 64.9, IW64.10	65.2, 65.6, 65.7, 65.8, 65.9, IW65.10	66.1, 66.5, 66.6, 66.7, 66.8, 66.9, IW66.10	67.1, 67.6, 67.7, 67.8, 67.9, IW67.10	68.1, 68.5, 68.7, 68.8, 68.9, IW68.10	69.1, 69.5, 69.7, 69.8, 69.9, IW69.10	70.1, 70.2, 70.4, 70.5, 70.6, 70.7, 70.8, 70.9, IW70.10

Lesson	71	72	73	74	75	76	77	78	79	80
Exercise	71.1, 71.3, 71.5, 71.6, 71.7, 71.8, 71.9, IW71.10	72.1, 72.4, 72.5, 72.6, 72.7, 72.8, 72.9, IW72.10	73.1, 73.3, 73.4, 73.5, 73.6, 73.7, 73.8, 73.9, IW73.10	74.1, 74.2, 74.4, 74.6, 74.7, 74.8, 74.9, IW74.10	75.1, 75.3, 75.5, 75.6, 75.7, 75.8, IW75.9	76.1, 76.2, 76.4, 76.6, 76.7, 76.8, IW76.9	77.2, 77.3, 77.5, 77.7, 77.8, IW77.9	78.1, 78.2, 78.4, 78.6, 78.7, 78.8, IW78.9	79.1, 79.2, 79.4, 79.5, 79.7, 79.8, IW79.9	80.1, 80.2, 80.4, 80.5, 80.6, 80.7, 80.8, IW80.9

Number and Operations in Base Ten (1.NBT)

Understand place value.

2. Understand that the two digits of a two-digit number represent amounts of tens and ones. Understand the following as special cases: a. 10 can be thought of as a bundle of ten ones — called a "ten." b. The numbers from 11 to 19 are composed of a ten and one, two, three, four, five, six, seven, eight, or nine ones. c. The numbers 10, 20, 30, 40, 50, 60, 70, 80, 90 refer to one, two, three, four, five, six, seven, eight, or nine tens (and 0 ones).

Lesson	41	42	43	44	45	46	47	48	49	50
Exercise	41.8, 41.10	42.5, IW42.11	43.2	44.4, 44.9	45.4, 45.9	46.3, 46.6	47.4	48.7, IW48.11	49.7, IW49.10	IW50.11

Lesson	51	52	53	54	55	56	57	58	59	60
Exercise	IW51.10	IW52.9	IW53.10	IW54.10	55.6, 55.9, IW55.11	56.5	IW57.11	58.7, IW58.10	59.10, IW59.11	IW60.11

Lesson	61	64	65	66	67	68	69	71	73	74
Exercise	IW61.10	64.7	65.7	66.6, IW66.10	67.3	68.2	69.3	IW71.10	73.3, IW73.10	74.2

Lesson	75	76	77	78	79	80
Exercise	75.3, IW75.9	76.2	77.5	78.4	79.4	80.5

Number and Operations in Base Ten (1.NBT)

Understand place value.

3. Compare two two-digit numbers based on meanings of the tens and ones digits, recording the results of comparisons with the symbols >, =, and <.

This standard is addressed in the following activities of the Student Practice Software:

- **Block 4:** Activity 4
- **Block 5:** Activity 6

Number and Operations in Base Ten (1.NBT)

Use place value understanding and properties of operations to add and subtract.

4. Add within 100, including adding a two-digit number and a one-digit number, and adding a two-digit number and a multiple of 10, using concrete models or drawings and strategies based on place value, properties of operations, and/or the relationship between addition and subtraction; relate the strategy to a written method and explain the reasoning used. Understand that in adding two-digit numbers, one adds tens and tens, ones and ones; and sometimes it is necessary to compose a ten.

Lesson	41	42	43	44	45	46	47	48	49	50
Exercise	41.1, 41.4, 41.6, 41.8, 41.9, 41.11, IW41.12	42.1, 42.3, 42.4, 42.6, 42.9, 42.10, IW42.11	43.1, 43.3, 43.4, 43.6, 43.8, 43.9, IW43.10	44.2, 44.3, 44.5, 44.6, 44.7, 44.9, 44.10, IW44.11	45.2, 45.3, 45.5, 45.6, 45.7, 45.8, 45.9, IW45.11	46.4, 46.5, 46.7, 46.8, IW46.10	47.1, 47.3, 47.5, 47.7, 47.8, 47.9, IW47.10	48.1, 48.2, 48.3, 48.5, 48.8, 48.9, IW48.11	49.4, 49.5, 49.6, 49.8, IW49.10	50.3, 50.5, 50.6, 50.7, 50.9, 50.10, IW50.11

Lesson	51	52	53	54	55	56	57	58	59	60
Exercise	51.1, 51.3, 51.4, 51.5, 51.6, 51.7, 51.8, IW51.10	52.1, 52.3, 52.4, 52.5, 52.6, 52.8, IW52.9	53.1, 53.2, 53.4, 53.5, 53.6, 53.9, IW53.10	54.1, 54.2, 54.3, 54.6, 54.7, 54.8, 54.9, IW54.10	55.1, 55.3, 55.5, 55.7, 55.8, IW55.11	56.1, 56.4, 56.7, 56.8, 56.9, IW56.10	57.2, 57.5, 57.8, 57.9, IW57.11	58.2, 58.6, 58.9, IW58.10	59.5, 59.9, IW59.11	60.1, 60.6, IW60.11

Lesson	61	62	63	64	65	66	67	68	69	70
Exercise	61.1, 61.5, IW61.10	62.1, 62.7, 62.9, IW62.10	63.1, 63.6, 63.7, IW63.10	64.1, 64.6, 64.7, IW64.10	65.2, 65.7, 65.8, IW65.10	66.1, 66.6, 66.7, 66.8, IW66.10	67.1, 67.6, 67.7, 67.9, IW67.10	68.1, 68.5, 68.8, IW68.10	69.1, 69.5, 69.7, IW69.10	70.2, 70.6, 70.8, IW70.10

Lesson	71	72	73	74	75	76	77	78	79	80
Exercise	71.1, 71.8, 71.9, IW71.10	72.1, 72.8, 72.9, IW72.10	73.1, 73.3, 73.8, IW73.10	74.1, 74.2, 74.4, IW74.10	75.3, 75.5, 75.8, IW75.9	76.1, 76.2, 76.6, IW76.9	77.5, 77.7, IW77.9	78.1, 78.4, 78.7, 78.8, IW78.9	79.1, 79.4, 79.5, IW79.9	80.1, 80.4, 80.5, IW80.9

Number and Operations in Base Ten (1.NBT)

Use place value understanding and properties of operations to add and subtract.

5. Given a two-digit number, mentally find 10 more or 10 less than the number, without having to count; explain the reasoning used.

Lesson	41	42	43	44	45	46	47	49	50	52
Exercise	41.9	42.3, IW42.11	IW43.10	IW44.11	IW45.11	IW46.10	47.3	IW49.10	50.5	52.1, 52.3

Lesson	53	54	55
Exercise	53.4	54.6	IW55.11

Number and Operations in Base Ten (1.NBT)

Use place value understanding and properties of operations to add and subtract.

6. Subtract multiples of 10 in the range 10–90 from multiples of 10 in the range 10–90 (positive or zero differences), using concrete models or drawings and strategies based on place value, properties of operations, and/or the relationship between addition and subtraction; relate the strategy to a written method and explain the reasoning used.

Lesson	54
Exercise	54.2

Measurement and Data (1.MD)

Measure lengths indirectly and by iterating length units.

1. Order three objects by length; compare the lengths of two objects indirectly by using a third object.

This standard is first addressed in **Lesson 93.**

Measurement and Data (1.MD)

Measure lengths indirectly and by iterating length units.

2. Express the length of an object as a whole number of length units, by laying multiple copies of a shorter object (the length unit) end to end; understand that the length measurement of an object is the number of same-size length units that span it with no gaps or overlaps. Limit to contexts where the object being measured is spanned by a whole number of length units with no gaps or overlaps.

This standard is addressed in the following activities of the Student Practice Software:

- **Block 3:** Activity 6
- **Block 6:** Activity 5

Measurement and Data (1.MD)

Measure lengths indirectly and by iterating length units.

3. Tell and write time in hours and half-hours using analog and digital clocks.

This standard is first addressed in **Lesson 91.**

Measurement and Data (1.MD)

Measure lengths indirectly and by iterating length units.

4. Organize, represent, and interpret data with up to three categories; ask and answer questions about the total number of data points, how many in each category, and how many more or less are in one category than in another.

This standard is first addressed in **Lesson 122.**

Geometry (1.G)

Reason with shapes and their attributes.

1. Distinguish between defining attributes (e.g., triangles are closed and three-sided) versus non-defining attributes (e.g., color, orientation, overall size); build and draw shapes to possess defining attributes.

Lesson	61	62	63	64	65	66	67	68	69	70
Exercise	61.2	62.4	63.3	64.2	65.4	66.4	67.5	68.4	69.2, 69.4	70.3

Lesson	71	72	73	74	75	76	77
Exercise	71.2	72.2	73.3	74.3	75.2	76.3	77.1

Geometry (1.G)

Reason with shapes and their attributes.

2. Compose two-dimensional shapes (rectangles, squares, trapezoids, triangles, half-circles, and quarter-circles) or three-dimensional shapes (cubes, right rectangular prisms, right circular cones, and right circular cylinders) to create a composite shape, and compose new shapes from the composite shape.

Lesson	69	70	75	76	77	78	79	80
Exercise	69.4	70.3	75.2	76.5	77.6	78.5	79.6	IW80.9

Geometry (1.G)

Reason with shapes and their attributes.

3. Partition circles and rectangles into two and four equal shares, describe the shares using the words halves, fourths, and quarters, and use the phrases half of, fourth of, and quarter of. Describe the whole as two of, or four of the shares. Understand for these examples that decomposing into more equal shares creates smaller shares.

This standard is first addressed in **Lesson 117.**

Standards for Mathematical Practice and Connecting Math Concepts

Connecting Math Concepts: Comprehensive Edition is a six-level series that is fully aligned with the Common Core State Standards for kindergarten through fifth grade. As its name implies, *Connecting Math Concepts* is designed to bring students to an understanding of mathematical concepts by making connections between central and generative concepts in the school mathematics curriculum, as defined by the Common Core State Standards (CCSS) for Mathematical Content. This document illustrates some of those connections as they relate to the eight CCSS for Mathematical Practices. Examples will be provided from each of the six levels of *Connecting Math Concepts (CMC)*, to illustrate how students engage in activities representative of the eight CCSS for Mathematical Practices.

MP1: MAKE SENSE OF PROBLEMS AND PERSEVERE IN SOLVING THEM.

CMC Level C (Grade 2): Comparison Story Problems

Students in the primary grades are just beginning to make sense of situations that can be expressed mathematically. Care must be taken to give them a way of thinking about situations that is not overly simplistic, resulting in a shallow understanding of the circumstance. For example, when considering two values in a statement such as *Sophia is 2 years older than Isabel,* students are likely to interpret the situation as implying addition because of the word *more.* Conversely, they are likely to interpret a statement such as *Isabel is 2 years younger than Sophia* as subtraction. The comparison statements do not imply either operation; they simply identify the larger and smaller of the two values being compared. Given one value, we can find the other by addition or subtraction. If the larger value is given, we subtract 2 (the difference) to find the smaller value. If the smaller value is given, we add the difference to find the larger value.

Students in *CMC* are taught to use this reasoning to make sense of word problems that compare. For example:

> Isabel is 2 years younger than Sophia. Isabel is 8 years old. How old is Sophia?

Rather than jumping into a solution attempt, students are taught to analyze and diagram the relationship between the two values, identify and substitute the value given in the problem, and base their solution strategy on the value that is not given. For this example, students first represent the relationship between the two values. Initial letters represent the values. The larger value is written at the end of a number-family arrow, with the smaller value and the difference shown on the arrow:

$$\xrightarrow[\quad 2 \quad]{\quad I \quad} S$$

The value given in the problem replaces the letter in the diagram:

$$\xrightarrow[\quad 2 \quad]{\quad \overset{8}{I} \quad} S$$

The smaller value is given, so we add to find Sophia's age:

$$\xrightarrow[\quad 2 \quad]{\quad \overset{8}{I} \quad} S \qquad \begin{array}{r} 2 \\ + 8 \\ \hline 1\,0 \end{array}$$

Here is the first part of the exercise from Level C Lesson 48, illustrating how the students apply the strategy.

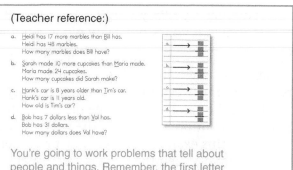

(Teacher reference:)

a. Heidi has 17 more marbles than Bill has.
   Heidi has 48 marbles.
   How many marbles does Bill have?

b. Sarah made 10 more cupcakes than Maria made.
   Maria made 24 cupcakes.
   How many cupcakes did Sarah make?

c. Hank's car is 8 years older than Tim's car.
   Hank's car is 11 years old.
   How old is Tim's car?

d. Bob has 7 dollars less than Val has.
   Bob has 31 dollars.
   How many dollars does Val have?

You're going to work problems that tell about people and things. Remember, the first letter of each name is underlined.

b. Touch problem A. ✔
• Heidi has 17 more marbles than Bill has.
  Say the sentence. (Signal.) *Heidi has 17 more marbles than Bill has.*
• What's the letter for Heidi? (Signal.) *H.*
• What's the letter for Bill? (Signal.) *B.*
  Heidi has 17 more marbles than Bill has.
• Make the family with two letters and the number. ✔
  (Display:)   [48:8A]

a.   _17___B_→H

Here's what you should have.

c. The next sentence in the problem says: Heidi has 48 marbles.
• Put a number for Heidi in the family. Then figure out how many marbles Bill has. Write that number in the family.
  (Observe students and give feedback.)

d. Check your work.
   (Add to show:)   [48:8B]

a.   _17__ ³¹Ƶ̷→¹̸/₁   ³¹48   48
                                −17
                                ‾‾‾‾
                                 31

Here's what you should have.
• The problem asks: How many marbles does Bill have? What's the answer? (Signal.) *31.*

from Lesson 48, Exercise 8

Note that three different types of problems are worked in this problem set:

The smaller-value unknown with "more" language (items a and c), the bigger-value unknown with "more" language (item b), and the bigger-value unknown with "less" language (item d). These types are described in Table 1 in the *Common Core State Standards for Mathematics* document (excerpt below).

Bigger Unknown	Smaller Unknown
(Version with "more"): Julie has three more apples than Lucy. Lucy has two apples. How many apples does Julie Have?	(Version with "more"): Julie has three more apples than Lucy. Julie has five apples. How many apples does Lucy Have?
(Version with "fewer"): Lucy has 3 fewer apples than Julie. Lucy has two apples. How many apples does Julie have? $2 + 3 = ?, 3 + 2 = ?$	(Version with "fewer"): Lucy has 3 fewer apples than Julie. Julie has five apples. How many apples does Lucy have? $5 − 3 = ?, ? + 3 = 5$

Table 1 from the *Common Core State Standards for Mathematics* describes 15 different addition-subtraction situations, all of which students can solve with variations of a number-family strategy by the end of CMC Level C. Foundation work for number families is illustrated in Standard MP7.

MP2: REASON ABSTRACTLY AND QUANTITATIVELY.

CMC Level F (Grade 5): Algebraic Translation

As students progress thorough the levels of *CMC* they make sense of quantities and their relationships through a variety of representations. The earliest representations are pictorial and countable, but by Level C, students frequently use more abstract letter representations for quantities, relationships, and units named in problem situations. By Level E, these problem situations are extended to include the four basic operations, ratio and proportion, and unit conversion.

Students in *CMC Level F* make sense of quantities and their relationships through extensive work with algebraic translation, which they apply to a variety of word-problem contexts, including fractions of a group and probability.

For example:

- 2/3 of the cats are awake. 6 cats are sleeping. How many cats are there?
- There are 8 marbles in a bag; 5 are green. If you take 24 trials at pulling a marble from the bag without looking, how many times would you expect to draw a marble that is not green?

Students are well prepared to *decontextualize* and manipulate symbols independently through initial work with solving letter equations. They also can *contextualize* by attending carefully to the details of the problem to be solved in order to discriminate which specific letter equation a given problem requires.

Early work equips students with the algebraic skills to solve letter equations of the form 2/5 R = M with a substitution for either letter.

Here are the solution steps when R = 10:

$$\frac{2}{5} R = M$$
$$\frac{2}{5} (10) = M$$
$$\frac{20}{5} = M = 4$$

Here are the solution steps when M = 90:

$$\frac{2}{5} R = M$$
$$\frac{2}{5} R = 90$$
$$\left(\frac{5}{2}\right) \frac{2}{5} R = 90 \left(\frac{5}{2}\right)$$
$$R = \frac{450}{2} = 225$$

Students are first taught the definition and application of a reciprocal and the equality principle that states the following: If we change one side of an equation, we must change the other side in the same way.

Here is part of an exercise from Lesson 38.

h. (Display:)   [38:3H]

$$\frac{5}{4} P = 10$$
$$1 P =$$

- Read the problem. (Signal.) *5/4 P = 10.* We have to figure out what 1 P equals.
- What do we change 5/4 into? (Signal.) *1.*
- So what do we multiply 5/4 by? (Signal.) *4/5.*
- What do we multiply the other side by? (Signal.) *4/5.*
  (Repeat until firm.)
  (Add to show:)   [38:3I]

$$\left(\frac{4}{5}\right) \frac{5}{4} P = \frac{10}{1} \left(\frac{4}{5}\right)$$
$$1 P =$$

i. (Point right.) Say the problem for this side. (Signal.) *10 × 4/5.*
- Raise your hand when you know the fraction answer. ✔
- What's the fraction? (Signal.) *40/5.*
  (Add to show:)   [38:3J]

$$\left(\frac{4}{5}\right) \frac{5}{4} P = \frac{10}{1} \left(\frac{4}{5}\right)$$
$$1 P = \frac{40}{5}$$

- Raise your hand when you know the number 1 P equals. ✔
- What does 1 P equal? (Signal.) *8.*
  (Add to show:)   [38:3K]

$$\left(\frac{4}{5}\right) \frac{5}{4} P = \frac{10}{1} \left(\frac{4}{5}\right)$$
$$1 P = \frac{40}{5} = \boxed{8}$$

j. Remember, multiply both sides by the reciprocal. Then figure out what the letter equals.

from Lesson 38, Exercise 3

Students have extensive practice with substitution and solving letter equations prior to the introduction of word problem applications. This level of proficiency enables them to manipulate symbols confidently once they have analyzed the problem and represented it with a letter equation.

Students first work with sentences that describe a fraction of a group, and write the letter equation. Here are two sentences and the corresponding equations from Lesson 46:

$\frac{3}{4}$ of the dogs were hungry.

$$\frac{3}{4} d = h$$

$\frac{1}{9}$ of the students wore coats.

$$\frac{1}{9} s = c$$

In subsequent lessons, students work the simplest type of word problem, where a number is given for one of the letters, and students solve for the other letter to answer the question the problem asks.

For example:

- 2/5 of the rabbits were white. There were 25 rabbits. How many white rabbits were there?

$$\frac{2}{5} r = w$$

$$\frac{2}{5}(25) = w = \frac{50}{5} = 10$$

$$\boxed{10 \text{ white rabbits}}$$

Here's part of an exercise from Level F Lesson 78, where students work a complete problem that asks two questions. They assess their initial solution and refer back to the problem to ascertain which question relates to the letter solution, and which question remains to be answered.

b. Problem A: 2/7 of the dogs were sleeping. 8 dogs were sleeping. How many dogs were awake? How many dogs were there?
- Write the letter equation. Replace one of the letters with a number. Stop when you've done that much. ✔
- Everybody, read the letter equation. (Signal.) 2/7 D = S.
- Read the equation with a number. (Signal.) 2/7 D = 8.
(Display:)   [78:4C]

a. $\frac{2}{7} d = s$

     $\frac{2}{7} d = 8$

Here's what you should have.
- Work the problem and write the unit name in the answer. Remember to simplify before you multiply. (Observe students and give feedback.)
c. Check your work.
- What does D equal? (Signal.) 28.
- Which question does that answer? (Signal.) How many dogs were there?
(Add to show:)   [78:4D]

a. $\frac{2}{7} d = s$

$$\left(\frac{7}{2}\right)\frac{2}{7} d = \overset{4}{\cancel{8}}\left(\frac{7}{\cancel{2}}\right)$$

$$d = 28$$

$$\boxed{28 \text{ dogs}}$$

Here's what you should have.
d. Now figure the answer to the other question. ✔ The other question is: How many dogs were awake?
- Everybody, say the subtraction problem you worked. (Signal.) 28 – 8.
- What's the answer? (Signal.) 20.
(Display:)   [78:4E]

a. $\frac{2}{7} d = s$

$$\left(\frac{7}{2}\right)\frac{2}{7} d = \overset{4}{\cancel{8}}\left(\frac{7}{\cancel{2}}\right)$$

$$d = 28$$

$$\begin{array}{r} 2\,8 \\ -\ \ 8 \\ \hline 2\,0 \end{array}$$

$$\boxed{28 \text{ dogs}} \quad \boxed{2\,0 \text{ awake dogs}}$$

Here's what you should have.

from Lesson 78, Exercise 4

On later lessons, students work more advanced problems where the problem gives a fraction for one part of the group but gives a number for the other part of the group. For example:

- 2/7 of the dogs were running. 20 dogs were *not* running. How many dogs were running? How many dogs were there in all?

Students must analyze the problem carefully to discriminate it from the earlier problem types that involve only two names (e.g., dogs and running dogs). The number given is for dogs *not running,* so rather than writing the basic equation 2/7 d = r, students write the complementary letter equation 5/7 d = n. (5/7 of the dogs were *not running.*)

Students first practice discriminating between sentence pairs that give two names and those that give three names.

For example:

- 3/8 of the children are boys. There are 15 boys. (2 names: 3/8 c = b)
- 3/8 of the children are boys. There are 15 girls. (3 names: 5/8 c = g)

Here is part of the exercise from Lesson 90, where students solve a complete problem involving three names.

g. (Display:)   [90:7C]

> $\frac{2}{7}$ of the dogs were running. 20 dogs were not running. How many dogs were running? How many dogs were there in all?

New problem: 2/7 of the dogs were running. 20 dogs were not running. How many dogs were running? How many dogs were there in all?
This problem asks two questions. You work it the same way you work the other problems with three names. You write the equation for the name that has a number and solve the equation. That answers one of the questions.

h. Write the equation and solve it. Write the answer with a unit name. Stop when you've done that much.
(Observe students and give feedback.)

i. I'll read the questions the problem asks. You'll tell me which question you can now answer:
How many dogs were running?
How many dogs were there in all?
- Which question can you answer? (Signal.)
How many dogs were there in all?
- What's the answer? (Signal.) *28 dogs.*

j. (Add to show:)   [90:7D]

> $\frac{2}{7}$ of the dogs were running. 20 dogs were not running. How many dogs were running? How many dogs were there in all?
>
> $$\frac{5}{7} d = n$$
> $$\left(\frac{7}{5}\right) \frac{5}{7} d = \overset{4}{\cancel{20}} \left(\frac{7}{\cancel{5}}\right)$$
> $$d = 28$$
>
> $$\boxed{28 \text{ dogs}}$$

Here's what you should have.

k. The other question is: How many dogs were running?
Now you can figure out how many dogs were running.
- Raise your hand when you can say the problem. ✔
- Say the problem. (Signal.) *28 – 20.*

l. Figure out the answer and write the unit name—running dogs. ✔
- Everybody, how many dogs were running? (Signal.) *8 running dogs.*

from Lesson 90, Exercise 7

Once students have set up the correct equation, they follow familiar solution steps. They figure out how many dogs there were in all (which is the answer to the second question). They then subtract to figure out the number of dogs that were running.

The strategy builds the habit of a coherent representation of the problem at hand (a familiar equation form), requires students to consider the units for the quantities represented by the letters as they solve the equation, consider both questions the problem asks, and then determine which has been answered and which remains to be answered.

Coherent representation is further strengthened in later applications to new domains such as probability. Students first learn to create a probability fraction based on the composition of the set. This fraction gives the likelihood of a particular object being drawn from the set on any given trial. The following examples from lesson 96 require students to construct probability fractions for given sets of objects and to construct a set of objects given the probability fraction.

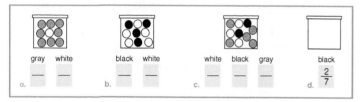

from Workbook Lesson 96, Part 1

For item a, students write the fraction 8/9 for gray and 1/9 for white. For item d, students draw 7 marbles, and shade 2 of them black.

With this background it is a simple transition to work problems that involve trials using the familiar equation form.

Here are two examples and the equations students work from Lesson 98.

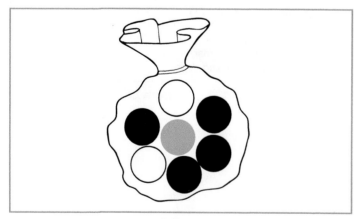

from Textbook Lesson 98, Part 2

- If you took trials until you pulled out four black marbles, about how many trials would you take?

$$\frac{4}{7} t = b$$

$$\frac{4}{7} t = 4$$

- If you take 98 trials, about how many black marbles would you expect to pull out?

$$\frac{4}{7} t = b$$

$$\frac{4}{7} (98) = b$$

Students already have the mathematical tools to solve the equations, and learning a wide range of applications for equation form builds flexibility in the application of those tools.

MP3: CONSTRUCT VIABLE ARGUMENTS AND CRITIQUE THE REASONING OF OTHERS.

CMC Level C (Grade 2): Column Addition/Subtraction

As students develop mathematically, they build a basis for constructing arguments to justify the mathematical procedures they use, or the steps they take to perform an algorithm. They are able to decide whether variations and deviations from the familiar progression of steps make sense. In a column addition or subtraction problem that involves regrouping, students should display an understanding of the underlying place value that conserves the numbers they are manipulating.

Students in *CMC Level C* work extensively with place value to build the conceptual basis for the steps they take in addition/subtraction regrouping problems, for example:

$$\begin{array}{r} \overset{1}{2}\,8 \\ +\ 4\,5 \\ \hline 7\,3 \end{array} \quad \text{or} \quad \begin{array}{r} \overset{8}{\cancel{9}}\overset{1}{2} \\ -\ 3\,6 \\ \hline 5\,6 \end{array}$$

Students are familiar with identifying the tens digit and ones digit of two-digit numbers and with the place value of each digit. When applied to column addition, the students simply write a 2-digit answer to the problem for the ones column as the "tens digit" and the "ones digit" in the appropriate order and in the appropriate column. Here's part of an early exercise from Lesson 21.

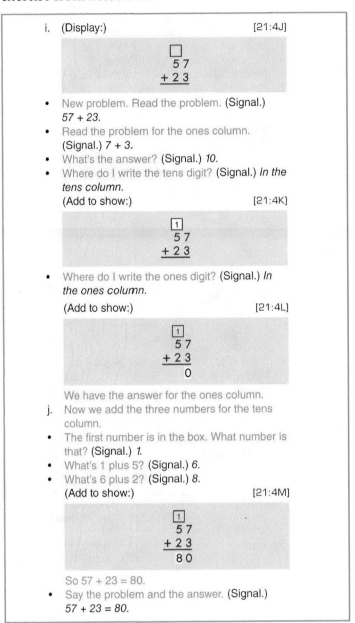

i. (Display:) [21:4J]

$$\square \\ 57 \\ +23$$

- New problem. Read the problem. (Signal.) *57 + 23.*
- Read the problem for the ones column. (Signal.) *7 + 3.*
- What's the answer? (Signal.) *10.*
- Where do I write the tens digit? (Signal.) *In the tens column.*
  (Add to show:) [21:4K]

$$\boxed{1} \\ 57 \\ +23$$

- Where do I write the ones digit? (Signal.) *In the ones column.*
  (Add to show:) [21:4L]

$$\boxed{1} \\ 57 \\ +23 \\ \hline 0$$

We have the answer for the ones column.
j. Now we add the three numbers for the tens column.
- The first number is in the box. What number is that? (Signal.) *1.*
- What's 1 plus 5? (Signal.) *6.*
- What's 6 plus 2? (Signal.) *8.*
  (Add to show:) [21:4M]

$$\boxed{1} \\ 57 \\ +23 \\ \hline 80$$

So 57 + 23 = 80.
- Say the problem and the answer. (Signal.) *57 + 23 = 80.*

from Lesson 21, Exercise 4

Having practiced this procedure to automaticity for more than 90 lessons, students are ready to discuss the steps they take and critique deviations from the procedure. Here is an exercise from Lesson 114.

a. (Display:) [114:4A]

$$46 \\ +39$$

I'm going to ask you a lot of questions about working this problem.
- Read the problem. (Signal.) *46 + 39.*
- Read the problem for the ones. (Signal.) *6 + 9.*
- What's the answer? (Signal.) *15.*
b. What if I just put 15 in the ones column?
  (Add to show:) [114:4B]

$$46 \\ +39 \\ \hline 15$$

- Can I do this? (Signal.) *No.*
- What's wrong with this? (Call on a student. Idea: *15 is too many to have in the ones column.*)
- What's the largest number we can have in the ones column? (Call on a student. *9.*)
c. The answer for the ones is 15. Everybody, say the place-value addition for 15. (Signal.) *10 + 5.*
- I can't write **both** digits in the ones column. Where do I write the **ones** digit of 15? (Signal.) *In the ones column.*
- Where do I write the **tens** digit of 15? (Signal.) *In the tens column.*
  (Change to show:) [114:4C]

$$46 \\ +39 \\ \hline 15$$

- (Point to **1.**) Can I write the tens digit here? (Signal.) *No.*
- Why not? (Call on a student. Idea: *You have to add it to the tens.*)
d. Where do I write it? (Call on a student. Idea: *At the top of the tens column.*)
  (Change to show:) [114:4D]

$$\overset{1}{4}6 \\ +39 \\ \hline 5$$

- Everybody, read the problem for the tens. (Signal.) *1 + 4 + 3.*
- Raise your hand when you know the answer. ✔
- What's 1 + 4 + 3? (Signal.) *8.*
  (Add to show:) [114:4E]

$$\overset{1}{4}6 \\ +39 \\ \hline 85$$

- Read the problem we started with and the answer. (Signal.) *46 + 39 = 85.*
- Say the place-value addition for the answer. (Signal.) *80 + 5 = 85.*

from Lesson 114, Exercise 4

A similar development is taught for subtraction. Before students work with column problems that require renaming the top number, student work on the conservation of two-digit values based on place value.

Here's an introductory exercise from Level C Lesson 26.

a. You're going to say the place-value fact for different numbers.
- Say the place-value fact for 53. (Signal.) *50 + 3 = 53.*
- Say the place-value fact for 29. (Signal.) *20 + 9 = 29.*
- Say the place-value fact for 70. (Signal.) *70 + 0 = 70.*
- Say the place-value fact for 71. (Signal.) *70 + 1 = 71.*
- Say the place-value fact for 17. (Signal.) *10 + 7 = 17.*
- (Repeat until firm.)
b. (Display:)   [26:5A]

$$30 + 6 = 36$$
$$40 + 8 = 48$$
$$70 + 1 = 71$$
$$50 + 3 = 53$$

c. (Point to **30.**) You're going to subtract 10 from this number.
- What's 30 – 10? (Signal.) *20.*
d. (Point to **40.**) You're going to subtract 10 from this number.
- Say the problem. (Signal.) *40 – 10.*
- What's the answer? (Signal.) *30.*

e. (Point to **70.**) You're going to subtract 10 from this number.
- Say the problem. (Signal.) *70 – 10.*
- What's the answer? (Signal.) *60.*
f. (Point to **50.**) You're going to subtract 10 from this number.
- Say the problem. (Signal.) *50 – 10.*
- What's the answer? (Signal.) *40.*
- (Repeat until firm.)
g. You've subtracted 10 from the tens number. You're going to add 10 to the ones number.
- (Point to **6.**) What's 10 + 6? (Signal.) *16.*
- (Point to **8.**) What's 10 + 8? (Signal.) *18.*
- (Point to **1.**) What's 10 + 1? (Signal.) *11.*
- (Point to **3.**) What's 10 + 3? (Signal.) *13.*
h. This time you're going to subtract 10 from the tens number and add that 10 to the ones number.
i. (Point to **30.**) What's 30 – 10? (Signal.) *20.* (Change to show:)   [26:5B]

$$30 + 6 = 36$$
$$20$$

- What's 10 + 6? (Signal.) *16.* (Add to show:)   [26:5C]

$$30 + 6 = 36$$
$$20 + 16 = 36$$

- Read the new place-value fact for 36. (Signal.) *20 + 16 = 36.*
j. (Display:)   [26:5D]

$$40 + 8 = 48$$

- (Point to **40.**) What's 40 – 10? (Signal.) *30.* (Add to show:)   [26:5E]

$$40 + 8 = 48$$
$$30$$

- What's 10 + 8? (Signal.) *18.* (Add to show:)   [26:5F]

$$40 + 8 = 48$$
$$30 + 18 = 48$$

- Read the new place-value fact for 48. (Signal.) *30 + 18 = 48.*

from Lesson 26, Exercise 5

Students then practice rewriting 2-digit numbers in isolation to show the "new fact." Here's part of the introduction from Lesson 33.

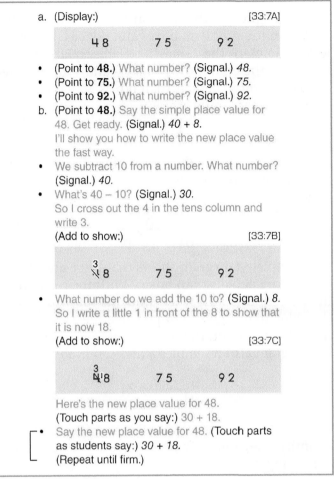

a. (Display:) [33:7A]

48   75   92

- (Point to **48**.) What number? (Signal.) *48.*
- (Point to **75**.) What number? (Signal.) *75.*
- (Point to **92**.) What number? (Signal.) *92.*
b. (Point to **48**.) Say the simple place value for 48. Get ready. (Signal.) *40 + 8.*
  I'll show you how to write the new place value the fast way.
- We subtract 10 from a number. What number? (Signal.) *40.*
- What's 40 − 10? (Signal.) *30.*
  So I cross out the 4 in the tens column and write 3.
  (Add to show:) [33:7B]

3
4̷8   75   92

- What number do we add the 10 to? (Signal.) *8.*
  So I write a little 1 in front of the 8 to show that it is now 18.
  (Add to show:) [33:7C]

3
4̷18   75   92

Here's the new place value for 48.
(Touch parts as you say:) *30 + 18.*
- Say the new place value for 48. (Touch parts as students say:) *30 + 18.*
(Repeat until firm.)

from Lesson 33, Exercise 7

Given this background and practice that continues for 15 lessons, students have a conceptual framework to make sense of and justify the steps in the subtraction algorithm, which begins on Lesson 43.

After the algorithm is established and practiced for 70 lessons, students also discuss and critique the steps in the algorithm and deviations from the procedure. Here is an exercise from Lesson 115.

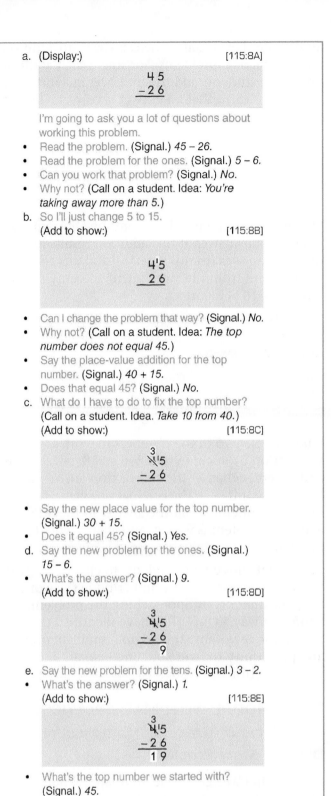

a. (Display:) [115:8A]

45
−26

I'm going to ask you a lot of questions about working this problem.
- Read the problem. (Signal.) *45 − 26.*
- Read the problem for the ones. (Signal.) *5 − 6.*
- Can you work that problem? (Signal.) *No.*
- Why not? (Call on a student. Idea: *You're taking away more than 5.*)
b. So I'll just change 5 to 15.
  (Add to show:) [115:8B]

4¹5
26

- Can I change the problem that way? (Signal.) *No.*
- Why not? (Call on a student. Idea: *The top number does not equal 45.*)
- Say the place-value addition for the top number. (Signal.) *40 + 15.*
- Does that equal 45? (Signal.) *No.*
c. What do I have to do to fix the top number? (Call on a student. Idea. *Take 10 from 40.*)
  (Add to show:) [115:8C]

3
4̷¹5
−26

- Say the new place value for the top number. (Signal.) *30 + 15.*
- Does it equal 45? (Signal.) *Yes.*
d. Say the new problem for the ones. (Signal.) *15 − 6.*
- What's the answer? (Signal.) *9.*
  (Add to show:) [115:8D]

3
4̷¹5
−26
9

e. Say the new problem for the tens. (Signal.) *3 − 2.*
- What's the answer? (Signal.) *1.*
  (Add to show:) [115:8E]

3
4̷¹5
−26
19

- What's the top number we started with? (Signal.) *45.*
- Say the problem we started with and the answer. (Signal.) *45 − 26 = 19.*

from Lesson 115, Exercise 8

Students in Level C also check answers to addition and subtraction problems. For single-digit addition problems, students add the numbers in a different order to confirm that an answer is correct. Given the problem and (incorrect) answer:

$$\begin{array}{r} 2 \\ 4 \\ + 9 \\ \hline 1\ 6 \end{array}$$

Students work the problem from the "bottom up" to check the answer. If this addition results in a different answer (in this case 15) students also work the problem from the "top down" to verify that the answer of 16 is incorrect.

Students also check answers to multi-digit problems using inverse operations. Given the problem and (incorrect) answer:

$$\begin{array}{r} 5\ 8 \\ + 2\ 1\ 8 \\ \hline 2\ 6\ 6 \end{array}$$

Students use two of the numbers to work a subtraction problem (either 266 – 218, or 266 – 58) and observe whether the subtraction answer is the third number in the original problem. If it is not, they conclude that the answer to the original addition problem is wrong, and they rework the problem to figure out the correct answer.

These strategies enable students to check and justify their answers, or evaluate and critique the work of other students by approaching the problem a "different way." Once they have identified that there is an error, reworking the original problem provides an opportunity to figure out and explain precisely where an error was made.

MP4: MODEL WITH MATHEMATICS.

CMC Level D (Grade 3): Multiplication/Division Story Problems

Students in *CMC Level D* learn to apply the mathematics they know to a range of word problems that represent situations arising in everyday life. This section will illustrate how students model problems solved by multiplication or division.

Students identify the important quantities in a situation and map their relationship with a number-family diagram showing one name that represents the product, and one name that represents a factor in a multiplicative relationship. This modeling parallels the work with addition/subtraction number families described in MP1 above.

Here's a basic problem:

> **Each box has 6 pencils. There are 24 pencils. How many boxes are there?**

Students represent the relationship between the items named in the problem: boxes and pencils. Initial letters represent the values. The larger quantity is written at the end of the number-family arrow. There are 6 times more pencils than boxes, so *p* is at the end of the number family arrow:

The smaller quantity *(b)* is written on the arrow, and the relationship number (6) is the first number in the diagram.

The completed diagram shows that there are six times more pencils than boxes. With the relationship modeled in this way, when students substitute for the value that is given (24 pencils), they see that they must divide to find the smaller quantity. (Note that the diagram also resembles a division problem).

Conversely, when students substitute for the smaller quantity, they multiply to find the larger quantity. For example:

> **Each box has 6 pencils. There are 50 boxes. How many pencils are there?**

The relationship that students map is the same as above:

Students substitute for b:

This diagram translates into a multiplication problem to figure out the larger quantity.

$$\begin{array}{r} 2\ 4 \\ \times\ \ 6 \\ \hline \end{array}$$

By carefully mapping the situation and relating the quantities before deciding which operation to perform, students can reliably solve problems that typically cause confusion for students. In the latter problem, students may read the word *each* and decide that the problem involves multiplication or division, but they are tempted to divide because the numbers given in the problem are from a familiar division fact (24 ÷ 6). By mapping the relationship before they work the problem, students in *CMC* are much less likely to make this mistake.

Students write number families from sentences for several lessons to establish the mapping strategy. They are then ready to work complete problems. Here's part of the exercise from Level D Lesson 72, where students work the following problems:

- Each dime is worth 10 cents. Tom had 8 dimes. How many cents did he have?
- Every chair has 4 legs. There are 12 legs. How many chairs are there?

b. Touch problem A in your textbook. ✔
    I'll read it: Each dime is worth 10 cents. Tom had 8 dimes. How many cents did he have?
- Raise your hand when you know which sentence tells about each or every. ✔
- Read that sentence. (Signal.) *Each dime is worth 10 cents.*
- Is the big number dimes or cents? (Signal.) *Cents.*
    Yes, there are more cents than dimes.
- Write two letters and a number in the number family.
    (Observe students and give feedback.)
    (Display:)   [72:6J]

Here's what you should have.
c. The problem gives another number for dimes.
- Look at the problem. Raise your hand when you know the number. ✔
- What's the number? (Signal.) *8.*
- Write the number for dimes in the family. ✔
    (Add to show:)   [72:6K]

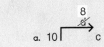

Here's what you should have.
d. Do you multiply or divide to find the missing number? (Signal.) *Multiply.*
- Start with the 2-digit number and say the problem you'll work. (Signal.) *10 × 8.*
- Write the problem next to the number family. Work the problem. Write the unit name in the answer.
    (Observe students and give feedback.)
- You worked the problem 10 times 8. What's the whole answer? (Signal.) *80 cents.*
    Yes, Tom had 80 cents.
e. Check your work.
    (Add to show:)   [72:6L]

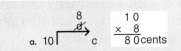

Here's what you should have.

f. Count down 4 lines on your lined paper. Write the letter B and a multiplication number family arrow. ✔
- Touch problem B. ✔
    I'll read the problem: Every chair has 4 legs. There are 12 legs. How many chairs are there?
- Raise your hand when you know which sentence tells about each or every. ✔
- Read that sentence. (Signal.) *Every chair has 4 legs.*
- Write two letters and a number in the family.
    (Observe children and give feedback.)
    (Display:)   [72:6M]

Here's what you should have.
g. The problem gives a number for one of the letters.
    Look at the problem. Raise your hand when you know the number. ✔
- What's the number? (Signal.) *12.*
- Is that number for chairs or legs? (Signal.) *Legs.*
- Write that number in the family. ✔
    (Add to show:)   [72:6N]

$$b. \ 4 \overset{c}{\underset{\cancel{x}}{\longrightarrow}} {\overset{12}{}}$$

Here's what you should have.
h. Say the problem you'll work to find the number of chairs. (Signal.) *12 divided by 4.*
- Write the problem next to the number family. Work the problem. Write the unit name in the answer.
    (Observe students and give feedback.)
- Read the problem and the whole answer. (Signal.) *12 divided by 4 = 3 chairs.*
    Yes, there were 3 chairs.
i. Check your work.
    (Add to show:)   [72:6O]

$$b. \ 4 \overset{c}{\underset{\cancel{x}}{\longrightarrow}} {\overset{12}{}} \quad \begin{array}{r} 3\,\text{chairs} \\ \hline 4\,\overline{)1\,2} \end{array}$$

Here's what you should have.

from Lesson 72, Exercise 6

Note that the same strategy accommodates money problems that convert a coin value into cents and vice versa. (For example: Tom had 80 cents in dimes. How many dimes did he have?)

Later in Level D, students learn to work multiplication and division problems that use the word "times." Here are two items and the student work from Lesson 91.

a. The yellow snake was 9 times as long as the red snake.
The red snake was 10 inches long.
How long was the yellow snake?

b. There were 6 times as many spoons as forks.
There were 66 spoons.
How many forks were there?

Item b results in a division problem. Again, representing the related values in the problem before choosing the operation reduces the likelihood that students will multiply instead of divide because the problem involves the word "times."

Another real-life application for the number-family mapping taught in Level D is unit conversion. Students make a number family for measurement facts. For example:

1 foot equals 12 inches.

Students identify that there are more inches than feet, so the number family relationship is:

Here are the facts that student map in Lesson 110:

- 1 week equals 7 days.
- 1 gallon equals 4 quarts.
- 1 quarter equals 25 cents.
- 1 pound equals 16 ounces.

Here's the first part of the exercise.

b. You're going to make number families for measurement facts.
- Touch and read fact A. (Signal.) *1 week = 7 days.*
- Are there more weeks or days? (Signal.) *Days.*
- So which is the big number? (Signal.) *Days.*
- Make the family for the fact about weeks and days.
(Observe students and give feedback.)
c. Check your work.
(Display:)   [110:4A]

Here's what you should have.
7 and weeks are small numbers. Days is the big number.
d. Read fact B. (Signal.) *1 gallon = 4 quarts.*
- Are there more gallons or quarts?
(Signal.) *Quarts.*
- So which is the name for the big number in the family? (Signal.) *Quarts.*
- Make the family.
(Observe students and give feedback.)
e. Check your work.
(Display:)   [110:4B]

Here's what you should have.
4 and gallons are small numbers. Quarts is the big number.

from Lesson 110, Exercise 4

On later lessons, students work complete problems. For example:

- 1 nickel equals 5 cents. How many nickels is 150 cents?
- 1 day equals 24 hours. How many hours is 9 days?

Students divide to find the number of nickels in the first item, and multiply to find the number of hours in the second item. Traditionally these problem types cause confusion for elementary students because they are counterintuitive. To solve the problems correctly, students *multiply* when the problem asks about the *smaller* unit, and *divide* when the problem asks about the *larger* unit. The number-family analysis provides students in *CMC* with a consistent and reliable way to tackle these and the full range of other situations that call for multiplication or division.

MP5: USE APPROPRIATE TOOLS STRATEGICALLY.

CMC Level E (Grade 4): Geometry

CMC teaches students self-reliance through the extensive use of conceptual strategies that can be applied using paper and pencil. Students generate solutions to a wide range of problems presented in the Textbook and Workbook. Tasks often move systematically from a workbook (more highly structured) to a textbook setting as students develop the conceptual tools to tackle problems more independently. For example, initial work with area and perimeter appears in the Workbook and only requires students to write the multiplication or addition problem. Here is the Workbook part and student work from Lesson 8.

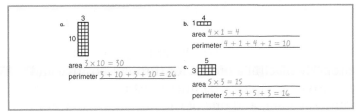

from Workbook Lesson 8 AK, Part 2

In later lessons, students respond to diagrams shown in the Textbook and work much more advanced problems. Below is a set of examples from Lesson 93 Textbook. For some problems, students find the area of a rectangle. For others they find the length of a side of the rectangle. They show their answers with appropriate linear or square units.

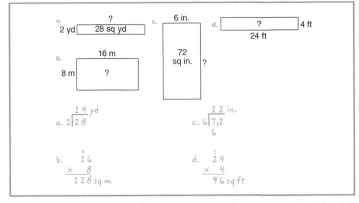

from Textbook Lesson 93 AK, Part 2

The work with area and perimeter culminates with story problems for which students sketch a rectangle to represent the problem. For example:

> a. A farmer wants to put a fence around a rectangular garden. The garden is 132 yards wide and 58 yards long. How many yards of fencing does the farmer need to put a fence around that garden?
>
> b. Jim has enough paint to cover 100 square feet of a wall. The wall he wants to paint is 8 feet tall. If he has just enough paint to cover the wall, how long is the wall?

from Textbook Lesson 126, Part 4

Here is part of the exercise.

> b. Read problem A. (Call on a student.) *A farmer wants to put a fence around a rectangular garden. The garden is 132 yards wide and 58 yards long. How many yards of fencing does the farmer need to put a fence around that garden?*
> - Tell me the length of the longest side of the rectangle. Get ready. (Signal.) *132 yards.*
> - What's the length of the other side? (Signal.) *58 yards.*
> - Make a sketch of the rectangle on your lined paper with the sides labeled. Make sure you label the longest side of your sketch with the biggest length.
> (Observe students and give feedback.)
> (Display:) [126:6A]

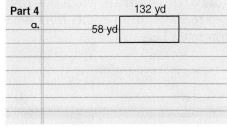

> Here's a sketch of the garden.
> - The question asks: How many yards of fencing does the farmer need to put a fence around that garden? Does that question ask about the area or perimeter? (Signal.) *Perimeter.*
> - Start with 132 and say the problem for finding the perimeter. Get ready. (Signal.) *132 + 58 + 132 + 58.*
> c. Read problem B. (Call on a student.) *Jim has enough paint to cover 100 square feet of a wall. The wall he wants to paint is 8 feet tall. If he has just enough paint to cover the wall, how long is the wall?*
> - The length of one of the sides of the wall is given. What's that length? (Signal.) *8 feet.*
> - Does the problem give another length or the area of the rectangle? (Signal.) *The area.*
> - What's the area of the rectangle? (Signal.) *100 square feet.*
> - Make a sketch of the rectangle on your lined paper with a side labeled and the area labeled.
> (Observe students and give feedback.)
> (Display:) [126:6B]

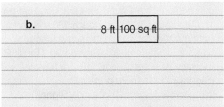

> Here's a sketch of the wall.
> - The question asks: How long is the wall? Say the problem for finding the length of the wall. Get ready. (Signal.) *100 ÷ 8.*
> d. Write the problem for each word problem below each rectangle and figure out the answer.
> (Observe students and give feedback.)

from Lesson 126, Exercise 6

Students work the addition or division problem and write the answer with a unit name.

Students are also familiar with tools appropriate to their grade, such as a protractor and ruler, which they use to explore angles, rays, line segments, and lines. They use a protractor first to measure given angles and then to construct angles. They use a ruler to complete the angle constructions and to graph straight lines on a coordinate system.

The protractor is introduced in Level E on Lesson 121. Before students physically use a protractor, they work with diagrams that show protractors properly placed. Students are often confused by the two rows of numbers on the protractor, so care is taken to teach students how to discriminate when one row or the other is used to measure an angle. Here is part of the introduction from Lesson 121.

a. In the last lesson, you learned about line segments and when they intersect.
• Think of a triangle. Are the sides of a triangle lines or line segments? (Signal.) *Line segments.*
• Do the line segments **intersect** at the corners or in the middle of a triangle? (Signal.)
  At the corners.
  (Repeat until firm.)
b. You're going to learn about a tool that's used to measure angles. The tool is called a protractor. Say **protractor.** (Signal.) *Protractor.*
• What's the name of a tool used to measure angles? (Signal.) *(A) protractor.*
(Display:)   [121:5A]

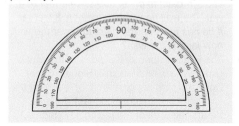

Here's a picture of a protractor. There are two sets of number scales on this protractor. The tens numbers on the inside scale start at zero here (**touch 180 over zero**) and go to 180 here (**touch zero over 180**). The tens numbers on the outside scale start at zero here (**touch zero over 180**) and go to 180 here (**touch 180 over zero**).
• Both number scales are the same for one number. What number? (Signal.) *90.* (Touch 90.)
• Everybody, what do you use a protractor for? (Signal.) *To measure angles.*
• What's the name of this tool? (Signal.) *(A) protractor.*

c. (Add to show:)   [121:5B]

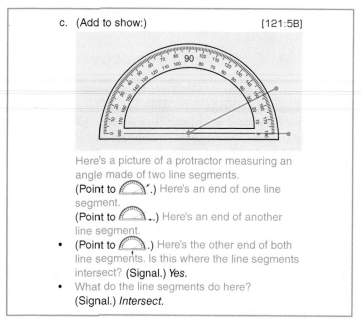

Here's a picture of a protractor measuring an angle made of two line segments.
(Point to ⌂.) Here's an end of one line segment.
(Point to ⌂.) Here's an end of another line segment.
• (Point to ⌂.) Here's the other end of both line segments. Is this where the line segments intersect? (Signal.) *Yes.*
• What do the line segments do here? (Signal.) *Intersect.*

from Lesson 121, Exercise 5

Similarly, students learn the critical features of properly placing a protractor through positive and negative examples. The lines must intersect at the center marker, and one of the lines must go through zero. This instruction ensures that students are able to use the tool accurately. Here are the examples from Lesson 123.

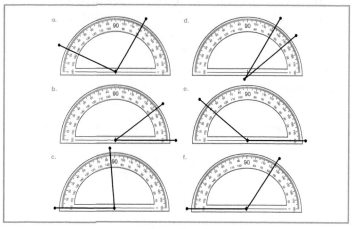

from Textbook Lesson 123, Part 2

Having measured angles for several lessons, students are well prepared to construct angles. Here is part of the exercise from Lesson 128.

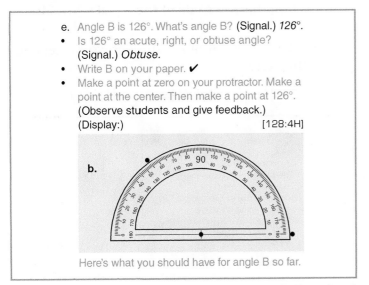

e. Angle B is 126°. What's angle B? (Signal.) *126°.*
• Is 126° an acute, right, or obtuse angle? (Signal.) *Obtuse.*
• Write B on your paper. ✔
• Make a point at zero on your protractor. Make a point at the center. Then make a point at 126°. (Observe students and give feedback.) (Display:)   [128:4H]

b.

Here's what you should have for angle B so far.

from Lesson 128, Exercise 4

Students also work with rays, lines, and line segments in the Practice Software. Here are three examples of a task students complete (using a line-drawing tool) to discriminate between a ray, a line, and a segment:

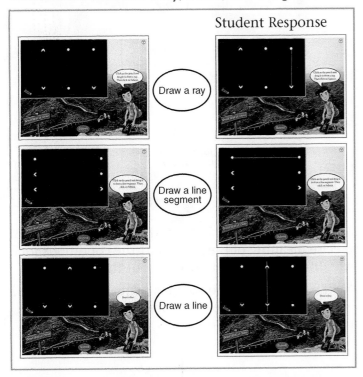

Student Response

Draw a ray

Draw a line segment

Draw a line

Following each example, students then make a line, line segment, or ray on the computer screen that is parallel or perpendicular to the first object they made.

Students also use a ruler to graph lines on the coordinate system. Here's the function table and coordinate system from Lesson 103.

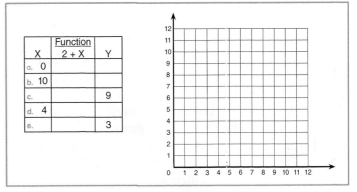

	X	Function 2 + X	Y
a.	0		
b.	10		
c.			9
d.	4		
e.			3

from Workbook Lesson 103, Part 1

Students plot two points, draw the line, and inspect the line to complete the missing X or Y values in the table. Here is the table and coordinate system showing the student work.

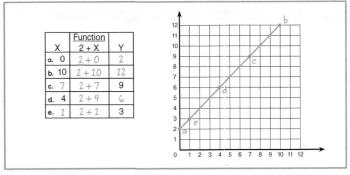

	X	Function 2 + X	Y
a.	0	2 + 0	2
b.	10	2 + 10	12
c.	7	2 + 7	9
d.	4	2 + 4	6
e.	1	2 + 1	3

from Workbook Lesson 103 AK, Part 1

Students' development of meaningful pencil-and-paper strategies, their use of grade-level appropriate tools, and their exposure to related concepts in computer-based activities deepens their understanding of the concepts, and prepares them for more advanced applications in later grades.

MP6: ATTEND TO PRECISION.

CMC Level A (Kindergarten): Equality and Equations.

Precise communication in mathematics is necessary at every grade level, beginning in kindergarten. In *CMC Level A*, students are taught to understand the meaning of the mathematical symbols they use, including the equal sign. Students in *CMC Level A* learn a concise and clear definition of equality: "You must have the same number on one side of the equals that you have on the other side." This rule is applied first in the simplest context: 4 = ■ or ■ = 7. Here's part of the introduction:

a. (Write on the board:) [19:6A]

  =

- (Point to =.) Everybody, what's this? (Touch.) *Equals.*
- Is equals a number? (Touch.) *No.*
  Right. It isn't.
- Here's the rule about equals. (Point to the space on the left.) You must have the same number on this side of the equals (point to the right side of the equals) and on this side of the equals.
b. Listen. (Point left.) If we have 4 on this side of the equals, (point right) we must have 4 on this side of the equals.
- (Point left.) Listen: If we have 4 on this side of the equals, (point right) how many must we have on this side of the equals? (Touch.) *4.*
  Yes, 4.
c. (Point right.) If we have 10 on this side of the equals, (point left) how many must we have on this side of the equals? (Touch.) *10.*
- (Point left.) If we have 15 on this side of the equals, (point right) how many must we have on this side of the equals? (Touch.) *15.*
  (Repeat steps b and c until firm.)
d. (Write to show:) [19:6B]

  4 = □

- This says **4 equals box.** What does it say? (Touch.) *4 equals box.*
- There's a number on one side of the equals. What number? (Signal.) *4.*
  Yes, 4.
e. (Point to 4.) If we have 4 on that side of the equals, (point to box) how many must there be on this side of the equals? (Touch.) *4.*
  Yes, that's the number that goes in the box.
- (Write to show:) [19:6C]

  4 = 4

  (Point to 4 = 4.) Now this says (touch each symbol as you read) 4 equals 4.
- What does it say? (Touch symbols.) *4 equals 4.*
f. (Write on the board:) [19:6D]

  □ = 2

  This says **box equals 2.**
- There's a number on one side of the equals. What number? (Signal.) *2.*
- (Point to 2.) If there are 2 on this side of the equals, (point to box) how many must there be on this side of the equals? (Touch.) *2.*
  Yes, that's the number that goes in the box.
- (Write to show:) [19:6E]

  2 = 2

- (Point to 2 = 2.) What does this say now? (Touch symbols.) *2 equals 2.*

from Lesson 19, Exercise 6

By Lesson 25 students have learned to discriminate whether sides are equal by counting lines on one side of the equal sign and comparing that number with a number on the other side of the equal sign. If the sides are not equal, they cross out the equal sign.

Here are 3 examples from the student worksheet and part of the exercise:

$$5 = |||||| \qquad |||||||| = 9 \qquad |||| = 4$$

There are equals in the row next to the duck. A number is on one side of each equals, and lines are on the other side. You'll cross out an equals if the sides are not equal.
- Touch the first equals. ✔
  You'll touch and count the lines.
- Finger over the first line. ✔
- Get ready. (Tap 6.) *1, 2, 3, 4, 5, 6.*
- How many lines? (Signal.) *6.*
- Touch the number on the other side. ✔
- What number? (Signal.) *5.*
  It says 5 equals 6.
- So are the sides equal? (Signal.) *No.*
  So you cross out that equals.
c. Put your pencil on the big ball and make one cross-out line.
  (Observe children and give feedback.)
  (Teacher reference:)

  ✍ 5 ≠ ||||||

d. Touch the next equals. ✔
  You'll touch and count the lines.
- Finger over the first line. ✔
- Get ready. (Tap 7.) *1, 2, 3, 4, 5, 6, 7.*
- How many lines? (Signal.) *7.*
- Touch the number on the other side. ✔
- What number? (Signal.) *9.*
  It says 7 equals 9.
- So are the sides equal? (Signal.) *No.*
- Make one cross-out line through the equals.
  (Observe children and give feedback.)
e. Touch the last equals. ✔
  You'll touch and count the lines.
- Finger over the first line. ✔
- Get ready. (Tap 4.) *1, 2, 3, 4.*
- How many lines? (Signal.) *4.*
- Touch the number on the other side. ✔
- What number? (Signal.) *4.*
- What does it say? (Signal.) *4 equals 4.*
- So are the sides equal? (Signal.) *Yes.*
  The sides are equal, so you don't cross out the equals.

from Lesson 25, Exercise 8

Through these early exercises, the students learn the precise meaning of the symbol. If the conditions for equality are not met, they cross out the sign.

Next, students learn to complete equality statements, either by drawing lines for a numeral or by writing the numeral for the lines.

$$8 = \qquad ||| =$$
$$= ||||||| \qquad = 2$$

This work lays the foundation for the operations that students perform in addition and subtraction. Students learn to modify sides of an equation to make sides equal:

For example:

$$3 = |||||$$

Students cross out two lines to make the sides equal:

$$3 = |||\cancel{||}$$

Other items show lines crossed out:

$$||\cancel{|} = \qquad ||||||\cancel{|} =$$

Students record the number of lines not crossed out on the other side of the equals.

This work prepares students to incorporate the concept of equality into the subtraction strategy they will learn. For the problem $5 - 1 = \blacksquare$, students "start with 5 and take away 1."

They draw 5 lines and cross out 1 line:

$$5 - 1 = \blacksquare$$

$$||||\cancel{|}$$

They count the lines not crossed out and apply the equality principle: They must have the same number on one side of the equals that they have on the other side. They write 4 in the box to make the sides equal. As each new problem type is introduced for addition or subtraction, the meaning of the equal sign is consolidated as students use the symbol consistently and appropriately

This careful foundation guards against a lack of understanding often displayed by older students who struggle with algebra. These students fail to develop a clear understanding of the equality principle. They may consider the equal sign to be more of a "punctuation mark" in a math statement than a symbol representing a powerful mathematical concept. This causes difficulty when they begin algebra, as the equality principle is the basis for modifying each side of an equation in the same way to conserve equality. The work in early levels of *CMC* prepares students to understand and apply the equality principle in a variety of contexts, in *CMC* and beyond.

MP7: LOOK FOR AND MAKE USE OF STRUCTURE.
CMC Levels B–D (Grades 1–3): **Number family Addition/Subtraction Facts.**

Students are taught to recognize the structure and organization of the number system in all levels of *Connecting Math Concepts*. They use patterns and relationships to master the basic facts in all four operations. For example, rather than learning the 200 basic addition/subtraction facts as isolated items of information, students learn related sets of facts through the concept of the number family.

The number family $\underset{\longrightarrow 7}{5 \qquad 2}$ generates 4 facts:

$$5 + 2 = 7 \qquad 7 - 2 = 5$$
$$2 + 5 = 7 \qquad 7 - 5 = 2$$

Students refer to the largest of the three related numbers as the "big number," and the other two numbers as "small numbers." This standard vocabulary helps students see continuity in the number system as facts are systematically added to their repertoire. Seeing the same structure represented by the number family in many cases (small number + small number = big number; big number – small number = small number) helps them look for and make use of the same structure in new applications. Through verbal and written exercises, students can generate new facts from known facts because they have internalized the structure of the number family.

Here is the first part of an exercise from Level B Lesson 51 that illustrates how students make use of what they know from one family and apply it to a new, related family.

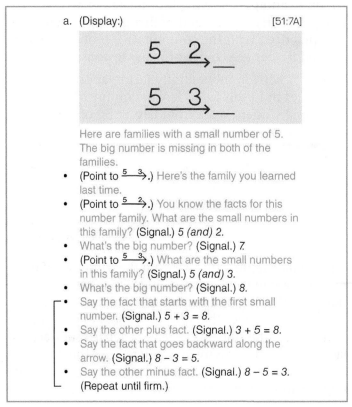

a. (Display:)   [51:7A]

$$\underset{\longrightarrow —}{5 \qquad 2}$$

$$\underset{\longrightarrow —}{5 \qquad 3}$$

Here are families with a small number of 5. The big number is missing in both of the families.
- (Point to $\overset{5 \quad 3}{\longrightarrow}$.) Here's the family you learned last time.
- (Point to $\overset{5 \quad 2}{\longrightarrow}$.) You know the facts for this number family. What are the small numbers in this family? (Signal.) *5 (and)* 2.
- What's the big number? (Signal.) *7.*
- (Point to $\overset{5 \quad 3}{\longrightarrow}$.) What are the small numbers in this family? (Signal.) *5 (and)* 3.
- What's the big number? (Signal.) *8.*
- Say the fact that starts with the first small number. (Signal.) *5 + 3 = 8.*
- Say the other plus fact. (Signal.) *3 + 5 = 8.*
- Say the fact that goes backward along the arrow. (Signal.) *8 – 3 = 5.*
- Say the other minus fact. (Signal.) *8 – 5 = 3.*
  (Repeat until firm.)

from Lesson 51, Exercise 7

The number family table below incorporates the 200 basic addition and subtraction facts in 55 families. Students in *CMC Levels B–D* are systematically introduced to related sets of facts from columns in the table (as in the example above), rows in the table (e.g., all facts in the top row involve a "small number" of 1) or the diagonal (which shows "doubles").

Number Family Table

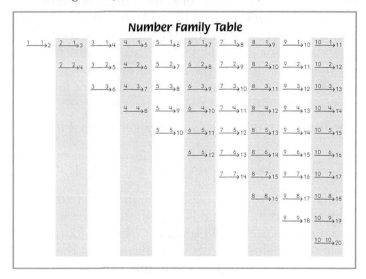

Here are the number-family items and student work from Level B Lesson 44 Independent Work.

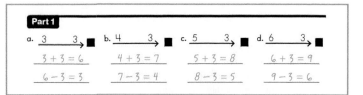

a.
$$6 \xrightarrow{\;2\;} 8$$
$6 + 2 = 8$
$2 + 6 = 8$
$8 - 2 = 6$
$8 - 6 = 2$

b.
$$3 \xrightarrow{\;2\;} 5$$
$3 + 2 = 5$
$2 + 3 = 5$
$5 - 2 = 3$
$5 - 3 = 2$

from Workbook Lesson 44, Part 6

Students are familiar with the number-family structure and generate the four related facts independently. By the end of first grade, students are well practiced with the commutative property, having applied it over the course of many lessons.

The work with number-family fact relationships continues in Levels C–E. Here are number-family items and student work from Lesson 30 of Level C.

Part 1

a.
$$3 \xrightarrow{\;3\;} \blacksquare$$
$3 + 3 = 6$
$6 - 3 = 3$

b.
$$4 \xrightarrow{\;3\;} \blacksquare$$
$4 + 3 = 7$
$7 - 3 = 4$

c.
$$5 \xrightarrow{\;3\;} \blacksquare$$
$5 + 3 = 8$
$8 - 3 = 5$

d.
$$6 \xrightarrow{\;3\;} \blacksquare$$
$6 + 3 = 9$
$9 - 3 = 6$

from Workbook Lesson 30, Part 1

The families are in sequence from the third row of the number-family table. For each item, students recognize that a missing big number requires addition (3 + 3, 4 + 3, 5 + 3, 6 + 3). Once the missing number in the family is identified, the students generate a related subtraction fact. This type of activity not only reduces the memory load for fact memorization, but it also prepares students for later work with inverse operations.

MP8: Look for and Express Regularity in Repeated Reasoning.
CMC Levels C–F: Renaming with Zeros
CMC Level E: Function Tables and the Coordinate System

In *CMC*, students work with repeated reasoning in a variety of situations. They learn to extend and apply what they have learned in one mathematical context to other, related contexts.

Subtraction: Renaming with Zeros

Students in several levels of the program apply the renaming strategy for the tens column to problems with zero in the minuend in an efficient and consistent way.

When renaming from the tens column, students learn a "new place value" for the top number. In the following example, 42 is 4 tens +2. It also equals 3 tens plus 12.

$$\begin{array}{r} \overset{3}{\cancel{4}}{}^{1}2 \\ -\ 7\,8 \end{array}$$

When a problem requires renaming from two columns, students apply the same reasoning and procedure:

402 is 40 tens plus 2. It also equals 39 tens plus 12.

$$\begin{array}{r} \overset{3\ 9}{\cancel{4}\cancel{0}}{}^{1}2 \\ -\quad 7\,8 \end{array}$$

This same reasoning and procedure can be applied to any number of columns:

$$\begin{array}{r} \overset{3\ 9\ 9}{\cancel{4}\cancel{0}\cancel{0}}{}^{1}2 \\ -\qquad 7\,8 \end{array}$$

$$\begin{array}{r} \overset{3\ 9\ 9\ 9}{\cancel{4}\cancel{0}\cancel{0}\cancel{0}}{}^{1}2 \\ -\qquad\quad 7\,8 \end{array}$$

The repeated reasoning provides students with an efficient algorithm, and a "shortcut" to solve any subtraction problem that requires renaming with zeros. In the last example above, students in *CMC* take one renaming step. In the example below, students not taught the method in *CMC* take four separate renaming steps to work the same problem.

$$\begin{array}{r} \overset{3\ 9\ 9\ 9}{\cancel{4}\,\cancel{0}\,\cancel{0}\,\cancel{0}}{}^{1}2 \\ -\qquad\quad 7\,8 \end{array}$$

This less efficient method increases the likelihood that students will make an error or become confused.

Function Tables and the Coordinate System

Students in *CMC Level E* apply repeated reasoning to rows of a function table. The function table shows three columns: the X value, the function, and the Y value. For example:

X	Function 3 + X	Y
a. 7		10
b. 0		3
c. 2		
d.		4
e. 5		

from Workbook Lesson 99, Part 3

This table format enables students to see that the same transformation from the X value to the Y value is repeated for each point they will plot on the coordinate system. They also learn that this type of repeated transformation results in points that lie on a straight line. Given two X and Y values, students make two points and plot the line. They then refer to the line to identify where other points lie on the line, based on either the x or y coordinate. Finally, they perform the calculation based on the function rule to verify the accuracy of the points they have identified. Here's part of the exercise from Lesson 99.

(Add to show:)   [99:3D]

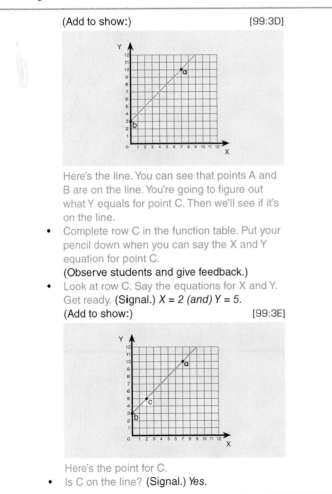

Here's the line. You can see that points A and B are on the line. You're going to figure out what Y equals for point C. Then we'll see if it's on the line.

• Complete row C in the function table. Put your pencil down when you can say the X and Y equation for point C.
  (Observe students and give feedback.)
• Look at row C. Say the equations for X and Y. Get ready. (Signal.) *X = 2 (and) Y = 5.*
  (Add to show:)   [99:3E]

Here's the point for C.
• Is C on the line? (Signal.) *Yes.*

d. We're going to figure out the missing X or Y value for points D and E by using the line.
• Look at row D in the function table and say the equation for Y. Get ready. (Signal.) *Y = 4.*
  To find the point, I go to Y equals 4 on the arrow and go across to the line.
  (Add to show:)   [99:3F]

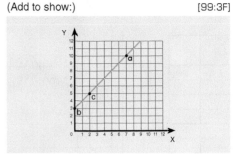

• What does X equal when Y equals 4? (Signal.) *1.*
  (Add to show:)   [99:3G]

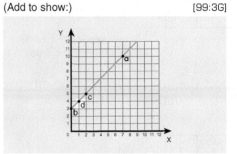

• Write what X equals for row D in the function table. ✔
f. Now we're going to check D to make sure it's right. The function for the table is 3 plus X.
• Raise your hand when you can say the problem for row D. ✔
• Say the problem for row D. (Signal.) *3 + 1.*
• What's the answer? (Signal.) *4.*
• Does Y equal 4 for row D? (Signal.) *Yes.*
  So the values for X and Y are right. Remember, these types of functions tell about all the points on a straight line.

from Lesson 99, Exercise 3

This exercise gives students the opportunity to apply repeated reasoning and also to evaluate the reasonableness of their results. If the point on the line does not correspond to the values in the table, they find the discrepancy and correct it.

On later lessons, students work with multiplication function tables. Here's an example from Lesson 102:

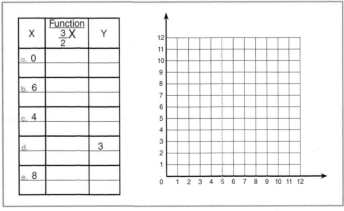

X	Function $\frac{3 X}{2}$	Y
a. 0		
b. 6		
c. 4		
d.		3
e. 8		

from Workbook Lesson 102, Part 3

Students deepen their understanding of functions and lines by repeating the same reasoning to these examples as they did to the add-subtract functions. They apply the function rule to figure out the y value for points a. and b., plot those points and draw the line.

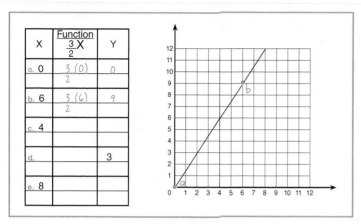

X	Function $\frac{3X}{2}$	Y
a. 0	$\frac{3(0)}{2}$	0
b. 6	$\frac{3(6)}{2}$	9
c. 4		
d.		3
e. 8		

from Workbook Lesson 102, Part 3

They inspect the line to identify the missing x or y values for the remaining points.

X	Function $\frac{3X}{2}$	Y
a. 0	$\frac{3(0)}{2}$	0
b. 6	$\frac{3(6)}{2}$	9
c. 4		6
d. 2		3
e. 8		12

from Workbook Lesson 102, Part 3

Finally, students evaluate their answers for points c through e by multiplying each X value in the table by 3/2.

SUMMARY

For each mathematical concept or strategy introduced in *CMC*, students progress through a careful sequence of examples, move from highly structured activities to independent practice over several lessons, and revisit earlier concepts as they learn new applications or integrate elements of previously taught concepts. The *Comprehensive Edition* of *CMC* includes strategically designed instructional activities that fully meet the Common Core State Standards for Mathematical Content and the Standards for Mathematical Practice in kindergarten through fifth grade.